AF862254

Annual Reports in Inorganic and General Syntheses – 1974

ACADEMIC PRESS RAPID MANUSCRIPT REPRODUCTION

Annual Reports in Inorganic and General Syntheses—1974

Edited by

Kurt Niedenzu

Department of Chemistry
University of Kentucky
Lexington, Kentucky

Hans Zimmer

Department of Chemistry
University of Cincinnati
Cincinnati, Ohio

Academic Press

New York San Francisco London 1975

A Subsidiary of Harcourt Brace Jovanovich, Publishers

COPYRIGHT © 1975, BY ACADEMIC PRESS, INC.
ALL RIGHTS RESERVED.
NO PART OF THIS PUBLICATION MAY BE REPRODUCED OR TRANSMITTED IN ANY FORM OR BY ANY MEANS, ELECTRONIC OR MECHANICAL, INCLUDING PHOTOCOPY, RECORDING, OR ANY INFORMATION STORAGE AND RETRIEVAL SYSTEM, WITHOUT PERMISSION IN WRITING FROM THE PUBLISHER.

ACADEMIC PRESS, INC.
111 Fifth Avenue, New York, New York 10003

United Kingdom Edition published by
ACADEMIC PRESS, INC. (LONDON) LTD.
24/28 Oval Road, London NW1

LIBRARY OF CONGRESS CATALOG CARD NUMBER: 72-9991

ISBN 0–12–040703–5

PRINTED IN THE UNITED STATES OF AMERICA

CONTENTS

PREFACE

This third volume of the "Annual Reports in Inorganic and General Synthesis" is organized along the lines used for the two preceding volumes. Again we have encouraged the authors to use synthetic aspects as their primary guideline for the arrangement and preservation of the information; however, an occasional deviation to include structural or mechanistic features seemed to be justified in order to reflect the particular features of a given element's chemistry.

The second edition also was received favorably by the scientific community. Again the reviewers stressed the need for an index. Consequently, a brief index without cross-referencing is provided. Since a very large number of new compounds are described, only an attempt to list representative compounds of a given type of species was made. Ms. Anita Wojtowicz assisted in preparing this index. Dr. Phyllis Kaplan and Dr. John Wasson deserve my special appreciation for agreeing on an extremely short notice to prepare the chapters on Pt-metals and Lanthanides, respectively. The chapter on nitrogen did not arrive in time to be included in this volume.

Professor Niedenzu spent 1974-1975 abroad on a sabbatical leave, consequently, I am solely responsible for this, the third issue, of ARIGS.

Hans Zimmer

SIMPLE AND COMPLEX METAL HYDRIDES OF MAIN GROUPS I, II and III

R.O. Bach

Manager, Research Department
Lithium Corporation of America
Bessemer City, North Carolina 28016

SODIUM HYDRIDE

The reaction of magnesium turnings with molten NaOH at 400°C occurs along two different paths (1):

1) $Mg + NaOH \longrightarrow MgO + NaH$

or

2) $Mg + 2NaOH \longrightarrow MgO + Na_2O + H_2$

Under the conditions of this experiment approximately 80% of the magnesium is consumed according to equation 1), the rate of consumption proceeds according to a psuedo zero order. A concentration of up to 0.41 wt.% of NaH in NaOH is observed. The possible effect of the presence of 0.4% Na_2CO_3 in the molten NaOH is not taken into consideration by the author.

A Czechoslovakian group of hydride researchers reports further progress in the direct synthesis of NaH. Ullrych (2) uses a two liter stirred laboratory autoclave containing up to 400 steel balls of 8 mm diameter. The temperature range explored is 230° to 330°, the hydrogen pressure is varied from 5 to 25 Kp/cm^2. The comparison between hydrogen, hydrogen containing 0.5 vol.% CO and hydrogen with the addition of 1.5 to 2.2 m mole of methanol per gram atom of sodium shows a substantial initial increase in the rate of reaction with CO and methanol. Carbon monoxide has superior thermal stability. As expected, the ratio increases with higher temperature and increasing H_2 pressure. Reaction rate studies show that the process is controlled by diffusion through the NaH_S/Na_L interphase, rather than by the activation energy.

The usefulness of CO as a catalyst is demonstrated by the large, specific surface of the NaH obtained (3). The reaction rate and the purity of the NaH decrease with further increasing CO amounts. A 96% conversion of Na to NaH can be obtained in eight hours at about 320°C and H_2 pressures of 120 atmospheres.

The role of CO as a catalyst is described (4) through the constants of the Arrhenius equation. The rate of the reaction is zero order with respect to liquid sodium in the absence of CO. The progress of the reaction up to 80% completion is linear, temperature and pressure controlling the slope. The presence of CO seems to establish a reaction rate of second order with respect to liquid sodium. The essential difference is that the specific surface of the NaH product is substantially increased while the reaction rate actually decreases (5): with 3% CO in H_2 the specific surface of NaH is about $5m^2/g$, while with 0.25% CO it is in the order of magnitude of cm^2/g. The role of other additives (6) on the specific surface of the NaH product is explored with branched alcohols and alcanoic acids. All agents yielding a product of 1.55 m^2/g surface area or more are considered useful.

The rate of reaction of hydrogen with liquid sodium and liquid sodium alloys depends linearly on the pressure. The activation energy is estimated to be 69 kJ/mole for sodium (7).

BERYLLIUM HYDRIDE

In 1962 and subsequent years, a number of U.S. patents were filed for the manufacture of BeH_2 with the intent of using this compound as a rocket propellant. These patents were issued in 1974. Three main objectives can be recognized in the research efforts contained in these patents:

Development of alternate methods of preparing a high purity product with improved yield and economy;

Densification for improved loading capacity in propellant applications, resulting in identification of several crystal phases;

Stabilization of the rather labile compound by means of adding agents during or after its preparation.

A. New Methods for Amorphous BeH_2

A process starting from Be halides, alkoxides or beryllium halide-ether adducts (8) consists of reacting these with dialkyl aluminum hydride in toluene at reflux temperatures for up to 1-1/2 hours. After filtering, the product is heated, under high vacuum, to 140°-145°C for 2-3 hours. The purity of the product ranges from 40-80 wt.%, the balance being aluminum alkyl or aluminum alkyl halide.

The pyrolisis of beryllium alkyls is a well-known method for obtaining BeH_2. A two-stage process combines the alkylation of $BeCl_2$ and the subsequent pyrolitic degradation (9). Trialkyl aluminum or sodium tetra-alkylaluminate are reacted with beryllium chloride bis(diethyletherate) at up to 50°C in low boiling aromatic solvents. After removal of the low boiling constituents, the mixture is heated in hydrocarbon oils or kerosene to between 190 and 220°C. The solid residue after isolation consists in BeH_2 of a purity up to 72%.

Thermal decomposition of adducts of BeH_2 with tertiary amines (see below) leads to a product having a purity of up to 93.5% (10). The preferred temperature range is from 180 to 220°C because, below 180°, amine is returned and because above 220°C BeH_2 decomposes rapidly. This serves as a method for purification of BeH_2 as it reacts selectively with tertiary amines.

Another method of purification claims to up-grade 58 wt.% BeH_2 to a product containing 88 wt.% (11). Beryllium chloride in the presence of ether reacts with BeH_2 to form a solution of chloroberyllium hydride which can be filtered from insoluble impurities. The solution is evaporated and the solid recovered, washed with dimethyl sulfide and stored in high vacuum at 100°C.

A two step reaction according to the following equations is carried out (12):

$$1)\quad 2\ NaBEt_3H + BeCl_2 \xrightarrow{Et_2O} 2\ NaCl + Be(BEt_3H)_2$$

$$2)\quad Be(BEt_3H)_2 \xrightarrow{\text{Boiling Xylene}} BeH_2 + 2\ BEt_3$$

The first step leads to a solution of $Be(BEt_3H)_2$ in ether, which, after filtration from NaCl, is concentrated to 40% of its initial volume. This concentrated solution is dropped into boiling xylene in step 2. Ether, xylene and triethylboron distill from the mixture, when kept above 120°C. The product analyzed 94.4 mole % BeH_2, the remainder being triethylboron.

B. Crystalline BeH_2

Two patent applications filed in 1964 (13), (14) describe two crystalline varieties of BeH_2. Beryllium hydride obtained by the methods described above or by the pyrolitic decomposition of dialkyl Be has a density of up to 0.69 g/cc depending on its purity. If placed under pressure of up to 900,000 psi at temperatures up to 300°C, two crystalline modifications can be obtained. This is claimed as a fusion process of densification. The higher the pressure and the higher the temperature, the denser the material obtained and the more pronounced its crystallinity. Increased density of neutral hydrides by fusion is claimed for all members of Groups II-A, II-B, III-A and III-B, but specific claims are restricted to BeH_2. Temperatures range from 135°C to 210°C while the fusion temperature of BeH_2 is established to be around 125°C under a pressure of 75,000 psi. Pressures range up to 900,000 psi and densities of up to .88 g/cc are reported. X-ray diffraction reveals the strongest d value at 3.30A.

In the thermal decomposition of di-tert butyl Beryllium in presence of butyllithium or lithium tetrahydridoaluminate (14), lithium hydride containing BeH_2 is obtained. After the samples are subjected to high pressures from 100,000 to 500,000 psi at temperatures from 150°C to 300°C, high density crystalline composites are obtained consisting from 80 to 99.5 mole percent BeH_2. When butyllithium is used, x-ray diffraction reveals the strongest line at d= 3.78A, the density is 0.78 g/cc. When $LiAlH_4$ is used, the density of the composite is 0.79 g/cc and the strongest x-ray diffraction line is at d=3.38A.

The crystal phase having the strongest line at d = 3.38A is identified as pertaining to the hexagonal system with the following lattice parameters: a = b = 4.14 - 4.20A and c = 6.76A with a calculated density of .82 g/cc (15).

The very high pressures and relatively long heating times are reduced by using alkali and alkaline earth fluorides as crystallization additives (16). Pressures of 50 to 100,000 psi at temperatures of 180 to 220° are sufficient to obtain the hexagonal variety of lithium-doped BeH_2, with densities of up to .80 g/cc.

C. Stabilization

BeH_2 prepared in the manner described above for the use as a rocket fuel exhibits "erratic and unpredictable" behavior when placed in propellant formulations (17). It also is quite susceptible to hydrolysis by boiling water. It was found that the suppression of hydrolysis is equivalent to a stable BeH_2 in propellant formulations, - thus hydrolysis testing is an accepted measure of quality control. One hour in water at 100°C hydrolyzes 15.5% of the BeH_2. Hydrolysis is stopped or greatly reduced by the addition of .05 to .2 wt.% (based on BeH_2) of following: alkali or alkaline earth salts of carboxylic acids, alkali salts or oxy acids of B, Si, P, S and group VI-B elements, alcohols, phenols and salts of alkylbenzene sulfonic acids.

D. Tertiary Amine Adducts of BeH_2

BeH_2 reacts with tertiary amines in four to ten days at temperatures of 130°-160°C using a rotating monel metal bomb (10). One to one adducts are obtained with trimethylamine, N-methyl pyrrolidine (M.P. 86.0-86.5°C), N, N-dimethylbenzylamine (M.P. 101-103°C) and also with N-ethyl pyrrolidine, N-methylpiperidine, dimethyl-N-dodecylamine and dimethyl-N-octylamine. Most of these are well crystallized compounds. The N, N, N', N' tetramethylethylenediamine adduct is polymeric. $BeH_2 \cdot NMe_3$ can also be obtained by the following reaction sequence:

$$2\ Et_2AlH + Et_2Be \longrightarrow Be[AlEt_3H]_2$$

$$+ NMe_3 \longrightarrow BeH_2\cdot NMe_3 + 2\ Et_3AlNMe_3$$

By ballmilling together BeH_2, triethylamine or N-methyl morpholine, the corresponding adducts are obtained.

E. <u>Diethyleneglycoldimethylether Adduct of BeH_2</u>

When solutions of $NaAlH_4$ and $Be(BH_4)_2$ in diethyleneglycol dimethylether are mixed, a white precipitate is obtained (18). After filtration and drying under reduced pressure at about 83°C, the composition corresponds approximately to:

$$BeH_2\cdot 0.22\ C_6H_{14}O_3$$

ALUMINUM HYDRIDE

A substantial number of U.S. patents were applied for in 1962 and subsequent years covering the preparation of AlH_3. This parallels the efforts for BeH_2 described above and is intended for the use as a rocket fuel. The patents issued in 1974 cover the following clearly discernable issues:

Preparation of Nonsolvated AlH_3;

Preparation of newly identified Crystalline Phases of AlH_3;

Purification and Stabilization of AlH_3.

A. <u>Nonsolvated AlH_3</u>

The reaction of gaseous HCl with an alkali tetrahydriodoaluminate in an ether solvent (19) leads to precipitation of an alkali chloride and an ether solution of AlH_3. After evaporation of the ether solution, the residue consists of ether solvated AlH_3 and excess $MAlH_4$. The finely ground solid is then heated for several hours at 75°C under vacuum. The desolvated product is washed with ether to remove the excess tetrahydridoaluminate, the examples specifying $LiAlH_4$. The residue is dried under high vacuum at room temperature for up to 8 hours. The final product is AlH_3 with a few percent of impurities containing Li, Cl, C and possibly O. The same reaction can also be carried out using $LiBH_4$ (20) or by removing the ether solvent by dropping the AlH_3 containing solution into a large volume of aromatic solvents, such as toluene. These are program heated from 40° to about 100°C. The ether distills off and AlH_3 precipitates. Most of the product is obtained in granular form. Instead of HCl, one can use also metallic chlorides (2) such as $HgCl_2$, Hg_2Cl_2, $FeCl_3$, $ZnCl_2$, and CuCl. The heavy metal precipitates together with the alkali chloride while hydrogen is evolved. After filtration, the ethereal AlH_3 solution is worked up as described above.

Sodium tetrahydridoaluminate when reacted with $AlCl_3$ in ether in the ratio of 3:1 will yield an ether solution of AlH_3 which is stabilized by the addition of a small quantity of $LiAlH_4$ (22). The work-up is essentially the same as above with the exception that the temperature of desolvation is raised to 100°C for about forty minutes. This method can also be carried out with lithium tetrahydridoaluminate (23). Instead of the complex hydrides one can also use lithium hydride with a small amount of $LiAlH_4$, reacting it with $AlCl_3$ in ether (24). The product has a similar composition, the recoveries are somewhat lower.

AlH_3 of high purity is prepared by the thermal decomposition of a tertiary amine -AlH_3 adduct in presence of LiH as a catalyst (25). An ether solution containing 66.2 g $Et_3N{:}AlH_3$ and 2.96 g $LiAlH_4$ is vacuum distilled to remove the ether. The solid residue is then decomposed at 50°C and 2-5 mm Hg. The amine is distilled off and the AlH_3 residue is heated to 90°C under a sharp vacuum to remove residual adduct. The final step is the removal of impurities and catalyst by washing with ether.

B. Crystallized Phases of AlH_3

The aluminum hydride preparations so far described are known to be somewhat unstable, decomposing even at room temperatures. The structure of these preparations as characterized by x-ray diffraction analysis (26) has its strongest absorption line at d= 2.88A. A preparation following approximately the techniques described in (23) leads to such a composition which is called gamma-aluminum hydride. When such a product is heated for another 14 to 15 hours under reduced pressure at 65°C a product of a different crystal structure diffraction is obtained, called alpha-aluminumhydride. Its strongest line is observed at d=3.22A, the structure is hexagonal, the calculated density is 1.49 g/cc and the unit cell dimensions are a=4.44A and c=5.89A. This material is stable at room temperature and when stored for 135 days it shows no decrease in reactivity. It appears useful as a solid rocket propellant. Another crystal phase of AlH_3 is obtained when the reaction between $LiAlH_4$ and $AlCl_3$ is carried out in the presence of $LiBH_4$, - it is called beta-aluminum hydride. The preparation of such solution is described in (27). When the desolvated product is heated at 65°C for 2 hours, the x-ray powder diffraction data shows the most intense line to be at d=5.25A. This product has a rhombohedral structure with a cell dimension of about a=3.18A and c=15.60A. There is no appreciable decomposition after 26 days at ambient temperature. It also is classified as a suitable rocket fuel. If this product is heated for an additional six hours, it is converted into the alpha-phase.

A solution of AlH_3 obtained by the known reaction of $AlCl_3$ with $LiAlH_4$ in ether is fed into boiling benzene or benzene diethylether mixtures (28). The molar ratios of the solution of $LiAlH_4$:AlH_3:$LiBH_4$ are about 1:8:1. When the reflux temperature is maintained at 76.5 to 77°C by addition of appropriate amount of Et_2O to the benzene, the AlH_3 solution being fed deposits hexagonal aluminum hydride of a particle size larger than 10 microns and a density of above 1.4 g/cc. When the addition is carried out at 80°C, gamma-aluminum hydride is obtained. A continuous reactor system eliminates the use of excess diethylether (29).

The decomposition of tri(aluminum hydride)-trimethylamine (see below) at a temperature of about 80°C under high vacuum for about 3.5 hours leads to alpha-aluminum hydride in a yield of up to 20% (30).

C. Purification and Stabilization of AlH_3

Usually the crystallized AlH_3 is obtained as a mixture of submicron and macrocrystalline particles. The latter are preferred for the application as a rocket fuel. Aluminum chloride mono-diethyletherate reacts with the submicron fraction preferentially, thereby solubilizing it as aluminum chlorohydride in ether. The reaction is carried out at 70-80°C for about 15 minutes. The residue is macrocrystalline alpha-aluminum hydride (31).

When AlH_3 is prepared according to the procedure described in (19) and Hg is added during the desolvating process, a product of increased thermal stability is obtained (32). A product containing .02% mercury, when subjected to heating to 100°C for 24 hours, loses about 5% of its activity. A similar product not containing mercury under the same test conditions loses 95% of its activity.

Certain sulphur containing aromatic compounds, such as phenothiazine and 2-mercaptobenzothiazole (33) seem to slow down thermal degradation of AlH_3. They are added during the drying step as toluene solutions. The reaction between $(CH_3)_2SiCl_2$ and lithium tertahydridoaluminate in diethylether leads to AlH_3 containing solutions. If these solutions are worked up according to the procedures described in (28), a silicol containing AlH_3 product is obtained with a substantially increased thermal stability. For instance a silicol containing material decomposes only 1.76% after ten days at 60°C. Untreated alpha-aluminum hydride decomposes rapidly on the 5th day under these conditions (34).

When $AlCl_3$ is reacted with $LiAlH_4$ and $LiBH_4$ in ether in the presence of small amounts of $MgCl_2$ (35), Mg is incorporated in the AlH_3 obtained. This product contains between 1 and 1.5% Mg. It shows the x-ray diffraction pattern of hexagonal alpha-aluminum hydride. The slightly expanded lattice constants point to a solid solution of Mg. The standard test for propellant application is the Taliani (35) test: the standard of acceptance being a maximum of 1% decomposition in 10 days at 60°C.

The Mg containing AlH_3 is treated with an aqueous buffer solution (pH 6-8) of potassium acid phosphate for about 15 minutes. After washing and drying, the product shows vastly superior stability. Products which originally could not pass the Taliani test are acceptable after the treatment. Magnesium containing AlH_3 is superior to ordinary alpha-aluminum hydride.

D. Tri(aluminumhydride)-trimethylamine $(AlH_3)_3{:}NMe_3$

When liquid $AlH_3{\cdot}NMe_3$ is heated to about 85°C and a partial vacuum is placed on the system, keeping it under reflux for 20 to 30 minutes, a white solid separates out. The liquid phase is then removed by vacuum. The solid has the composition of $(AlH_3)_3{\cdot}NMe_3$ and is not soluble in benzene or diethyl-ether. It decomposes rapidly at 135°C without melting. The x-ray diffraction pattern showed the most prominent absorption line at d=9.75A (30).

E. Aluminum Chlorohydride

$AlCl_3$, NaCl and KCl are mixed in quantities to provide on a mole percent basis the eutectic $AlCl_3$ 66, NaCl 20 and KCl 14%. 5 g of finely divided MgH_2 are mixed with 15 g of the eutectic in a N_2 atmosphere for 30 minutes at 90°C. The mass is extracted with Et_2O. The solution, after filtration, is concentrated and reduced to dryness. The composition corresponds to an aluminum chlorohydride etherate of the following elemental analysis: C 19.21, H 4.19, Al 12.40, Mg 1.78 and Cl 50.8% (36).

HYDRIDOALUMINATES

A direct synthesis from the elements is carried out under hydrogen pressure (37). The novel aspect consists in the fact that the synthesis is carried out in the absence of any solvent and the products are obtained as liquids at high temperatures. $NaAlH_4$ is obtained by melting together Na and Al in a ratio of 1:1.1 to 1:1.3 under 300 to 420 atmospheres of hydrogen and at temperatures of 200 to 240°C. Yields are as high as 90%. At lower pressures and temperatures, mixtures of Na_3AlH_6 and Al are obtained. $KAlH_4$ requires a temperature of 280°C and a hydrogen pressure of up to 285 atmospheres giving a yield of 92%. $CsAlH_4$ is obtained with a yield of 90% using pressures of 330 atmospheres at 280°C. The tetrahydridoaluminates are obtained after cooling as solid masses, obviously solidified from the molten state. All these reactions are complete within 2-3 hours. The synthesis of $LiAlH_4$ was not successful.

The reaction of $MgCl_2$ with $NaAlH_4$ and $LiAlH_4$ (38) in ether leads to a variety of double complex hydrides etherates.

The standard heat of formation of $(CH_3)_4NAlH_4$ is -15.16 kcal mole^{-1} (39).

Li_3AlH_6 is obtained when aluminum hydride triethylamine adduct is reacted with butyllithium (40), according to the following equation:

$$2\ AlH_3{\cdot}NEt_3 + 3\ BuLi \longrightarrow Li_3AlH_6 + AlBu_3 + NEt_3$$

REFERENCES

1. G.A. Vorob'ev, V.L. Kubasov (USSR) ZH. NEORG.KHIM. 1974, 19 (3), 339-42 (Russ).
2. J. Ullrych (Hutn.Fak., Vys. SK. Banske, Ostrava, Czech.) SB. VED.PR.VYS. SK. BANSKE OSTRAVE, RADA HUTN. 1973, 19 (2), 225-37 (Czech).
3. V. Prochazka, P. Kriz, J. Tlaskal (Ustav Anorg. Chem. Ceskoslov. Akad Ved, Rez/Prague) CHEM.PRUM. 1973, 23 (12), 615-20 (Czech).
4. V. Prochazka, M. Nedved (Cesk.Akad.Ved, Prague-Rez, Czech.) COLLECT. CZECH. CHEM. COMMUN. 1973, 38(10), 2845-9 (Eng.).
5. V. Prochazka, M. Nedved (Cesk.Adak.Ved, Prague-Rez, Czech.) COLLECT. CZECH. CHEM. COMMUN. 1973, 38(10), 2850-4 (Eng.).
6. J. Subrt, V. Prochazka (Ustav Anorg. Chem., Cesk.Akad.Ved, Rez/ Prague, Czech) CHEM.PRUM. 1973, 23 (7), 345-9 (Czech).
7. M.R. Hobdell, A.C. Whittingham (Berkeley Nucl.Lab., Cent. Electr. Generating Board, Berkeley, Engl.). LIQUIDALKALI METALS, PROC. INT. CONF. 1973 (Pub. 1973), 5-12 (Eng.).
8. P. Kobetz, R.T. Laran (Ethyl Corp.) U.S. 3,832,456 27 Aug.1974.
9. R.J. Larau, P. Kobetz, R.W. Johnson, Jr. (Ethyl Corp.) U.S. 3,830,906 20 Aug. 1974).
10. L.H. Shepherd, G. TerHaar (Ethyl Corp.) U.S. 3,808,307 30 April 1974.
11. J.H. Murib, S. Schoss, C.A. Bonecutter (National Distillers and Chemicals Corp.) U.S. 3,786,128 15 Jan. 1974.
12. L.H. Shepherd (Ethyl Corp.) U.S. 3,845,199 29 Oct. 1974.
13. L.M. Niebylski (Ethyl Corp.) U.S. 3,784,682 8 Jan. 1974
14. P. Kobetz, L.M. Niebylski (Ethyl Corp.) U.S. 3,784,423 8 Jan. 1974.
15. G.J. Brendel (Ethyl Corp.) U.S. 3,816,193 11 June 1974.
16. L.M. Niebylski (Ethyl Corp.) U.S. 3,808,063 30 April 1974.
17. W.B. Kirsch (Ethyl Corp.) U.S. 3,812,237 21 May 1974.
18. P.F. Reigler, Lz.F.Lamoria (Dow Chemical Co.) U.S. 3,832,407 27 August 1974.
19. T.C. Kraus, J.A. Scruggs, S.I. Trotz (Olin) U.S. 3,767,781 23 Oct. 1973.
20. T.C. Kraus, D.J. Mangold (Olin) U.S. 3,758,677 11 Sept. 1973.
21. T.C. Kraus, J.A. Scruggs, S.I. Trotz (Olin) U.S. 3,821,360 28 May 1974.

22. T.C. Kraus (Olin) U.S. 3,816,606 11 June 1974.
23. W.M. King (Aerojet General) U.S. 3,810,974 14 May 1974.
24. T.C. Kraus (Olin) U.S. 3,769,385 30 Oct. 1973).
25. J.H. Murib (National Distillers) U.S. 3,764,666 9 Oct. 1973.
26. N.E. Matzek, D.F. Musinski, (Dow) U.S. 3,819,819 25 June 1974.
27. G.H. Mattson, (Pollution Control Products) U.S. 3,819,808 25 June 1974.
28. P.F. Reigler, Lz.F. Lamoria (Dow) U.S. 3,822,342 2 July 1974.
29. R.D. Daniels, J.A. Snover (Dow) U.S. 3,819,335, 25 June 1974.
30. C.B. Roberts, F.M. Brower (Dow) U.S. 3,826,820, 30 July 1974.
31. B.L. York, K. Terada (Dow) U.S. 3,819,337, 25 June 1974.
32. S.I. Trotz (Olin) U.S. 3,785,890, 15 Jan. 1974.
33. A.E. Ardis, F.S. Natoli (U.S. Army) U.S. 3,801,707, 2 April 1974.
34. C.B. Roberts (Dow) U.S. 3,803,082, 9 April 1974.
35. C.B. Roberts (Dow) U.S. 3,821,044, 28 June 1974.
36. D.L. Schmidt, A.L. Daniels, R.V. Petrella (Dow) U.S. 3,812,244 21 May 1974.
37. T.N. Dymova, N.G. Eliseeva, S.I. Bakum, Yu.M. Dergachev, (Inst. Obshch. Neorg.Khim. im. Kurnakova, Moscow, USSR) DOKL. AKAD. NAUK SSSR 1974, 215 (6), 1369-72 (Russ).
38. K.N. Semenenko, B.M. Bulychev, K.B. Bitsoev (Mosk.Gos.Univ., Moscow, USSR). VESTN.MOSK.UNIV.KHIM. 1974, 15(1), 74-7 (Russ).
39. K.N. Semenenko, T.S. Il'ina, A.P. Savchenkova (Mosk.Gos.Univ, im. Lomonsova, Moscow, USSR), ZH.FIZ.KHIM. 1974, 48(1), 228 (Russ).
40. R.O. Bach, D.K. Hoffman (Lithium Corp.) CHIMIA 1974, 28 (2) 70 (Eng.).

BORON

L. M. Frenzel

Department of Chemistry
University of New Orleans
New Orleans, Louisiana 70122

GENERAL

Reviews of the mass spectra of boron compounds (1), the preparation of organoboranes and their use in an organic synthesis (2), and metalloboranes and the theory of metal-boron bonding (3) were published in 1974. H. C. Brown elaborated the use of thexylborane as a highly versatile reagent for organic synthesis by hydroboration methods (4). New boron hydrides and sulfur boranes were studied for protein-binding capabilities so that they might be developed for neutron capture cancer therapy (5,6,7).

NOVEL PROCEDURES

Tetraphenylborate functioned as a nuclear magnetic resonance shift reagent for aminosulfonium cations (8). Brilliant green-tetrafluoroborate was the basis for a liquid state ion selective electrode for tetrafluoroborate (9) while methylene blue-tetrafluoroborate was developed for the spectrophotometric determination of boron in condensed water solutions (10).

A discharge-flow reactor connected to a time-of-flight mass spectrometer was used to study the reaction of atomic oxygen with diborane(6) without an explosion (11). Concentric cylinders which were maintained at temperatures of 165°C and -78°C provided an apparatus which was applied to the synthesis of B_2Cl_4 and <u>nido</u>-$C_2B_6H_{10}$ (12, 13).

INORGANIC BORON DERIVATIVES

Zirconium tetraborate was prepared from zirconium chloride oxide and boric acid. It was found to have a formula of $[ZrOB_4O_7]_4 \cdot 4\ H_2O$ (14). The first preparation of boron phosphate from ortho-phosphoric acid and boric acid in aqueous solution was described (15). The following new ternary borides were identified in three phase system diagrams: $YbCo_3B_2$, $PrCo_2B_2$, $HoCo_2B_2$(16); Cr_3NiB_6, $Cr_2Ni_3B_6$(17); and $ThMoB_4$ (18).

METAL-BORON COMPOUNDS

This section cites metal derivatives of boranes or carboranes which are sigma- or pi-bonded. In general they are prepared under inert atmosphere conditions in organic solvents and are subsequently precipitated, sublimed, distilled, or separated by chromatography techniques. Three new thiaboranes, 2,2′-$(1\text{-}B_9H_8S)_2$, 2,6′-$(1\text{-}B_9H_8S)_2$, and 6,6′-$(1\text{-}B_9H_8S)_2$ (19), and new nido- and closo- arsacarboranes, 7-$B_{10}H_{12}As^-$ (20), 1,2-$B_{10}H_{10}As_2$ (20), 3-Br-3-As-1,2-$C_2B_9H_{11}$ (21), and $[(CH_3)_2As]_2C_2B_9H_{11}$ (21), have been isolated. Diborane(6) reacts with potassium methylselenide to form μ-methylselenodiborane(6), μ-$CH_3SeB_2H_5$ (22).

Direct insertion of iron pentacarbonyl or hapto-cyclopentadienyldicarbonylcobalt into 1,5-$C_2B_3H_5$, 1,6-$C_2B_4H_6$, or 2,4-$C_2B_5H_7$ led to new structurally novel small cage systems with six vertex cages, $(CO)_3FeC_2B_3H_6$ and $(\eta\text{-}C_5H_5)CoC_2B_3H_5$, or seven vertex cages, $(CO)_6Fe_2C_2B_3H_5$ and $(\eta\text{-}C_5H_5)_2Co_2C_2B_3H_5$ (23). The first closo- metallocarborane in which the metal is bound to a B_4 face, $(\eta\text{-}C_5H_5)NiB_{10}CH_{11}$, was prepared from $B_{10}H_{12}CN(CH_3)_3$, $NiBr_2 \cdot 2C_2H_4(OCH_3)_2$, and NaC_5H_5 (24). New cobalt carboranes included 8,8′-$CH_3S(C_2B_9H_{10})_2Co$ in which a sulfur atom links the two carborane cages as a inter-ligand bridge (25); $(\eta\text{-}C_5H_5)_2Co_2C_2B_5H_7$ and $(\eta\text{-}C_5H_5)_2Co_2C_2B_7H_9$ which were formed by polyhedral expansion of smaller metallocarboranes (26); and $(C_2B_3H_5)Co_2(\eta\text{-}C_5H_5)_2$, $(C_2B_4H_8)Co(\eta\text{-}C_5H_5)$, and $(C_2B_3H_7)Co(\eta\text{-}C_5H_5)$ which are triple decker or double sandwich compounds (27).

The first univalent nickel and cobalt tetrahydridoborate complexes, $Ni(BH_4)[P(C_6H_5)_3]_3$ and $Co(BH_4)[P(C_6H_5)_3]_3$ were reported (28). Formally three-coordinate neutral rhodium(I) complexes contained 1,2- or 1,7-dicarba-closo- dodecaborane as the boron ligand (29). Rhodium and iridium complexes of B_6H_{10} were found to contain a three center-two electron bond between the metal and the 4,5 bridging position (30). The first neutral heptaborane derivative, $Fe(CO)_4B_7H_{11}$, was isolated and it was postulated that the iron moiety provided electrophilic stabilization to the heptaborane species (31). Tridentate $B_9H_{13}^{-2}$ derivatives of manganese and rhenium were structurally similiar to $B_{10}H_{14}$ derivatives in that there were two metal-H-B bridging hydrogen bonds and one metal-boron sigma-bond interaction (32). Oxidative addition of tris(triphenylphosphine)iridium chloride with 1,2-$C_2B_{10}H_{12}$ in toluene led to 3 - $[(P(C_6H_5)_3)_2Ir\ HCl]$ - 1,2-$C_2B_{10}H_{11}$ (33).

The first apically substituted hexaborane(10) derivative, 1-$(CH_3)_3MB_6H_9$, where M=Si or Ge, was isolated from the reaction of H_2BCl and $(CH_3)_3MB_5H_7$ (34). Germacarboranes bonded to other metal atoms were prepared from $[1,2\text{-}B_{10}H_{10}CHGe]^-$ and iron,molydenum, chromium, or tungsten carbonyl derivatives (35,36). New tris(trialkyltin)borates were obtained by distillation from a boric acid and bis(trialkyltin)oxide reaction solution (37).

ORGANOBORANES

Sterically overcrowded aminoboranes were produced from trichloroborane and acridane in order to study compounds with nonplanar ground states (38). Other sterically hindered aminoboranes which were isolated were tris-(diorganoamino)boranes, produced from trifluoroborane and lithio amines (39), and (trimethylsilylalkylamino)(dimethylamino)phenylborane (40). The thermal stability of the hindered derivatives varied with the size of the alkyl group.

Hydroboration: Sodium tetrahydridoborate was utilized in the reduction of (a) alkenes to produce mono-alcohols (41), (b) sterically hindered pyridinium salts to produce 1,2-dihydropyridines (42), (c) 4,5-dihydro-1,2-oxazolium tetrafluoroborates to form stereospecific tetrahydro-1,2-oxazoles (43), (d) imine ethers to mono-alkyl primary aromatic amines (44), and (e) indoles to indolines (45). Lithium dimesitylborohydride bis(dimethoxyethane) etherate was developed as a new crystalline reagent for the quantitative stereoselective reduction of mildly hindered ketones (46). Borane-methyl sulfide was found to be more stable and more versatile as a hydroboration-oxidation reagent for alkenes than are tetrahydrofuran solutions of borane(3) (47). Borane(3) reduced 1,5-cyclooctadiene to 9-borabicycyclo[3.3.1]nonane (48) and reductively cleaved acetals and ketals to form ethers or hydroxy ethers in 90% yields (49). Arylmercurial halides and borane(3) followed by addition of water furnished a synthetic route to arylboronic acids, $ArB(OH)_2$ (50). Pinacol and tetrakis(dimethoxyboryl)methane yielded $C[BO_2C_2(CH_3)_4]_4$ and, after further reaction, the triborylmethide ion which was subsequently converted to alkene diboronic esters or methanetriboronic esters (51). Triorganoboranes reacted with α,α'-dibromoketones in the presence of base to produce ketones (52); they also added across multiple bonds in unsaturated organo-nitrogen compounds to form derivatives of the type $CH_2C(CH_3)C(R)NBR_2$ (53).

Carboranes: The dilithiosalt of 2,4-dicarba-closo-heptaborane(7) was reacted with bis(trifluoromethyl)iodophosphine to yield 2-bis(trifluoromethyl)phosphino-2,4-dicarba-closo-heptaborane(7) (54). Naptho derivatives of 1,2-dicarba-closo-dodecaborane(12) exhibited properties of low aromaticity with poor conjugation effects (55, 56). The pyrolysis of nido-2-B_5CH_9 at 250°C formed closo-1,7-$B_{10}C_2H_{12}$ with no evidence for 1,2-$B_{10}C_2H_{12}$ or 1,12-$B_{10}C_2H_{12}$ (57). Oxidation of 7,8-dicarba-nido-undecaborate(12) with an aqueous solution of iron(III) chloride followed by pyrolysis at 120°C yielded the new compound, 1,2-dicarba-closo-decaborane-(10) (58). Other new carboranes which were prepared are 8-bromo-1,2-dicarba-closo-dodecaborane(12) (59), 3-OH-1,8-$(CH_3)_2$-1,8-$B_9C_2H_8$ (60), 3,7-$(OH)_2$-1,8-$(CH_3)_2$-1,8-$B_9C_2H_7$ (60), and $(CH_3)_4C_4B_8H_8$ which is a byproduct from the preparation of iron metallocarboranes (61).

Boranes: Triborane(7) carbonyl decomposed at 0°C to form a mixture of several boranes from which white solid 1,2-bis(carbonyl)diborane(4) was isolated (62). Electrochemical oxidation of $B_{10}H_{10}^{-2}$ derivatives in acetonitrile produced substituted $B_{20}H_{18}^{-2}$ ions (63). Tridecaborane(19) formed in the pyrolysis of B_6H_{10} (64) while $B_{14}H_{22}$ and $B_{16}H_{20}$ were isolated from the reaction of B_6H_{10} with B_8H_{12} under different conditions (65). The solubilities of $B_{10}H_{14}$ in benzene and toluene at various temperatures were studied; a eutectic mixture of 20% mole $B_{10}H_{14}$ and 80% mole benzene was observed at -7.0°C (66).

RING SYSTEMS

A series of 2-substituted-4H-1,3,2-benzodioxaborins, $H_2COB(R)OC_6H_4$, with aryl, amine, ether and sulfur substituents were prepared from phenyl boroxine or triorganoboranes and o-hydroxybenzyl alcohols (67). Amine-borane derivatives of 3,4-dihydro-2H-1,3-benzoxazines produced [4H]-1-oxa-3-azonia-2-boratanaphthalenes upon heating (68). Lithium trialkylalkynyl-borates, $[R_3'B\text{-}C{\equiv}C\text{-}R]^- Li^+$, reacted with acetyl chloride to form 2-oxa-3-borolenes, $RBC(R')_2C(R)C(CH_3)O$, which are distillable liquids (69). New air stable five and six member hetereocycles, $H(R)BOC(O)CH_2N(R')_2$ and $H(R)BOC(O)CH_2CH_2N(R')_2$, were investigated for base strengths and related to other compounds such as glycine and 3-aminopropionic acid (70).

Two large borazines (Fig. 1, 2) were obtained from the cyclization of trisorganoborates or tris(dialkylamino)boranes and amines (71, 72). The

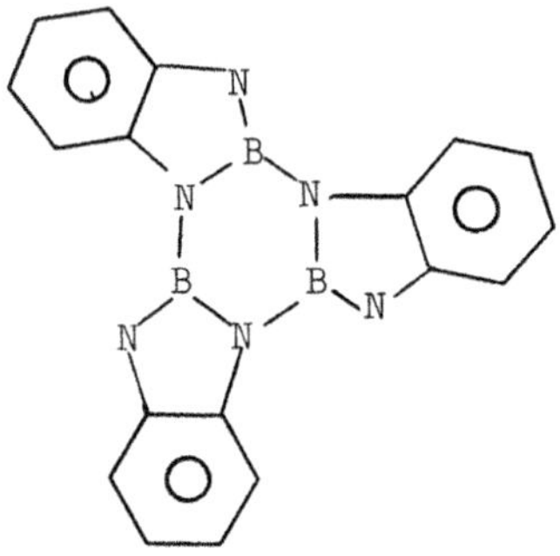

Figure 1

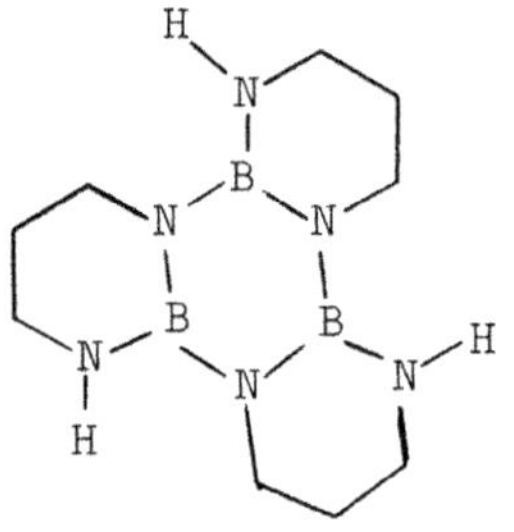

Figure 2

application of substituent effects allowed the preparation of pure 2,4-dichloroborazine (73). Bis(μ-dimethylamino)triborane(9) was synthesized by a stepwise boron-nitrogen chain buildup (74). In the process, $H_2BN(CH_3)_2B(H)_2N(CH_3)_2B(H)_2B(H)_2H$, was isolated and characterized (74).

The reaction of various aminobenzonitriles with trihaloboranes yielded amine-halogenboranes or there was an insertion of the nitrile group into a boron-halogen bond to form the 1,3,2-diazaboranaphthalene ring (75). Yellow liquid 1,3-diphenyl-2,4,5-trimethyl-1,3,2-diazaboroline was prepared from methyldibromoborane and $C_6H_5N[C(CH_3)]_2NC_6H_5$ (76). Other new heterocycle systems included $\overline{SB(R)N(R')P(R'')N(R')B}(R)$ (77), [$\overline{P(R)_2N(R')P(R)_2N(R')B(R'')N}(R')$]$^+$ X^- (77), and $Y\overline{BS(CH_2)_3O}$, where Y = Ph, OR, NHR, NR_2, SR (78).

FOUR COORDINATE BORON

<u>BX_3 Compounds:</u> Methyldichlorophosphine formed 1:1 adducts with BCl_3, BBr_3, and BI_3 but not with BF_3; borane(3) yielded a mixture of products (79). Trichlorophosphine did not form adducts with BF_3 or BCl_3 (80) while dimethylchlorophosphine and trimethylphosphine formed adducts with BCl_3 and BF_3 (80). In another study, BI_3 and BBr_3 were reacted with $(CH_3)_3P$, $(CH_3)_2C_6H_5P$, $(C_6H_5)_3PO$, $CH_3(C_6H_5)_2P$, and $(CH_3)_3PS$ (81). These results agreed with those of P. M. Boorman who investigated the complexes of BCl_3, BBr_3, and BI_3 with R_3PS and R_3PSe (82); no BF_3 complexes were found in these studies. Trimethylarsine and trimethylstibine formed 1:1 adducts with BBr_3 and BI_3 (83). The trimethylstibine adducts slowly decomposed at room temperature while the trimethylarsine adducts were stable indefinitely in an inert atmosphere. Trimethylamine-trihaloborane exchanged halogen with trihaloborane in solution to yield mixed halide adducts, $(CH_3)_3NBX_2Y$ (84). The reaction proceeds whether or not the heavier halogen is originally on the free BY_3. Other mixed halogen-hydrogen borane adducts were prepared with N,N,N',N'-tetramethylethylene diamine (85) and trimethylamine (86) borane adducts by exchange reactions with HX or MX.

<u>BX_4^- Compounds:</u> Two new aluminium tetrahydridoborates were isolated and studied for reduction properties: $H_2B[N(CH_3)_2]_2Al(BH_4)_2$ (87) and $(C_4H_9)_4NAlX_2(BH_4)_2$, where X = Cl, Br (88). Barium tetrahydridoborate, $Ba(BH_4)_2$, resulted from the double decomposition of barium iodide and lithium tetrahydridoborate in tetrahydrofuran (89). In a new preparative method, direct gaseous fluorination of BH_4^- and $LiAlH_4$ yielded $NaBF_4$, KBF_4, or $LiAlF_4$ in approximately 100% yield (90). Trichloroborane was reacted with KCl, RbCl, CsCl, $(CH_3)_4NCl$, or $(C_2H_5)_4NCl$ in chloroform and trichloroborane as solvents to form $MBCl_4$ in yields ranging from 50% to 97% (91). Silver tetrafluoroborate formed solid complexes with oxaziridines (92), cyclohexanone (93), or acetophenones (93) while maintaining the $AgBF_4$ moiety.

REFERENCES

1. R.H. Cragg and A.F. Weston, J. ORGANOMETAL. CHEM., <u>67</u>, 161 (1974).
2. K. Smith, CHEM. SOC. REVS., <u>3</u>, 443 (1974).
3. N.N. Greenwood and I.M. Ward, CHEM. SOC. REVS., <u>3</u>, 231 (1974).
4. E. Negishi and H.C. Brown, SYNTHESIS, 77 (1974).

5. H.S. Wong, E.I. Tolpin, and W.N. Lipscomb, J. MED. CHEM., 17, 785 (1974).
6. E. I. Tolpin, H.S. Wong, and W.N. Lipscomb, J. MED. CHEM., 17, 792 (1974).
7. R.L. Sneath, A.H. Soloway, and A.S. Day, J. MED. CHEM., 17, 796 (1974).
8. G.P. Schiemenz and P. Klemm, ORG. MAG. RES., 6, 276 (1974).
9. A.G. Fogg, A.S. Pathan, and D.T. Barns, ANAL. LETT., 7, 545 (1974).
10. A. Isozaki and S. Utsumi, BULL. CHEM. SOC. JAPAN, 47, 2891 (1974).
11. C.W. Hand and L.K. Derr, INORG. CHEM., 13, 339 (1974).
12. J.P. Brennan, INORG. CHEM., 13, 490 (1974).
13. T.J. Rielly and A.B. Burg, INORG. CHEM., 13, 1250 (1974).
14. Y.A. Afanasev, A.I. Ryabinin, and V.P. Eremin, PROC. ACAD. SCI., USSR, CHEM. SECT., 215, 147 (1974).
15. D. Ashboren, ISRAEL J. CHEM., 12, 831 (1974).
16. P. Rogl, MONATSH. CHEM., 104, 1623 (1973).
17. E. Lugscheider, O. Knotek, and H. Reimann, MONATSH. CHEM., 105, 80 (1974).
18. P. Rogl and H. Nowotny, MONATSH. CHEM. 105, 1082 (1974).
19. W.R. Pretzer and R.W. Rudolph, J. C. S., CHEM. COMMUN. 629 (1974).
20. J.L. Little, S.S. Pao, and K.K. Sugathan, INORG. CHEM. 13, 1752 (1974).
21. H.D. Smith, Jr. and M.F. Hawthorne, INORG. CHEM., 13, 2312 (1974).
22. J.J. Mielcarek and P.C. Keller, J. AMER. CHEM. SOC., 96, 7143 (1974).
23. V.R. Miller, L.G. Sneddon, D.C. Beer, and R.N. Grimes, J. AMER. CHEM. SOC., 96, 3090 (1974).
24. C.G. Salentine, R.R. Rietz, and M.F. Hawthorne, INORG. CHEM., 13, 3025 (1974).
25. J. Plesek, S. Hermanek, and Z. Janousek, CHEM. IND. (LONDON), 108 (1974).
26. W.J. Evans and M.F. Hawthorne, INORG. CHEM. 13, 869 (1974).
27. R.N. Grimes, et.al., INORG. CHEM., 13, 1138 (1974).
28. D.G. Holah, A.N. Hughes, B.C. Hui, and K. Wright, CAN. J. CHEM., 52, 2990 (1974); INORG. NUCL. CHEM. LETT. 10, 427 (1974).
29. S. Bresdale and B. Longate, INORG. CHEM., 13, 539 (1974).
30. A. Davidson, D.D. Traficante and S.S. Wreford, J. AMER. CHEM. SOC., 96, 2802 (1974).
31. O. Hollander, W.R. Clayton, and S.G. Shore, J. C. S., CHEM. COMMUN., 604 (1974).
32. J.W. Lott and D.F. Gaines, INORG. CHEM., 13, 2261 (1974).
33. E.L. Hoel and M.F. Hawthorne, J. AMER. CHEM. SOC., 96, 6770 (1974).
34. D.F. Gaines, S. Hildebrandt, and J. Ulman, INORG. CHEM., 13, 1217 (1974).
35. T. Yamamoto and L.J. Todd, J. ORGANOMETAL. CHEM., 67, 75 (1974).
36. G.S. Wikholm and L.J. Todd, J. ORGANOMETAL. CHEM., 71, 219 (1974).
37. S.K. Mehrotra, G. Strivastava, and R.C. Mehrotra, J. ORGANOMETAL. CHEM., 65, 361 (1974).
38. J. Casanova and M. Geisel, INORG. CHEM., 13, 2783 (1974).
39. W.R. Purdum and E.M. Kaiser, J. INORG. NUCL. CHEM., 36, 1465 (1974).
40. R.H. Neilson and R.L. Wells, INORG. CHEM., 13, 480 (1974).
41. V. Hach, SYNTHESIS, 340 (1974).

42. R.E. Lyle and C.B. Boyce, J. ORG. CHEM., 39, 3708 (1974).
43. A. Cerri, C. deMicheli, and R. Gandolfi, SYNTHESIS, 710 (1974).
44. R.A. Crochet, Jr. and C.D. Blanton, Jr., SYNTHESIS, 55 (1974).
45. G.W. Gribble, et.al., J. AMER. CHEM. SOC., 96, 7812 (1974).
46. J. Hooz, S. Akiyama, et.al., J. AMER. CHEM. SOC., 96, 274 (1974).
47. C.F. Lane, J. ORG. CHEM., 39, 1437 (1974).
48. H.C. Brown, E.F. Knights, and C.G. Scouter, J. AMER. CHEM. SOC., 96, 7765 (1974).
49. B. Fleming and H.I. Bolker, CAN. J. CHEM., 52, 889 (1974).
50. S.W. Breuer and F.G. Thorpe, TETRAHEDRON LETT., 3719 (1974).
51. D.S. Matteson, R.A. Davis, and L.A. Hagelee, J. ORGANOMETAL. CHEM., 69, 45 (1974); D.S. Matteson, P.B. Tripathy, op.cit. 69, 53 (1974); P.S. Matteson, M. Farve, op.cit. 69, 63 (1974).
52. R.H. Prager and J.M. Tippett, AUST. J. CHEM., 27, 1457 (1974).
53. A. Meller and W. Gerger, MONATSH. CHEM., 105, 684 (1974).
54. L. Maya and A.B. Burg, INORG. CHEM., 13, 1522 (1974).
55. D.S. Matteson and R.A. Davis, INORG. CHEM., 13, 859 (1974).
56. L.I. Zakharkin, A.V. Kazautsev, and B.T. Ermaganbetov, J. GEN. CHEM., USSR, 44, 220 (1974).
57. R.R. Rietz and M.F. Hawthorne, INORG. CHEM., 13, 759 (1974).
58. J. Plesek and S. Hermanek, COLL. CZECH. CHEM. COMMUN., 39, 821 (1974).
59. V. Gregor and J. Stuchlik, COLL. CZECH. CHEM. COMMUN., 38, 3623 (1973).
60. G.D. Merle and F.R. Scholer, INORG. CHEM., 13, 2256 (1974).
61. W.M. Maxwell, V.R. Miller, and R.N. Grimes, J. AMER. CHEM. SOC., 96, 7116 (1974).
62. J. Rathke and R. Schaeffer, INORG. CHEM., 13, 760 (1974).
63. A.P. Schmitt and R.L. Middaugh, INORG. CHEM., 13, 163 (1974).
64. J. Rathke, D.C. Moody and R. Schaeffer, INORG. CHEM., 13, 3040 (1974).
65. J. Rathke and R. Schaeffer, INORG. CHEM., 13, 3008 (1974).
66. K.G. Myakishev, I.S. Posnaya, and V.V. Volkov, RUSS. J. INORG. CHEM., 19, 761 (1974).
67. R.H. Cragg and M. Nazery, J. C. S., DALTON TRAN., 162 (1974).
68. R.S. Lyle and D.A. Walsh, J. ORGANOMETAL. CHEM., 67, 363 (1974).
69. M. Naruse, T. Tomita, K. Utimoto, and H. Nazaki, TETRAHEDRON, 30, 835 (1974).
70. N.E. Miller, INORG. CHEM., 13, 1459 (1974).
71. S.H. Dandegaonker and A.S. Mane, J. IND. CHEM. SOC., 50, 622 (1973).
72. R.H. Cragg and M. Nazery, INORG. NUCL. CHEM. LETT., 10, 481 (1974).
73. O.T. Beachley, Jr. and T.R. Durkin, INORG. CHEM., 13, 1768 (1974).
74. P.C. Keller, J. AMER. Chem. SOC., 96, 3078 (1974).
75. A. Meller, W. Maringgele, and G. Maresch, MONATSH. CHEM., 105, 637 (1974).
76. L. Weber and G. Schmid, ANGEW. CHEM. INT. ED., 13, 467 (1974).
77. H. Noth and W. Tinhof, CHEM. BER., 107, 3806 (1974).
78. R. H. Cragg and M. Nazery, J. C. S., DALTON TRAN., 1438 (1974).
79. R.M. Kren, M.A. Mathur, and H.H. Sisler, INORG. CHEM., 13, 174 (1974).
80. R.T. Markham, E.A. Dietz, Jr., and D.R. Martin, J. INORG. NUCL. CHEM., 36, 503 (1974).
81. M.L. Denniston and D.R. Martin, J. INORG. NUCL. CHEM., 36, 1467 (1974).

82. P.M. Boorman and D. Potts, CAN. J. CHEM., 52, 2016 (1974).
83. M.L. Denniston and D.R. Martin, J. INORG. NUCL. CHEM., 36, 2175 (1974).
84. B.W. Benton and J.M. Miller, CAN. J. CHEM., 52, 2866 (1974).
85. M.G. Hu and R.A. Geanangel, INORG. CHIM. ACTA, 10, 83 (1974).
86. P.S. Bratt, M.P. Brown, and K.R. Seddon, J. C. S., DALTON TRAN., 2161 (1974).
87. P.C. Keller, J. AMER. CHEM. SOC., 96, 3073 (1974).
88. L.V. Titov, L.A. Gavrilova, L.N. Erofeev, and V.Y. Rosolovskii, BULL. ACAD. SCI., U.S.S.R., CHEM. DIV., 23, 5 (1974).
89. V.I. Mikheeva and L.N. Tolmacheva, RUSS. J. INORG. CHEM., 19, 665 (1974).
90. R.J. Lagow and J.L. Margrave, INORG. CHIM. ACTA, 10, 9 (1974).
91. K.V. Titova, I.P. Vavilova, and V.Y. Rosolovskii, RUSS. J. INORG. CHEM., 18, 597 (1973).
92. D.R. Crist, G.J. Jordan, and J.A. Hashmall, J. AMER. CHEM. SOC., 96, 4927 (1974).
93. D.R. Crist, Z. Hsich, G.J. Jordan, F.P. Schinco, and C.A. Maciorowski, J. AMER. CHEM. SOC., 96, 4932 (1974).

ALUMINUM, GALLIUM, INDIUM AND THALLIUM

B.H. Freeland and D. G. Tuck

Department of Chemistry,
University of Windsor,
Windsor, Ontario N9B 3P4, Canada.

INTRODUCTION

This review follows the general pattern of previous treatments (1,2). Main group preparative chemistry continues to offer problems no less than those encountered in other parts of the periodic table.

SYNTHETIC METHODS

Hydride Complexes: A new route to the alkali metal fluoroborates and fluoroaluminates (3) involves direct fluorination of the corresponding hydride complex.

The tetra n-butylammonium salts of the AlH_4^- and GaH_4^- ions have been prepared (4) from $K[MH_4]$ and $(nBu)_4NBr$ in THF. Aluminum trihalides react with $[nBu_4N]$ $[BH_4]$ in benzene to form $[nBu_4N][AlX_4]$ and $[nBu_4N][AlX_2(BH_4)_2]$ (5). These products are also formed initially in THF, but further reaction then produces, *inter alia*, $[nBu_4N][B_2H_7]$. In a study of the course of this reaction, the formation of the same products from $[Bu_4N][Al(BH_4)_3Br]$ in THF has been investigated *via* Al^{27} NMR (6).

Diethyl ether provides a convenient solvent for the preparation of mixed borohydride-chloride complexes of Al (7), isolated as the etherates;

$$LiBH_4 + AlCl_3 \xrightarrow{Et_2O} Al(BH_4)Cl_2.OEt_2$$

$$AlCl_3 + 2Al(BH_4)_3.OEt_2 \longrightarrow 3Al(BH_4)_2Cl.OEt_2$$

A series of compounds of unusual stoichiometry has been isolated (8) from the reactions of manganese iodides with trimethyl- or triethylaluminum in THF:

$$MnI_2 + Et_3Al \longrightarrow Mn(H)I\ (THF)_{1.5}$$

$$MnI_2 \xrightarrow{I_2} MnI_{10},(THF)_n \xrightarrow{AlR_3} MnAl_2R_4I_4(THF)_4$$

$$\rightleftharpoons MnI_2.(THF)_n + R_2AlI.THF$$

(R = Me, Et)

A hydride complex, presently formulated as $Mn_2AlH_4I_6(THF)_8$, has also been isolated.

An infrared study (9) of the reaction of AlH_3 with $BeCl_2$, produced no evidence to support earlier claims for the presence of species such as $LiAl_2H_7$. The reaction proceeds in stepwise fashion.

$$BeCl_2 \xrightarrow{AlH_3} H_2AlCl + HBeCl \xrightarrow{AlH_3} H_2AlCl + H_2Be$$

The use of $LiAlH_4$ to produce GeH_4 from $GeCl_4$ has also received attention (10); the protonic intermediate $H^+GeCl_3^-$ is suggested. $SnCl_4$ behaves analogously.

The reaction of $Ca[AlH_4]_2$ with analine yields crystals of calcium-N-phenyliminodialane (11). On the basis of vibrational spectra and ebulliometric results, the following structure is postulated.

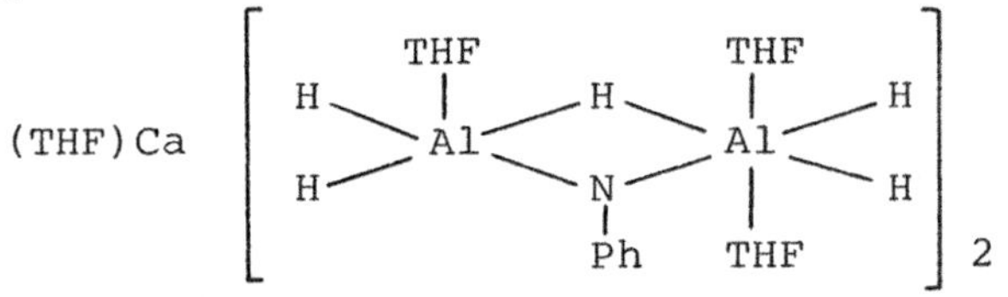

A study of the triethylamine adducts of $AlH_{3-x}Cl_x$, (x = 0-3), indicates (12), on the basis of PMR data, that the Lewis acid character of Al increases as Cl is substituted for H. The x-ray crystal structure determination (13) of the $AlCl_3.NH_3$ adduct indicates a very strong Al-N linkage (1.996Å). These initially formed adducts react further with ammonia to produce species containing the $[Al(NH_3)_6]^{3+}$ ion (14). The trimethylamine adduct of AlH_2I undergoes a nucleophilic displacement reaction with $Me_2NCH_2CH_2NMe_2$(L) to form the four coordinate ionic complex $[AlH_2L]I$ (15). In contrast $Me_3N.AlHBr_2$ reacts with L to form the neutral five coordinate complex $[AlHBr_2L]$.

The N-methyl(diethanolamino)gallane dimer has been prepared by reaction N-methyldiethanolamine with trimethylamine-gallane in benzene (16).

$$MeN(CH_2CH_2OH)_2 + Me_3N.GaH_3 \longrightarrow MeN(CH_2CH_2O)_2GaH + 2H_2\uparrow + Me_3N$$

An x-ray structure of this complex has identified units which are dimerized in a novel manner through bridging oxygen atoms producing a distorted trigonal bipyramidal arrangement about each five coordinate gallium atom.

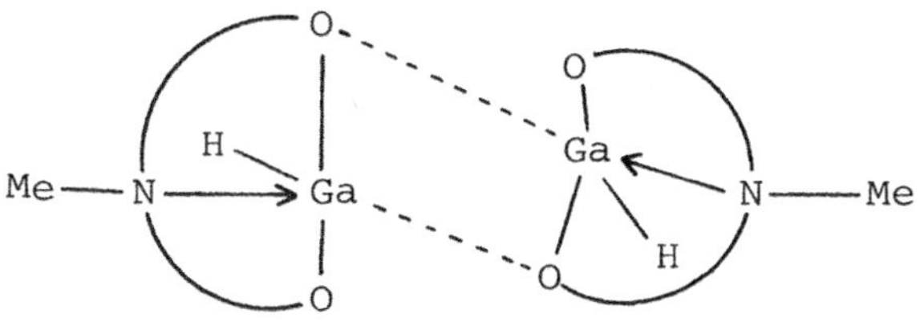

Sesqui(alkoxyalkoxy)aluminum hydrides have been prepared by the treatment of the sodium or lithium salts of $[AlH_2(OQOR)_2]^-$ with the diethyl ether adduct of $AlCl_3$ in benzene (17).

Activated Metals: Several recent papers (18,19) on the synthesis of alkyl- and aryl- metal halide derivatives of groups III metals report the use of an activated metal, prepared by the treatment of the corresponding group III trihalide with potassium or sodium metal in xylene. Indium prepared in this way reacts with iodine in xylene to form InI. While so far used only to prepare known derivatives such as Me_2InI, $PhAlBr_2$ and Ph_2AlBr, this method appears to offer some potential for new syntheses.

Alkyl and Aryl Compounds: The reaction of alkaline earth metals and Et_3Al to form $M[AlEt_4]_2$ compounds (M = Ca, Sr, Ba) has been examined in detail (20).

$$8AlEt_3 + 3M^o \longrightarrow 3M[AlEt_4]_2 + 2Al^o$$

The yields from these reactions are excellent when trialkylaluminum is employed as the solvent medium. Diethyl ether and THF strongly inhibit this reaction while diglyme accelerates it (21). In the latter case, the product formed is the bis(diglyme) adduct, which can also be prepared by the addition of diglyme to $M[AlEt_4]_2$ in benzene or hexane.

A simple method of producing Ph_3Al involves the dehalogenation of phenylaluminum chlorides (22).

$$3Ph_2AlCl + 2AlR_3 \longrightarrow 2AlPh_3 + 3AlR_2Cl$$

$$Ph_3Al_2Cl_3 + 2AlR_3 \longrightarrow AlPh_3 + 3R_2AlCl$$

(R = Me, Et)

The volatile R_2AlCl and nonvolatile Ph_3Al are readily separated by fractional distillation. This method eliminates the tedious separation of Ph_3Al from the solid salts of alkali or alkaline earth metals required in other reaction schemes.

Only one metal-carbon bond in the trialkyl compounds of Al, Ga, In and Tl is cleaved by SO_3 (23). The dialkylmetal-alkylenesulphonate derivatives formed may also be obtained by treating the trialkyl compounds with alkylene sulphonic acids. The Al and Ga compounds are dimeric or trimeric in benzene, with bridging $R(O)SO_2$ groups. The In and Tl compounds are freely soluble in water, dissociating into R_2M^+ cations and RSO_3^- anions.

The product isolated from the reaction of Ph_3In and SO_2, previously considered to be the o-benzenesulphinatodiphenyl-indium, by insertion of SO_2 into an In-C bond, is now formulated as the 1:1 adduct, $Ph_3In(O_2S)$ (24). The proposed structure is,

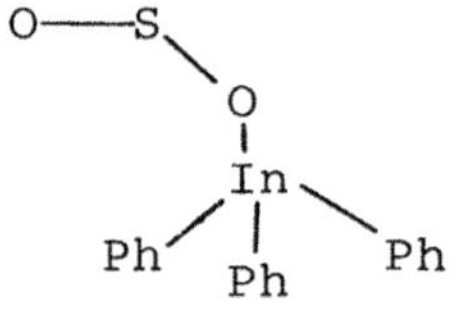

The intermediate product in the C-methylation reaction of dimethylaluminumacetylacetonate with trimethylaluminum has now been isolated (25), confirming that opening of the chelate ring occurs.

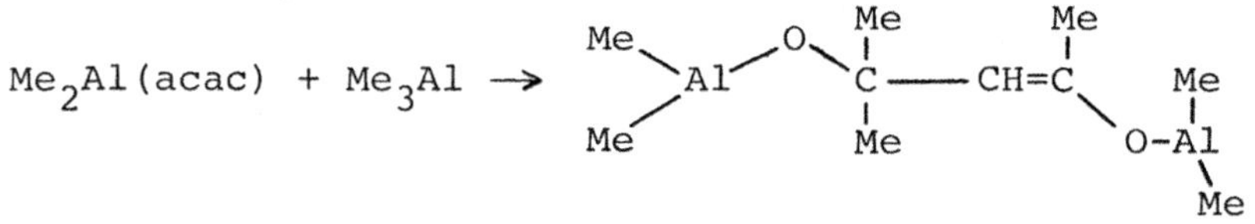

Me_2AlCl behaves analogously, but with $MeAlCl_2$, C-methylation does not occur. The initial step in the reaction is postulated as the formation of the Lewis acid-base adduct;

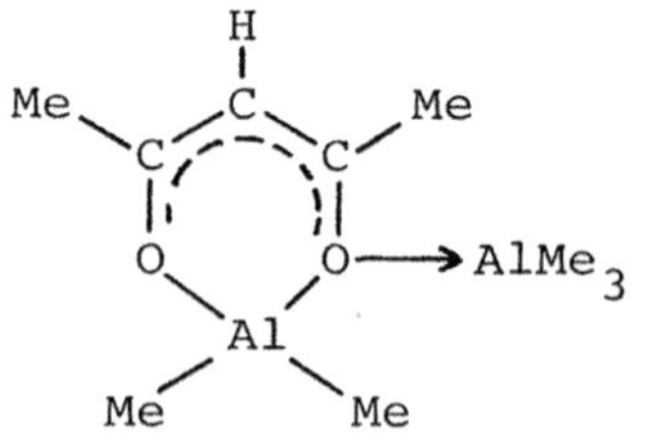

Similar intermediates have been proposed in the reaction of $Co(acac)_3$ with $AlEt_3$ (26). MeAl(Cl)(acac) and $Cl_2Al(acac)$ do not undergo C-methylation with those aluminum alkyls studied (25).

Tris(trimethylsilylmethyl)aluminum, prepared from $Hg(CH_2SiMe_3)$ and aluminum foil (27), is a colourless, pyrophoric liquid which appears to exist as a mixture of monomer and dimer in benzene at room temperature. Reaction with monomethylamine yields the dimeric amido derivative $[(Me_3SiCH_2)AlNHMe]_2$. A variety of trimethylsilylmethyl derivatives of Tl have also been prepared by metathesis (28). Unlike the simple dialkylthallium chlorides, for which polymeric structures containing linear C-Tl-C units predominate in the solid state, all of these silyl derivatives exist as halogen bridged dimers.

PERFLUOROPHENYL COMPOUNDS:

Perfluorophenyl compounds of Al have been prepared by the following route (29):

$$ClAlH_2 + C_6F_5Li \xrightarrow{Et_2O} AlH_{3-x}(C_6F_5)_x.nL$$

$$(X = 1,2; \quad n = 1,2) \qquad (L = Et_2O,\ NMe_3)$$

The dioxane adduct of $(C_6F_5)InCl_2$ has been shown to react with bidentate ligands such as bipy and 1,10-phenanthroline forming 1:1 adducts in which indium is five coordinate (30). Pyridine, DMSO or Ph_3P cause the dioxane adduct to decompose:

$$C_6F_5InCl_2.diox \xrightarrow{B} InCl_3.B_n + (C_6F_5)_3In$$

The tris(perfluorophenyl)In.diox reacts with quinol-8-ol to form the oxinates; $(C_6F_5)_nIn(ox)_{3-n}$, (n = 1,2).

ORGANOMETALLIC

Reactions of alkyl- or arylmetal halides: Bis(dimethylaluminum)sulphate has been prepared by the reaction of Me_2AlCl and Na_2SO_4 in cyclohexane (31). This material is soluble in both benzene and pentane.

Methylenedialuminumtetrabromide, $H_2C(AlBr_2)_2$, which is easily obtained from the reaction of Al metal with CH_2Br_2 (32), reacts with aldehydes in THF to form methylenic olefins in good yield (33):

$$RCHO + H_2C(AlBr_2)_2 \longrightarrow RCH{=}CH_2$$

Under similar conditions ketones do not yield the corresponding olefins, but the reaction of $LiCH_2AlBr_2$ from $H_2C(AlBr_2)_2$ and BrLi, does form methylenic olefins on reaction with fluorenone and decanol.

Dimethyl- or diphenylthallium halides react with the disodium salts of dianionic tetradentate ligands with S, N and

and O donor atoms to form anionic complexes, $[R_2TlL]^-$, (34), viz.;

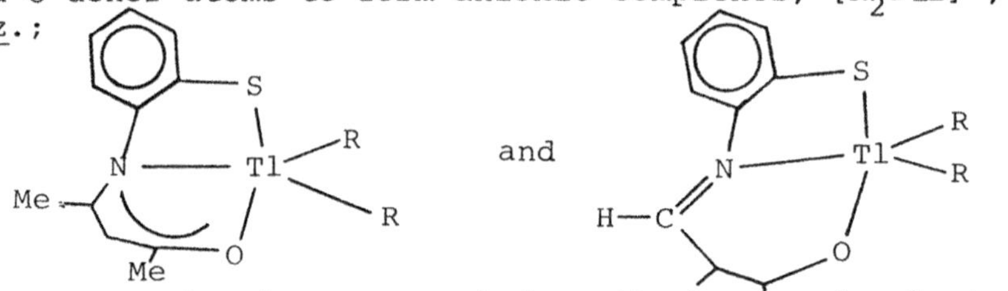

A structure has been suggested on the basis of mol.wt., IR and NMR for bis(chloromethyl)alumoxane (35), formed on the partial hydrolysis of the diethylether or benzonitrile complexes of Me_2AlCl.

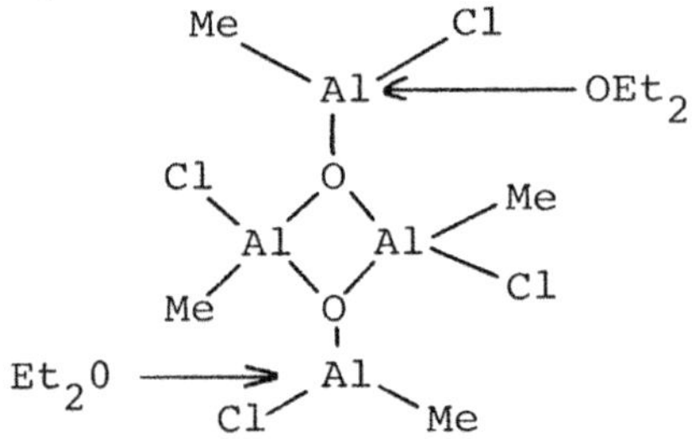

Cyclic derivatives have been formed via an addition complex on reaction of diacetyldianil (DAD) with alkylaluminum halides (36).

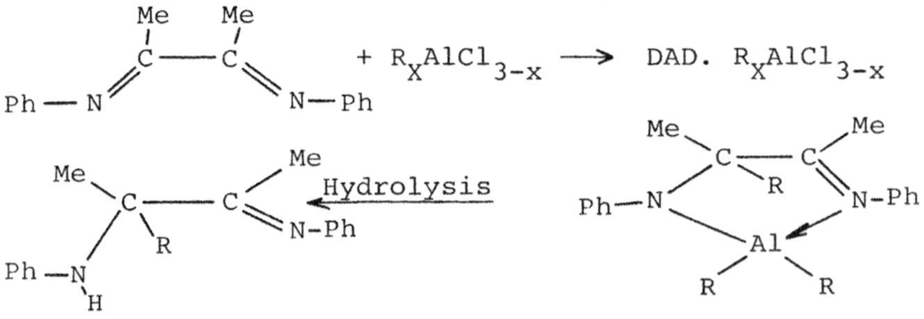

Alkane Elimination Reactions: Bis(dimethylaluminum)sulphide has been prepared by the reaction of Me_3Al and liquid H_2S (37). The compound forms a benzonitrile complex and may form both a mono- and a bis- adduct with pyridine.

The low pressure distillation of the crude product from the $Ni(acac)_2$ catalysed 1,4-addition of Me_3Al to mesityl oxide yields Me_2Al(Z)-4,4-dimethylpent-2-en-2-olate, which is dimeric in benzene (38). At ca. 100-150° this compound isomerizes to form Me_2Al(E)-4,4-dimethylpent-2-en-2-olate which may be separated into a dimeric distillate and a trimeric residue.

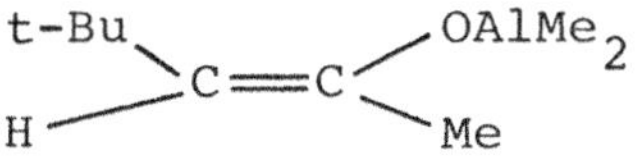

(Z)-enolate dimer
mp 63°

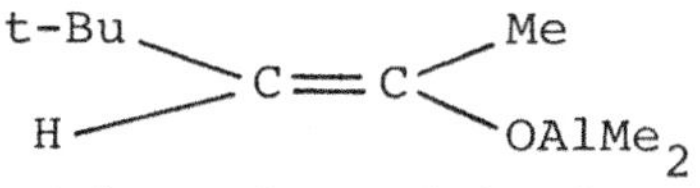

(E)-enolate (a) dimer mp 27
(b) trimer(liquid)

The enolates form monomeric 1:1 complexes with bases such as Me_3NO, HMPA, Ph_3PO and pyridine, and also undergo condensation reactions with acetaldehyde and benzaldehyde to form complexes, some of which appear to contain five coordinate aluminum.

Triethylaluminum reacts with dialkyl-N-methylphosphoramidates to form the corresponding diethylaluminum derivatives in high yield (39).

$$(RO)_2P(O)NHMe + Et_3Al \xrightarrow[80^{\circ}]{C_6M_6} (RO)_2P(O)N(Me)AlEt_2 + C_2H_6\uparrow$$

Tetraalkylphosphorodiamidic-N-methylamide behaves analogously. These compounds are dimeric in benzene when R is small. The introduction of bulky groups results in a lowering of the degree of association. The infrared spectra of these compounds have been rationalized in terms of structures in which $(p\text{-}d)\pi$ bonding between P and N is involved.

In the reaction of $(RO)_2P(O)N(Me)AlEt_2$ with aromatic aldehydes or phenylisocyanate the corresponding imino compounds were obtained by an addition-elimination process.

The alkane elimination route has again proven useful for the preparation of a series of phosphoric and phosphinic acid derivatives of group III metal dialkyls (40).

$$Me_3M + HX \xrightarrow[\text{or } Et_2O \text{ or pentane}]{C_6H_6} Me_2MX \qquad (M = Al, Ga, In, Tl)$$

$$(X = O_2PF_2,\ O_2PCl_2,\ O_2PH_2,\ OSPMe_2,\ S_2PMe_2)$$

The ligands behave as bidentate donors and the compounds are formulated as monomeric, dimeric or trimeric ring systems.

Diborane, B_2H_6, reacts readily with $Al(NMe_2)_3$ in diethyl ether to form, inter alia, $H_2B(NMe_2)_2Al(BH_4)_2$, for which the following structure has been proposed (41).

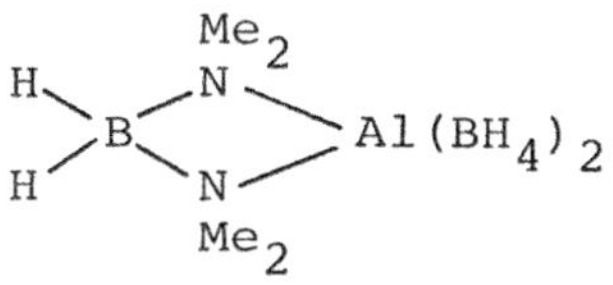

Bis(dimethylgallium)oxalate has been prepared from Me_3Ga and oxalic acid (42). The x-ray crystal structure shows two fused coplanar five membered GaO_2C_2 rings.

Cyclic dimeric dialkylgallium and -indium derivatives of N-methylacetamide have been prepared, from R_3M (43).

$$MeNHAc + R_3M \longrightarrow [R_2MNMeAc]_2 \quad (R = Me, Et;\ M = Ga, In)$$

Aluminum trialkyls, triaryls and dialkyl- or diaryl chlorides react with salicaldoxime or dimethylglyoxime (44) to form dimeric products, e.g.,

R_3Al + salicaldoxime →

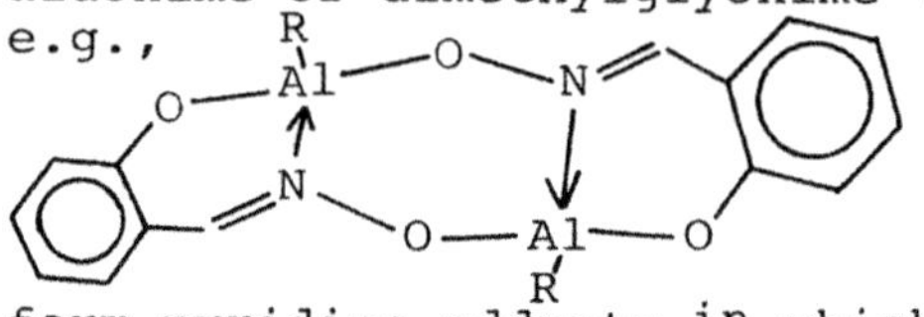

These complexes also form pyridine adducts in which the Al atoms are presumably five coordinate. Salicaldoximato metal (Ni,Pd) complexes react analogously with $Al(iso\text{-}Bu)_3$ (45).

A variety of indium(III) carboxylates have been prepared from In metal or $InMe_3$ and the anhydrous acid (46).

$$InMe_3 + RCO_2H \longrightarrow In(O_2CR)_3$$

$(R = H, Me, Et, n\text{-}Pr, Me_2CH, Me_3C)$

Hydrolysis of the tertiary-butyl compound forms a hydroxy derivative $In(O_2CR)_2OH$, for which eight coordinate indium has been proposed.

The thermal decomposition of $AlMe_3.NH_2Me$ at 215° in toluene results in the formation of a polymeric species $(MeAlNMe)_n$. Cryoscopic measurements in benzene suggested that the molecular formula is octameric, but x-ray crystallography confirms a heptameric cluster (47). The molecule has a cage structure, each Al and each N is four coordinate, bound to one methyl group and three atoms in the cage.

A study has been made of the stereoisomers of cyclo-di-μ-tert-butylamino bis(ethylbromoaluminum),$[Et(Br)AlN(H)CMe_3]_3$, formed on the thermal decomposition of $Et_2(Br)Al.NH_2CMe_3$ (48).

Triphenylaluminum reacts with the enol forms of 1,3-dicarbonyls to produce chelate complexes, $Ph_2Al(O_2C_3R_2)$ with the elimination of benzene (49). These compounds also form adducts with 1,10-phen and bipyridyl.

<u>Miscellaneous Complexes with O, N, S and P ligands:</u> A series of Tl(I) hexafluoroacac complexes with bipy, 1,10-phen and $Me_2NC_2H_4NMe_2$ ligands have been prepared and their IR and NMR spectra discussed (50). The reaction between indium monohalides and acetylacetone produces mixtures of $In(acac)_3$ and $InX_2(acac)$ (51); the latter form $InX_2O_2N_2$

adducts with ligands such as bipy and phen. Tris(acac)-gallium, prepared as orange crystals from Na(acac) and $GaCl_3$ in methanol, has been the subject of an x-ray crystallographic investigation (52). In related work, anhydrous $GaCl_3$ reacted with the sodium salt of 8-hydroxy-2-methylquinoline in methanol to form green crystals of chlorobis(8-hydroxy-2-methylquinolinato)gallium(III) (53). X-ray diffraction reveals that this compound is an example of the rare five coordinate state in Ga(III).

Aluminum and gallium trihalides react with trichloroacetonitrile to form 1:1 N-bonded adducts in contrast to the reactions of BX_3 derivatives, which add across the CN triple bond (54).

The treatment of porphyrins with Tl(III) acetate or trifluoroacetate affords the corresponding Tl(III) chelates, which undergo ligand exchange on chromatography columns to yield the corresponding aquoporphinatohydroxides (55). The complexes are readily demetallated on treatment with acid, or by reduction of the Tl(III) to Tl(I). Spectral results suggest that the metal centre is situated out of the plane of the porphyrin ring.

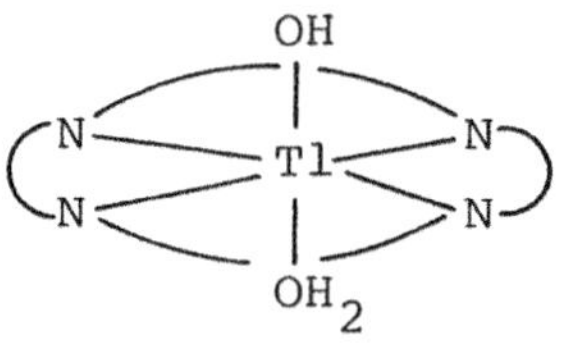

$AlCl_3$ dissolves readily in hot PCl_3, but compound formation does not occur (56). In contrast, $AlCl_3$ and PF_3 under pressure form monomeric $F_3P.AlCl_3$, which undergoes halogen exchange to PCl_3 and AlF_3. There is some evidence for interaction between $AlMe_3$ and PF_3, but no reaction occurs between $AlMe_3$ (or $AlCl_3$) and CO.

Dimethylaluminumamides, $Me_2Al\,NR(Ph)$, have been shown to react with CS_2 to form compounds, $Me_2AlS_2CN(Ph)R$ (57).

The crystal structure of thallousdimethyldithiocarbamate (58), has established a dimeric molecular formula, $[Tl(S_2CNMe_2)]_2$ and a structure based on seven coordinate Tl.

Aluminum- and galliumbisulphate complexes have been prepared by the controlled hydrolysis of the corresponding chlorosulphate (59). Gallium and indium perchlorates form addition compounds with thioxane oxide and tetramethylene sulphoxide (60); coordination to the central metal involves the sulphoxide oxygen.

Diethylaluminum dimethylamide and diethylaluminum ethane thiolate, Et_2AlX, (X = NMe_2 or SEt) react with diketene, by aryl oxygen bond cleavage, followed by a 1,3-hydrogen shift to give the corresponding diethylaluminumacetoacetic acid dimethylamide (a) or acetothioloacetic ethylacid ester (b) both of which have chelate structures (61).

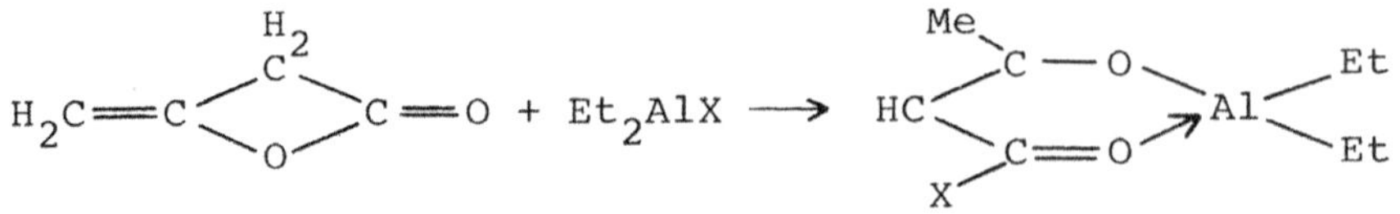

(a) X = NMe_2 (b) X = SEt

Similar complexes are formed by insertion on reaction with ketenimines (62).

METAL-NITROGEN RING SYSTEMS

The reactions of MMe_3 or $Me_3N.MMe_3$ (M = Ga, Al) with pyrazole result in pyrazolyl derivatives of Al and Ga (63). The compounds are volatile, tricyclic dimers. The dideuterio (pyrazol-1-yl) gallane dimer has been the subject of an x-ray crystallographic study (64). In all cases the Ga_2N_4, six membered ring system exists in the boat form.

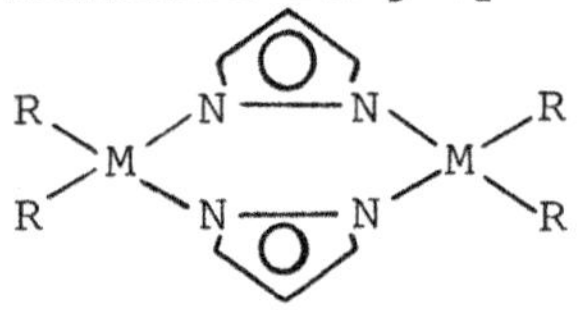

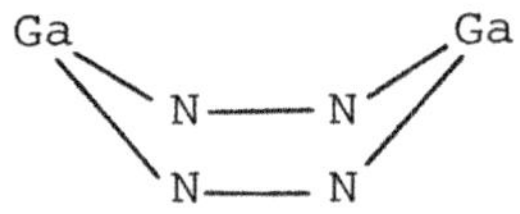

HALOGEN BRIDGED COMPOUNDS:

Analogues of Ziegler-Natta complexes with aluminum-halogen-metal bridges continue to attract interest.

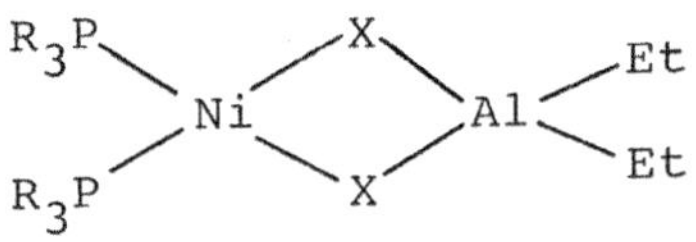

complexes have been prepared from $(R_3P)_2NiX_2$ and $AlEt_3$ (65).

($X = NO_3^-$, SCN^-, Cl^-, Br^-, I^-)

An x-ray crystallographic study (66) has confirmed a chlorine bridge in the zirconium complex.

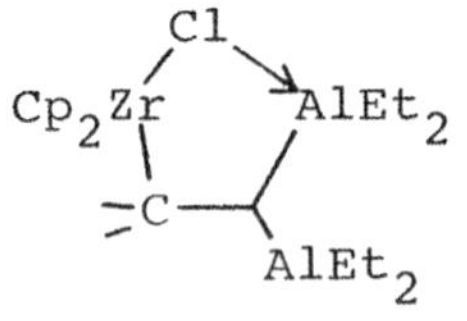

Kovar et al., have developed convenient syntheses for the preparation of base free R_nGaCl_{3-n} compounds (R = n-butyl; n = 1,3) (67). n-Butyllithium in benzene is added to a solution of $GaCl_3$ in benzene in the required stoichiometry. This report also includes experimental details and a description of apparatus used for the preparation of $GaCl_3$ from Ga and Cl_2. The reaction involves the intermediate formation of $Ga[GaCl_4]$. Cryoscopic molecular weight measurements in benzene indicate the formation of dimers with $RGaCl_2$ and R_2GaCl although $(n\text{-}Bu)_3Ga$ is reported to be monomeric. Chlorine bridged structures are proposed.

On the basis of vibrational spectra the products $R_3In_2X_3$, obtained from the reaction of indium metal and alkyl halides, RX, (R = Me, Et, n-Pr, n-Bu; X = Br,I), are now formulated as true sesquihalides rather than intimate mixtures of $RInX_2$ and R_2InX (68). These sequihalides provide a convenient route to the dialkyl indium halides by reaction with KBr or KI followed by sublimation of the R_2InX.

$$R_3In_2X_3 + KX \longrightarrow R_2InX + K^+ \; RInX_3^-$$

A continuing investigation of the insertion reaction of InBr into RBr compounds (69) has now resulted in the isolation of several intermediate species, some of which are formulated as the subhalides, In_7Br_9 and In_5Br_7. The proposed reaction sequence is:

$$9InBr + 3RBr \longrightarrow \underset{(yellow)}{In_7Br_9} + RInBr_2 + R_2InBr$$

$$8InBr + 2RBr \longrightarrow In_7Br_9 + R_2InBr$$

$$In_7Br_9 + R_2InBr + 6RBr \longrightarrow 8RInBr_2$$

$$In_7Br_9 + 2RBr \longrightarrow \underset{(white)}{In_5Br_7} + 2RInBr_2$$

$$In_5Br_7 + 3RBr \longrightarrow 3RInBr_2 + In_2Br_4$$

An intermediate with an ionic formulation has also been proposed.

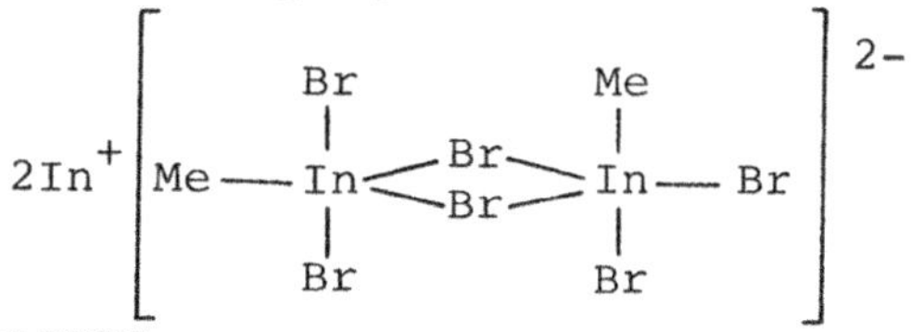

HALIDE AND PSEUDOHALIDE COMPLEXES:

The formation of tetracoordinate monoanions of group III metals with halides and pseudohalides continues to be an active field with structural studies commonplace, e.g., $K[MeAlCl_3]$ (70), $K[Al_2Me_6F].C_6H_6$ (71), $K[R_3GaF]$ and $Me_4N[(R_3Ga)_2X]$ (72) and $Me_4N[Me_3Tl(CN)]$ and $Me_4N[Me_3Tl)_2F]$ (73).

The thermal decomposition of the benzene and toluene complexes, Rb or $K[Al_2Me_6N_3]$, yields the corresponding alkali metal tetramethylaluminate and dimethylaluminumazide (74), according to the equation:

$$M[Al_2Me_6N_3].\ Arene \xrightarrow{180^\circ} M[AlMe_4] + Me_2AlN_3$$

(M = Rb, K)

The azido complex has an unexpected thermal stability.

The formation of an (isopropylideneamino) complex in the thermal decomposition of $K[Al_2Me_6(SCN)]$;

$$K[Al_2Me_6(SCN)] \longrightarrow [Me_2AlNCMe_2]_2$$

suggests the possibility of a "head-to-tail" bridged reaction intermediate (75).

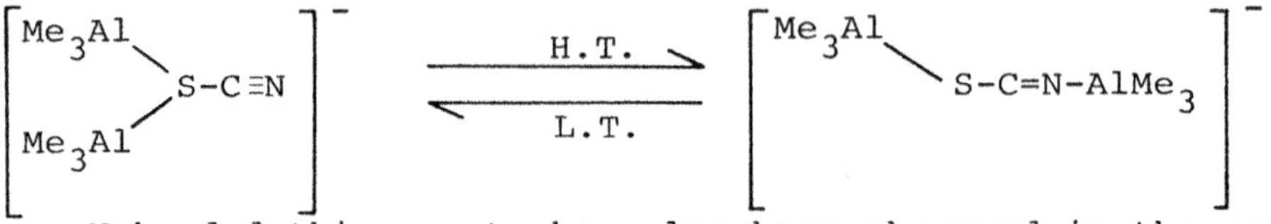

N-bonded thiocyanate has also been observed in the complex $[Me_2Tl][Me_3Al(NCS)]$ (76); with the K^+ counterion, thiocyanate is S-bonded (77).

The $[Me_2Tl][AlMe_3(NCS)]$ salt has been prepared from TlSCN and $AlMe_3$.

$$3TlSCN + 3AlMe_3 \rightarrow [Me_2Tl][AlMe_3NCS] + 2Tl + 2Me_2AlSCN$$

The coordination sphere about the Tl atom consists of the two methyl groups and four sulphur atoms in an essentially octahedral array. This coordination of sulphurs with the Tl may provide the driving force for the stabilization of the N-bonded thiocyanate on aluminum.

Thermal decomposition has provided a route to the trimethylhydroaluminate ion which has also been the subject of an x-ray crystallographic analysis (78).

$$K[AlMe_3SiH_3] \longrightarrow K[AlMe_3H] + (SiH_2)_n$$

Ligand exchange between $GaMe_3$ and $K_2[Hg(CN)_4]$ provides a straightforward synthesis of the dimethyldicyanogallate ion (79).

$$K_2[Hg(CN)_4] + 2GaMe_3 \longrightarrow 2K[Me_2Ga(CN)_2] + HgMe_2$$

COMPLEXES WITH Π ACID LIGANDS

Kumar et al., have continued their work on the preparation of cyclopentadienyl thallium compounds with the synthesis of Cp_2TlCl and $(Indenyl)_2TlCl$ from $(pyH)_2TlCl_5$ and the appropriate sodium salt (80). The products react in THF with the sodium salts of a variety of pseudohalide derivatives to form involatile, insoluble Cp_2Tl derivatives (81).

$$Cp_2TlCl + NaX \xrightarrow{THF} Cp_2TlX \qquad (X = N_3^-, NCO^-, CN^-, NCS^-)$$

Cp_2TlBH_4 and $(Ind)_2TlBH_4$ have been prepared similarly (82); the vibrational spectra indicate that the BH_4 group is bonded to the metal via two hydrogen bridges. Utilizing the same dipyridium $[TlCl_5]$ salt as starting material these workers have also obtained thallium cyclooctatetrene complexes by the reaction with the dipotassium salt of COT (83).

$$(pyH)_2TlCl_5 + K_2(COT) \xrightarrow{THF} (COT)TlCl$$

Treatment of this product with potassium metal yields the potassium salt of a COT complex of Tl(I). The vibrational spectra are consistent with an ionic formulation for the Tl(III) complex while a covalent bonding scheme is suggested for the Tl(I) material. A mixed oxidation state complex, $(COT)_2Tl^{-}Tl^{+}$, has also been prepared from the reaction of (COT)TlCl and the sodium salt of COT in THF (84).

Cyclopentadienylindium(I) has proven to be a useful starting material for the preparation of a number of In(III) complexes. It reacts with iodine in $CHCl_3$ to form $CpInI_2$ which reacts further with iodide ion to form the anion $[CpInI_3]^-$ (85). Adducts of $CpInI_2$ with 2,2'-bipyridyl and 1,10-phenanthroline have also been prepared. Reaction of $CpInI_2$ with Cp_3In yields the unstable Cp_2InI.

Tris(allyl)boron is an effective exchange reagent for the preparation of allyl-aluminum derivatives on reaction with Me_3Al or Et_3Al (86).

In the reductive Friedel-Crafts reaction, the $AlCl_4^-$ ion once again appears to play an important role in stabilizing a metal complex with a π-arene ligand (see (1),ref.74). The anions occupy bridging positions in the x-ray crystal structure of $\pi\text{-}C_6H_6Pb(AlCl_4)_2 \cdot C_6H_6$ (87).

Alkenyl derivatives of Al, Ga and In have been shown to undergo intra molecular cyclization reactions (88). A mechanism based on the interaction between the metal centre and the π-system of the olefin has been proposed.

METAL CARBONYL COMPLEXES:

Attention continues to be focused on complexes of group III elements and transition metal carbonyls. Lewis acid-base adduct formation between carbonyl oxygens and group III halides now appears to have an important effect on the rearrangements of carbonyl groups in polynuclear metal carbonyls (89). In a related case, a series of rhenium dinitrogen complexes interact with $InCl_3$ with elimination of the dinitrogen ligand (90). Occasionally, an oxidation-reduction reaction occurs resulting in the formation of the corresponding cationic dinitrogen complex and $InCl_2$ (although the latter was not isolated).

Thallium complexes of metal carbonyls have been prepared by a variety of routes and with numerous metal carbonyl moieties (91,93). For example, $Tl + Co_2(CO)_8$, $Tl + Hg[Co(CO)_4]_2$, $TlNO_3 + Na\ Co(CO)_4$, and $Tl + Tl[Co(CO)_4]_3$ all react to form $Tl[Co(CO)_4]$ (91,92). In addition $[Tl(OEt)_4]$ reacts with $HMn(CO)_5$ to form the corresponding $-Mn(CO)_5$ derivative (92) and Tl + $[CpM(CO)_3]_2$ (M=Cr,Mo) readily forms the tris(metalcarbonyl) derivatives (93).

The redox balance between $Tl[Co(CO)_4]_3$ and $Tl[Co(CO)_4]$ has also been studied (91). The crystal structure of the latter compound has shown it to be essentially ionic although some degree of metal-metal bonding appears to be present in solution. Further it readily forms a complex ion $Tl[Co(CO)_4]_2^-$ suggesting a Lewis acid behaviour for $Tl[Co(CO)_4]$.

In a continuing investigation of the reactions of group III halides with cobalt carbonyls, complexes with the formulae, $MCo_2Br_4(CO)_6$, (M = Ga,In) have been identified for which the ionic formulation

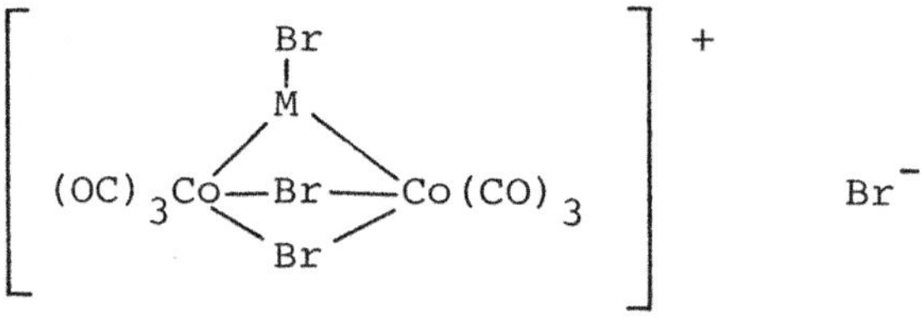

has been suggested (94).

The reactions between gallium or indium metal and manganese carbonyls or their mercury compounds, or pyrolysis of $In[Mn(CO)_5]_3$, yield compounds of the stoichiometry $M_2Mn_4(CO)_{18}$ (95,96). The proposed structure is

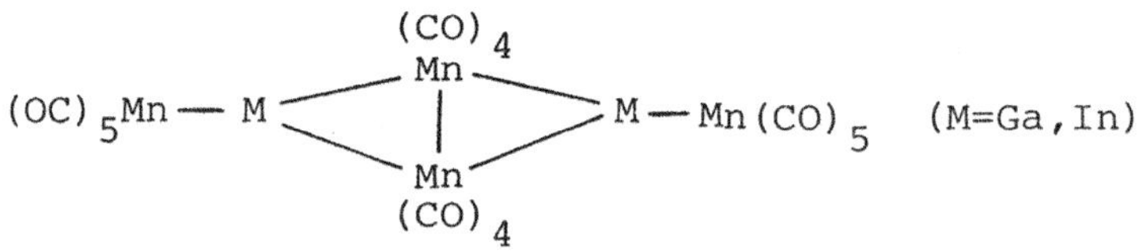

These materials react with bases (e.g.pyridine) to form adducts in which the base is apparently coordinated to the heteroatom. Ionic species such as:

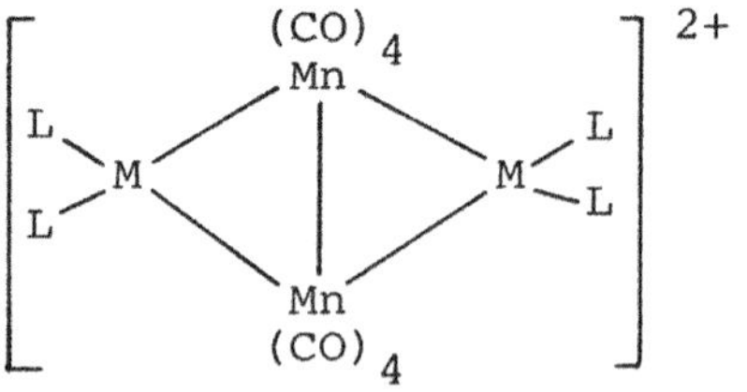

(M = Ga,In; L = py)

are also believed present.

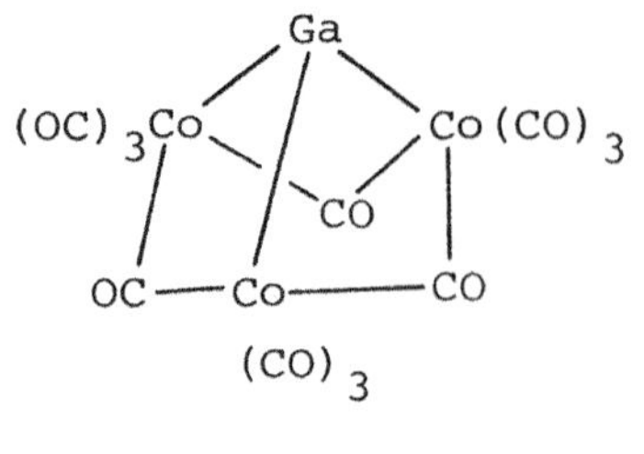

The product $GaCo_3(CO)_{12}$, for which the open cluster structure is proposed, on the basis of ms and ir, results from the reaction of $NaCo(CO)_4$ with $GaBr_3$ in benzene (97).

As part of a continuing study of the stabilization of short lived molecules as ligands in coordination complexes (98), Schmid has replaced a carbonyl group in $Fe(CO)_4Br_2$ with the alumylene, $AlNMe_2$, which is formally a derivative of Al(I). The resulting complex is dimeric, with $AlNMe_2$ units occupying bridging positions between the irons.

METAL-METAL BONDED COMPOUNDS:

The treatment of $GaBr_3$ with Ga metal at 180° results in a compound with the stoichiometry $GaBr_{1.5}$ (99). An ionic formulation, $2Ga^+ \; Ga_2Br_6^{2-}$, is proposed, on the basis of its Raman spectrum. For the similar gallium(II) complex $[n\text{-}Pr_4N]_2[Ga_2Br_6]$ an x-ray diffraction study has revealed a Ga-Ga bond length of 2.419Å (100).

In a study of the system NH_4F/InF_3, phases corresponding to the ions $In_3F_{14}^{5-}$, $In_3F_{11}^{2-}$, $In_3F_{10}^{3-}$ and $In_2F_7^-$ in addition to those of InF_4^- and InF_6^{3-} (101).

In the reaction of diethylthallium chloride with mercury, Et_2Hg_2 is formed at the mercury surface (102). There is also evidence for the formation of the intermediate dimetallic cation, $[EtHgTlEt]^+$

$$Et_2Tl^+ + Hg^0 \longrightarrow EtTlHgEt^+ \xrightarrow{Hg^0} EtHgHgEt + Tl^+$$

The silylaluminum complex $(Ph_3Si)_2AlEt.2LiBr.2THF$ reacts with tolan in hot toluene to eliminate LiBr and form the THF adducts $(Ph_3Si)_2AlEt.THF$ and $Ph_3Si\ AlEt_2.THF$ (103). On the basis of hydrolysis products, which include $Ph_3SiCPh{:}CHPh$, an intermediate formulated $(Ph_3Si)_2AlEt.C_2Ph_2.THF$ is postulated.

Diethyl ether is displaced and a 1:1 adduct formed between Me_3In and $SbEt_3$ when $Me_3In.OEt_2$ and $SbEt_3$ react at elevated temperatures (104).

Dialkylphosphido lithium reacts with $AlCl_3$ to form the quarternary salt $LiAl(PMe_2)_4$ (105).

REFERENCES

1. B.H. Freeland and D.G. Tuck, ANN. REP. INORG. GEN. SYNTHESIS, 2, 18 (1974).
2. J.P. Oliver, ANN. REP. INORG. GEN. SYNTHESIS, 1, 55 (1973).
3. R.J. Lagow and J.L. Margrave, INORG.CHIM. ACTA, 10, 9 (1974).
4. S.I. Bakum and S.F. Ereshko, IZV. AKAD. NAUK. SSSR, Ser. Khim., 2138 (1974).
5. L.V. Titov, L.A. Garilova, L.N. Erofeev and Y.Y. Rosolovski, IZV. AKAD. NAUK. SSSR, Ser. Khim., 10 (1974).
6. L.V. Titov, E.R. Eremin, L.N. Erofeev and V.V. Rosolovskii, IZV. AKAD. NAUK. SSSR, Ser. Khim., 7 (1974).
7. K.N. Semenenko, Y.I. Malov and V.N. Fokin, IZV. AKAD. NAUK. SSSR, Ser. Khim., 2179 (1973).
8. A. Yamamoto, K. Kato and S. Ikeda, J. ORGANOMETAL. CHEM., 60, 139 (1973).
9. E.C. Ashby, P. Claudy and R.D. Schwartz, INORG. CHEM., 13, 192 (1974).
10. N.S. Nametkin, V.K. Korolev, O.V. Kuz'min and K.I. Kobrakov, DOKL. AKAD. NAUK. SSSR, 215, 1385 (1974).
11. S. Cucinella, G. Dozzi and A. Mazzei, J. ORGANOMETAL. CHEM., 63, 17 (1973).
12. V.V. Belov, G.N. Boiko, B.M. Bulychev, Y.I. Malov and K.N. Semenenko, IZV. AKAD. NAUK. SSSR, Ser. Khim., 2171 (1974).
13. M. Hargittai, I. Hargittai and V.P. Spiridona, J. CHEM. SOC., CHEM. COMMUN., 750 (1973).
14. K.N. Semenenko, S.P. Shilkin, O.V. Kravchenko and V.B. Polyakova, IZV. AKAD. NAUK. SSSR, Ser. Khim., 1455 (1974).
15. K.R. Skillern and H.C. Kelly, INORG. CHEM., 13, 2802 (1974).
16. S.J. Rettig, A. Storr and J. Trotter, CAN. J. CHEM., 52, 2206 (1974).
17. B. Casensky, O. Kriz and H. Kadlecova, Z. CHEM., 13, 436 (1973).
18. R.D. Rieke and L. Chao, SYN. INORG. METAL-ORG. CHEM., 4, 101 (1974).
19. L.C. Chao and R.D. Rieke, J. ORGANOMETAL. CHEM., 67, C64 (1974).
20. L.L. Ivanov, S.Y. Zavizion and L.I. Zakharkin, SYN. INORG. METAL-ORG. CHEM., 3, 323 (1973); ZH. OBSHCH. KHIM., 43, 2254 (1973).
21. L.I. Ivanov, S.Y. Zavizion and L.I. Zakharkin, ZH. OBSHCH. KHIM., 43, 2265 (1973).
22. R. Streck, J. ORGANOMETAL. CHEM., 71, 181 (1974).
23. H. Olapinski, J. Weidlein and H.D. Hausen, J. ORGANOMETAL. CHEM., 64, 193 (1974).
24. A.T.T. Hsieh and G.B. Deacon, J. ORGANOMETAL. CHEM., 70, 39 (1974).

25. S. Pasynkiewicz and K. Dowbor, J. ORGANOMETAL. CHEM., 78, 49 (1974).
26. S. Pasynkiewicz, A. Pietrzykowski and K. Dowbor, J. ORGANOMETAL. CHEM., 78, 55 (1974).
27. J.Z. Nyathi, J.M. Ressner and J.D. Smith, J. ORGANOMETAL. CHEM., 70, 35 (1974).
28. S. Numata, H. Kurosawa and R. Okawara, J. ORGANOMETAL. CHEM., 70, C21 (1974).
29. R. Uson and F. Iranzo, REV. ACAD. CIENC. EXACTAS, FIS-QUIM. NATUR. ZARAGOZA, 28, 481 (1974).
30. G.B. Deacon and J.C. Parrott, AUST. J. CHEM., 27, 2547 (1974).
31. H. Olapinski and J. Weidlein, Z. NATURFORSCH, TEIL B. 29, 114 (1974).
32. M.R. Ort and E.H. Mottus, J. ORGANOMETAL. CHEM., 50, 47 (1973).
33. A. Bongini, D. Savoia and A. Umani-Ronchi, J. ORGANOMETAL. CHEM., 72, C4 (1974).
34. L. Pellerito, R. Cefalù and G. Ruisi, J. ORGANOMETAL. CHEM., 63, 41 (1973).
35. M. Boleslawski, S. Pasynkiewicz, A. Minorska and W. Hryniow, J. ORGANOMETAL. CHEM., 65, 165 (1974).
36. A. Giezynski, S. Pasynkiewicz, and A. Serwatowska, J. ORGANOMETAL. CHEM., 69, 345 (1974).
37. M. Boleslawski, S. Pasynkiewicz, A. Kunicki and J. Smola, J. ORGANOMETAL. CHEM., 65, 161 (1974).
38. E.A. Jeffery, A. Meisters and T. Mole, J. ORGANOMETAL. CHEM., 74, 365, 373 (1974).
39. K. Urata, K. Itoh and Y. Ishii, J. ORGANOMETAL. CHEM., 76, 203 (1974).
40. B. Schaible, W. Haubold and J. Weidlein, Z. ANORG. ALLG. CHEM., 403, 289 (1974).
41. P.C. Keller, J. AMER. CHEM. SOC., 96, 3073, 3078 (1974).
42. H.D. Hausen, K. Mertz and J. Weidlein, J. ORGANOMETAL. CHEM., 67, 7 (1974).
43. H.U. Schwerring and J. Weidlein, CHIMIA, 27, 335 (1973).
44. N. Voiculescu, REV. ROUM. CHIM., 18, 1919 (1973).
45. N. Voiculescu and I. Popescu, REV. ROUM. CHIM., 18, 1911 (1973).
46. W. Lindel and F. Huber, Z. ANORG. ALLG. CHEM., 408, 167 (1974).
47. P.B. Hitchcock, G.M. McLaughlin, J.D. Smith and K.M. Thomas, J. CHEM. SOC., CHEM. COMMUN., 934 (1973).
48. R.E. Bowen and K. Gosling, J. CHEM. SOC., DALTON TRANS., 964 (1974).
49. D. Giurgiu and I. Popescu, REV. ROUM. CHIM., 18, 1963 (1973).
50. D.E. Fenton and R. Newman, J. CHEM. SOC., DALTON TRANS., 655 (1974).

51. J.G. Contreras and D.G. Tuck, J. CHEM. SOC., DALTON TRANS., 1249 (1974).
52. K. Dymock and G.J. Palenik, ACTA CRYST., B, 30,1364(1974).
53. K. Dymock and G.J. Palenik, J.CHEM.SOC., CHEM. COMMUN., 884 (1973).
54. G.W.A. Fowles, D.A. Rice and N. Rolfe, J.INORG.NUCL.CHEM., 36, 31 (1974).
55. R.J. Abraham, G.H. Barnett and K.M. Smith, J.CHEM.SOC., PERKIN TRANS., 2142 (1973).
56. E.R. Alton, R.G. Montemayor and R.W. Parry, INORG. CHEM., 13, 2267 (1974).
57. K. Wakatsuki, Y. Takeda and T. Tanaka, INORG. NUCL. CHEM., LETT., 10, 383 (1974).
58. P. Jennische and R. Hesse, ACTA CHEM. SCAND., 27,3531(1973).
59. B. Vandorpe and M. Drache, C.R.ACAD.SCI., Ser.C., 277, 1121 (1973).
60. G. Vincentini and W.N. Dehima, AN. ACAD. BRASIL. CIENC., 45, 219 (1973).
61. K. Urata, K. Itoh and Y. Ishii, J. ORGANOMETAL. CHEM., 66, 229 (1974).
62. K. Urata, K. Itoh and Y. Ishii, J. ORGANOMETAL. CHEM., 63, 11 (1973).
63. A. Arduini and A. Storr, J. CHEM. SOC., DALTON TRANS., 503 (1974).
64. D.F. Rendle, A. Storr and J. Trotter, J. CHEM. SOC., DALTON TRANS., 2252 (1973).
65. F.K. Schmidt, V.V. Saraev, G.M. Larin, Y.G. Lipovich and L.V. Mironova, IZV. AKAD. NAUK. SSSR, Ser. Khim., 2136 (1974).
66. W. Kaminsky, J. Kopf and G. Thirase, JUSTUS LIEBIGS ANN. CHEM., 9, 1531 (1974).
67. R.A. Kovar, G. Loaris, H. Derr and J.O. Callaway, INORG. CHEM., 13, 1476 (1974).
68. M.J.S. Gynane and I.J. Worall, J. ORGANOMETAL CHEM., 81, 329 (1974).
69. I.G. Watterworth and I.J. Worall, J. ORGANOMETAL. CHEM., 81, 23 (1974).
70. J.L. Atwood, D.C. Hrncir and W.R. Newberry, CRYST. STRUCT. COMMUN., 3, 615 (1974).
71. J.L. Atwood and W.R. Newberry, J.ORGANOMETAL CHEM., 66, 15 (1974).
72. I.L. Wilson and K. Dehnicke,J.ORGANOMETAL.CHEM., 67, 229 (1974).
73. T. Ehemann and K. Dehnicke, J.ORGANOMETAL. CHEM., 71, 191 (1974).
74. J.L. Atwood and D.C. Hrncir, J.ORGANOMETAL CHEM., 61, 43 (1973).
75. S.K.Seale and J.L.Atwood, J.ORGANOMETAL CHEM.,73,27(1974).
76. S.K.Seale and J.L.Atwood, J.ORGANOMETAL CHEM.,64,57(1974).
77. F.Weller,I.L.Wilson and K.Dehnicke, J.ORGANOMETAL. CHEM., 30, C1 (1971).

78. G.Hencken and E.Weiss, J.ORGANOMETAL.CHEM., 73,35(1974).
79. T.Ehemann and K.Dehnicke, J.ORGANOMETAL. CHEM., 64, C33 (1974).
80. N.Kumar, B.L.Kalsotra and R.K. Multani, J.INORG.NUCL. CHEM., 35, 4295 (1973).
81. N. Kumar, B.L. Kalsotra and R.K. Multani, J.INORG.NUCL. CHEM., 36, 1157 (1974).
82. N. Kumar, R.K. Multani and B.L. Kalsotra, J.ORGANOMETAL. CHEM., 59, 105 (1973).
83. N. Kumar and R.K. Multani, J. ORGANOMETAL.CHEM., 63, 47 (1973).
84. N. Kumar and R.K. Sharma, CHEM. IND. (LONDON), 261 (1974).
85. J.G. Contreras and D.G.Tuck,J.ORGANOMETAL.CHEM., 66, 405 (1974).
86. G. Weigand and K.H. Thiele, Z. ANORG. ALLG. CHEM., 405, 101 (1974).
87. A.G. Gash, P.F. Rodesiler and E.L. Amma, INORG. CHEM., 13, 2429 (1974).
88. T.W. Dolzine and J.P. Oliver, J. ORGANOMETAL. CHEM., 78, 165 (1974).
89. J.S. Kristoff and D.F. Shriver, INORG. CHEM., 13, 499 (1974).
90. D.J. Darensbourg and D. Madrid, INORG.CHEM., 13,1532(1974).
91. D.P. Schussler, W.R.Robinson and W.F. Edgell, INORG. CHEM., 13, 153 (1974).
92. J.M. Burlitch and T.W. Theyson, J. CHEM. SOC., DALTON TRANS., 828 (1974).
93. H. Behrens, J. Ellermann, P. Merbach and E. Weps, Z. NATURFORSCH., TEIL B, 29, 469 (1974).
94. G. Schmid and V.Baetzel, J. ORGANOMETAL. CHEM., 81, 321 (1974).
95. H.J. Haupt and F. Neumann, J. ORGANOMETAL. CHEM., 74, 185 (1974).
96. H. Preut and H.J. Haupt, CHEM.BER., 107, 2860 (1974).
97. W. Kelbfus, J. Kiefer and K.E. Schwarzhans, Z. NATURFORSCH., TEIL.B, 28, 503 (1973).
98. G. Schmid, ANGEW.CHEM.INT.ED. (ENGL.), 11,1034 (1972).
99. M.Wilkinson and I.J. Worall, INORG. NUCL. CHEM. LETT., 10, 747 (1974).
100, H.J.Cumming, D.Hall and C.E. Wright, CRYST.STRUCT. COMMUN., 3, 107 (1974).
101. J.C. Champarnaud-Mesjard, J.Grannec and B. Gaudreau, C.R. ACAD.SCI., Ser.C, 278, 171 (1974).
102. K.P. Butin, V.V.Strelets, I.P. Beletskaya and O.A. Reutov, J. ORGANOMETAL. CHEM., 64, 189 (1974).
103. G.A. Razuvaev, I.V. Lomakova and L.P. Stepovik, ZH. OBSHCH, KHIM., 43, 2416 (1973).
104. L.N. Nemirovskii, B.I. Kozyrkin, A.F. Lanstov, B.G. Gribov and I.M. Skvortsov, DOKL. AKAD. NAUK. SSSR, 214, 590 (1974).
105. G.Fritz and H.Schaefer, Z.ANORG.ALLG.CHEM., 406, 167 (1974).

SILICON CHEMISTRY

J. Y. Corey

Department of Chemistry
University of Missouri-St. Louis
St. Louis, Missouri 63121

GENERAL

The majority of the synthetic studies on silicon compounds tend to concentrate in the organosilicon area and many deal with predominantly organic problems. For this reason the synthetic methods outlined in this chapter were restricted to those transformations that occurred at the silicon center and were selected from those articles abstracted in both the Inorganic and Organometallic sections of Chemical Abstracts. Emphasis is placed on bond formation reactions and each section is headed by the group of elements that forms the bond to silicon. Although this is a convenient classification it was considered appropriate to deal with polysilanes and their reactivity as a group.

Several reviews pertaining to particular types of silicon derivatives have included information on synthetic methods. The areas reviewed include, carbosilanes(1), polysilanes(2), peroxides(3), silicon-sulfur derivatives(4), silicon-transition metal compounds(5), and molecular rearrangements(6). The annual survey of organosilicon chemistry for 1973 has been conveniently divided into sections concerning synthesis and reactivity(7), applications to organic synthesis(8), bonding and structure(9) and reaction mechanisms(10).

SILICON-HYDROGEN

Reduction side products occasionally accompany reactions of organosilicon compounds (SiOR, SiF, SiCl) with saturated Grignard reagents. Such reduction reactions with the Grignard reagent predominate at room temperature in ether when a metal catalyst is added. Conversions of SiOR, SiF and SiCl to SiH with $^{i}PrMgBr-Cp_2TiCl_2$ occur in approximately 100% yield in a 4 hour reaction time with high stereospecificity (93-100% retention)(11). This reactivity is similar to that of $LiAlH_4$. Other Grignard reagents were also reported to give reduction in the presence of Cp_2TiCl_2. This mild reduction reaction may also be used to convert bulky alkoxy groups to SiH, however, longer time periods are required. Use of other metal catalysts has also been explored, but yields of reduction product vary(12). Longer reaction periods are required with EtMgX-Ni(PØ$_3$)$_2$Cl$_2$ and reactions of SiF and SiOMe proceed with retention and SiCl with inversion(12).

SILICON-GROUP IV

Organometallic Reagents: One of the most commonly used methods of Si-C bond formation remains the coupling of RMgX (or RLi) with an SiCl, SiF or SiOR derivative. The influence of solvent on such coupling reactions has been examined employing the optically active organosilicon substrates, $R_1R_2R_3SiX$ (X = OMe, F). The rate of the coupling reaction increases with the basicity of solvent (Et_2O < THF < DME) and the stereochemical course of the reaction changes from inversion (Et_2O) to retention (THF, DME) (13). In coupling reactions of $R_1R_2Si^*$(OMe)(OMn) involving two different alkoxides, the smaller alkoxy group is preferentially substituted (retention) by $R_{(sat)}$MgX or C_6H_5MgX in Et_2O, THF or DME but the bulkier alkoxy group is substituted (inversion) by unsaturated or benzyl Grignard reagents (14). For cyclic derivatives the smaller alkoxy group is replaced but significant amounts of disubstituted products are observed.

An improved method of synthesis of ethynylsilanes was reproted. The coupling of Me_3SiX (X = Cl,F) with Mg and Li derivatives of RC≡CH proceeds in THF but not in Et_2O or hydrocarbon solvents. Highest yields were obtained with lithium acetylides and the silyl fluoride (15). The coupling reaction of sodium acetylide with ArR_2SiCl occurs in xylene-nitrobenzene (4 hours, 50-60°) to give 42% yield of product(16). Reactions of the ethynyl group bonded to silicon could be performed without cleavage of the Si-C bond and provide a convenient entry into functionalized organosilicon compounds (16).

A new synthesis of silicon substituted acids based on the β-effect has been developed from reaction of $R_1R_2Si(CH_2Cl)Cl$ with two equivalents of $NaCH(CO_2Et)_2$ in hydrocarbon solvents. Treatment of the reaction product with alkali resulted in cleavage of the $Si-CH(CO_2Et)_2$ linkage and isolation of $[HOOCCH_2CH_2-SiR_1R_2]_2O$. When $R_1=R_2=CH_3$ the lactone was isolated (50%) (17).

Trialkylsilyl substituted derivatives of biphenyl were prepared from both m-, and p-biphenyl Grignard and lithium reagents. Better yields were reported for the Grignard reaction and also for the m-biphenylyl derivatives (18).

A Wurtz coupling reaction of adamantylchloride(bromide) and $SiCl_4$ with Na in boiling cyclohexane provided tetraadamantylsilane, Ad_4Si (18%) (19). Ad_4Si is described as a chemically inert material and was suggested as a ^{29}Si NMR reference.

In a unique coupling reaction, $(Me_3Si)_2CBrLi$ reacts with Me_2SiCl_2 in Hex-THF-Me_2O at -115° to give 36% of the disilabutane product (20).

SiCl/Li/THF and SiCl/Mg/HMPT: Although silyl Grignard reagents are uncommon, the combination Me_3SiCl/Li/THF or Me_3SiCl/Mg/HMPT reacts in the manner of an active organometallic when unsaturated organic derivatives such as α-ethylenic aldehydes (21), α-acetylenic carbonyl derivatives (22), esters (23), tetralin (24), polyenes (25,26,27), and vinylsilanes (28), are present. Typical reactions are summarized in eq. (1).

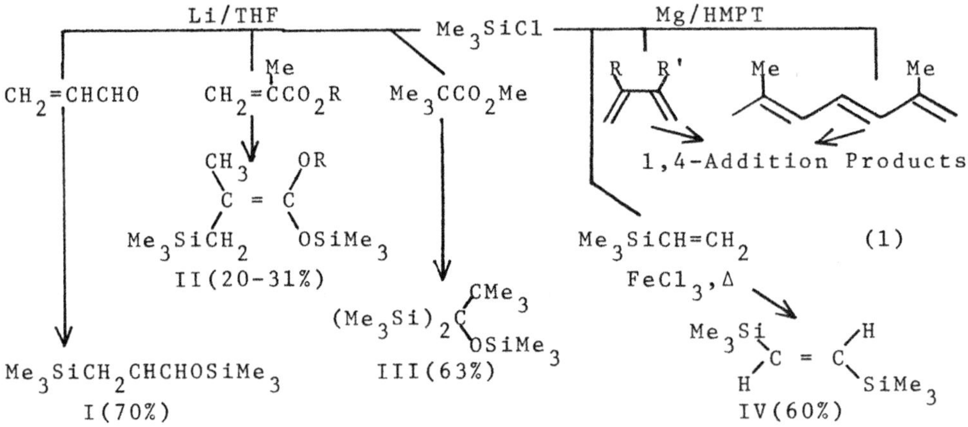

(1)

Reaction of Me_3SiCl with Me_3CCO_2Me is the first example of silylation of a saturated ester and may provide an entry into gem-disilyl derivatives (23). Derivatives such as I and II are easily hydrolyzed and thus provide an entry into carbon-functional organosilicon compounds. Reactions of vinylsilanes with Me_3SiCl and excess magnesium in the presence of $FeCl_3$ provide high yields of tetrasilylsubstituted ethanes (28). Although these silylation reactions have normally involved the Me_3Si-group it should not in principal be limited to this group.

The 1,4-addition reaction with polyenes has been effectively employed as a synthetic route to silacyclopentene derivatives under mild conditions (25), eq. (2). Higher yields of

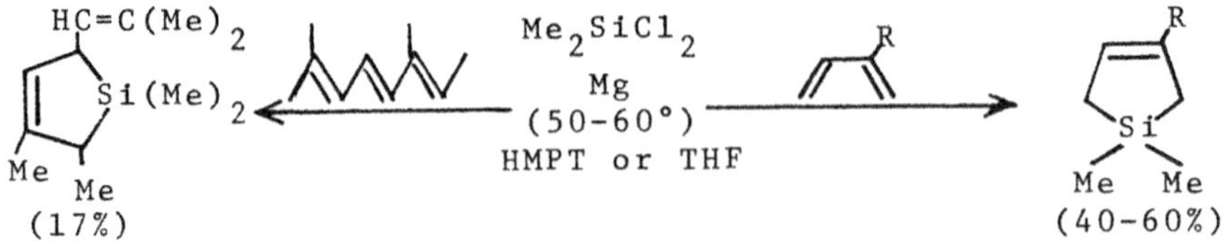

(2)

cyclic product were reported for reaction of hexatriene in the presence of Cu_2Cl_2 (26). The effect of solvent on the ring closure reaction has also been studied (27). Previous methods employed Na/THF and produced lower yields of the silacyclopentene.

In contrast to other routes which require low temperatures partial silylation of gem-polychlorinated compounds with Me_3SiCl/Mg/HMPT can be accomplished under mild conditions (30-60°, 15 to 40 hours) to give 10 to 80% yields of coupled product (29). The α-chlorosilyl derivatives thus produced may be converted to α-functional silylated compounds. Complete silylation of gem-polychlorinated compounds also occurs in good yield (29).

Hydrosilylation: Hydride addition of SiH to olefins remains an important synthetic route to formation of organosilicon compounds. Recent studies have focused on new catalysts, the mode of addition, reactions with dienes and asymmetric syntheses.

Reactions of SiH with allyl derivatives, $RCH_2CH=CH_2$ (R=F, CF_3CO_2, and NR_1R_2), in the presence of H_2PtCl_6 afford $Cl_3Si(CH_2)_3F$ (93%) (30), $CF_3CO_2(CH_2)_3SiCl_2Me$ (up to 52%) (31) and $(EtO)_3Si(CH_2)_3NR_1R_2$ (major product in varying yields)(32). Selective hydrosilylation of styrene under mild conditions (room temperature) can be determined by choice of catalyst. In the presence of $Ni(CO)_4$, 88-97% yields of α-silyl adduct are produced, and in the presence of H_2PtCl_6-$P(C_6H_5)_3$ the β-silyl adduct was obtained in 69-91% yields (33). Previously reported routes were less efficient and produced mixtures of isomers(34).

Hydrosilylation is also useful for synthesis of silicon heterocycles from $CH_2=CH-(CH_2)_n-Si(Me)_2H$ in the presence of H_2PtCl_6. When n=0 or 1, no ring closure products were observed and when n=2 to 6, mixtures of two cyclic isomers were obtained. The largest ring size produced by this route was the silacycloheptane ring. The maximum yield of ring-closed products was observed for n=4.

Addition of SiH to butadienes produces products of 1,4-hydrosilylation (80-95%) (35 - 38). A variety of metal catalysts have been studied for the hydrosilyation of dienes and include $(p\text{-}RC_6H_4CN)_2PdCl_2$ (36), $[(C_6H_5)_3P]_4Pd$ (37), $M(acac)_n/AlEt_3$ (M=Ni,Fe,Co) (38), and $M(CO)_6/h\nu$ (M=Cr,Mo,W) (39). α-ω-Dienes apparantly undergo a stepwise addition of SiH (40).

Formation of a Pt-coated resin for use in hydrosilylation reactions has been described as well as a procedure for recovery of the Pt (41).

Reactions of silicon dihydrides as hydrosilylation reagents are also currently under study. Asymmetric hydrosilyation of ketones in the presence of chiral metal catalysts has been reported. The optical purity of the alkoxysilane so produced was measured by conversion to a triorganosilane of known specific rotation. With the chiral catalyst $[Rh\{\underline{R}-C_6H_5CH_2PMeC_6H_5\}_2H_2L_2]^+ClO_4^-$ (43) (L=solv.),63-86% optically active alkoxide was produced and up to 28% optical purity in the product silane. The catalyst system $[(C_8H_{14})_2RhCl]_2$ plus (+) or (-) diop produced silane end products of up to 46% optical purity (42).

Reaction of H_2SiCl_2 with acetylenes in the presence of Pt catalyst proceeds via cis addition to gave both terminal (major product, 34-56%) and internal olefins (10-15%) as well as bis-(trans-dialkenyl)dichlorosilanes (6-30%). A large excess of H_2SiCl_2 minimizes formation of bis adducts (44).

Polysilanes: Disilanes can be produced from the coupling of chlorosilanes with Mg or Na in THF (yields vary from 38-59% with Mg and 0-75% with Na) (45). Although symmetrical disilanes are usually produced from this reaction, unsymmetrical disilanes and trisilanes are reported in small yields from the reaction of mixtures of chlorosilanes with one chlorosilane in large excess. A similar approach has been utilized to form Si_{18} and Si_{24} chains by reaction of $Me_3Si(SiMe_2)_4SiMe_2Cl$ and $Cl(SiMe_2)_6Cl$ with potassium in benzene solvent (46). When dichlorosilanes are coupled with alkali metals, cyclic polysilanes are obtained with 5 or more ring members. If bulky organic groups are also bonded to silicon reasonable yields (25% crude) of cyclotetrasilanes can be realized (47).

In a unique coupling reaction the branched chain polysilane, $[(Me_3Si)_3Si]_2$ was produced in 92% yields from peroxide coupling of $(Me_3Si)_3SiH$ at 130° (48), where previously reported Wurtz coupling gave low yields. This route may be applicable only if bulky substituents are present on silicon.

A useful preparation of an optically active disilane from photolysis of a disilylmercurial has been reported (48a). Reaction of R_3Si^*Li with R_3Si^*Cl gave a mixture of diastereomeric meso and d, ℓ disilanes (49).

α-ω-Dichloromethylpolysilanes, $ClSiMe_2(SiMe_2)_xSiMe_2Cl$, react with MeMgI at 30° to give monosubstituted products (major) with only small amounts of disubstitution. Yields of monosubstituted product increase with increasing chain length (90% monosubstituted product for the hexasilane) (50). Linear polysilanes, $H(SiH_2)_nH$ (n=3-5) give both mono- and disubstituted terminal products with BuLi (51).

Cyclic polysilanes are cleaved by alkali metals or halogens to give difunctional reagents, such as $X(SiR_2)_4X$, which have been used as intermediates in the preparation of the 5-membered heterocycles SiSiQSiSi, Q=RB, O, RN (52) [but not $Q=CR_2$ (53)].

The $-Si_x-$ (x=2,3) unit can be incorporated simultaneously into an organic heterocycle (54,55,56,57) employing methods described in the previous section, eq. (3). These new heterocycles

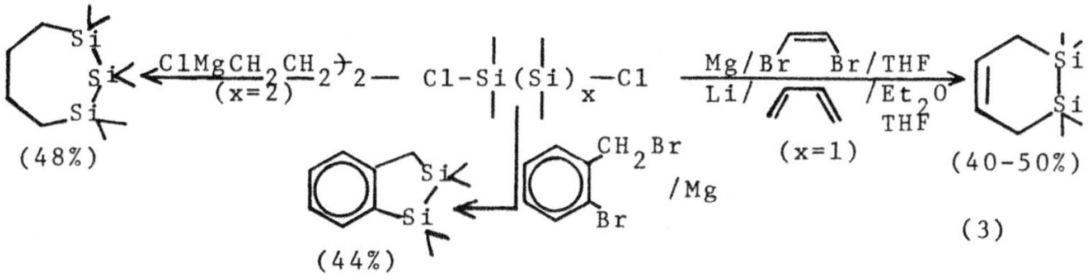

(3)

are O_2 sensitive and usually react with air.

Reactions of direct process residues, $Me_nSi_2Cl_{6-n}$ with HMPT or H_3PO_4 causes 65-80% cleavage of SiSi bonds and formation of $MeSiCl_3$ and Me_2SiCl_2 as major products. In the presence of $(NH_4)_3PO_4 \cdot 3H_2O$, $MeHSiCl_2$ becomes the major product (58). Direct process residues also react with organic chlorides in the presence of metal catalysts to produce Me_nRSiCl_{3-n} (n= 0,1,2) (59). Both these results suggest limitations in chemical transformations of disilanes utilizing HMPT as a solvent or with regard to transformations which require metal catalysis.

New synthetic methods have been developed for preparation of fluorinated disilanes. Hexamethoxydisilane was converted to Si_2F_6 by BF_3 in 76% yield (60). Pentafluorodisilane has been isolated from the decomposition of the insertion product, SiF_2HSiF_2Br, produced from reaction of SiF_2 with HBr (61).

Silicon-Germanium: Difluorogermylene inserts into the SiCl bond to generate derivatives of the type $(C_6H_5)_xSi(GeF_2Cl)_{4-x}$ (63).

SILICON-GROUP V

Silicon-Nitrogen: The SiCl bond is rapidly converted to SiN at ≈0° by reaction with amines containing an NH bond, whereas the corresponding CCl to CN conversion usually requires the lithiated amine and higher temperatures. The difference in reactivity is well illustrated in the synthesis of chelating ligands of the type $Me_2XSiMe_2CH_2X'Me_2$ (X=X'=N,P; X=N,X'=P; X=P, X'=N), from the precursor $Me_2Si(CH_2Cl)Cl$ (63).

Current investigations have focused on the controlled amination of one SiCl bond in either a $SiCl_2$ derivative or derivatives that contain two different SiCl bonds. This transformation can be performed successfully at 0° under high dilution in petroleum ether. Yields of partial amination are quite high from Me_2SiCl_2 [90% yield of $(ClSiMe_2)_2NR$ from $MeNH_2$] (64) but in the range of 51-69% for partial amination of $R'N(SiMe_2Cl)_2$ to give $R'N(SiMe_2Cl)(SiMe_2NR_2)$. In the majority of the studies reported the reaction which produced the SiN bond also involved the formation of cyclic products with ring sizes varying from 5 to 8. The general synthetic principle involves the condensation of a preformed HN(R)-X-N(R)H unit with R_2SiCl_2 or ClSi-X-SiCl units but hydrazines and 1,2-dichlorodisilanes have also been employed in these condensation reactions. The majority of the ring systems that have been reported that exemplify these sythetic principles (and involve reaction at silicon) are indicated below as a function of ring size. Only the elements that form the ring are listed. The two NH groups in Schiff

Ring Size	Elements in Ring
5	Si_2NC_2(66), Si_2N_2P(67), Si_2N_3(68)
6	Si_2NC_3(66), Si_2N_2OS(69), $Si_2N_2C_2$(70), Si_2N_3S(69), Si_2N_3S(71), Si_2N_4(68), Si_3NO_2(72), Si_3N_2O(72), Si_3N_3(73), Si_4N_2(71)
7	SiN_4C_2(74), $Si_2N_3S_2$(71), $Si_3N_2O_2$(68) Si_3N_4(68)
8	$Si_2N_2P_2O_2$(75), Si_4N_4(66,76)

base complexes of Co and Ni also react with α-ω-dichloroorganosilicon compounds (78). Spirocyclic derivatives have also been prepared from reaction of diamines with $SiCl_4$ (79).

Exchange reactions can also be used to form SiN bonds. Transammination of the $Si(NEt_2)_2$ group with organic diamines produces cyclic derivatives with ring sizes from 5 to 8 (74,79). Transsilylation reactions, i.e. replacement of one silyl group attached to N by another silyl group is a useful method for synthesis of silicon functional silazanes in high yield. An example is shown in eq. (4). Other chlorosilanes undergo the

$$Me_3SiNRSiMe_3 \xrightarrow{Me_2SiCl_2} \underset{(74\%)}{Me_2SiClNRSiMe_3} \xrightarrow{Me_2SiCl_2} \underset{(73\text{-}87\%)}{Me_2Si\!\!\rightarrow_2 NR} \qquad (4)$$

exchange reaction (71,80,81) and a unique exchange reaction of SnN to SiN has also been reported to form Si-containing SN

8-membered rings (82,83).

The SiN bond has been cleaved by $COCl_2$ (84), $AlCl_3$ (85) and $SOCl_2$ (86) and CO_2 inserts between the Si and N (87).

Silicon-Phosphorus: Compounds containing Si-P bonds have been prepared by methods similar to those described in the previous sections. Coupling reactions of Me_3SiCl and $PCl_n(^tBu)_{3-n}$ in the presence of Mg in THF generated 41-64% yields of silylphosphines (88). Higher yields of silylphosphines are obtainable from $LiAl(PMe_2)_4$ (formed in diglyme from Me_2PLi and $AlCl_3$) and $H_{3-x}Me_xSiX$ (89). Cyclic polyphosphines react with K to give K_2PR which couples with Me_3SiCl to give the disilylphosphine (64%) (90).

Redistribution reactions of $MeSiH(PH_2)_2$ with halosilanes produce phosphines of the general formula, $X_{3-(n+m)}Me_nH_mSiPH_2$ but yields are frequently low (91).

Silicon-Arsenic: The exchange of Si for Sn by reaction of Me_3SnAsH_2 and F_3SiBr in n-Bu_2O at -25° gave F_3SiAsH_2 (70%) (92).

SILICON-GROUP VI

Silicon-Oxygen: Reaction of SiCl with an OH grouping normally leads to formation of an SiO bond. Further condensation reactions to polysiloxanes may also occur. The controlled methanolysis of Me_2SiCl_2 in the gas phase in the presence of $ZnCl_2$ afforded the cyclic polysiloxanes $[Me_2SiO]_x$, x=3(3%), 4(70%), and 5(22%) (93). The cyclic pentamer is not normally obtainable in such high yields in solution. Cyclosiloxanes with different SiR_1R_2 units can be prepared from condensation of silane diols and silicon dihalides at 0-5° in ether in the presence of an HCl scavenger (94). In an extraordinary case of slow, controlled hydrolysis with undried N_2 as the source of moisture, the cyclic trimer, $[MeSi(Cl)CH_2]_3$ afforded the cage siloxane product (96%) (95).

Diols, such as o-$HOC_6H_4CH_2OH$ react with R_2SiCl_2 in benzene to form the cyclic derivative when Et_3N is added as an HCl scavenger (96). In absence of Et_3N only polymers were obtained. Salicylic acid derivatives react in a similar way (97,98). The SiN bonds in $R_2Si(NEt_2)_2$ are converted to cyclic derivatives containing SiO bonds by reaction with glycols (99).

The SiH bond can be converted to SiOH in the presence of base. If the ring size is appropriate, silalactones may be synthesized by generation of a silanol (from SiH) in a derivative which already contains an acid functional group. The method has only been explored for hydrosilyl derivatives of toluic acids (100). The SiH bond in trialkylsilanes reacts with alcohols in the presence of F^- in CH_3CN to give Et_3SiOR (55-95%) (101).

The formation of SiO bonds also may occur from addition of SiH to a carbonyl group. Fluoride ion catalyzes the addition of Et_3SiH to ArCHO to produce the alkoxides, Et_3SiOCH_2Ar (101). Me_3CN adds to both aldehydes and ketones in the presence of KCN-crown ether (102), a transformation that may be related to SiH addition to carbonyls.

Dialkylsilanes, R_2SiH_2, add to carbonyls in the presence of metal catalysts but in α- and β-hydroxy substituted carbonyl compounds, R_2SiH_2 reacts only at the hydroxy group (103). An asymmetric synthesis of alkoxysilanes has been developed from reaction of $R_1R_2SiH_2$ with R*OH or ROH in the presence of the rhodium catalyst, $[(C_8H_{14})_2RhCl]_2$ + (+) or (-)diop. The optical purity of the products were ascertained by conversion of the alkoxide product, $R_1R_2Si(H)OR$ to $R_1R_2R_3SiH$. When achiral alcohols were used the optical purity of the products varied from 3-19%, but ranged from 13-56% for the chiral alcohols studied (104).

The reaction of R_3SiCl with NH_3 in Et_2O at 0° followed by addition of 98% H_2O_2 provided peroxide derivatives, $R_3SiOOSiR_3$ (40-50%) (105).

Complexes with coordination numbers greater than 4 have been obtained from reactions of several different silicon reagents with an appropriate chelating agent in nonprotonic solvents. $Si(OAc)_4$ reacts with Schiff bases in anhydrous benzene to afford complexes of stoichiometry, $Si(OAc)_3L$, $Si(OAc)_2L_2$ and $Si(OAc)L_3$ (L=base) (106). A coordination number of 7 is proposed for the later complex. Disilazanes react with 8-hydroxyquinoline in toluene to generate the complexes R_3SiL and R_2SiL_2 (107). $SiCl_4$ reacts with 3 equivalents of N-arylhydroxylamine in $CHCl_3$ to give $[SiL_3]^+HCl_2^-$ (108).

<u>Silicon-Sulfur</u>: Symmetrical disilthianes, $(R_3Si)_2S$, are prepared from the reaction of H_2S with R_3SiCl compounds (109, 110). In the presence of an HCl scavenger yields of up to 82% of disilthiane have been realized. If bulky substituents or phenyl groups are bonded to silicon, silylthiols, R_3SiSH are obtained (58-92%) (110). Unsymmetrical disilthianes, $R_3SiSSiR'_3$, are obtained from reaction of R_3SiSH with R'_3SiCl in the presence of Et_3N (110). Reaction of H_2S and Me_2SiCl_2 provides hexamethylcyclotrisilthiane (109).

An improved synthesis of silylthioethers has been developed employing imidazole catalyzed reactions of RSH with $(Me_3Si)_2NH$

to give $RSSiMe_3$(50-93%) (111).

Hexamethyldisilthiane reacts with both organic and mineral acids to give high yields of esters (cf. $(Me_3SiO)_2SO_2$ from reaction with H_2SO_4) (112), and with metal oxides to give $Me_3SiO-SiMe_3$ (113).

Silicon-Selenium: In an approach similar to that reported for formation of silylphosphines, reaction of $LiAl(SeMe)_4$ with bromosilanes in diglyme affords the derivatives, $H_xSi(SeMe)_{4-x}$ (x=2,3) (114).

SILICON-GROUP VII

Silanes, R_3SiH, may be chlorinated in high yields by copper(I) chloride in acetonitrile (115), or by chlorobenzene in the presence of Ni (150°) (116).

A unique method of fluorination of alkoxysilanes has been reported using BF_3 as the fluorinating agent (117).

Cleavage of the SiSi bond in Br_6Si_2 occurs in the presence of o,o'-bipyridine and the new complex, $[(bipy)_3Si]Br_4$, was formed (118).

SILICON-TRANSITION METAL BOND

A σ-bond between Si and a transition metal element may be formed by a variety of methods but the most common approach appears to be the coupling of a chlorosilane with a sodium carbonylmetallate such as $Na[M(CO)_xC_5H_5]$ (M=Fe,Mo,W) or $Na[Mn(CO)_5]$ (symbolized as Na[M]), as exemplified in eq. (5) (119,120,121).

$$Na[M] + SiX_{4-(n+m)}(CH_3)_nH_m \xrightarrow{C_6H_{12}} M\text{-}SiX_{3-(n+m)}(CH_3)_nH_m \ (22\text{-}92\%)$$

$$Na[M] + Me_5Si_2X \longrightarrow Me_5Si_2Mn(CO)_5 \ (26\%) \qquad (5)$$

Chlorosilacyclobutane also has been shown to undergo such coupling reactions (120,123). The SiH bond in these transition metal complexes are readily halogenated by CX_4 in high yields without cleavage of the SiM bond (120). The SiCl bonds in the complexes can also be converted to SiF by reaction with $AgBF_4$ (121,122).

Another entry into the formation of the SiM bond is the

reaction of a silicon hydride with a metal carbonyl containing a M-M bond (123,124,125,126) as indicated in eq. (6). Silicon

$$Mn_2(CO)_{10} \quad + \quad Me_5Si_2H \longrightarrow Me_5Si_2Mn(CO)_5 \quad (35\%) \qquad (6)$$

hydrides replace M-C sigma bonds or ethylene in phosphine complexes of Pt(II) (127,128) and Ni(II) (129), eq. (7).

$$\underline{cis}\text{-}[Me_2PtL_2] \quad + \quad (C_6H_5)_3SiH \longrightarrow \underline{cis}\text{-}[PtH\{Si(C_6H_5)_3\}L_2]$$
(L=phosphine) (7)

Since complications frequently accompany reactions of metallate anions with silyl chlorides, an alternate route to <u>cis</u>-$Fe(CO)_4(SiMe_3)_2$ utilizing photolysis of $Hg(SiMe_3)_2$ in the presence of an iron carbonyl has been devised (130).

Derivatives that may be described as π-silacyclopentadiene complexes of metals have only recently been prepared (131,132, 133), eq. (8).

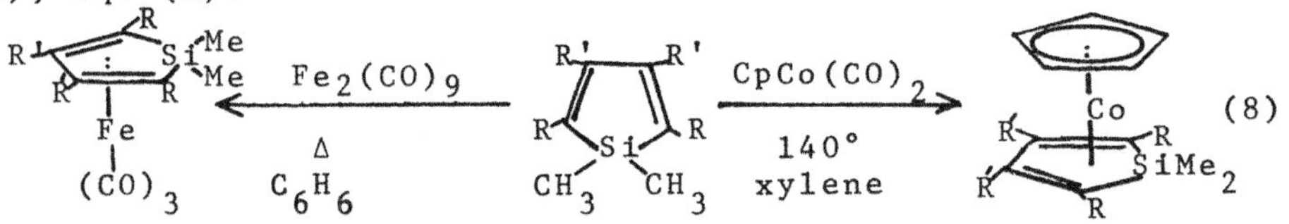

(8)

An unusual reaction of the Ni complex, $Ni(bipy)(SiCl_2R)_2$, with acetylenes may provide a useful new route to 1,2-disila-substituted ethylenes (134).

MOLECULAR REARRANGEMENTS

Several rearrangement reactions which may prove of synthetic utility in future investigations have been studied. Examples of such rearrangements are shown in eq. (9) for anionic one atom migrations (135,136,137) and in eq. (10) for a free-radical three-atom migration (138). A rearrangement resulting

$$C_6H_5CH_2ESiMe_3 \begin{cases} \xrightarrow[2)\ H^+]{1)\ {}^tBuLi/THF} C_6H_5CH(SiMe_3)OH \quad [E=O] \\ \xrightarrow[2)\ RX]{1)\ BuLi/THF} C_6H_5CH(SiMe_3)SR \quad [E=S] \end{cases} \qquad (9)$$

$$ArMe_2Si(CH_2)_2CH_2X \xrightarrow[{}^tBuO)_2]{{}^nBu_3SnH} HSiMe_2(CH_2)_3Ar + ArSiMe_2(CH_2)_2CH_3 \qquad (10)$$

from generation of a cationic site α to silicon has been employed in formation of silicon heterocycles, eq. (11) (139).

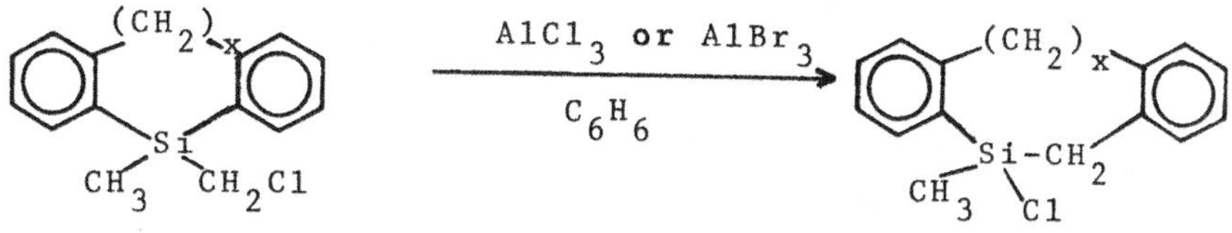

(11)

REFERENCES

1. G. Fritz, FORTSCHR. CHEM. FORSCH., 50, 43 (1974).
2. E. Hengge, FORTSCHR. CHEM. FORSCH., 51, 1 (1974).
3. D. Brandes and A. Blaschette, J. ORGANOMETAL. CHEM., 78, 1 (1974).
4. A. Haas, SULFUR ORG. INORG. CHEM., 1, 3 (1971).
5. F. Höfler, FORTSCHR. CHEM. FORSCH., 50, 129 (1974).
6. A. G. Brook, ACCTS. CHEM. RES., 7, 77 (1974).
7. R. F. Cunico, J. ORGANOMETAL. CHEM., 83, 65 (1974).
8. S. S. Washburne, J. ORGANOMETAL. CHEM., 83, 155 (1974).
9. P. J. Jones, J. ORGANOMETAL. CHEM., 83, 213 (1974).
10. O. W. Steward, J. ORGANOMETAL. CHEM., 83, 265 (1974).
11. R. J. P. Corriu and B. Meunier, J. ORGANOMETAL. CHEM., 65, 187 (1974).
12. R. J. P. Corriu and B. Meunier, J. ORGANOMETAL. CHEM., 60, 31 (1973).
13. R. J. P. Corriu and B. J. L. Henner, J. ORGANOMETAL. CHEM., 71, 393 (1974).
14. G. Chauviere, R. Corriu, A. Kpoton and G. Lanneau, J. ORGANOMETAL. CHEM., 73, 305 (1974).
15. O. Florensova, L. I. Volkova, Yu. V. Maroshin, Yu. G. Kryazhev, ZH. OBSHCH. KHIM., 43, 1992 (1973).
16. I. A. Shikhiev, D. S. Dzhafarov, S. F. Karaev, ZH. OBSHCH. KHIM., 43, 1994 (1973).
17. G. S. Smirnova, D. N. Andreev and E. S. Lobanova, ZH. OBSHCH. KHIM., 44, 782 (1974).
18. H. Zimmer, M. A. Barcelon and W. R. Jones, J. ORGANOMETAL. CHEM., 63, 133 (1973).
19. R. M. G. Roberts, J. ORGANOMETAL. CHEM., 63, 159 (1973).
20. D. Seyferth and J. L. Lefferts, J. AMER. CHEM. SOC., 96, 6237 (1974).
21. J. Dunogues, A. Ekouya, N. Duffaut and R. Calas, J. ORGANOMETAL. CHEM., 66, C36 (1974).
22. G. Merault, P. Gourgeois, J. Dunogues, and N. Duffaut, J. ORGANOMETAL. CHEM., 76, 17 (1974).
23. J. Dunogues, A. Ekouya, R. Calas, J.-P. Picard and N. Duffaut, J. ORGANOMETAL. CHEM., 66, C39 (1974). J.-P. Picard, J. Dunogues and R. Calas, J. ORGANOMETAL. CHEM.,

77, 167 (1974).
24. R. Calas, J. Dunogues, J. P. Pillot and N. Ardoin, J. ORGANOMETAL. CHEM., 73, 211 (1974).
25. J. Dunogues, B. Arreguy, C. Brian, R. Calas and F. Pisciotti, J. ORGANOMETAL. CHEM., 63, 119 (1973).
26. L. Birkofer and D. Pruestel, J. ORGANOMETAL. CHEM., 60, C1 (1973).
27. G. Manuel, P. Mazerolles and G. Cauquy, SYN. REACTIV. INORG. METAL.-ORG. CHEM., 4, 133 (1974).
28. J. Dunogues, J. P. Pillot, N. Duffaut and R. Calas, C.R. ACAD. SCI., SER. C, 278, 467 (1974).
29. J. Dunogues, E. Jousseaume, and R. Calas, J. ORGANOMETAL. CHEM., 71, 377 (1974).
30. V. B. Puchnarevic, J. Vcelak, M. G. Voronkov and V. Chavlovsky, COLLECT. CZECH. CHEM. COMMUN. 39, 2616 (1974).
31. Z. V. Belyakova, M. G. Pomerantseva and S. A. Golubtsov, ZH. OBSHCH. KHIM., 44, 1780 (1974).
32. Z. V. Belyakova, Z. V. Belikova, V. N. Bochkarev, M. S. Yamova, and S. A. Golubtsov, ZH. OBSHCH. KHIM., 44, 1484 (1974).
33. M. Čapka, P. Svoboda and J. Hetflejš, COLLECT. CZECH. CHEM. COMMUN., 38, 3830 (1973).
34. A. I. Nogaideli, L. I. Nakaidze and V. S. Tskhovrebashvili, ZH. OBSHCH. KHIM., 44, 1763 (1974).
35. J. V. Swisher and H. H. Chen, J. ORGANOMETAL. CHEM., 69, 83 (1974).
36. V. Vaisarova, J. Langová, J. Hetflejš, G. Oehme and H. Pracejus, Z. CHEM., 14, 64 (1974).
37. J. Tsuji, M. Hara and K. Ohno, TETRAHEDRON, 30, 2143 (1974).
38. M. F. Lappert, T. A. Nile and S. Takahashi, J. ORGANOMETAL. CHEM., 72, 425 (1974).
39. M. Wrighton and M. A. Schroeder, J. AMER. CHEM. SOC., 96, 6235 (1974).
40. A. A. Oswald, J. AMER. CHEM. SOC., DIV. PET. CHEM. PREPR., 19, 155 (1974).
41. V. O. Reikhsfel'd, V. N. Vinogradov and N. A. Filippov, ZH. OBSHCH. KHIM., 43, 2216 (1973).
42. R. J. P. Corriu and J. J. E. Moreau, J. ORGANOMETAL. CHEM., 64, C51 (1974).
43. T. Hayashi, K. Yamamoto and M. Kumada, TETRAHEDRON LETT., 331 (1974).
44. R. A. Benkeser and D. F. Ehler, J. ORGANOMETAL. CHEM., 69, 193 (1974).
45. P. Gerval, E. Frainnet, G. Lain and F. Moulines, BULL. SOC. CHIM. FR., 1548 (1974).
46. W. G. Boberski and A. L. Allred, J. ORGANOMETAL. CHEM., 71, C27 (1974).
47. M. Biernbaum and R. West, J. ORGANOMETAL. CHEM., 77, C13 (1974).
48. M. Ishikawa, A. Nakamura and M. Kumada, J. ORGANOMETAL. CHEM., 59, C11 (1973).

48a. C. Eaborn, R. A. Jackson, D. J. Tune and D. R. M. Walton, J. ORGANOMETAL. CHEM., 63, 85 (1973).
49. M. Lequan and Y. Besace, J. ORGANOMETAL. CHEM., 61, C23 (1973).
50. W. G. Boberski and A. L. Allred, J. AMER. CHEM. SOC., 96, 1244 (1974).
51. F. Feher and R. Freund, INORG. NUCL. CHEM. LETT., 10, 569 (1974).
52. E. Hengge and D. Wolfer, J. ORGANOMETAL. CHEM., 66, 413 (1974).
53. P. K. Sen, T. B. Brennan, and H. Gilman, J. INDIAN CHEM. SOC., 51, 561 (1974).
54. P. Mazerolles, M. Joanny and G. Tourrou, J. ORGANOMETAL. CHEM., 60, C3 (1973).
55. L. Birkofer and W. Weniger, CHEM. BER., 106, 3595 (1973).
56. F. P. Tsui and G. Zon, J. ORGANOMETAL. CHEM., 70, C3 (1974).
57. H. Sakurai, Y. Kobayashi, and Y. Nakadaira, J. AMER. CHEM. SOC., 96, 2656 (1974).
58. R. Calas, J. Dunogues and G. Deleris, J. ORGANOMETAL. CHEM., 71, 371 (1974).
59. W. H. Atwell and G. N. Bokerman, Fr. Demande, 2,163,579. CHEM.ABSTR., 80, 3625v (1974).
60. E. Hengge and S. Waldhoer, Z. NATURFORSCH., TEIL B, 29, 437 (1974).
61. J. F. Bald, Jr., K. G. Sharp and A. G. MacDiarmid, J. FLUORINE CHEM., 3, 433 (1973).
62. J. Satge, P. Riviére and A. Boy, C. R. ACAD. SCI., SER. C., 278, 1309 (1974).
63. J. Grobe and G. Heyer, J. ORGANOMETAL. CHEM., 61, 133 (1973).
64. H. J. Wismar and U. Wannagat, MONATSH. CHEM., 104, 1465 (1973).
65. U. Wannagat and D. Labuhn, MONATSH. CHEM., 104, 1457 (1973).
66. K. Andrianov, G. V. Kotrelev, A. M. Kononov and I. M. Prudnik , DOKL. ACAD. NAUK SSSR, 216, 1041 (1974).
67. O. J. Scherer, W. Glaessel and R. Thalacker, J. ORGANO-METAL. CHEM., 70, 61 (1974).
68. U. Wannagat and J. M. Schlingmann, Z. ANORG. ALLG. CHEM., 406, 7 (1974).
69. U. Wannagat and L. Labuhn, Z. ANORG. ALLG. CHEM., 402, 147 (1973).
70. V. F. Mironov, V. D. Sheludyakov and E. S. Rodionov, ZH. OBSHCH. KHIM., 44, 1496 (1974). V. D. Sheludyakov, E. S. Rodionov and V. F. Mironov, ZH. OBSHCH. KHIM., 44, 1044 (1974).
71. O. J. Scherer and G. Wolmershaeuser, Z. NATURFORSCH. TEIL B., 29, 444 (1974).
72. Z. Lasocki and M. Witekowa, SYNTH. REACT. INORG. MET.-ORG. CHEM., 4, 231 (1974).
73. U. Wannagat and D. Labuhn, MONATSH. CHEM., 104, 1453 (1973); ibid., 105, 209 (1974).

74. G. S. Gol'din, L. S. Baturina, A. G. Kucher and N. L. Ivanova, ZH. OBSHCH. KHIM., 44, 566 (1974).
75. A. M. Pinchuk, A. M. Khmaruk and T. V. Kovalevskaya, ZH. OBSHCH. KHIM., 44, 460 (1974).
76. F. Hoefler and D. Wolfer, Z. ANORG. ALLG. CHEM., 406, 19 (1974).
77. K. A. Andrianov, A. B. Zachernyuk and E. A. Zhdanova, DOKL. AKAD. NAUK SSSR, SER. KHIM., 214, 325 (1974).
78. U. Wannagat and M. Schlingman, Z. ANORG. ALLG. CHEM., 406, 312 (1974).
79. G. S. Gol'din and L. S. Baturina, ZH. OBSHCH. KHIM., 43, 2245 (1973).
80. J. P. Mooser, H. Nöth and W. Tinhof, Z. NATURFORSCH, TEIL B., 29, 166 (1974).
81. H. Nöth, W. Tinhof, and T. Taeger, CHEM. BER., 107, 3113 (1974).
82. H. W. Roesky and H. Wiezer, ANGEW. CHEM., INT. ED., 13, 146 (1974).
83. H. W. Roesky and H. Wiezer, CHEM. ZTG., 97, 661 (1973).
84. V. F. Mironov, V. D. Sheludyakov and E. S. Rodionov, ZH. OBSHCH. KHIM., 44, 1502 (1974).
85. N. S. Nametkin, V. M. Vdovin, V. N. Karel'skii, B. V. Kacharmin and E. D. Babich, IZV. AKAD. NAUK SSSR, SER. KHIM., 2149 (1973).
86. D. A. Armitage and A. W. Sinden, J. INORG. NUCL. CHEM., 36, 993 (1974).
87. V. F. Mironov, V. D. Sheludyakov and E. S. Rodionov, ZH. OBSHCH. KHIM., 44, 2557 (1974).
88. H. Schumann and L. Rösch, CHEM. BER., 107, 854 (1974).
89. G. Fritz and H. Schäfer, Z. ANORG. ALLG. CHEM., 406, 167 (1974).
90. M. Baudler and A. Zarkadas, CHEM. BER., 106, 3970 (1973).
91. G. Fritz and H. Schäfer, Z. ANORG. ALLG. CHEM., 407, 295 (1974).
92. R. Demuth, Z. NATURFORSCH. TEIL B., 29, 42 (1974).
93. K. A. Andrianov, M. A. Ezerets, I. I. Khazanov, V. G. Ser-ov and V. T. Kaufman, IZV. AKAD. NAUK SSSR, SER. KHIM., 1894 (1973).
94. A. A. Zhdanov, K. A. Andrianov, T. V. Astapova and V. G. Vinogradova, IZV. AKAD. NAUK SSSR. SER. KHIM., 2628 (1973).
95. J. H. Burk and W. A. Kriner, J. ORGANOMETAL. CHEM., 63, C1 (1973).
96. R. H. Cragg and M. Nazery, J. ORGANOMETAL. CHEM., 71, 225 (1974).
97. J. Lukasiak, A. Radecki and H. Lamparczk, ROCZ. CHEM., 48, 891 (1974).
98. J. Lukasiak, A. Radecki and J. Halkiewicz, ROCZ. CHEM., 48, 1099 (1974).
99. A. I. Nogaideli, R. Sh. Tkeshelashvili and N. P. Makhara-shvili, ZH. OBSHCH. KHIM., 44, 119 (1974).
100. J. M. Wolcott and F. K. Cartledge, J.ORG.CHEM., 39, 2420

(1974).
101. M. Deneux, I. C. Akhrem, D. V. Avetissian, E. I. Myssoff, and M. E. Volpin, BULL. CHEM. FR., 2638 (1973).
102. D. A. Evans and L. K. Truesdale, TETRAHEDRON LETT., 4929 (1973).
103. I. N. Lapkin, T. N. Povarnitsyna and T. Yu. Denisova, ZH. OBSHCH. KHIM., 44, 123 (1974).
104. R. J. P. Corriu and J. J. E. Moreau, TETRAHEDRON LETT., 4469 (1973).
105. V. A. Yablokov, A. N. Sunin and A. V. Ganyushkin, ZH. OBSHCH. KHIM., 44, 287 (1974).
106. R. N. Prasad and J. P. Tandon, ACTA CHIM., 80, 447 (1974).
107. M. Wada, T. Suda and R. Okawara, J. ORGANOMETAL. CHEM., 65, 335 (1974).
108. T. Seshadri, J. INORG. NUCL. CHEM., 36, 519 (1974).
109. D. A. Armitage, M. J. Clark, A. W. Sinden, J. N. Wingfield, E. W. Abel and E. J. Louis, INORG. SYNTH., 15, 207 (1974).
110. E. A. Lebedev, D. V. Fridland and V. O. Reikhsfel'd, ZH. OBSHCH. KHIM., 44, 294 (1974).
111. R. S. Glass, J. ORGANOMETAL. CHEM., 61, 83 (1973).
112. E. P. Lebedev, V. O. Reikhsfel'd and V. A. Baburina, ZH. OBSHCH. KHIM., 44, 787 (1974).
113. E. P. Lebedev, V. A. Baburina and V. O. Reikhsfel'd, ZH. OBSHCH. KHIM., 44, 1212 (1974).
114. G. K. Barker, J. E. Drake and R. T. Hemmings, J. CHEM. SOC., DALTON TRANS., 450 (1974).
115. R. J. Cunico, SYNTH. REACT. INORG. METAL.-ORG. CHEM., 4, 23 (1974).
116. Yu. I. Khudobin, N. A. Andreeva, N. P. Kharitonov and M. G. Voronkov, IZV. AKAD. NAUK., USSR, SER. KHIM., 856 (1974).
117. F. Hoefler and R. Jannach, INORG. NUCL. CHEM. LETT., 10, 711 (1974).
118. D. Kummer and H. Koester, Z. ANORG. ALLG. CHEM., 402, 297 (1973).
119. W. Malisch and M. Kuhn, CHEM. BER., 107, 979 (1974).
120. W. Malisch and M. Kuhn, CHEM. BER., 107, 2835 (1974).
121. W. Malisch and P. Panster, J. ORGANOMETAL. CHEM., 64, C5 (1974).
122. T. J. Marks and A. M. Seyam, INORG. CHEM., 13, 1624 (1974).
123. B. K. Nicholson, J. ORGANOMETAL. CHEM., 72, 211 (1974).
124. C. S. Cundy and M. F. Lappert, J. ORGANOMETAL. CHEM., 57, C72 (1973).
125. A. P. Hagen, L. McAmis and M. A. Stewart, J. ORGANOMETAL. CHEM., 66, 127 (1974).
126. L. Vancea and W. A. G. Graham, INORG. CHEM., 13, 511 (1974).
127. C. Eaborn, A. Pidcock and B. Ratcliff, J. ORGANOMETAL. CHEM., 66, 23 (1974).
128. C. Eaborn, T. N. Metham and A. Pidcock, J. ORGANOMETAL. CHEM., 63, 107 (1974).
129. Y. Kiso, K. Tamao and M. Kumada, J. ORGANOMETAL. CHEM., 76, 95 (1974).
130. W. Jetz and W. A. G. Graham, J. ORGANOMETAL. CHEM., 69, 383 (1974).
131. H. Sakurai and J. Hayashi, J. ORGANOMETAL. CHEM., 63, C10 (1973).
132. W. Fink, HELV. CHIM. ACTA, 57, 167 (1974).
133. J. C. Brunet, J. Bertrand and C. Lesenne, J. ORGANOMETAL. CHEM., 71, C8 (1974).
134. Y. Kiso, K. Tamao and M. Kumada, J. ORGANOMETAL. CHEM., 76, 105 (1974).

135. A. Wright and R. West, J. AMER. CHEM. SOC., 96, 3222 (1974).
136. A. Wright and R. West, J. AMER. CHEM. SOC., 96, 4692 (1974).
137. A. G. Brook and J. M. Duff, J. AMER. CHEM. SOC., 96, 4692 (1974).
138. J. W. Wilt and W. K. Chwang, J. AMER. CHEM. SOC., 96, 6194 (1974).
139. E. A. Francis and J. Y. Corey, J. ORGANOMETAL. CHEM., 61, C20 (1973).

Germanium, Tin, and Lead

R.D. Joyner

Department of Chemistry
Ball State University
Muncie, Indiana 47306

GENERAL

Reviews and books covering the chemistry of the group IVA elements this year include: 1972 annual surveys of germanium (1), tin (2), and lead (3); a review of germanium, tin, and lead (4); a review of germanium and its compounds in organic chemistry (5); unusual stereochemistry of compounds of silicon and germanium (6); a review of organometallic derivatives of hydroxylamines and oximes (7); a review of metalloboroxanes (8); a monograph on the redox chemistry of tin and lead porphyrins (9); organo-group IV peroxides (10); two studies of the behavior of organotin compounds toward SO_2 (11, 12); a review of fluxional group IVA organometallic compounds (13); cycloalkanes containing heterocyclic germanium, tin, and lead (14); and a new quarterly journal devoted entirely to main group IV elements and their compounds (15).

MX_4 COMPOUNDS (X's not necessarily the same)

Metal-carbon bonded compounds. A series of 4-oxopentyltin compounds was prepared by a novel use of Grignard reactions (16). Other syntheses include β-chlorovinyl compounds of tin and germanium (18),metallated enamines prepared by hydrostannylation (19), reactions of trialkylstannanes with ketene (20), preparation of methanesulfonates of tin(II) and tin(IV) (21), methyl and ethyl esters of 2-germaacetic acid (22), the preparation of tetramethyllead from trimethylaluminum and lead sulfide (23), a series of pentafluorophenylgermanium hydrides (24), synthesis and reactivities of N-(2-triphenylstannylethyl)amines (25), a series of trifluoromethylphenyltin(IV) compounds (26), sulfur containing organometallic compounds of tin and germanium (27), organolead derivatives of diazoalkanes (28), an improved yield of triphenylvinyllead (29), and 24 new organogermanium compounds in a study of radical phenyl migrations (30).

Metal-nitrogen bonded compounds. Tri-tert-butylstannylamine was claimed as the first primary stannylamine (31). Other preparations include: carbodiimides of Ge and Sn (32); reaction of $[(CH_3)_3Sn]_3N$ with methylchlorosilanes to substitute a silicon moiety for one $(CH_3)_3Sn$ group (33); preparation of germylhydrazines by hydrogermylation of nitrogen-nitrogen double bonds (34); and formation of organolead(IV) azido complex anions (35).

Metal-oxygen and metal-sulfur bonded compounds. Reactions in this section include: a mechanistic study of sulfur dioxide insertion into Sn-C bonds (36); the preparation of digermthiane $(H_3Ge)_2S$(17); synthesis of chelates of general formula,

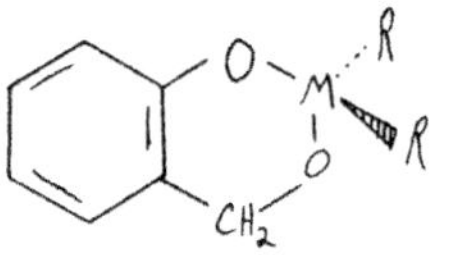

(37);

formation of several cyclic derivatives by reaction of butyltin triisopropoxide with alkanolamines (38); a new preparation of oxagermacyclopentanes and -pentenes (39); reaction of organic peroxides with stannyllithium compounds to prepare tin alkoxides (40); preparation of Sn(IV) compounds from tin(II) alkoxides (41); the preparation of tin(II) alkoxides and phenoxides (42); synthesis of tributylgermanium oximates (43); preparation of di- and triphenylstannyl esters of aromatic dithio acids (44); formation of the Sn-S bond by heating thioesters with $(C_2H_5)_3SnOCH_3$ (45); and the preparation of oxadiazaboroles which contain Sn-O-B linkages (46).

Metal-halogen bonded compounds. High yield syntheses of $(CH_3)_2Ge(H)X$ (X=halogen, psuedohalogen, or acetate) were reported with a normal coordinate analysis assignment of all fundamental vibrational frequencies (47). Other preparations include: $GeBr_2$ from $Ge(OH)_2$ and HBr (48); chlorogermane and iodogermane from GeH_4 (49); bromo- and dibromogermane from GeH_4 and Br_2 (50); organotin chlorides and bromides from ICl or IBr and tetraorganotins (51); alkyltin halides from Sn and alkyl halides (52); organotrifluorostannanes (53); (silylmethyl)trihalostannanes, $H_3SiCH_2SnX_3$ (54); fluorogermane from bromogermane and PbF_2 (55); $H_3GeNCNGeH_3$ from fluorogermane and $(CH_3)_3SiNCNSi(CH_3)_3$ (55); the synthesis of organofluorogermanes (56); and the preparation of tin(IV) chloride alkoxides (57).

Reactions of MX_4 compounds. The activation of Grignard reagents by catalytic quantities of Ni(II) complexes in the formation of Ge-C bonds was studied further (58). An extension of the known catalytic activity of octacarbonyldicobalt in reactions of organosilanes included similar studies with organostannanes and -germanes (59). Other studies include: a strong preference of F for Si rather than comparable C or Ge centers (60); the interaction of organotin halides with phosphorous ylids (61); reactions of bis(tributylstannyl) mercaptoethanolate with carbon disulfide, isothiocyanates, and thionecarbonates (62); oxidative addition reactions of organotin(II) compounds (63); transmetallation reactions with n-butyl and methyllithium (64) and $CH_3(R)NCH_2Li$ (65); disproportionation of hexaorganodistannanes in liquid SO_2(66); the cleavage of tin allyl bonds as a route to organostannoxanes (67); stereochemical studies of nucleophilic substitution of chloro- and alkoxygermanes by germyllithium reagents (68) and of reactions of alkoxygermanes with organolithium and Grignard reagents (69); reaction of organogermazanes to give organo(oxinato)germanes (70); hydrostannylation under γ-radiation (71); insertion reactions of difluorogermylene (72); synthesis of diarylmercury from dialkyldiaryltin (73); cyclodestannylation (74); and cycloaddition reactions of trimethyllead diazoacetic ester with activated oelfins and acetylenes (75).

METAL-METAL BONDED COMPOUNDS

Organometallic compounds. Compounds of the type $(CH_3)_3Ge\text{-}Ge(CH_3)_2X$ were prepared and studied. Changes in X did not markedly effect the stability of the Ge-Ge bond (76). In a separate study Ge-Ge bond reactivity was enhanced in hexakis(pentafluorophenyl)digermane (77). An optically active digermane was prepared (78). Syntheses and reactions of triethylgermyllithium, -sodium, -potassium, -rubidium, and -cesium were studied in weakly solvating solvents (79). Additional Sn-Cd-Sn and Sn-Hg-Sn compounds were prepared (80). The compound bis[tris(trimethylsilylmethyl)stannyl] cadmium was reported (81).

Coordination Compounds with ligands containing group IVA elements. Reactions with $SnCl_2$ provided Rh-$SnCl_3$ (82) and Pt-$SnCl_3$ (83) moieties, as well as $SnCl_2$ insertion into Fe-C bonds (84, 85). A study of olefin isomerization included an $SnCl_3^-$ complex of Pt as catalyst (86). Several other Sn(II) or Ge(II) reactions were reported to form transition metal complexes (87-90). Several complexes which have Ge or Sn bonded to a transition metal were reported (91-98). Most of these complexes were carbonyl, phosphine, or cyclopentadienyl bonded complexes. Several of the complexes were germanium or tin bridged bis-complexes (89, 99-102) including a stable stannane (101) and two metal cluster compounds (102, 103). Alkali metal derivatives of metal carbonyls reacted with $(CH_3)_3SnCH_2I$ to give tin-carbon bond cleavage followed by substitution of $(CH_3)_3Sn$ for the alkali metal (104). Novel carbene complexes were prepared from triphenyl- and trimethylgermylpentacarbonylmanganese (105). Cleavage reactions of some group IV-transition metal bonds were studied (106). Titanium complexes of the type

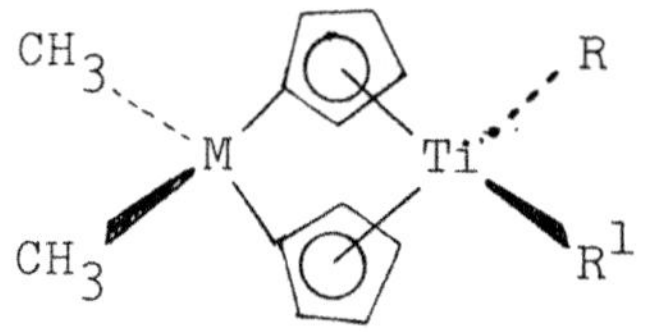

(M = Ge or Si) were claimed (107). Organolead(IV) adducts to some nickel(II) complexes were described (108).

COORDINATION COMPOUNDS OF Ge, Sn, Pb

Complexes of tin(II) (109, 110) and of germanium(II) (111) were reported. A five coordinate tin compound was claimed (112). The structure assignment was based upon dipole moment and proton nmr data. Polymers with the central atom of a macrocycle in the main chain included hemiporphyrazinato germanium compounds (113).

Coordinating groups used in forming other complexes include: several tridentate ligands (114); hexamethylphosphoric triamide (HMPA) (115); tetracycanoethylene and 7,7,8,8-tetracyanoquinodimethane (116); tetracyano-p-quinodimethane(TCNQ) (117); diacetyl-bis-benzoylhydrazone (118); aryl- and alkylimines (119); isooctyl thioglycolate (120); tetradentate Schiff bases (121); carboxylic acid anhydrides (122); bis(2-mercaptoethyl)ether (123); bis(2-mercaptoethyl)sulfide (124); both mono- and dicyclopentadienyls (125, 127); and mono-, di-, and triethanolamines (126). Seven new anionic triphenyltin complexes were reported (128). The compound bis[N,N,-bis(trimethylsilyl)amino]tin(II) was reported to be a stabilized Sn(II) compound (129).

MISCELLANEOUS COMPOUNDS AND REACTIONS

Icosahedral germacarboranes (130) and apically substituded B_6H_{10} derivatives with $(CH_3)_3Ge$ and -Si as the apex (131) were reported. Syntheses of bis(pentaboranyl)-

group IVA compounds were reported (132). 1,3-Bis(boryloxy) tetraalkyldistannoxanes were prepared (133). Syntheses of α-heterosubstituted nitrosamines included α-Sn substitution (134). Transfer of methyl groups from $(CH_3)_3Ga$ to Sn, Ge, and Si films was observed (135). Syntheses were given for tetraalkylammonium trichlorogernamates and -trichlorostannates (136). The synthesis is reported for germylphosphine (137). The first 1,3-distannacyclobutane was reported (138). Compounds of the general formula $(C_6H_5)_4Sn_2[OOCM(C_6H_5)_3]_2$ were reported (139). γ-Hydroxymetal hydrides of Si, Ge, and Sn were described (140).

Several divalent germanium reactions were reported. A rapid route to Ge(II) compounds using hypophosphite in methanol was reported (141). Several papers involved germylenes and germylgermylenes as intermediates (142-145).

REFERENCES

1. B. Pant, J. ORGANOMETAL. CHEM., 68,221 (1974).
2. P. Harrison, J. ORGANOMETAL. CHEM., 79,17 (1974).
3. L. Willemsens, J. ORGANOMETAL. CHEM., 62,285 (1973).
4. P. Harrison and P. Hubberstey, INORG. CHEM. MAIN-GROUP ELEM., 1,143 (1973).
5. E. Bulten and G. Van der Kerk, NEW USES GERMANIUM 1974, 51. CHEM. ABST., 81,130206e.
6. C. Glidewell, INORG. CHIM. ACTA., REV., 7,69 (1973).
7. A. Singh, V. Gupta, G. Srivastava, and R. Mehrotra, J. ORGANOMETAL. CHEM., 64,145 (1974).
8. S. Mehrotra, G. Srivastava, and R. Mehrotra, J. ORGANOMETAL, CHEM., 73,277 (1974).
9. D. Whitten, T. Meyer, F. Hopf, J. Ferguson and G. Brown, ANN. N.Y. ACAD. SCI., 206,516 (1973).
10. D. Brandes and A. Blachette, J. ORGANOMETAL. CHEM., 78,1 (1974).
11. U. Kunze, E. Lindner, and J. Koola, J. ORGANOMETAL. CHEM., 57,319 (1973).
12. U. Kinze and J. Koola, J. ORGANOMETAL, CHEM., 80,281 (1974).
13. R. Larrabee, J. ORGANOMETAL. CHEM., 74,313 (1974).
14. B. Pant, J. ORGANOMETAL. CHEM., 66,321 (1974).
15. M. Gielen; Ed., "Reviews on Silicon, Germanium, Tin and Lead Compounds", Scientific Publications Division, Freund Publishing House Ltd., P.O.B. 35010, Tel-Aviv, Israel.
16. S. Abbas and R. Poller, J. CHEM., SOC. - DALTON, 1974, 1769.
17. D. Rankin, INORG. SYNTH., 15,182 (1974).
18. A. Nesmeyanov and A. Borisov, IZV. AKAD. NAUK SSSR, SER. KHIM., 1974, 1667. CHEM. ABST., 81,105651p.

19. M. Kazankova, T. Zverkova, M. Levin, and I. Lutsenko, ZH. OBSHCH KHIM., 44,230 (1974). CHEM. ABST.,80,96113q.
20. M. Kazankova, T. Zerkova, A. Lutsenko, and I. Lutsenko, ZH. OBSHCH. KHIM., 44,229 (1974). CHEM. ABST., 80,96115s.
21. R. Paul, V. Kapila, and S. Sharma, INDIAN J. CHEM., 12,651 (1974).
22. K. Strom and W. Jolly, J. ORGANOMETAL. CHEM., 69,201 (1974).
23. M. Boleslawski, S. Pasynkiewicz, and A. Kunicki, J. ORGANOMETAL. CHEM., 73,193 (1974).
24. M. Bochkarev, L. Maiorova, S. Korneva, L. Bochkarev, and N. Vyazankin, J. ORGANOMETAL. CHEM., 73,229 (1974).
25. Y. Sato, Y. Ban, and H. Shirai, J. ORG. CHEM., 38,4373 (1973).
26. M. Barnard, P. Smith, and R. White, J. ORGANOMETAL CHEM., 77,189 (1974).
27. R. Taylor and J. Wardell, J. ORGANOMETAL. CHEM., 77,311 (1974).
28. R. Gruning and J. Lorberth, J. ORGANOMETAL. CHEM., 78,221 (1974).
29. G. Barbiere and G. Borelli, SYN. REACTIV. INORG. METAL-ORG. CHEM., 4,139 (1974).
30. H. Sakurai, I. Nozue, and A. Hosomi, J. ORGANOMETAL, CHEM., 80,71 (1974).
31. H. Goetze, ANGEW. CHEM., 86,104 (1974). CHEM. ABST., 80,108609w.
32. I. Matsuda, K. Itoh, and Y. Ishii, J. ORGANOMETAL. CHEM., 69,353 (1974).
33. H. Roesky and H. Wiezer, CHEM. BER.,107,3186 (1974). CHEM. ABST., 81,16910g.
34. K. Linke and H. Goehausen, CHEM. BER, 106,3438 (1973). CHEM. ABST., 80,27334a.
35. N. Bertazzi, G. Alonzo, G. Stocco, and L. Pellerito, J. ORGANOMETAL. CHEM., 80,225 (1974).
36. J. Koola and U. Kunze, J. ORGANOMETAL. CHEM., 77,325 (1974).
37. R. Cragg and M. Nazery, J. ORGANOMETAL. CHEM., 71,225 (1974).
38. D. Gaur, G. Srivastava, and R. Mehrotra, J. ORGANOMETAL. CHEM., 65,195 (1974).
39. J. Barrau, M. Massal, and J. Satge, J. ORGANOMETAL. CHEM. 71,C45 (1974).
40. T. Basalgina, O. Kruglaya, G. Kalinina, and N. Vyazankin, IZV. AKAD. NAUK. SSSR, SER. KHIM., 1973, 2776. CHEM. ABST., 80,83150z.
41. I. Wakeshima and I. Kijima, J. ORGANOMETAL. CHEM., 76,37 (1974).
42. I. Wakeshima and I. Kizima, NIPPON KAZAKU KAISHI, 1974, 536. CHEM. ABST., 80,145645a.

43. A. Singh, A. Rai, and R. Mehrotra, J. ORGANOMETAL. CHEM., 57,301 (1973).
44. S. Kato, T. Kato, T. Yamuchi, Y. Shibahashi, E. Kakuda, M. Mizuta, and Y. Ishii, J. ORGANOMETAL. CHEM., 76,215 (1974).
45. M. Voronkov, R. Mirskov, O. Ishchenko, V. Chernova, and S. Sitnikova, ZH. OBSHCH. KHIM., 44,228 (1974). CHEM. ABST., 80,96114r.
46. A. Goel and V. Gupta, J. ORGANOMETAL. CHEM., 77,183 (1974).
47. G. Barker, J. Drake, and R. Hemmings, CAN. J. CHEM., 52,2622 (1974).
48. O. Zakolodyazhnaya, R. Magunov, and I. Kovalevskaya, ZH. NEORG. KHIM., 19,2269 (1974). CHEM. ABST., 81,144805m.
49. S. Cradock, INORG. SYNTH., 15,161 (1974).
50. M. Swiniarski and M. Onyszchuk, INORG. SYNTH., 15,157 (1974).
51. A. Folaranmi, R. McLean, and N. Wadibia, J. ORGANOMETAL. CHEM., 73,59 (1974).
52. H. Matschiner, R. Voigtlander, and A. Tzschach, J. ORGANOMETAL. CHEM., 70,387 (1974).
53. V. Shiryaev, L. Makhalkina, F. Huzmina, V. Krylov, V. Osipov and V. Mironov, ZH. OBSHCH. KHIM., 43,2232 (1973). CHEM. ABST., 80,48116w.
54. V. Mironov, V. Shiryaev, V. Yankov and N. Gladchenko, U.S.S.R. 396,340, 29 Aug. 1973. CHEM. ABST. 80,15075d.
55. S. Cradock, INORG. SYNTH., 15,164 (1974).
56. J. Bulkowski, DISS. ABST. INT. B, 35,679 (1974). CHEM. ABST., 81,136215d.
57. N. Yoshino, Y. Kondo, and T. Yoshino, SYN. INORG. METAL-ORG. CHEM., 3,397 (1973).
58. F. Carre and R. Corriu, J. ORGANOMETAL. CHEM., 74,49 (1974).
59. G. Bradley and S. Stobart, J. CHEM. SOC - DALTON, 1974, 264.
60. S. Pace and J. Riess, J. ORGANOMETAL, CHEM., 76,325 (1974).
61. J. Buckle and P. Harrison, J. ORGANOMETAL. CHEM., 77,C22 (1974).
62. S. Sakai, H. Niimi, and Y. Ishii, J. ORGANOMETAL. CHEM., 72,103 (1974).
63. R. Bos, E. Bulten, and J. Noltes, J. ORGANOMETAL. CHEM., 67,C13 (1974).
64. D. Seyferth and T. Jula, J. ORGANOMETAL. CHEM., 66,195 (1974).
65. D. Peterson and J. Ward, J. ORGANOMETAL. CHEM., 66,209 (1974).
66. U. Kunze and H. Volker, J. ORGANOMETAL. CHEM., 66.C43 (1974).

67. V. Peruzzo and G. Tagliavini, J. ORGANOMETAL. CHEM., 66,437 (1974).
68. F. Carre and R. Corriu, J. ORGANOMETAL. CHEM., 65,349 (1974).
69. F. Carre and R. Corriu, J. ORGANOMETAL. CHEM., 65,343 (1974).
70. M. Wada, F. Suda, and R. Okawara, J. ORGANOMETAL. CHEM., 65,335 (1974).
71. V. Lopatina, N. Sheverdina, V. Chernoplekova, and K. Kocheshkov, DOKL. AKAD. NAUK, SSSR, 213,846 (1973). CHEM. ABST., 80,83157g.
72. J. Satze, P. Riviere and A. Boy, C. R. ACAD. SCI., SER. C., 278,1309 (1974). CHEM. ABST., 81,91659h.
73. O. Syutkina, E. Panov, V. Lodochnikova, and K. Kocheshkov, DOKL. AKAD. NAUK SSSR,, 216,333 (1974). CHEM. ABST., 81,91673h.
74. D. Peterson, M. Robbins, J. Hansen, J. ORGANOMETAL. CHEM. 73,237 (1974).
75. R. Gruning and Jorg Lorberth, J. ORGANOMETAL. CHEM., 69,213 (1974).
76. A. Andy and J. Thayer, J. ORGANOMETAL. CHEM., 76,339 (1974).
77. M. Bochkarev, G. Razuvaev, N. Vyazankin, and O. Yu. Semenov, J. ORGANOMETAL. CHEM., 74,C4 (1974).
78. F. Carre and R. Corriu, J. ORGANOMETAL. CHEM., 73,C49 (1974).
79. E. Gladyshev, N. Vyazankin, E. Fedorova, L. Yuntila, and G. Razuvaev, J. ORGANOMETAL. CHEM., 64,307 (1974).
80. B. Fedot'ev, O. Kruglaya, and N. Vyazankin, IZV. AKAD. NAUK SSSR, SER. KHIM., 1974, 713. CHEM. ABST., 81, 13617d.
81. G. Kalinina, O. Krugloya, B. Petrov, E. Shchupak, and N. Vyazankin, ZH. OBSHCH. KHIM., 43,2224 (1973).
82. W. Hill and C. McAuliffe, INORG. CHEM., 13,1524 (1974).
83. Y. Kukushkin, P. Antonov, K. Dubonos and L. Konovalov, ZH. NEORG. KHIM., 18,3017 (1973). CHEM. ABST., 80,55422w.
84. B. Cole, J. Cotton, and D. McWilliam, J. ORGANOMENTAL. CHEM., 64,223 (1974).
85. C. Magatti and W. Giering, J. ORGANOMETAL. CHEM., 73,85 (1974).
86. D. Bingham, D. Webster, and P. Wells, J. CHEM. SOC. - DALTON, 1974, 1514.
87. T. Kruck, F. Becker, H. Breuer, K. Ehlert, and W. Rother, Z. ANORG. ALLG. CHEM., 405,95 (1974). CHEM. ABST., 81,32655f.
88. H. Brunner and S. Loskot, Z. NATURFORSCH., TEIL B, 28,314 (1973). CHEM. ABST., 80,96148e.
89. T. Kruck, H. Breuer, and F. Becher, Z NATURFORSCH. TEIL B, 29,55 (1974). CHEM. ABST., 81,37626g.

90. A. Cornwell, P. Harrison, and J. Richards, J. ORGANOMETAL. CHEM., 76,C26 (1974).
91. A. Nesmeyanov, K. Anisimov, N. Kolobova, and V. Khandozhko, ZH. OBSHCH. KHIM., 44,1287 (1974). CHEM. ABST. 81,57704j.
92. T. Kruck and W. Molls, Z. NATURFORSCH. TEIL B, 29,198 (1974). CHEM. ABST. 81,72064u.
93. J. Collman, D. Murphy, E. Fleischer, and D. Swift, INORG. CHEM., 13,1 (1974).
94. K. Mann, W. Morrison, Jr., and D. Hendrickson, INORG. CHEM., 13,1180 (1974).
95. A. Nesmeyanov, K. Anisimov, N. Kolobova, and V. Khandozhko, ZH. OBSHCH. KHIM., 44,1079 (1974). CHEM. ABST., 81,44804s.
96. H. Schumann, H. Breunig, and U. Frank, J. ORGANOMETAL. CHEM., 60,279 (1973).
97. H. Clark and A. Rake, J. ORGANOMETAL. CHEM., 74,29 (1974).
98. S. Stobart, INORG. SYNTH., 15,174 (1974).
99. R. Adams, F. Cotton, and B. Frenz, J. ORGANOMETAL. CHEM., 73,93 (1974).
100. M. Basato, J. Fawcett, and A. Poe, J. CHEM. SOC. - DALTON, 1974, 1350.
101. P. Hackett and A. Manning, J. ORGANOMETAL. CHEM., 66,C17 (1974).
102. K. Bos, E. Bulten, and J. Noltes, J. ORGANOMETAL. CHEM., 71,C52 (1974).
103. C. Garner and R. Senior, INORG. NUCL. CHEM. LETT., 10,609 (1974).
104. R. King and K. Hodges, J. ORGANOMETAL. CHEM., 65,77 (1974).
105. M. Webb, R. Stewart, Jr., and W. Graham, J. ORGANOMETAL. CHEM., 59,C21 (1973).
106. R. Bichler, H. Clark, B. Hunter, and A. Rake, J. ORGANOMETAL. CHEM., 69,367 (1974).
107. H. Kolpf and W. Kahl, J. ORGANOMETAL. CHEM., 64,C37 (1974).
108. L. Pellerito, R. Cefalù, and A. Gianguzza, J. ORGANOMETAL. CHEM., 70,C27 (1974).
109. D. Perry and R. Geanangel, J. INORG. NUCL. CHEM., 36,205 (1974).
110. S. Copinathan, C. Gopinathan, and J. Gupta, INDIAN J. CHEM., 11,1067 (1973). CHEM. ABST., 80,90518d.
111. O. Nefedov, S. Kolesnikov, and I. Rogozhin, IZV. AKAD. NAUK SSSR, SER. KHIM., 1973, 2824. CHEM. ABST. 80,77719a
112. A. Tzschach and K. Poenicke, Z ANORG. ALLG. CHEM., 404,121 (1974). CHEM. ABST., 80,133566n.
113. G. Meyer and D. Woehrl, MAKROMOL. CHEM., 175,715 (1974). CHEM. ABST. 81,152673g.

114. L. Pellerito, R. Cefalu, A. Silvestri, F. DiBianca, R. Barbieri, H. Haupt, H. Preut, and F. Huber, J. ORGANOMETAL. CHEM., 78,101 (1974).
115. K. Elegbede and R. McLean, J. ORGANOMETAL. CHEM., 69,405 (1974).
116. A. Cornwell, P. Harrison, and J. Richards, J. ORGANOMENTAL. CHEM., 67,C43 (1974).
117. J. Richards and P. Harrison, J. ORGANOMETAL. CHEM., 64,C3 (1974).
118. R. Cefalu, F. Magzio, L. Pellerito, and V. Romano, INORG. NUCL. CHEM. LETT., 20,529 (1974).
119. V. Kogan, A. Egorov, O.Osipov, V. Sokolov, ZH. OBSHCH. KHIM., 43,2719 (1973). CHEM. ABST. 80,77780p.
120. C. Stapfer and R. Herber, INORG. NUCL. CHEM. LETT., 10,161 (1974).
121. N. Biradar and V. Mahale, REV. ROUM. CHIM., 18,2077 (1973). CHEM. ABST. 80,90542g.
122. J. Johnsborg and K. Taugbol, J. INORG. NUCL. CHEM., 36,3861 (1974).
123. R. Engler, Z. ANORG. ALLG. CHEM., 407,35 (1974). CHEM. ABST., 81,72038p.
124. R. Engler, Z. ANORG. ALLG. CHEM., 406,74 (1974). CHEM. ABST. 81,72036m.
125. K. Kocheshkov, N. Zemlyanskii, N. Kolosova, A. Azizov, and Yu Ustynyuk, IZV. AKAD. NAUK SSSR, SER. KHIM., 1974, 1208. CHEM. ABST., 81,49762m.
126. T. Srivastava, D. Rupainwar, and J. Gaur, J. INORG. NUCL. CHEM., 36,733 (1974).
127. K. Kocheskov, N. Zemlyanskii, N. Kolosova, A. Azizov, O. Subbotin, and Yu. Ustynyuk, IZV. AKAD. NAUK SSSR, SER. KHIM., 1974, 737. CHEM. ABST., 81,13616c.
128. J. Feiccabrino and E. Kupchik, J. ORGANOMETAL. CHEM., 73,319 (1974).
129. C. Schaeffer and J. Zuckerman, J. AM. CHEM. SOC., 96,7160 (1974).
130. G. Wikholm and L. Todd, J. ORGANOMETAL. CHEM., 71,219 (1974).
131. D. Gaines, S. Hildebrandt, and J. Ulman, INORG. CHEM., 13,1217 (1974).
132. D. Gaines, and J. Ulman, INORG. CHEM., 13,2792 (1974).
133. S. Mehrotra, G. Srivastava, and R. Mehrotra, J. ORGANOMETAL. CHEM., 65,367 (1974).
134. D. Seebach and D. Enders, J. MED. CHEM., 17,1225 (1974).
135. D. Schlyer and M. Ring, J. ORGANOMETAL. CHEM., 71,C25 (1974).
136. G. Parshall, INORG. SYNTH., 15,222 (1974).
137. A. Norman, D. Wingeleth, and C. Heil, INORG. SYNTH., 15,177 (1974).
138. D. Seyferth and J. Lefferts, J. AM. CHEM. SOC., 96,6237 (1974).

139. G. Plazzogna, V. Peruzzo, and G. Tagliavini, J. ORGANOMETAL. CHEM., 66,57 (1974).
140. M. Massol, J. Barrau, J. Satze and B. Bouyssieres, J. ORGANOMETAL, CHEM., 80,47 (1974).
141. P. Poskozim and C. Guengerich, INORG. CHEM., 13,241 (1974).
142. J. Satge and G. Dousse, J. ORGANOMETAL. CHEM., 61,C26 (1973).
143. P. Riviere, J. Satge, and D. Soula, J. ORGANOMETAL. CHEM., 72,329 (1974).
144. P. Riviere, J. Satgé, G. Dousse, M. Riviere-Bandet, and C. Couret, J. ORGANOMETAL. CHEM., 72,339 (1974).
145. P. Riviere, J. Satge, and D. Soula, J. ORGANOMETAL. CHEM., 63,167 (1973).

PHOSPHORUS

H.-D. Ebert and E. Lindner
Lehrstuhl für Anorganische Chemie II
der Universität Tubingen
D-7400 Tübingen 1
Germany

INTRODUCTION

A number of review articles dealt with the latest developments in the area of phosphorus chemistry. Specifically the following subjects were reviewed: The role of trivalent P-compounds in organic chemistry[1] reactions of trivalent P-compounds with polyhalogen compounds,[2] syntheses and reactions of methyldichloro- and dimethylchlorophosphines and the practical applications of the resulting methylphosphorus compounds,[3] molecular structure of oxyphosphoranes,[4] synthesis, chemical and physical properties of cyclic halogen free P(V)-compounds having O-P-O groups,[5] chemical properties of a number of phosphorous, phosphonons, and phosphinous acids and their complexing abilities with alkali, alkaline earth and transition metals,[6] and reaction of coordinated pnictogen donors.[7]

ELEMENTAL PHOSPHORUS

The chemiluminescence of phosphorus when subjected to moist air was reinvestigated.[8] The vapor pressure of commercial grade red phosphorus as well as that of a particular monoclinic modification in dependency of temperature was measured.[9] The diffusion of phosphorus in melting gallicum[10] and silicon[11] was discussed.

PHOSPHIDES

KMnP was obtained by heating the elements in a closed system to 800°; it crystallizes according to X-ray structure analysis in a layer type lattice.[12] The reaction of red phosphorus with rubidium yielded Rb_4P_6 as a black compound besides an unidentified red phosphide.[13] Metallation of $(CH_3)_2PH$ with butyllithium and subsequent reaction with $AlCl_3$ led to the formation of $LiAl[P(CH_3)_2]_4$.[14] Numerous papers deal with galliumphosphides[15-22] and indiumphosphides.[9,22,23] Through vacuum evaporation amorphous SiP-films were obtained.[24] FeP_2, CoP_3 and CoPS can function as electrocatalysts in the hydrogen electrode.[25] Treatment of Eu_2O_3 with phosphorus yields EuP_2.[26]

PHOSPHORANES

In a gasphase reaction CH_3-PF_4, $(CH_3)_2PF_3$, and $(CH_3)_3PF_2$ reacted with $(CH_3)_3SiH$ to give CF_3-PF_3H, $(CF_3)_2PF_2H$, and $(CF_3)_3PH_2$. CF_3-PF_3H converts in bromotrifluoromethane solution at -70° in the presence of $(CH_3)_3SiH$ into $CF_3-PF_2H_2$.[27] A new method for the preparation of organofluorophosphanes of the type R_nPF_{5-n} was reported.[28] Syntheses of $(C_3F_7)_3PCl_2$, $(C_3F_7)_2P(CF_3)Cl_2$, and $C_3F_7P(CF_3)_2Cl_2$ were accomplished.[29] Compounds of the general structure $(C_6H_5)_nPCl_{5-n}$ (n=1-3) react in CH_3CN with H_2N-SO_3H with formation of $(C_6H_5)_nPCl_{3-n}N-SO_2-Cl$ and $(C_6H_5)_nPOCl_{3-n}$. The product of the reaction with n=2 decomposes during work-up.[30] Finally the reaction according to equ. (1)[31] and the formation of the cyclic phosphorane I should be mentioned.[32]

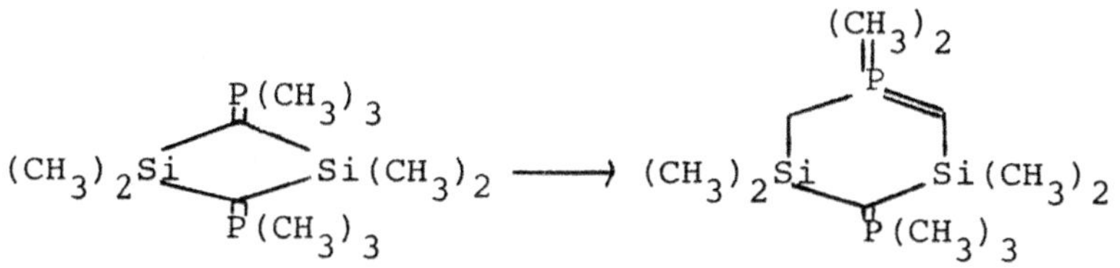

(1)

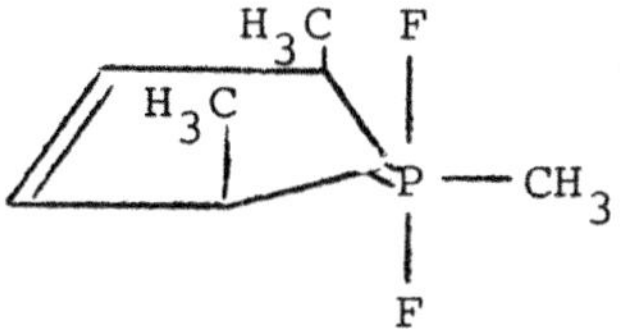

I

PHOSPHINES

Good yields of phosphines can be obtained by reacting $(C_6H_5O)_3P$ with alkylmagnesium iodides.[33] P_2F_4 reacts with $F_2P-CH_2-CH=CH_2$ with formation of $F_2P-CH_2-C(PF_2)H-CH_2-PF_2$ which reacts with BF_3 to give a variety of complexes.[34] The acidity and tautomeric behavior of β-ketodiphosphonium salts were measured and discussed.[35] Hexamethylenetetramine reacts with $P(CH_2OH)_3$ to phosphatriazaadamantane. This compound gives with CH_3I the corresponding $NH_4^{\oplus}$ salt and with H_2O_2 its phosphine oxide.[36] $PF_2(CH=CH_2)$ and $PF_2(CH=CH_2)\cdot BH_3$ were also prepared.[37] Reacting $BrPF_2$ with $(CF_3S)_2Hg$ leads to formation of CF_3SPF_2 in high yields; this compound oxidizes at -78° in the presence of O_2 to $CF_3-S-P(O)F_2$. Analogously difluorophosphines of the type XPF_2 react with O_2 at -78° to yield the corresponding phosphonyl compounds (X=Cl, Br, SCN, $N(CH_3)_2$).[38] $(CH_3)_3Si-C_6H_4-PCl_2$ and $(CH_3)_3SiCl$ are obtained by a reaction of $(CH_3)_3Si-C_6H_4-Si(CH_3)_3$ with PCl_3.[39] $(CH_3)_3P(O)H$ was obtained from $(CH_3)_2PCl$ and water.[40] The phosphine oxide of equ. (2) underwent ringclosure when reacted with P_2S_5; the compound was converted into its sulfone by treatment with $CH_3C(O)OOH$.[41]

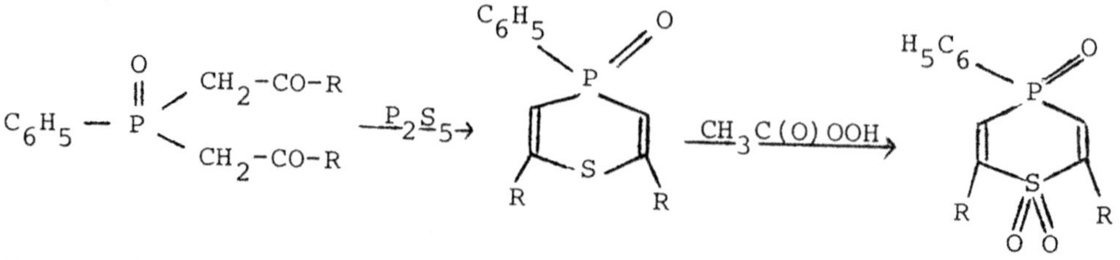

R = $-C(CH_3)_3$; $-C_6H_5$

(2)

PHOSPHORUS-BORON-COMPOUNDS

Co-reduction of BBr_3 and PCl_3 by H_2 leads to the compounds BP and B_6P as well as mixtures.[42] Reaction of $H_2P(BH_3)_2Na$ with HCl in etheral solution at -96° yields μ-phosphinodiborane, $(\mu\text{-}H_2PB_2H_5)_n$; its thermolysis leads to formation of diborane and polymeric phosphinodiborane.[43] Synthesis and spectroscopical investigations of dimethylchlorophosphinoborane was reported.[44] It was found that BCl_3, BBr_3 and BI_3 give with CH_3PCl_2 1:1 complexes. Diborane reacts with CH_3PCl_2 only to mixtures. Halogen exchange involving CH_3PCl_2 and BBr_3 or BI was observed.[45] The existence of $(CH_3)_2PCl_2 \cdot BF_3$ was demonstrated.[46] The reaction of B_5H_8Li with $(CF_3)_2PCl$ leads to formation of 2-$(CF_3)_2P\text{-}B_5H_8$.[47] The synthesis of $(CF_3)_2P\text{-}CB_5H_5C\text{-}R$ was reported [R=H, CH_3, $P(CF_3)_2$].[48] The following 1:1 adducts were prepared and NMR-spectroscopically investigated: $(CH_3)_{3-n}(C_6H_5)_nPBX_3$, $(CH_3)_3P(Z)BX_3$ and $(C_6H_5)_3P(O)BI_3$, (X=I,Br; n=0,1,2; Z=Se,S).[49]

PHOSPHORUS-SILICON-COMPOUNDS

The possibility of substituting silicon by phosphorus in the zeolithe synthesis was discussed.[50] Phosphorussilicate films on glass were obtained on Si-substrates by co-oxidation of SiH_4 and PH_3 at 350°.[51] F_3SiPH_2 can be prepared by reaction of $BrSiF_3$ with either $MAl(PH_2)_4$ (M=Li,Na)[52] or with $(CH_3)_3SnPH_2$ or $(CH_3)_3SnPD_2$ respectively.[53] Reactions leading to the formation of silylphosphines of the types $H_{3-x}(CH_3)_x$-$SiP(CH_3)_2$[14] and $X_{3-(n+m)}(CH_3)_nH_mSiPH_2$ (X=F,Cl,Br were investigated.[54] $(CH_3)_{3-n}H_nSiPH_2$ (n=0-3) reacts with $LiP(C_2H_5)_2$ in diglyme in the ratio of 1:2 yielding $(CH_3)_{3-n}H_nSiPh_2$; its reaction with CH_3Cl gives $(CH_3)_{3-n}$=$H_nSiP(CH_3)_2$. A reaction of $(CH_3)_{3-n}H_nSiPH_2$ with $LiPHCH_3$ gorms $(CH_{3-n}H_nSiPHLi$ which disproportionates upon heating.[55] Cyclic sila-, germa, and stannaphospholanes were prepared.[56] Compounds of the composition R_xECl_2 react with alkalimetal phosphides either according to equ. 3 or by a metal-halogen exchange. Cyclic or polymeric metal phosphides[57] were obtained by reacting $C_6H_5\text{-}P(H)\text{-}CH_2)_n\text{-}P(H)C_6H_5$ with $Zn(C_2H_5)_2$ or Al $(C_2H_5)_3$. Ethane is split off in these reactions. According to NMR analysis complete random distribution of trimethylelement groups is observed if $[(CH_3)_3Si]P$ is allowed to react with $(CH_3)_3GeCl$ and $(CH_3)_3SnCl$.[58]

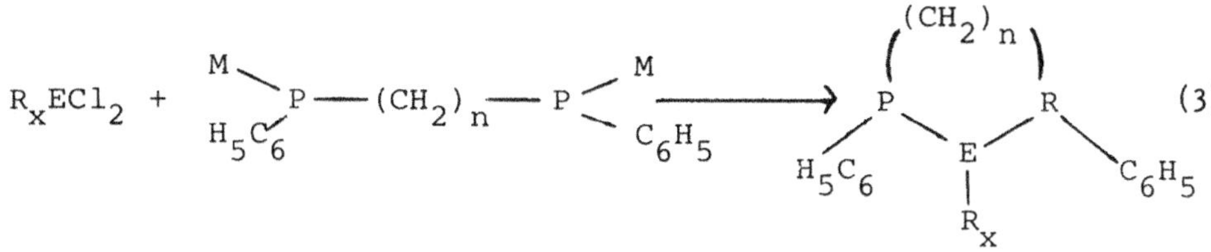

(3)

M = Na, K, Li; ER_x = $Si(CH_3)_2$, $Sn(C_2H_5)_2$, AlC_2H_5, AlC_6H_5; n = 3
ER_x = $Si(CH_3)_2$, CH_2, PC_6H_5; n = 2

PHOSPHORUS-NITROGEN COMPOUNDS

Cyclic phosphazenes: $N_3P_3F_5NH_2$ forms at 100° with $ClSO_4NCO$ the isocyanate $N_3P_3F_5NCO$ which adds $(CH_3)_2NH$ and yields the urea derivative $N_3P_3F_5NHC(O)N(CH_3)_2$. Analogously reacts $N_3P_3F_5Br$ with AgSCN. The resulting isothiocyanate also adds $(CH_3)_2NH$ and yields the thiourea derivative $N_3P_3F_5NHC(S)N(CH_3)_2$. Photochlorination of $N_3P_3F_5NCS$ affords $N_3P_3F_5$-N=CCl_2. Compounds of the type $N_4P_4F_7NCX$ (X=O, S) were obtained similarly.[59] Fluorination of $N_3P_3Cl_5N(CH_3)_2$ was discussed.[60] Reaction of $N_3P_3Cl_6$ with $HN(C_2H_5)_2$ leads theoretically to twelve isomers, nine of which were isolated and characterized.[61] Dimethylamine reacts with $N_3P_3Cl_6$ to IIA. In the presence of SbF_3, however, IIb is formed.[62]

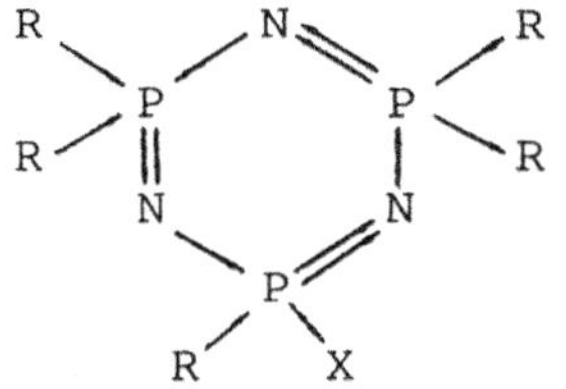

II

R = $N(CH_3)_2$
IIa X = Cl, IIb X = F;

Reaction of $[(CH_3)_2N]_2PCl$ at -70° with excess ammonia yields $N_3P_3H_3[N(CH_3)_2]_3$.[63] The reaction of $N_3P_3Cl_4(NH_2)R$ (R=Cl, NH_2) with $[(CH_3)_3Si]_2NH$ was investigated.[64] Transformations according to equ. 4 were described.[65]

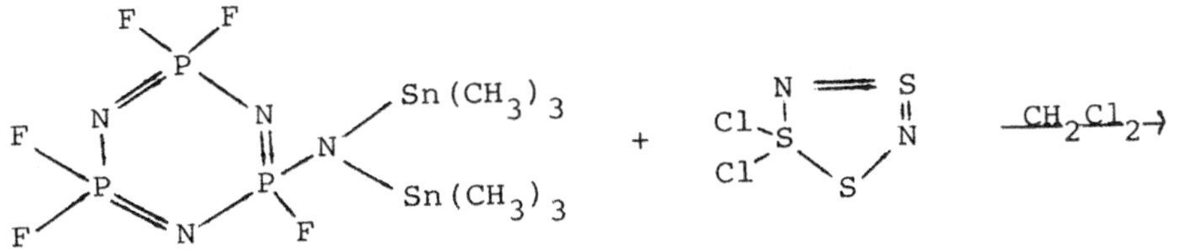

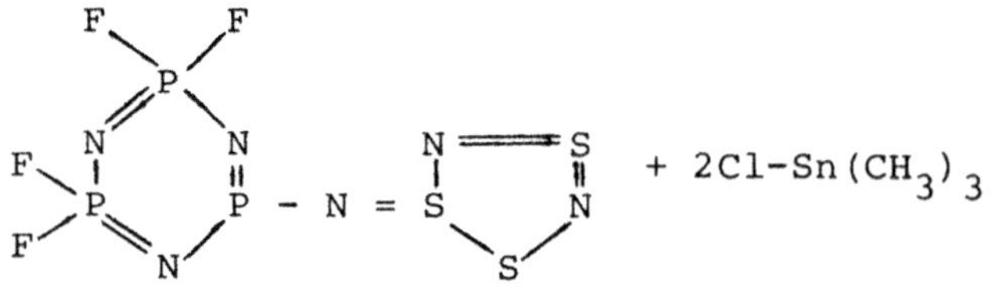

(4)

A Friedel-Crafts reaction was accomplished involving $N_3P_3Cl_{6-n}R_n$ (R=N-piperidyl) and benzene.[66] The phosphazenes which are formed between $N_3P_3Cl_6$ and 2,3-dihydroxynaphthalene give with certain organic molecules clathrates.[67] Under attack of $NaNH_2$ the chlorine atoms in $N_3P_3(OR)_5Cl$ (R=aryl) is substituted by an NH_2-group; with PCl_5 it affords $N_3P_3(OR)_2$-N=PCl_3.[68] The fluorination reaction according to equ. 5 leads in dependency of solvent acid ratio of the reactants to the following non-geminal mono-, di- (4 isomers), tri- (3 isomers), tetra- (3 isomers), and pentafluoro derivatives.[69]

$$P_4N_4[N(CH_3)_2]_8 \xrightarrow{SbF_3} P_4N_4F_n[N(CH_3)_2]_{8-n} \qquad (5)$$

Finally the preparation of oligomeric dimethylaminocyclophosphazene was reported.[70]

Heterocyclic Phosphazenes: A five-membered siliconorganic cyclophosphazene of structure III was prepared.[71] In refluxing benzene PCl_5 reacts with N≡C-N=C(R)NH_2 to the phosphatriazenes IV.[72] Nitriles of the general composition X-CH_2-CN react with R-C(Cl)$_2$-N=PCl_3 initially to linear cyanoethylenes which partially cyclize under the influence of the generated HCl to the phosphazenes V. The course of the reaction depends largely on the substituent R.[73]

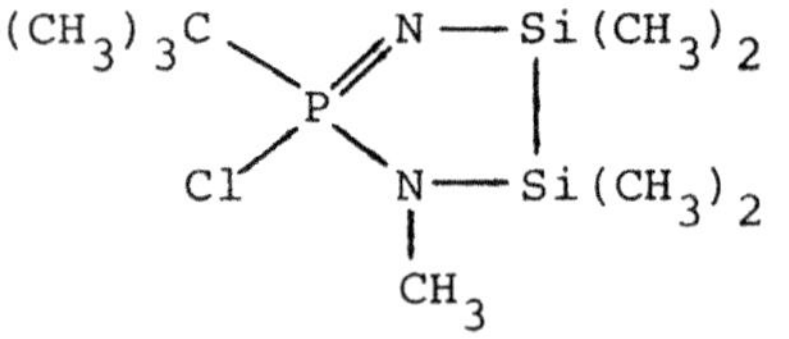

III

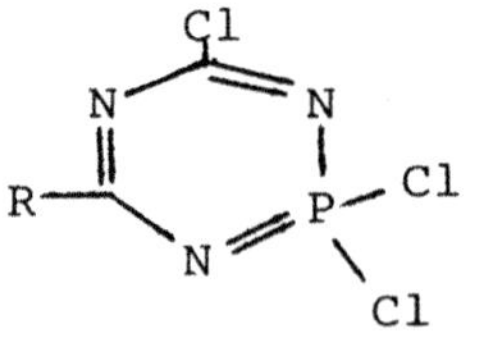

IV

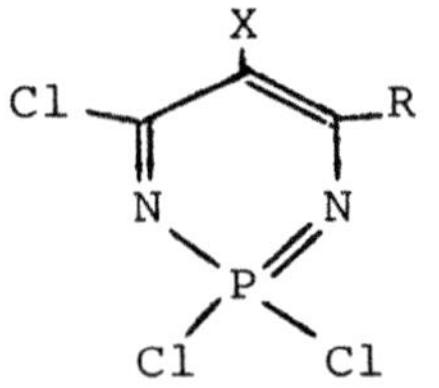

V

Thermolysis of Cl-SO_2-N=P(Cl_2)-N(CH_3)-Si(CH_3)$_3$ yields the S-containing phosphazene VI.[74] Ammonia or $(CH_3)_3$SiNRR' aminate the dichlorophosphazene according to equ. 6.[75]

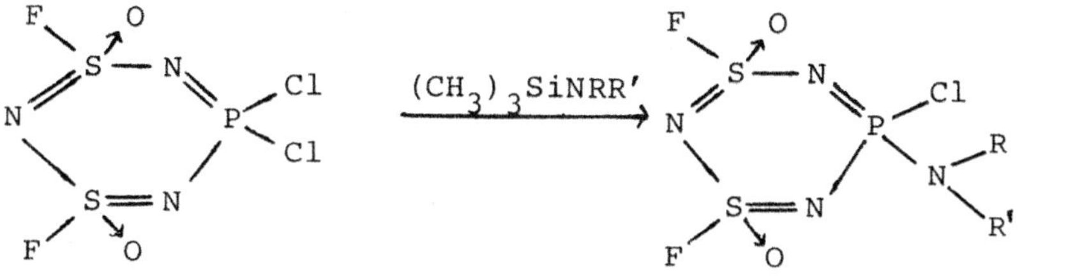

(6)

R,R′ = H, CH_3, C_2H_5

R auch $Si(CH_3)_3$

A number of additional six, as well as seven, and eight-membered heterocyclic phosphazenes were described.[76] The eight-membered ring compound $(NPCl)_3NSOCl$ was isolated from a reaction between $[Cl_3P{=}N{-}PCl_2{=}N{-}PCl_3]^{\oplus}$-$[PCl_6]^{\ominus}$ and $SO_2(NH_2)_2$.[77] $RSiCl_3$ and $(R'O)_2P(O)NCl_2$ react to VII.[78]

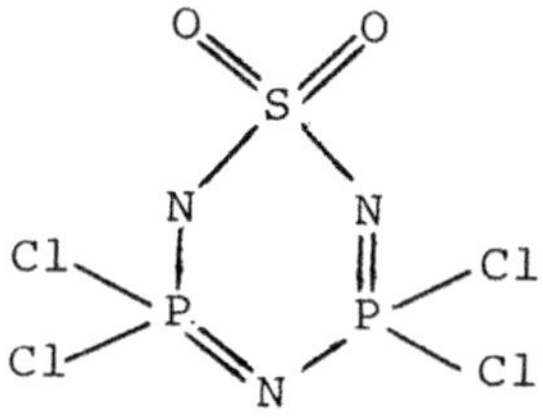

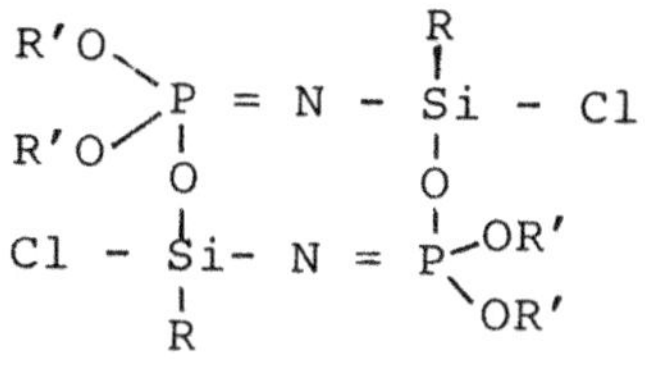

R = Cl, CH_3

R′= C_2H_5,

C_3H_7

VI VII

Oligomeric and polymeric phosphazene derivatives are formed when $(CH_3NPCl_3)_2$, $[CH_3NP(C_6H_5)Cl_2]_2$, and $(CH_3NPF_3)_2$ were decomposed.[79]

Linear Phosphazenes: A reaction between KSCN and PCl_5 yields $Cl_3C{-}N{=}PCl_3$ which combines with p-R-C_6H_4-SO_2-NH_2 to yield p-R-C_6H_4-SO_2-N=C(Cl)-N=PCl_3. Thermolysis of the reaction products yields compound VIII.[80] The reaction of $Hg(NSOF_2)_2$ with $ClSO_2NPCl_3$ proceeds to give F_2OSN-SO_2-$NPCl_3$ and $HgCl_2$.[81] The compounds $F_2P(R){=}N{-}C(CF_3)$-$(CH_3)N{=}C(CF_3)_2$ ($R{=}C_2H_5$, C_6H_5) also were synthesized.[82] A quantitative yield of $Cl_3P{=}N{-}POCl_2$ is obtained from interaction of $(Cl_2OP)_2NH$ with PCl_5.[83] Halogen derivatives of silicon or germanium and also carboxylic acid chlorides react with $(CCl_3)_2P(Cl)NH$ to yield $(CCl_3)_2P(Cl)NR$.[84] Reaction of phosphites or tris(dimethylamino)phosphine with $(CF_3)_2C{=}N{-}N{=}C(CF_3)_2$ results in formation of $RR'R''P{=}N{-}C(CF_3)_2{-}N{=}C(CF_3)_2$.[85] The preparation of R-NH-$P(C_6H_5)_2$=N-$P(C_6H_5)_2$-NH_2 from $RN[P(C_6H_5)_2]$ and chloroamine was reported.[86] Preparation, substitution reactions, and degradations of phosphazocyclotri(thiazene) compounds have been discussed.[87] Reaction of $(R_3Si)_2N{-}P{=}N{-}SiR_3$ with R_3SiN_3 initially gives an adduct which converts to IX($R{=}CH_3$).[88]

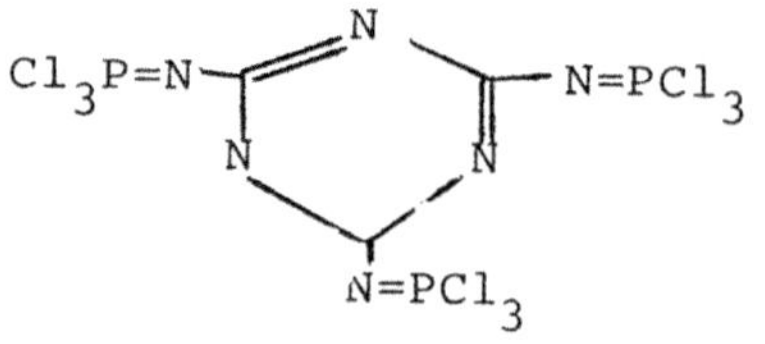

VIII

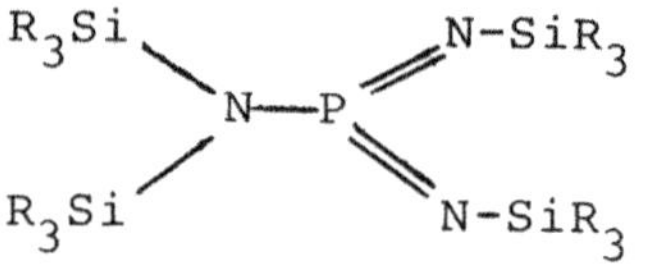

IX

Upon heating, the phosphazene X reacts with selenium according to equ.7.[89]

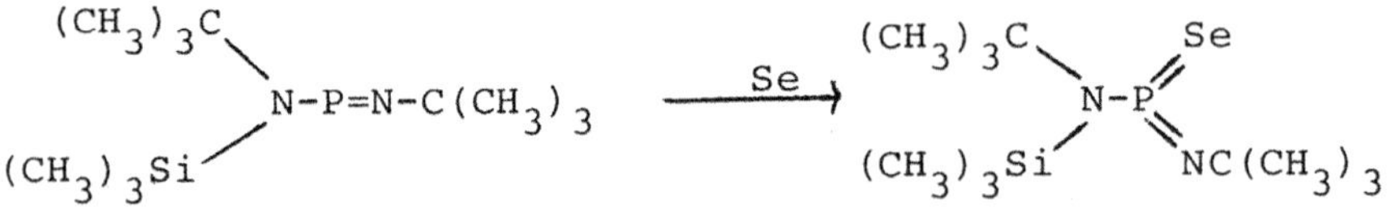

(7)

X

Soluble acrylic polymeric phosphazenes with branched chains and molecular weights between 1400 and 2000 were obtained through a reaction between Li-phenyl and $N_3P_3Cl_6$.[90]

Cyclic Phosphazanes and Rings with P-N-Bonds:

Phosphorus containing amines of the type $(PCl_2)_2NR$ and $(PCl_2X)_2NR'$ respectively react with $H_2NC(CH_3)_3$ to the cyclic derivatives XI and XII; as by-products six- and eight-membered systems are also formed ($R=CH_3$, C_2H_5, $C(CH_3)_3$; $R'=CH_3$, C_2H_5; $X=O,S$).[91] Compound XIII is obtainable through a reaction of PCl_5 or PF_5 with CH_3NH_2.[92] In the diazadiphosphetidine XIII ($X=F$) up to 4 F-atoms can be substituted by CH_3 or CH_3O-groups.[93]

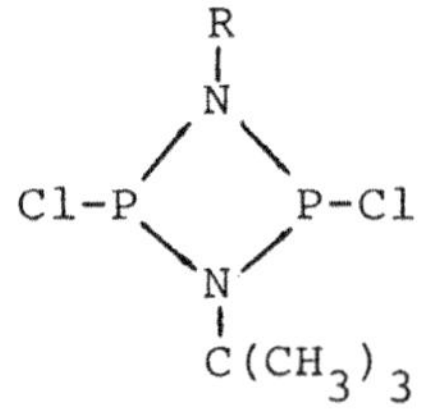

XI

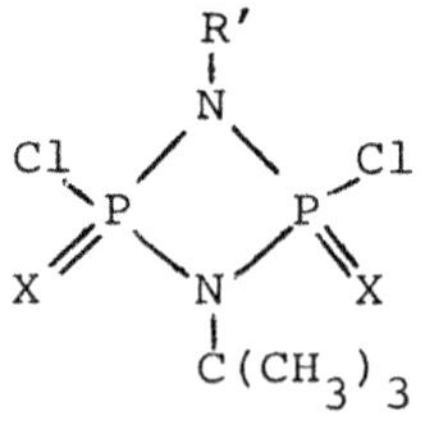

XII

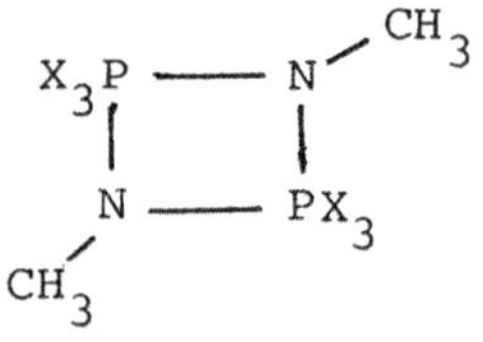

XIII

Reaction of XIII with $[LiN(CH_3)-CH_2-]_2$ leads to formation of XIV. Equation 8 describes as similar reaction.[94] The synthesis of diazadiphosphet-

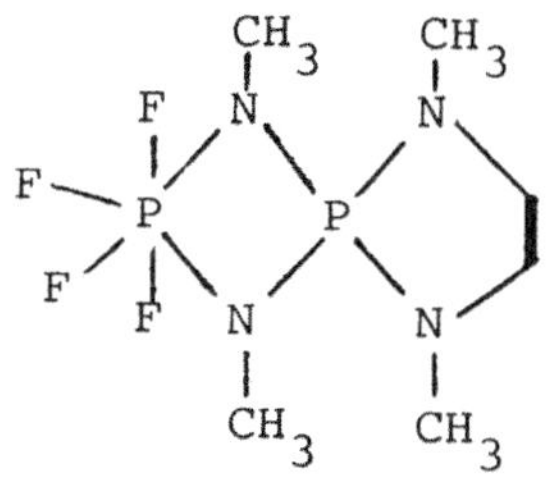

XIV

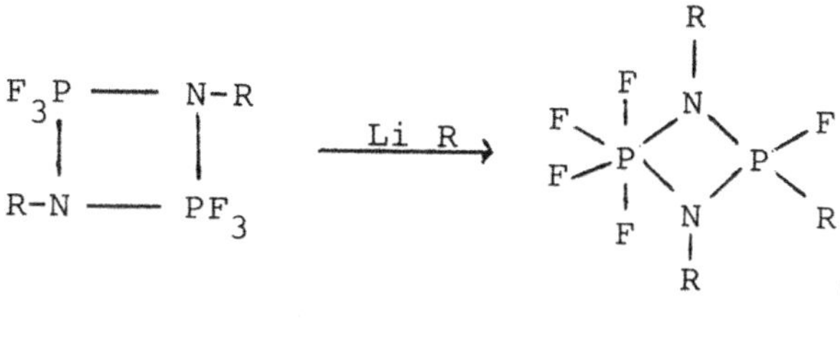

(8)

$R = C(CH_3)_3$

idiimines was accomplished.[95] In addition a dizadiphosphetidine substituted by a perfluoropinakoline group was made.[96] A tetrameric oligomer probably is obtained from the monomeric $C_2H_5N{=}P(Cl)_2{-}N(C_2H_5)_2$; its structure probably is represented by XV.[97]

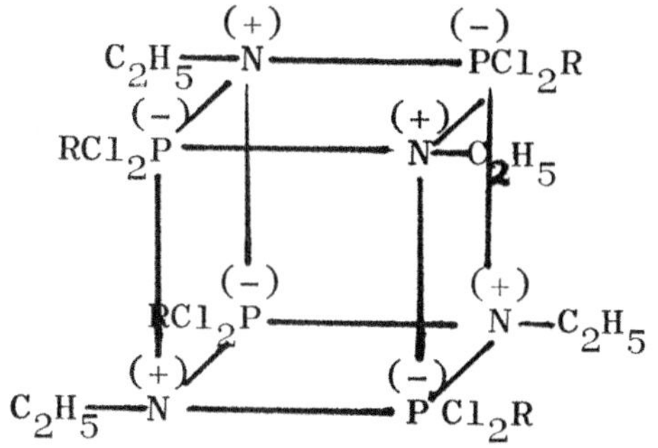

XV

N-metallated triazaboradecaline reacts with $(ClRP)_2NR'$ or $[Cl_2P(O)]_2NCH_3$ respectively via a ring closure to a phosphaborazine of type XVI. The heterocycle XVII containing a B_2N_3P-system could be prepared by a cleavage of a SiN-bond by $CH_3P(O)Cl_2$ of the linear BNSi-compound XVIII.[98]

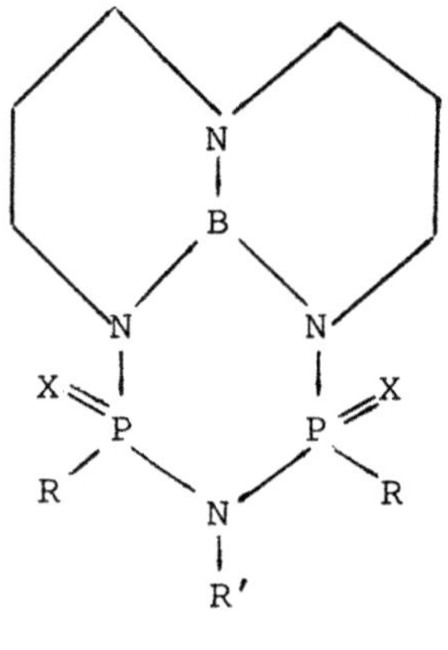

XVI

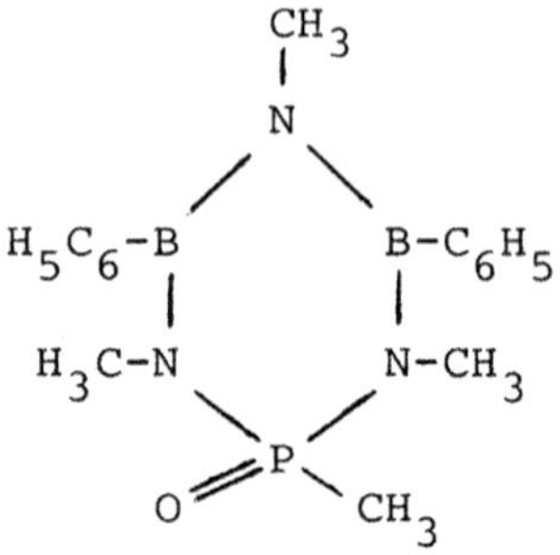

XVII

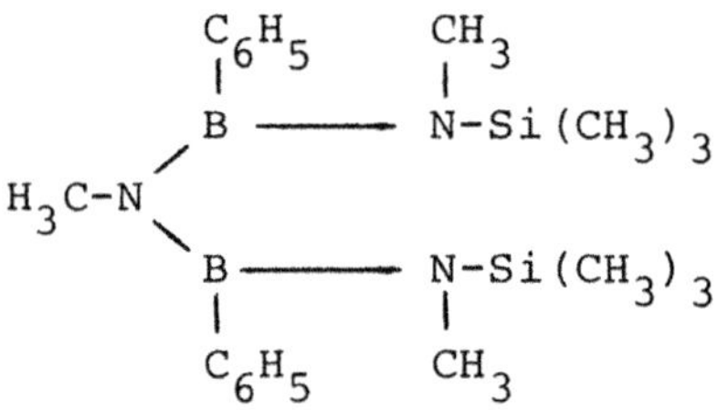

XVIII

R = CH_3, C_6H_5, Cl; R′ = CH_3, C_6H_5; X = O or no substituent

General methods for the synthesis of five- and six-membered rings with P-N-bonds were investigated.[99] The reaction of CH_3O-C(O)C≡CO(O)-CH_3 with RR'C=N-N=P$(C_6H_5)_3$ leads via 1,4-addition to XIX (R=R'=C_6H_5, -$CH_2C_6H_5$; R=H, CH_3, C_2H_5, R'=C_6H_5).[100] Five-membered cyclophosphazenes were obtained by a reaction between aminophosphines and chlorosilanes or metallated aminosilane and dichlorophosphines in accordance with equ. 9.[71]

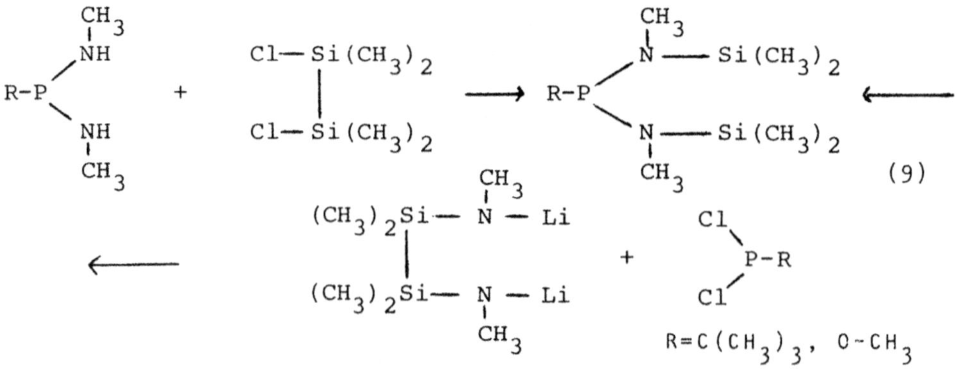

(9)

The synthesis of two novel ring systems XX and XXI were also reported.[101]

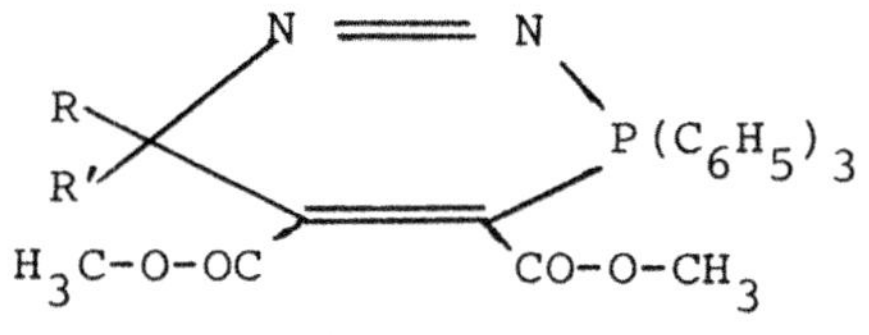

XIX

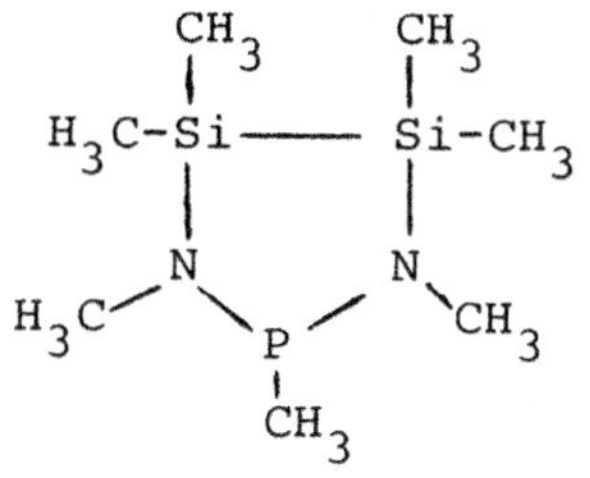

XX

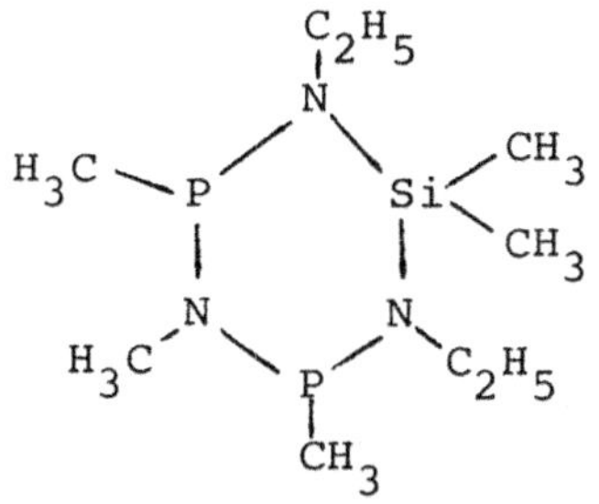

XXI

$SO_2(NPCl_3)_2$ reacts with alcohols ROH ($R=CH_3, C_2H_5, C_3H_7, C_6H_5$) to yield XXII. Acyl migration converts it to XXIII. This compound reacts with $[(CH_3)_3Si]_2NCH_3$ to the cyclic compounds XXIV, equ. 10.[102]

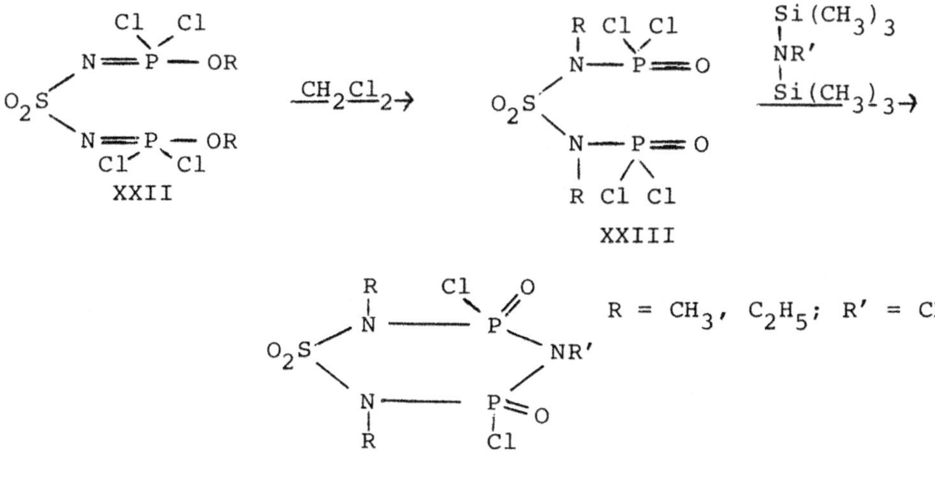

XXII XXIII XXIV

$R = CH_3,\ C_2H_5;\ R' = CH_3$ (10)

$[C_6H_{22}P(S)O]_3$ and $(CH_3)_2NCN$ react to yield XXV.[103] A ring system having either structure XXVI or XXVII was obtained from $(CH_3)_3Si\text{-}N{=}S{=}N\text{-}Si(CH_3)_3$ and PF_5 or $PF_4C_6H_5$ respectively.[104]

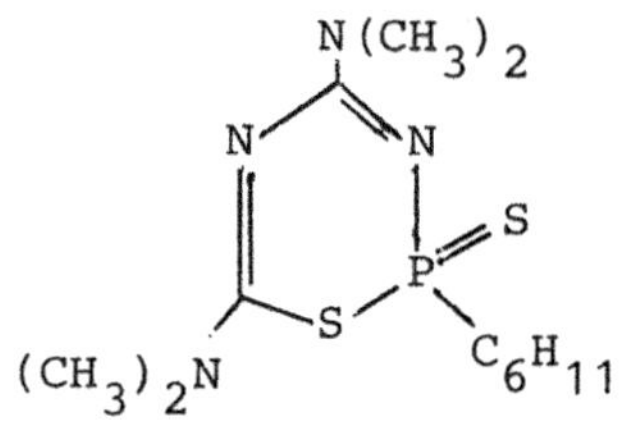

XXV

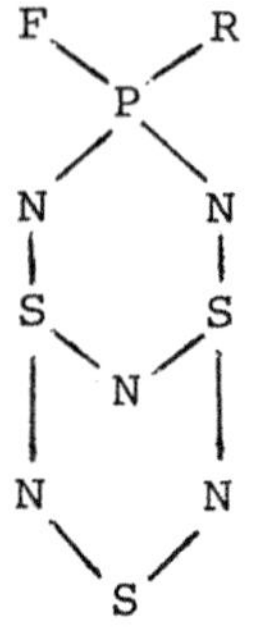

XXVI

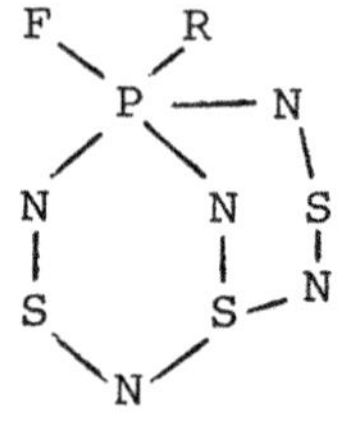

XXVII

R = C_6H_5, F

Linear Phosphazanes: $Cl_3P{=}N\text{-}P(O)Cl_2$ reacts with formic acid to afford $Cl_2P(O)\text{-}NH\text{-}P(O)Cl_2$; with NH_3, $(CH_3)_2NH$ or $(C_2H_5)_2NH$ the corresponding tetramides were obtained.[105] $[(CH_3)_2N]_2PCl$ furnishes with $AlCl_3$ an ionic 1:1 adduct. Analogously prepared were $[(R_2N)_2P]^{\oplus}[PF_6]^{\ominus}$ and $[(R_2N)_2P]^{\oplus}$-$[B_2F_7]^{\ominus}$.[107] The inorganic ringsystems XXVIII were also synthesized.[108]

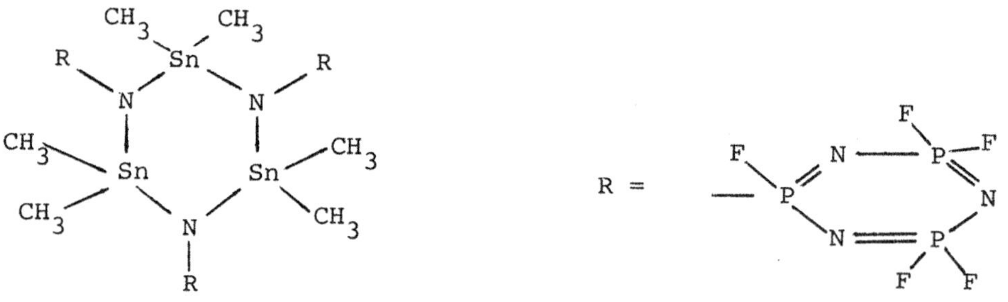

XXVIII

Reaction between CF_3PI_3 and secondary amines furnishes $CF_3P(NR_2)_2$.[109] The compounds $(CF_3)_3P(Cl)\text{-}N(CH_3)_2$, $(CF_3)_3P[N(CH_3)_2]_2$, $(CF_3)_2Cl_2P\text{-}N(CH_3)_2$, and $CF_3P(X)[N(CH_3)_2]_2$ (X=O, S) were obtained by aminolysis of the appropriate chlorides.[110] Vacuum pyrolysis at 100° of $(CF_3)_3P[N(CH_3)_2]_2$ leads to formation of $(CF_3)_2PF[N(CH_3)_2]_2$ whereas at 215° $(CF_3)PF_2[N(CH_3)_2]_2$ was isolated. Alkaline hydrolysis yields instead $CF_3P(O)[N(CH_3)_2]_2$ and $CF_3PO_3^{\ominus\ominus}$.[111] The relative reactivity of $[(CH_3)_2N]_2P\text{-}C{\equiv}C\text{-}R$ (R=CH_3, $C(CH_3)_3$, C_6H_5) in 1,3-dipolar cycloadditions was investigated.[112] $R_2N\text{-}P{=}NR$ is directly available from a reaction between $LiNR_2$ and PCl_3 or PBr_3 (R=$Si(CH_3)_3$).[113] R_2NCN reacts with $PXCl_3$ (R=CH_3, C_2H_5, C_4H_9; X=O, S) to form $R_2N\text{-}C(Cl){=}N\text{-}P(X)Cl_2$.[114] Through condensation of RPF_4 (R=C_2H_5, C_6H_5) with $LiN{=}C(CF_3)_2$ at -196° $F_3P(R)\text{-}N{=}C(CF_3)_2$ is obtained.[82] $H_2N\text{-}CH_2\text{-}CH{=}CH_2$ reacts with P_2F_4 to afford $F_2P\text{-}NH\text{-}CH_2\text{-}CH{=}CH_2$.[34] The reaction of $RO-PCl_2$ with $Cl_2C=N-SCl$ yields $RO-P(S)Cl-N=CCl_2$, which reacts with chlorine to $Cl_2P(O)-N=CCl_2$ (R=alkyl).[115] Subjecting $Cl_2P-N(R)-CRR'-C(O)OC_2H_5$ to chlorine attack results in formation of $Cl_2P(O)-NR-CR'=CCl-OC_2H_5$ (R=R'=alkyl).[116] The preparation of $(RO)_2P-N(CH_3)-P(OR)_2$ (R=CH_3, C_2H_5) was reported.[117] In the presence of sec. or tert. amines $(R_2N)_2P-OH$ reacts with sulfur affording ammonium salts of the type $[(R_2N)_2POS]^{\ominus}$ $[R_nNH_{4-n}]^{\oplus}$ (n=2,3)(R=CH_3, C_2H_5).[118] Under the catalytic influence of HCl PCl_5 and C_6H_5CN furnish $[C_6H_5-CCl=N-PCl_3]^{\oplus}[PCl_6]^{\ominus}$ and $[C_6H_5-C(Cl)=N-P(Cl)_2-N=C(Cl)-C_6H_5]^{\oplus}$. Linear polymers were obtained from a HCl-catalyzed reaction of PCl_5 and 1,4-dicyanobenzene; the reaction of these polymers with SO_2 was investigated.[119]

Azides and Hydrazine Derivatives: Reaction of $[(C_6H_5)_{4-n}PCl_n]^{\oplus}[SbCl_6]^{\ominus}$ with NaN_3 give the azides $[(C_6H_5)_{4-n}P(N_3)_n]^{\oplus}[SbCl_6]^{\ominus}$ (n=1, 2, 3) in nearly quantitative yields.[120] RR'PCl reacts with $H_2N-N(CH_3)_2$ to afford $RR'-P-NH-N(CH_3)_2$. These compounds convert after treatment with CCl_4 into $RR'P(Cl)=N-N(CH_3)_2$ (R=CH_3, $C(CH_3)_3$, R'=$C(CH_3)_3$).[121] Interaction of $ZNH-NH_2$ with Br-Z leads to formation of Z-NHNH-Z (Z=$N_3P_3F_5$).[122] Condensation of $R-C_6H_4-CO-NH-P(O)Cl_2$ with $p-(H_2N-NHSO_2-C_6H_4-CH_3)$ in the presence of $(C_2H_5)_3N$ produces the hydrazide XXIX.[123] Synthesis of the bicyclic system XXX was achieved by reaction of $[(CH_3)_2N]_3P$ with $CH_3-NHNH-CH_3$; XXXI was obtained by reacting the ring compound XXX with PCl_3, an excess of PCl_3 leads to formation of $Cl_2P-N(CH_3)N(CH_3)PCl_2$, treatment of this compound with $[(CH_3)_3Si]_2NCH_3$ reconverts it to XXX.

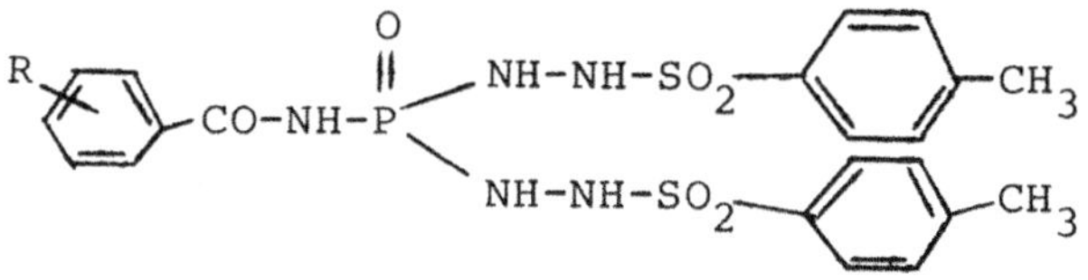

XXIX

R = H, p-Br, p-F, p-OCH_3, p-Cl, m-$OCH_2C_6H_5$;

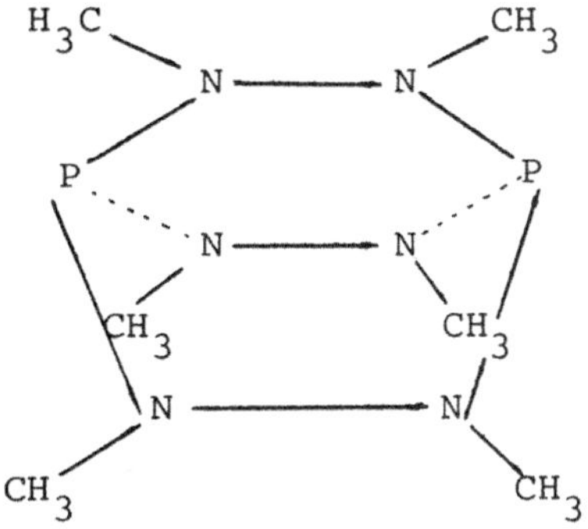

XXX

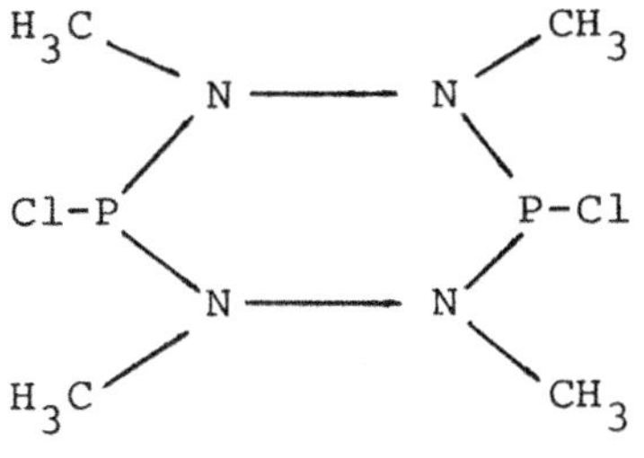

XXXI

COMPOUNDS WITH P-P-BONDS

Reaction of $(P\text{-}C_6H_5)_5$ with sulfur yields XXXII.[125] Electrolytic reduction of cyclophosphaalkanes at low temperature gives anion radicals.[126] The synthesis of $RP(OR)\text{-}P(O)(OR)_2$ ($R=C_4H_9$)[127] and $RR'P(S)\text{-}P(S)RR'$ ($R=C_4H_9$; $R'=CH=CH\text{-}C_6H_5$ was reported.[128] Attempts to synthesize a silylated iminophosphine were successful.[129]

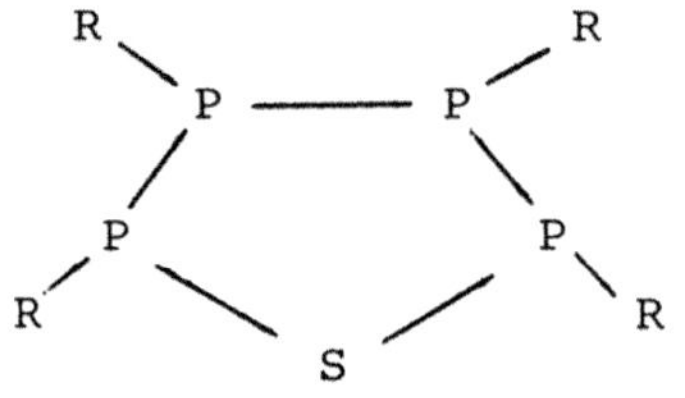

R = C_6H_5

XXXII

PHOSPHORUS COMPOUNDS WITH ELEMENTS OF THE 6th MAIN GROUP

Phosphorus Sulfides and Selenides: Upon heating the system P-S to 400-450° one obtains glasses within the range of 5-25 atom-% P-content, whereas in the system P-Se such phases are observed only at temperature of 600° with a P-range between 0-52 atom %.[130] A high temperature reaction between P_4S_3 and As_2S_3 leads to formation of $P_2As_2S_3$ which has a cage structure.[131] Reactions between NF_3 and P_4S_3 were investigated.[132] At 75 mm Hg Cl_2-pressure $FePSe_3$ is obtained in a gas phase reaction; it has a CdI_2-type structure.[133] The preparation of CoPY (Y=S, Se) was reported.[134]

Derivatives of Hypophosphorus Acid: The synthesis of simple phosphoric acid derivatives was described.[135] The anhydride $[C_6H_5P(O)CF_3]_2O$ was prepared from CF_3COOH and $C_6H_5PCl_2$, the free acid was obtained too.[136] The P-S-P-bridges in the anhydrides XXXIII and $[(C_6H_5)_2P(S)]_2S$ can be cleaved by $[(CH_3)_2N]S$.[137] $[(C_6F_3)_2P]_2Se$ was obtained from $(CF_3)_2PI$ and Ag_2Se as well as by interaction of Se at 100° on $(CF_3)_2P\text{-}P(CF_3)_2$, $(CF_3)_2P\text{-}P(CF_3)_2$; its hydrolysis gives $(CF_3)_2PH$, CF_3H, $(CF_3)_2POSe^{\ominus}$, $(CF_3)_2PSe_2^{\ominus}$, and $CF_3PHO_2^{\ominus}$.[138] $R_2P(S)X$ (X=Cl, Br; $R=CH_3, C_2H_5, (CH_2)_2CH_3$) reacts with $HN[Si(CH_3)_3]_2$ to $R_2P(S)\text{-}NH\text{-}Si(CH_3)_3$.[139] By melting the appropriate chloro- or bromophosphines with sulfur in the presence of $AlCl_3$, compounds of the type $(C_6H_5)_2P(S)X$ and $R(C_6H_5)P(S)X$ (X=F, Cl, Br; $R=CH_3$, C_2H_5, C_4H_9, C_6H_5) were obtained; the corresponding fluoro derivative could be made via a halogen exchange. The formation of XXXIV can be rationalized as the result of a cleavage of an unidentified intermediate according to equ. 11.[141] The synthesis of a dihypophosphite substituted by a dialkyltin group was reported.[142]

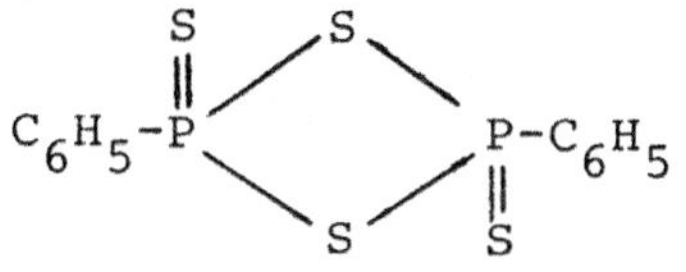

XXXIII

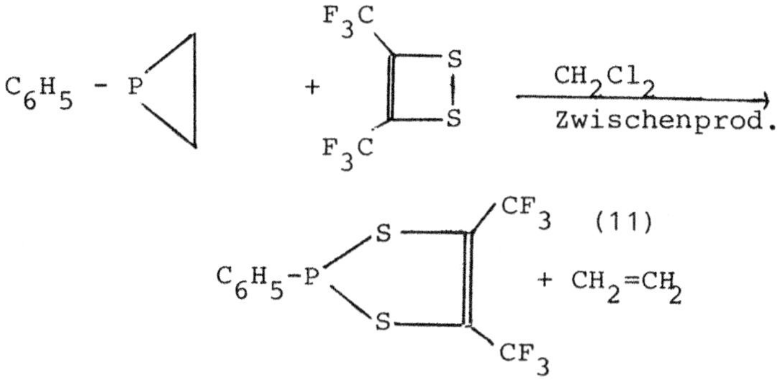

XXXIV

Derivatives of Phosphorous and Phosphonic Acids: Slow addition of $SnCl_4$ or $TiCl_4$ to cold H_3PO_3 resulted in $Sn(HPO_3)_2 \cdot \frac{1}{2}H_2O$ and $Ti(HPO_3)_2 \cdot H_2O$.[143] The phosphites $Zn_2H_2P_3O_9$ and $ZnP_2H_2O_5$ were accessible via a thermolysis reaction.[144] Without occurrence of side reactions $CH_3\overline{P(O)}(OCH_3)_2$ is converted by PCl_5 into $CH_3P(O)Cl_2$.[145] Similarly $(C_2H_5O)_2PCl$ is obtained from triethyl phosphite with 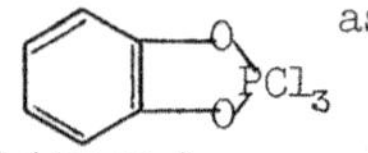 as chlorinating agent.[146] Optically active $(-)C_2H_5P(Se)(OH)(OC_2H_5)$ was prepared.[147] $(C_2H_5O)_2P$-O-$Si(CH_3)_3$ reacts with $(CF_3)_2CO$ to $(C_2H_5O)_2P(O)$-$C(CF_3)_2$-O-$Si(CH_3)_3$.[148] From a reaction between CS_2 and $(C_2H_5S)_2PN(CH_3)_2$ $(C_2H_5S)_2P$-S-CS-$N(CH_3)_2$ is obtained. In equ. 12 an additional CS_2 insertion reaction is described.[149]

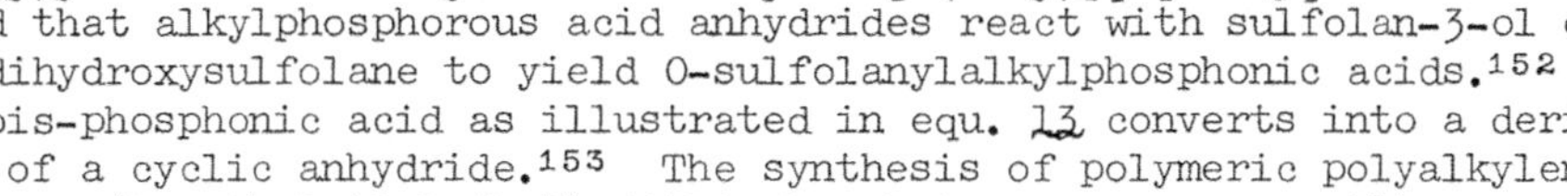

(12)

R, R' = H, CH_3, -CH=C(CH_3)-CH=CH Z = O, S;

Several alkoxymethylthiophorous acid derivatives have been synthesized.[150] $P(SCH_3)_3$ reacts with CH_3SCl and $SbCl_5$ to $[P(SCH_3)_4]^{\oplus}[SbCl_6]^{\ominus}$.[151] It was found that alkylphosphorous acid anhydrides react with sulfolan-3-ol or 3,4-dihydroxysulfolane to yield O-sulfolanylalkylphosphonic acids.[152] The bis-phosphonic acid as illustrated in equ. 13 converts into a derivative of a cyclic anhydride.[153] The synthesis of polymeric polyalkyleneglycol-S-(trialkylsilylethyl)-dithiophosphates was reported.[154]

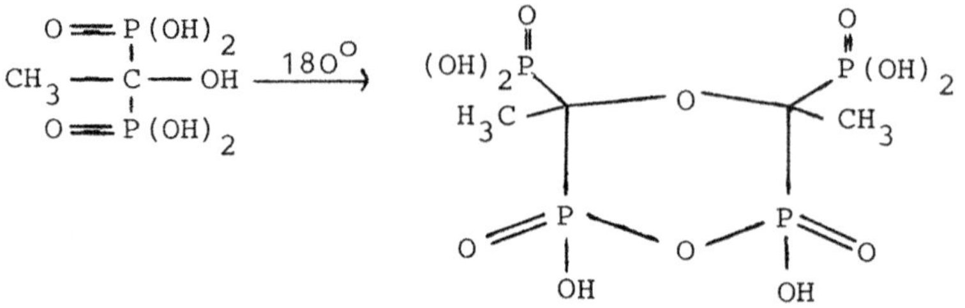

(13)

Orthophosphates and Derivatives of Orthophosphoric Acid: Potassium phosphate catalyzes the decomposition of H_2O_2. In several cases intermediates such as $K_3PO_4 \cdot nH_2O_2$ (n=1,2,3) and $K_4P_2O_7 \cdot 6H_2O_2$ were isolated.[155] An investigation of the products which result by heating carbamide with $NH_4H_2PO_4$ was started.[156] The crystal structure of hydrothermally synthesized $Ca_{18}Mg_2H_2(PO_4)_{14}$ resembles that of β-$Ca_3(PO_4)_2$.[157] The anodic dissolution of brass (60-90% Cu) in H_3PO_4 was investigated and discussed. Combination of solutions of mixtures of H_3PO_4-HNO_3 and $Zr(NO_3)_4$ results in formation of amorphous gel-like products which crystallize upon prolonged standing. The crystalline phase consists of $Zr(HPO_4)_2 \cdot H_2O$.[159] PBr_3/Br_2 and Br_2/CO mixture are best suited for chemical transport of all rare earth phosphates.[160] While determining the solubility of Y_2O_3 in H_3PO_4 or $H_4P_2O_7$ solutions, crystalline phosphates of the type $H_3[Y(PO_4)_2] \cdot 2H_2O$, $Y(H_2PO_4)_3 \cdot \frac{3}{2}H_2O$, $YHP_2O_7 \cdot 6H_2O$, and $Y(H_3P_2O_7)_3 \cdot 6H_2O$ precipitated.[161] An additional reaction between $[(CH_3)_3SiO]_3P{=}O$ and $(CH_3)_3SiX$ (H=F, Cl, Br, I, ClO_4, NO_3, SCN) leads to formation of phosphonium salts of the structure $[(CH_3)_3SiO]_4P^{\oplus}X^{\ominus}$. The iodide of this series is the end product of successive addition of $(CH_3)_3SiI$ to trimethyl phosphate; $[(CH_3)_3SiO]_nP(O)(OCH_3)_{3-n}$ were found to be intermediates.[162] The product of a reaction between $(RO)_2P(S)Cl$ and HNR_2' is $(RO)_2P(S)NR_2'$ (R=$(CH_2)_n$)CH_3, n=7,8,9; R'=C_2H_5, $(CH_2)_3CH_3$, $(CH_2)_2CH(CH_3)_2$).[162] Aminolysis or organylphosphoranes seems to be asuitable way for the preparation of a series of aza analogs of phosphorus oxygen acids.[164] UV-photolysis of monophenyl phosphate in aqueous solution gives rise to formation of orthophosphate, phenol and in smaller amounts phosphorous acid, benzene, as well as 2,2'-2,4'-, and 4,4'dihydroxybiphenyl.[165] Two novel compounds having phosphorus in two different valence states, $F_2P(S)SP(CF_3)_2$ and $(CF_3)_2P(S)SPF_2$ were obtained by reacting the acids F_2PS_2H or $(CF_3)_2PS_2H$ with $(CF_3)_2PN(CH_3)_2$ or $F_2PN(CH_3)_2$ respectively.[166]

Diphosphates and Polyphosphates: The rate of hydrolysis of the reaction

$$H_nP_2O_7^{4-n} + H_2O \longrightarrow 2H_{1+\frac{n}{2}}PO_4^{(2-n)\cdot\frac{1}{2}}$$

is greatly enhanced by $VO_2^{\oplus}$ and $H_2V_{10}O_{28}^{4\ominus}$ ions.[167] The stoichiometry of the reaction of molten zinc with $K_4P_2O_7$ was determined.[168] As a result of heating of mixtures of $(NH_4)_2H_2P_2O_7$ and carbamide or thiocarbamide respectively to 140° or 170° respectively, mixtures of ortho-, di-, tri-, tetra-, and traces of pentaphosphates were observed.[168] In the system

$AlCl_3$-$Li_4P_2O_7$-H_2O the solid phases $Al_4(P_2O_7)\cdot 12H_2O$ and $LiAlP_2O_7\cdot 5H_2O$ could be identified.[170] The co-oxidation of $POCl_3$ and $TiCl_4$ by oxygen proceeds according to equ. 15.

$$TiCl_4 + 2\ POCl_3 + \frac{5}{2}\ O_2 \longrightarrow TiP_2O_7 + 5\ Cl_2$$

Depending on the ratio of the starting materials Anatas was also found.[171] A report on $NaAlP_2O_7$, $Ga_4(P_2O_7)_3$, and $In(P_2O_7)_3$ and their hydrates appeared.[172] $KThP_3O_{10}$ was obtained from KH_2PO_4, ThO_2 and HPO_3 at 900° in a Pt-crucible.[173] The pyrophosphate XXXV was also synthesized.[174]

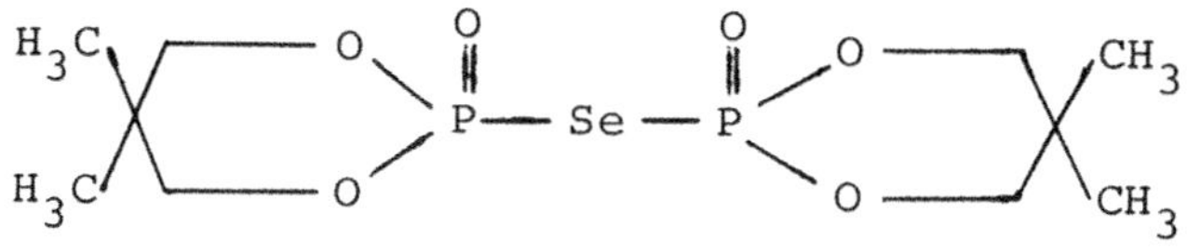

XXXV

Detailed investigations concerned with the thermal behavior and composition of basic calcium phosphates with a Ca/P ratio $\geqq \frac{3}{2}$ [175] or $> \frac{3}{2}$ [176] appeared. Zinc reacts with $CCl_3P(O)(OC_2H_5)_2$ via intermediate formation of a complex and splitting off C_2H_5Cl to polymers.[177]

Metaphosphates: Alkalimetaloctametaphosphates, $(MPO_3)_8$ (M=Li, Na, K, Rb, Cs), are obtainable either by ionic exchange with $(NaPO_3)_8$ or by a reaction between $Rb_4P_8O_{24}$ with alkalimetal sulfides.[178] The stoichiometry of the reaction of Zn with $NaPO_3$ in molten state was investigated.[168] The range of glass formation of the systems $NaPO_3$-LiF, $NaPO_3$-NaF, $Zn(PO_3)_2$-NaF, and $Al(PO_3)_3$-NaF were elucidated. The structures and the conductivity of the glass phases were determined.[179] End product of dehydration of $Ba(H_2PO_4)_2$ at 700° is the stable $Ba(PO_3)_2$.[180] The stability constants of complexes of trimeta-, tetrameta-, hexameta-, and octametaphosphate anions with Ca-, Sr, Zn-, and Co(II)- cations have been measured.[181] An investigation deals with the influence of water on the formation of the A- and B-types of $Al_4(P_4O_{12})_3$.[182] The reaction of ZrO_2 and HfO_2 with molten P_4O_{10} proceeds first to form the metastable phase MP_2O_7 (M=Zr, Hp) which above 430° or 470° respectively convert to the stable metaphosphates $M(PO_3)_4$.[183] A paper deals with condensed and cyclic phosphates.[184]

PHOSPHORUS-HALOGEN-COMPOUNDS

The radical anion $PF_5\cdot^{\ominus}$ is generated during γ-irradiation of NH_4PF_6 and KPF_6. The existence of the radical anion was demonstrated by ESR-spectroscopy.[185] $BrPF_2$ reacts with $(CF_3S)_2Hg$ to yield CF_3-S-PF_2. This compound reacts with oxygen at -78° to give CF_3-S-$P(O)F_2$. XPF_2 (X=Cl, Br, SCN, $N(CH_3)_2$) also can be oxidized to $XP(O)F_2$.[186] Reaction of XCl (X=F_2PO, F_2PS, CF_3S) with AgOCN leads to X-NCO. These compounds react with alcohols and mercaptans to carbanates and thiocarbamates of the type X-NH-CO-Y-R (R=CH_3, C_2H_5; Y=O, S).[187]

COMPLEXES WITH PHOSPHORUS LIGANDS

Bis[π-cyclopentadienyl]molybdenumdihydride, $(\pi\text{-}C_5H_5)_2MoH_2$, reacts with an excess of white phosphorus in toluene at 90° to yield $(\pi\text{-}C_5H_5)_2MoP_2H_2$.[188] Triphenylphosphine was used for the first time in certain transition metal complexes.[189] Tris-(aziridino)-phosphine or bis(aziridino)-morpholinophosphine (L) forms with $M(CO)_6$ (M=Cr, Mo) and with $(C_6H_5CN)_2M'Cl_2$ (M=Pd, Pt) complexes of the type $N_3P_3[N(CH_3)_2]_6$[191] and $N_6P_6[N(CH_3)_2]_{12}$[192] were also described. A nearly quantitative yield of polymeric phosphate complexes are the result of a reaction of methyltin- or methyllead chlorides. The penta- or hexacoordinated metal atoms are linked via O-P-O bridges.[192] $Na[Co(CO)_4]$ forms with PX_3 and $S=PX_3$ (X=Cl, Br) the cluster compounds $Co_2(CO)_6P_2$ and $Co_3(CO)_9PS$.[194] $Mo(CO)_5P_4S_3$ could be made from $Mo(CO)_6$ and P_4S_3 in cyclohexane.[195] Upon reaction of $(CH_3)_2P(S)$-NH-CS-R with Ni(II)-acetate in alcoholic solution a Ni(II)-SS'-chelate complex is formed.[196]

Numerous publications deal with the preparation, properties, and spectroscopical behavior of PF_3-complexes with vanadium,[197] cobalt[198, 199], nickel[200], rhodium,[201] and mercury.[202] Fluorine exchange reactions involving metaltrifluorophosphine complexes were investigated.[203-206] $C_6H_5PF_2$ forms with $Ni(II)_2$ and $Co(II)X_2$ (X=Br, I) water sensitive complexes of the type MX_2L_3. A disproportionation is observed with $CoCl_2$ and $CoBr_2$.[207] $Fe(CO)_4(PF_2Br)$ can be oxidized to $(CO)_4Fe\text{-}PF_2\text{-}O\text{-}PF_2\text{-}Fe(CO)_4$ by $AgMnO_4$, Ag_2O, Cu_2O, and $(C_4H_9)_2SnO$.[208] Interaction of PF_5 with $F_2PN[Si(CH_3)_3]_2$ leads to formation of $F_2P\text{-}N=PF_3$; $Mo(CO)_4(F_3P=N\text{-}PF_2)_2$ is a complex derived of this compound.[209] The ligands $[(CF_3)_2P]_2X$ (X=O, S, Se) react with $Fe(CO)_2(NO)_2$ and $Fe_2(CO)_9$ to the complexes $FeCO(NO)_2[(CF_3)_2PXP(CF_3)_2]$ and $Fe(CO)_4[(CF_3)_2PXP(CF_3)_2]$ as monodentate ligands coordinated <u>via</u> phosphorus. With $Fe_2(CO)_9$ and with X=Se the five-membered cyclic compound XXXVI is obtained. The complexes $\pi\text{-}C_5H_5Fe(CO)_2\text{-}P(CF_3)_2Fe(CO)[P(CF_3)_2(O)]C_2H_5\text{-}\pi$, $\pi\text{-}C_5H_5Fe(CO)_2P(X)(CF_3)_2$ and $\pi\text{-}C_5H_5Fe(CO)_2P(CF_3)_2$ also were prepared.[210] The last complex reacts with X_2 (X=Cl, Br, I) to give $[\pi\text{-}C_5H_5Fe(CO)_2P(CF_3)_2X](X_3)$.[211] Results of a comparison of the reactivity of $(C_6H_5)_3P=C=P(C_6H_5)_3$ and $(C_6H_5)_3P=CH_2$ towards metal carbonyls was published.[212] Also, some complexes involving phosphineacetylenes were prepared.[213,214]

XXXVI

REFERENCES

1. J. I. G. Cadogan and R. K. Mackie, CHEM. SOC. REV. 3, 87 (1974).
2. H. Teichmann, Z. CHEM. 14, 216 (1974).
3. H. Staendeke and H. J. Kleiner, ANGEW. CHEM. 85, 974 (1973).
4. F. Ramirez and I. Ugi, BULL. SOC. CHIM. FRANCE 453 (1974).
5. B. A. Arbuzov and N. A. Polezhaeva, USP. KHIM. 43, 933 (1974).
6. M. I. Kabachnik, T. Ya. Medved', N. M. Dyatlova and M. V. Rudomino, USP. KHIM. 43, 1554 (1974).
7. Ch. S. Kraihanzel, J. ORGANOMET. CHEM. 73, 137 (1974).
8. R. J. van Zee and A. U. Khan, J. AMER. CHEM. SOC. 96, 6805 (1974).
9. K. J. Bachmann and E. Buehler, J. ELECTROCHEM. SOC. 121, 835 (1974).
10. K. Kaneko, M. Ayabe, M. Dosen, K. Morizane, S. Usui and N. Watanabe, J. ELEKTROCHEM. SOC. 121, 556 (1974).
11. J. Middlehoek and J. Holleman, J. ELEKTROCHEM. SOC. 121, 132 (1974).
12. L. Linowsky and W. Bronger, Z. ANORG. ALLG. CHEM. 409, 221 (1974).
13. W. Schmettow, A. Lipka and H. G. von Schnering, ANGEW. CHEM. 86, 379 (1974).
14. G. Fritz and H. Schäfer, Z. ANORG. CHEM. 406, 167 (1974).
15. A. A. Bergh, R. H. Saul and C. R. Paola, J. ELECTROCHEM. SOC. 120, 1558 (1973).
16. L. R. Plauger, J. ELECTROCHEM. SOC. 121, 455 (1974).
17. M. Hegems, M. B. Panish and J. R. Arthur, J. CHEM. THERMODYN. 6, 157 (1974).
18. A. S. Jordan, A. R. von Neida, R. Caruso and C. K. Kim, J. ELECTROCHEM. SOC. 121, 153 (1974).
19. E. S. Yurova, E. M. Kistova, M. N. Kevorkov, M. A. Konstantinova, I. I. Shtogrin and V. N. Maslov, IZV. AKAD. NAUK. SSSR, NEORG. MATER. 10, 21 (1974).
20. T. Sugano and Y. Mori, J. ELECTROCHEM. SOC. 121, 113 (1974).
21. S. Yu. Ovchinikov, V. S. Sorokin and D. A. Yas'kov, IZV. AKAD. NAUK. SSSR, NEORG. MATER. 10, 234 (1974).
22. G. A. Antypas and R. L. Moon, J. ELECTROCHEM. SOC. 120, 1574 (1973).
23. M. G. Astles, F. G. H. Smith and E. W. Williams, J. ELECTROCHEM. SOC. 120, 1750 (1973).
24. E. P. Domashevskaya, Ya. A. Ugai, S. N. Miroshnichenko, M. S. Alfeeva and I. F. Nikonov, IZV. AKAD. NAUK. SSSR, NEORG. MATER. 10, 731 (1974).
25. K. Mund, G. Richter, R. Schulte and F. v. Sturm, BER. BUNSENGES. PHYS. CHEM. 77, 839 (1973).
26. K. E. Mironov and G. P. Brygalina, IZV. AKAD. NAUK. SSSR, NEORG. MATER. 10, 920 (1974).
27. J. W. Gilje, R. W. Braun and A. H Cowley, J. CHEM. SOC., CHEM. COMMUN. 813 (1973).
28. R. Appel and A. Gilak, CHEM. BER. 107, 2169 (1974).
29. A. N. Lavrent'ev, I. G. Maslennikov and E. G. Sochilin, ZH. OBSHCH. KHIM. 43, 2663 (1973).
30. D. E. Arrington, SYNTH. REACT. INORG. METAL-ORG. CHEM. 4, 107 (1974).
31. W. Malish and H. Schmidbaur, ANGEW. CHEM. 86, 554 (1974).
32. N. J. De'Ath, D. Z. Denney, D. B. Denney and C. D. Hall, PHOSPHORUS 3, 205 (1974).

33. W. Wolfsberger and H. Schmidbaur, SYNTH. REACT. INORG. METAL-ORG. CHEM. 4, 149 (1974).
34. E. R. Falardeau, K W. Morse and J. G Morse, INORG. CHEM. 13, 2333 (1974).
35. T. A. Mastryukova, Kh. A. Suerbaev, E. I. Matrosov, P. V. Petrovskii and M. I. Kabachnik, ZH. OBSHCH. KHIM. 43, 2613 (1973).
36. D. J. Daigle, A. B. Pepperman jun. and S. L Vail, J. HETEROCYCL. CHEM. 11, 407 (1974).
37. E. L. Lines and L. F. Centofanti, INORG. CHEM. 13, 1517 (1974).
38. G. H. Sprenger and J. M. Shreeve, J. FLUORINE CHEM. 4, 201 (1974).
39. K. Dey, J. INDIAN CHEM. SOC. 50, 224 (1973).
40. H.-J. Kleiner, LIEBIGS ANN. CHEM. 751 (1974).
41. M. S. Chatta, S. G. Quo, J. A. Kuczkowski and A. M. Aguiar, PHOSPHORUS 3, 65 (1973).
42. B. A. Savel'ev, V. N. Doronin, V. A. Krenev and V. I. Evdokimov, IZV. AKAD. NAUK. SSSR, NEORG. Mater. 10, 1538 (1974).
43. H. Hofstötter and E. Mayer, MONATSH. CHEM. 105, 712 (1974).
44. J. D. Odom, S. Riethmiller and J. R. Durig, J. INORG. NUCL. CHEM. 36, 1713 (1974).
45. R. M. Kren, M. A. Mathur and H. H. Sisler, INORG. CHEM. 13, 174 (1974).
46. R. T. Markham, E. A. Dietz jun., and D. R. Martin, INORG. NUCL. CHEM. 36, 503 (1974).
47. A. B. Burg, INORG. CHEM. 12, 3017 (1973).
48. L. Maya and A. B. Burg, INORG. CHEM. 13, 1522 (1974).
49. M. L. Denniston and D. R. Martin, J. INORG. NUCL. CHEM. 36, 1461 (1974).
50. R. M. Barrer and M. Liquornik, J. CHEM. SOC., DALTON TRANSACTIONS 2126 (1974).
51. N. Nagasima, H. Suzuki, K. Tanaka and S. Nishida, J. ELECTROCHEM. SOC., 121, 434 (1974).
52. G. Fritz, H. Schäfer, R. Demuth and J. Grobe, Z. ANORG. ALLG. CHEM. 407, 287 (1974).
53. R. Demuth, Z. NATURFORSCH., 29b, 42 (1974).
54. G. Fritz and H. Schäfer, Z. ANORG. ALLG. CHEM. 407, 295 (1974).
55. G. Fritz, H. Schäfer and W. Hölderich, Z. ANORG. ALLG. CHEM. 407, 266 (1974).
56. C. Couret, J. Escudie, J. Satgé and G. Redoules, C. R. ACAD. SCI., SER. C 279, 225 (1974).
57. K. Issleib and W. Böttcher, Z. ANORG. ALLG. CHEM. 406, 178 (1974).
58. H. Schumann, H.-J. Kroth and L. Rösch, Z. NATURFORSCH. 29b, 608 (1974).
59. H. W. Roesky and E. Janssen, Z. NATURFORSCH., 29b, 174 (1974).
60. B. Green, J. CHEM. SOC., DALTON TRANSACTIONS 1113 (1974).
61. W. Lehr and N. Rosswag, Z. ANORG. ALLG. CHEM. 406, 221 (1974).
62. P. Clare and D. B. Sowerby, J. INORG. NUCL. CHEM. 36, 729 (1974).
63. A. Schmidpeter and R. Roßknecht, CHEM. BER. 107, 3146 (1974).
64. A. T. Fields and Ch. W. Allen, J. INORG. NUCL. CHEM. 36, 1929 (1974).
65. H. W. Roesky and E. Janßen, CHEM.-ZTG. 98, 260 (1974).
66. S. Das, R. A. Shaw and B. C. Smith, J. CHEM. SOC., DALTON TRANSACTIONS 1610 (1974).
67. H. R. Allock and M. T. Stein, J. AMER. CHEM. SOC. 96, 49 (1974).

68. A. A. Volodin, V. V. Kireev, V. V. Korshak and A. A. Fomin, ZH, OBSHCH. KHIM. 43, 2206 (1973).
69. D. Millington and D. B. Sowerby, J. CHEM. SOC., DALTON TRANSACTIONS 1070 (1974).
70. S. N. Nabi and R. A. Shaw, J. CHEM. SOC., DALTON TRANSACTIONS 1618 (1974).
71. O. J. Scherer, W. Glässel and R. Thalacker, J. ORGANOMET. CHEM. 70, 61 (1974).
72. P. P. Kornuta and T. V. Kolodka, ZH. OBSHCH. KHIM. 43, 2573 (1973).
73. V. P. Kukhan' and A. P. Boiko, ZH. OBSHCH. KHIM. 44, 1247 (1974).
74. H. W. Roesky and W. Grosse-Böwing, Z. ANORG. ALLG. CHEM. 406, 260 (1974).
75. W. Heider, U. Klingebiel, T.-P. Lin and O. Glemser, CHEM. BER. 107, 592 (1974).
76. R. Appel and I. Ruppert, Z. ANORG. ALLG. CHEM. 406, 131 (1974).
77. C. Voswijk and J. C. van de Grampel, REC. TRAV. CHIM. PAYS-BAS 93, 120 (1974).
78. A. M. Pinchuk, A. M. Khmaruk and T. V. Kovalevskaya, ZH. OBSHCH. KHIM. 44, 460 (1974).
79. H.-G. Horn, Z. ANORG. ALLG. CHEM. 406, 199 (1974).
80. V. Ya. Semenii, A. . Boiko, G. F. Solodushchenko, N. A. Kirsanov and V. P. Kukhar', ZH. OBSHCH. SHIM. 44, 1251 (1974).
81. Ch. Jäckh and W. Sundermeyer, ANGEW. CHEM. 86, 442 (1974).
82. J. A. Gibson and R. Schmutzler, Z. NATURFORSCH. 29b, 441 (1974).
83. L. Riesel, M. Mauck and E. Herrmann, Z. ANORG. ALLG. CHEM. 405, 109 (1974).
84. E. S. Kozlov, S. N. Gaidamaka and L. I. Bobkova, ZH. OBSHCH. KHIM. 44, 1075 (1974).
85. K. Burger, W. Thenn and J. Fehn, CHEM. BER. 107, 1526 (1974).
86. D. F. Clemens and W. E. Perkinson, INORG. CHEM. 13, 333 (1974).
87. I. Ruppert, V. Bastian and R. Appel. CHEM. BER. 107, 3426 (1974).
88. E. Niecke and W. Flick, ANGEW. CHEM. 86, 128 (1974).
89. O. J. Scherer and N. Kuhn, J. ORGANOMET. CHEM. 78C 17 (1974).
90. M. Biddlestone and R. A. Shaw, PHOSPHORUS 3, 95 (1973).
91. G. Bulloch and R. Keat, J. CHEM. SOC., DALTON TRANSACTIONS 2010 (1974).
92. P. Haasemann and J. Goubeau, Z. ANORG. ALLG. CHEM. 408, 293 (1974).
93. R. K. Harris, M. I. M. Wazeer, O. Schlak and R. Schmutzler, J. CHEM. SOC., DALTON TRANSACTIONS 1912 (1974).
94. O. Schlak, R. Schmutzler, H.-M. Schiebel, M. I. M. Wazeer and R. K. Harris, J. CHEM. SOC., DALTON TRANSACTIONS 2153 (1974).
95. O. J. Scherer, P. Klusmann and N. Kuhn, CHEM. BER. 107, 552 (1974).
96. J. A. Gibson and G. V. Röschenthaler, J. CHEM. SOC., CHEM. COMMUN. 694 (1974).
97. A. M. Pinchuk and V. A. Kovenya, ZH. OBSHCH. SHIM. 44, 705 (1974).
98. H. Nöth and W. Tinhof, CHEM. BER. 107, 3806 (1974).
99. D. J. Collins, J. W. Hetherington and J. M Swan, AUST. J. CHEM. 27 1759 (1974).
100. I. Shahak and Y. Sasson, ISR. J. CHEM. 11, 729 (1973).
101. U. Wannagat, M. Schlingmann and H. Autzen, CHEM.-ZTG. 98, 111 (1974).
102. G. Schöning, U. Klingebiel and O. Glemser, CHEM. BER. 107, 3756 (1974).

103. L. Maier, HELV. CHIM. ACTA 56, 2490 (1973).
104. R. Appel, I. Ruppert, R. Milker and V. Bastian, CHEM. BER. 107, 308 (1974).
105. L. Riesel, H. H. Pätzmann and H.-P. Bartich, Z. ANORG. ALLG. CHEM. 404, 219 (1974).
106. M. G. Thomas, R. W. Kopp, C. W. Schultz and R. W. Parry, J. AMER. CHEM. SOC. 96, 2646 (1974).
107. R. Keat, J. CHEM. SOC., DALTON TRANSACTIONS 876 (1974).
108. H. W. Roesky and H. Wiezer, CHEM. BER. 107, 1153 (1974).
109. O. Adler and F. Kober, J. ORGANOMET. CHEM. 72, 351 (1974).
110. D. D. Poulin and R. G. Cavell, INORG. CHEM. 13, 2324 (1974).
111. D. D. Poulin and R. G. Cavell, INORG. CHEM. 13, 3012 (1974).
112. L. A. Tamm, V. N. Chistokletov, E. S. Turbanova and A. A. Petrov, ZH. OBSHCH. KHIM. 44, 1468 (1974).
113. O. J. Scherer and N. Kuhn, CHEM. BER. 107, 2123 (1974).
114. I. M. Kosinskaya, N. P. Pisanenko and V. I. Shevchenko, ZH. OBSHCH. KHIM. 44, 1264 (1974).
115. Zh. M. Ivanova, N. I. Gusar and Yu. G. Gololobov, ZH. OBSHCH. KHIM. 44, 538 (1974).
116. Yu. G. Gololobov, L. I. Nesterova and A. V. Kirsanov, ZH. OBSHCH. KHIM. 44, 457 (1974).
117. H. Binder and R. Fischer, CHEM. BER. 107, 205 (1974).
118. E. E. Nifant'ev and I. V. Shilov, ZH. OBSHCH. KHIM. 43, 2658 (1973).
119. E. Fluck and F. Horn, PHOSPHORUS 3, 59 (1973).
120. W. Buder and A. Schmidt, CHEM. BER. 106, 3812 (1973).
121. O. J. Scherer and W. Gick, Z. NATURFORSCH. 29b, 129 (1974).
122. H. W. Roesky and E. Janssen, Z. NATURFORSCH. 29b, 177 (1974).
123. I. P. Titarenko and L. D. Protsenko, ZH. OBSHCH. KHIM. 44, 753 (1974).
124. H. Nöth and R. Ullmann, CHEM. BER. 107, 1019 (1974).
125. M. Baudler, Th. Vakratsas, D. Koch and K. Kipker, Z. ANORG. ALLG. CHEM. 408, 225 (1974).
126. T. C. Wallace, R. West and A. H. Cowley, INORG. CHEM. 13, 182 (1974).
127. V. L. Foss, V. V. Kudinova, Yu. A. Veits and I. F. Lutsenko, ZH. OBSHCH. KHIM. 44, 1209 (1974).
128. N. A. Sukhorukova, B. V. Timokhim, E. F. Grechkin and L. M. Sergienko, ZH. OBSHCH. KHIM. 44, 456 (1974).
129. R. Appel and R. Milker, CHEM. BER. 107, 2658 (1974).
130. F. Keyder and D. Linke, Z. CHEM. 13, 480 (1973).
131. A.-M. Leiva, E. Fluck, H Müller and G. Wallenwein, Z. ANORG. ALLG. CHEM. 409, 215 (1974).
132. A. Tasaka and O. Glemser, Z. ANORG. ALLG. CHEM. 409, 163 (1974).
133. B. Taylor, J. Steger, A. Wold and E. Kostiner, INORG. CHEM. 13, 2719 (1974).
134. H. Nahigian, J. Steger, H. L. McK nzie, R. J. Arnott and A. Wold, INORG. CHEM. 13, 1498 (1974).
135. M. Finke and H.-J. Kleiner, LIEBIGS ANN. CHEM. 741 (1974).
136. P. Sartori and R. Hochleitner, Z. ANORG. ALLG. CHEM. 404, 161 (1974).

137. E. Fluck, G. Gonzales and H. Binder, Z. ANORG. ALLG. CHEM. 406, 161 (1974).
138. R. C. Dobbie and M. J. Hopkinson, J. FLUORINE CHEM. 3, 367 (1973/74).
139. H. Steinberger and W. Kuchen, Z. NATURFORSCH. 29b, 611 (1974).
140. M. Fild and Th. Stankiewicz, Z. NATURFORSCH, 29b, 206 (1974).
141. D. B. Denney and L. S. Shih, J. AMER. CHEM. SOC. 96, 317 (1974).
142. T. Chivers, J. H. G. van Roode, J. N. R. Ruddick and J. R. Sams, CAN. J. CHEM. 51, 3702 (1973).
143. K. A. Avduevskaya and V. S. Mironova, IZV. AKAD. NAUK SSSR, NEORG. MATER. 10, 1364 (1974).
144. M. Ebert and M. Pelikanova, MONATSH. CHEM. 105, 11 (1974).
145. H. Quast, M. Heuschmann and M. O. Abdel-Rahman, SYNTHESIS 490 (1974).
146. J. Gloede, M. Mikolajczyk, A. Lopusinski and J. Omelanczuk, J. PRAKT. CHEM. 316, 703 (1974).
147. I. A. Nuretdinov, N. A. Buina, E. V. Bayandina and F. G. Sibgatulina, IZV. AKAD. NAUK SSSR, SER. KHIM. 483 (1974).
148. A. N. Pudovik, T. Kh. Gazizov and A. M. Kibardin, ZH. OBSHCH. KHIM. 44, 1210 (1974).
149. H. Boudjebel, H. Gonçalves and F. Mathis, BULL. SOC. CHIM. FRANCE 1671 (1974).
150. D. W. Grisley jun. and K. Szabo, J. CHEM. ENG. DATA 19, 175 (1974).
151. H. Stoll and J. Goubeau, Z. ANORG. ALLG. CHEM. 406, 307 (1974).
152. O. N. Grishina and N. A. Andreev, ZH. OBSHCH. KHIM. 44, 706 (1974).
153. A. J. Collins, G. W. Fraser, P. G. Perkin and D. R. Russell, J. CHEM. SOC., DALTON TRANSACTIONS 960 (1974).
154. L. B. Brainin, A. M. Kabzan, R. A. Cherkasov and A. N. Pudovik, ZH. PRIKL. KHIM. 47, 1385 (1974).
155. G. A. Bogdanov, Z. N. Dymova and T. P. Sidorova, ZH. FIZ. KHIM. 48 249 (1974).
156. L. M. Poznyak, L. V. Kubasova, N. I. Mishin, Yu. Ya. Kharitonov and S. I. Vol'fkovich, IZV. AKAD. NAUK SSSR, NEORG. MATER. 10, 506 (1974).
157. R. Gopal, C. Calvo, J. Ito and W. K. Sabine, CAN. J. CHEM. 52, 1155 (1974).
158. S. Ya. Grilikhes and E. D. Levin, ZH. PRIKL. KHIM. 47, 921 (1974).
159. A. B. Degtyareva, I. A. Korshunov, N. G. Chernorukov, A. I. Shafiev, Yu. V. Efremov and G. N. Yakovlev, IZV. AKAD. NAUK SSSR, NEORG. MATER. 10, 1908 (1974).
160. V. P. Orlovskii, Kh. M. Kurbanov, B. S. Khalikov, V. I. Bugakov and I. V. Tananaev, IZV. AKAD. NAUK SSSR, NEORG. MATER. 10, 670 (1974).
161. I. V. Tananaev, E. G. Davitashvili, N. A. Dzhabishvili and M. V. Landija, IZV. AKAD. NAUK SSSR, NEORG. MATER. 9, 2174 (1973).
162. H. Schmidbaur and R. Seeber, CHEM. BER. 107, 1731 (1974).
163. V. I. Isagulyants, V. A. Parfenova, N. M. Medvedeva and G. I. Bolotova, ZH. PRIKL. KHIM. 47, 615 (1974).
164. B. Ross and K.-P. Reetz, CHEM. BER. 107, 2720 (1974).
165. N. Getoff and S. Solar, MONATSH. CHEM. 105, 241 (1974).
166. L. F. Doty and R. G. Cavell, INORG. CHEM. 13, 2722 (1974).
167. V. A. Leitsin, S. D. Grekov and V. G. Lobosh, ZH. PRIKL. KHIM. 46, 2757 (1973).

168. D. H. Kerridge and I. A. Sturton, INORG. CHIM. ACTA 7, 701 (1973).
169. D. M. Poznyak, L. V. Kubasova, N. I. Mishin, Yu. Ya. Kharitonov and S. I. Vol'fkovich, IZV. AKAD. NAUK SSSR, NEORG. MATER. 10, 1367 (1974).
170. L. F. Yastrebova and L. Yu. Alikberova, IZV. AKAD. NAUK SSSR, NEORG. MATER. 10, 174 (1974).
171. N. I. Vorob'ev, V. V Pechkovskii, V. P. Titov and V. I. Poloiko, IZV. AKAD. NAUK SSSR, NEORG. MATER. 10, 1693 (1974).
172. A. Muck, T. Hynie, J. Stejskal and B. Hájek, Z. CHEM. 14, 69 (1974).
173. Ž. Ružić-Toros, B. Kojić-Prodić, R. Liminga and S. Popović, INORG. CHIM. ACTA 8, 273 (1974).
174. D. S. Rycroft and R. F. M. White, J. CHEM. SOC., CHEM. COMMUN. 444 (1974).
175. A. Delay, C. Friedli and P. Lerch, BULL. SOC. CHIM. FRANCE 828 (1974).
176. A. Delay, C. Friedli and P. Lerch, BULL. SOC. CHIM. FRANCE 839 (1974).
177. K. Brandt, W. Walczyk and Z. Jedliński, J. ORGANOMET. CHEM. 70, 309 (1974).
178. U. Schülke and N. N. Chudinova, IZV. AKAD. NAUK SSSR, NEORG. MATER. 10, 1697 (1974).
179. G. I. Veksler, K. K. Evstrop'ev and B. S. Kondrat'eva, IZV. AKAD. NAUK SSSR, NEORG. MATER. 10, 171 (1974).
180. G. P. Nikolina, IZV. AKAD. NAUK SSSR, NEORG. MATER. 10, 857 (1974).
181. G. Kura, S. Ohashi and S. Kura, J. INORG. NUCL. CHEM. 36, 1605 (1974).
182. M. Tsuhako, K. Hasegawa, T. Matuso, I. Motooka and M. Kobayashi, CHEM. LETT. 573 (1973).
183. A. V. Lavrov, L. S. Guzeeva, P. M Fedorov and I. V. Tananaev, IZV. AKAD. NAUK SSSR, NEORG. MATER. 10, 851 (1974).
184. Th. Goonek, J R. van Wazer, R. A. Kleps and T. C. Myers, INORG. CHEM. 13, 2337 (1974).
185. S. P. Mishra and M. C. R. Symons, J. CHEM. SOC., CHEM. COMMUN. 279 (1974).
186. G. H. Sprenger and J. M. Shreeve, J. FLUORINE CHEM. 4, 401 (1974).
187. S. R. O'Neill and J. M. Shreeve, J. FLUORINE CHEM. 3, 361 (1973/74).
188. J. C. Green, M. L. H. Green and G. E. Morris, J. CHEM. SOC. CHEM. COMMUN. 212 (1974).
189. M. Baudler and M. Bock, ANGEW. CHEM. 86, 124 (1974).
190. R. B. King and O. von Stetten, INORG. CHEM. 13, 2449 (1974).
191. H. R. Allcock, E. C. Bissell and E. T. Shawl, Inorg. Chem. 12, 2963 (1973).
192. R. J. Dudley, B. J. Hathaway, P. G. Hodgson, P. C. Power and D. J. Loose, J. CHEM. SOC., DALTON TRANSACTIONS 1005 (1974).
193. K. Dehnicke, R. Schmitt, A.-F. Shihada and J. Pebler, Z. ANORG. ALLG. CHEM. 404, 249 (1974).
194. A. Vizi-Orosz, G. Palyi and L. Marko, J. ORGANOMET. CHEM. 60, C25 (1973).
195. A. W. Cordes, R. D. Joyner, R. D. Shores and E. D. Dill, INORG. CHEM. 13, 132 (1974).

196. J. Bödeker and H. Zärtner, Z. CHEM. 14, 56 (1974).
197. Th. Kruck and H.-U. Hempel, ANGEW. CHEM. 86, 233 (1974).
198. M. A. Cairns and J. F. Nixon, J. ORGANOMET. CHEM. 74, 263 (1974).
199. M. A. Cairns and J. F. Nixon, J. CHEM. SOC., DALTON TRANSACTIONS 2001 (1974).
200. S. J. Cyvin, Z. ANORG. ALLG. CHEM. 403, 193 (1974).
201. J. F. Nixon, B. Wilkins and D. A. Clement, J. CHEM. SOC., DALTON TRANSACTIONS 1993 (1974).
202. P. A. W. Dean and D. G. Ibbott, INORG. NUCL. CHEM. LETTERS 11, 119 (1974).
203. Th. Kruck, M. Höfler and H. Jung, CHEM. BER. 107, 2133 (1974).
204. Th. Kruck, H. Jung, M. Höfler and H. Blume, CHEM. BER. 107, 2145 (1974).
205. Th. Kruck, H. Jung, M. Höfler and H. Blume, CHEM. BER. 107, 2156 (1974).
206. Th. Kruck, G. Mäueler and G. Schmidgen, CHEM. BER. 107, 2421 (1974).
207. O. Stelzer, CHEM. BER. 107, 2329 (1974).
208. W. M. Douglas, R. B. Johannesen and J. K. Ruff, INORG. CHEM. 13, 371 (1974).
209. G.-V. Röschenthaler and R. Schmutzler, Z. NATURFORSCH. 29b, 436 (1974).
210. R. C. Dobbie and M. J. Hopkinson, J. CHEM. SOC., DALTON TRANSACTIONS 290 (1974).
211. R. C. Dobbie and P. R. Mason, J. CHEM. SOC., DALTON TRANSACTIONS 2439 (1974).
212. W. C. Kaska, D. K. Mitchell, R. F. Reichelderfer and W. D. Korte, J. AMER. CHEM. SOC. 96, 2847 (1974).
213. A. J. Carty, G. Ferguson, H. N. Paik and R. Restivo, J. ORGANOMET. CHEM. 74, C14 (1974).
214. S. Jacobson, A. J. Carty, M. Mathew and G. J. Palenik, J. AMER. CHEM. SOC. 96, 4330 (1974).

ARSENIC, ANTIMONY AND BISMUTH

L. C. Duncan

Department of Chemistry
Central Washington State College
Ellensburg, Washington 98926

REVIEWS

Several comprehensive review type articles concerned with the preparative chemistry of arsenic, antimony, and bismuth have been published and then abstracted in the past year. Those reviews included the annual surveys covering the chemistry of As (1), Sb (2), and Bi (3) for the year 1972. A review concerning the elements of group V (4) and another directed toward the organometallic compounds of group V (5) appeared. The chemistry of metal complexes with phosphorous, arsenic, and antimony ligands (6) were also reviewed. Finally, acetylenic organoarsenic compounds (7) were the subject of another review.

ARSENIC

ARSENIC (III)

Arsenic bound to carbon: The tertiarycyanoarsine p-$MeC_6H_4AsEtCN$ (8) was prepared and studied as a synthetic intermediate. Di-p-tolylalkyl-(aryl)arsines (9) and alkylarylaminophenylarsines (10) (ie) Et(p-MeC_6H_4) $AsC_6H_4NH_2$ as well as di-α-napthylarsines $(\alpha\text{-}C_{10}H_7)_2AsR$ (11) have been described. The isonitrile $Ph_2As(CH_2)_2NC$ (12) was reported by King and Efraty who prepared Cr complexes using that material as a ligand. The potential ligands $(Ph_2AsCH_2CH_2)_2PPh$ (13) and cis-1,3-bis(dimethylarsino)-2,2,4,4-tetramethyl cyclobutane (14) were synthesized and described. Monocyclopentadienyl arsines (I) with C-As σ bonds were prepared and studied by Jutzi and Kuhn (15). A study of the selective cleavage of

(I) 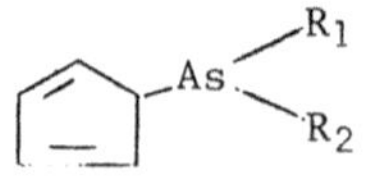

arsenic-phenyl bonds with HI (16) II, and a study of transformations of tertiaryarsine oxides (17) III appeared.

(II) $(CH_3)_2AsC_6H_5 \xrightarrow{HI} (CH_3)_2AsI$

(III) $(C_6H_5)_2EtAs) \xrightarrow{RX} (C_6H_5)_2AsOR$

Cyclic compounds: The reaction of catecholtrichlorophosphorane with AsF_3 yields 2-chloro-1,3,2-benzodioxaarsole (18) IV. The oxarsenium

(IV) 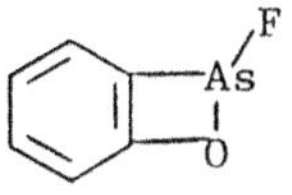

bromide 4,4-diphenyl-2,6-dimethyl-1,4-oxarsenium bromide (19) (V) was prepared and reported by Horner and Samaan. Maerkel and co-workers

(V) 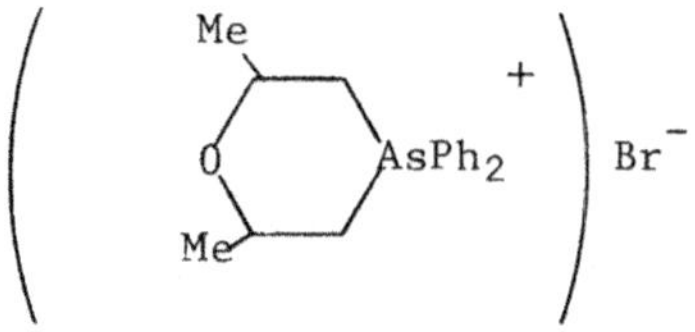

(20)(21) obtained several arsadiazoles (VI) and (VII) through the reactions of selected phenylhydrozones with $AsCl_3$. They also reported

(VI) 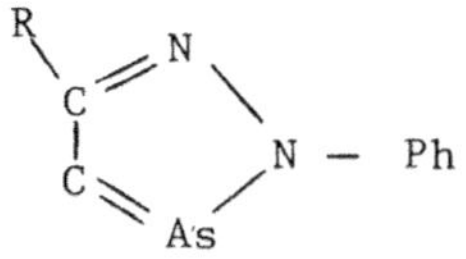

(VII) 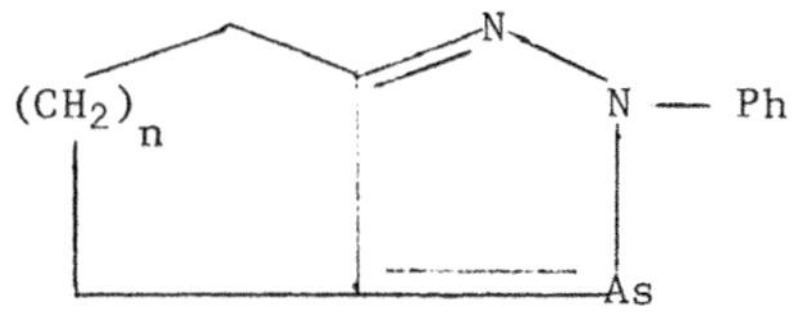

arsabenzene preparative routes through displacement of Sn by As (22) (VIII), via ring enlargement of arsoles (23) (IX), and via thermolyses of arsolyacetates (24) (X).

(VIII) 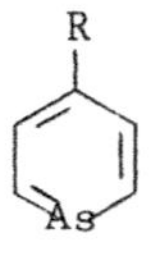

(IX) 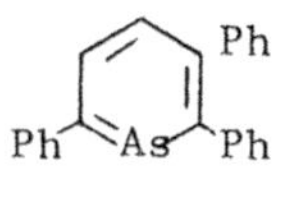

(X) 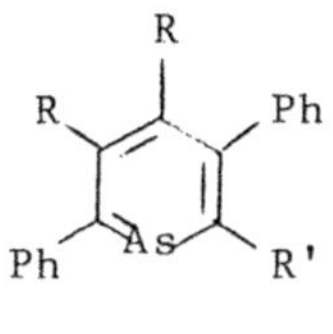

Arsenic compounds containing boron: Arsacarboranes (25)(26) and As containing boranes (27)(28) have been prepared, characterized, and reported.

Compounds with As-Si bonds: Reports of trimethylsilylbis(trifluoromethyl)arsine (29) and trifluorosilylarsine (30) preparations have appeared.

Compounds with As-N or As-P bonds: Alkoxy arsenic (V) nitrides were studied (31) and are reported to be mixtures of trimeric, tetrameric, pentameric, and hexameric units (XI). Kober and co-workers prepared,

(XI)

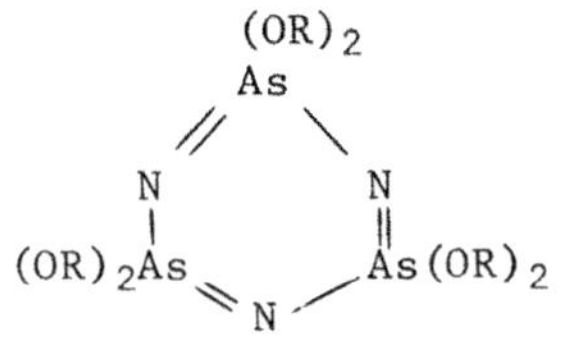

characterized and then studied the reactions of several bis(dialkylamino)(perfluoromethyl)arsines (32) and dimethylarsinoamines (33)(34). The five membered ring compound (XII) was obtained as a product of the reaction of 1,2-bis(methylamino)tetramethyldisilane with $AsCl_3$ or $SbCl_3$ (35). Diphosphinetrifluoromethylarsine $CF_3As(PH_2)_2$ was obtained on

(XII)

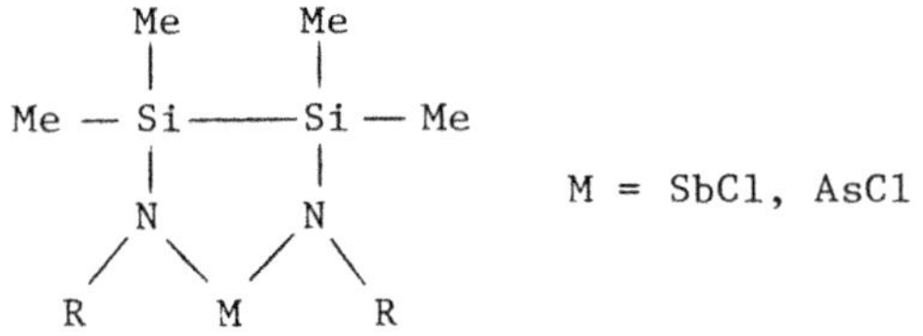

reaction of H_3SiPH_2 with CF_3AsI_2 (36). The preparation of a phosphonium chloride with a P-As bond (XIII) was described (37).

(XIII)

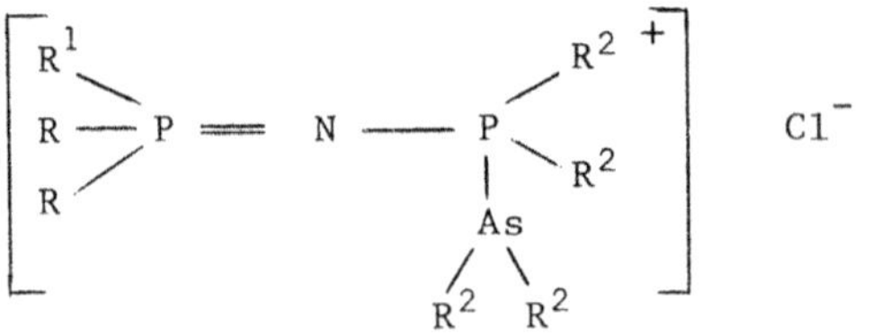

Compounds containing As-O or As-S bonds: Kober and co-workers reported the preparation of fluoroarsenic acid esters (38) as well as the preparation of several O-(dimethylarsino)oximes (39). Engler described the preparation of Bis(2-mercaptoethyl)sulfide and bis(2-mercaptoethyl) ether complexes containing arsenic, antimony, and bismuth (40)(41) (ie) $X(CH_2CH_2S)_2MCl$ (where X = O, S; M = As, Sb, Bi).

Compounds containing As-Hg bonds: Tris(methylmercury)-arsine $As(HgMe)_3$ and tetrakis(methylmercury)arsonium salts $[As(HgCH_3)_4]X$ were reported (42).

Misc: The synthesis and reactions of arsenides, stibides and bismuthides have been reported (43).

ARSENIC (V)

Kendurkar and Tewari prepared (p-Nitrobenzylidene)triphenylarsinane (44) and studied its' reactions with carbonyl compounds. Triarylarsinimes $Ph_3As = N\text{-}X$ were prepared via two routes (45)(46) by Cadogan and Gosney. The N.M.R. of the complexes formed between boron trifluoride and trialkylarsine oxide was studied (47). A synthetic route to pentamethylarsorane $(CH_3)_5As$ (48) was published. The IR and Raman spectra of Ph_5As and Ph_5Sb were reexamined; the arsenic compound is trigonal bipyramidal while the antimony compound shows a square pyramid structure (49). The synthesis of trimethylmethoxyarsonium iodide has been reported (50).

ANTIMONY

Antimony (III): Labeled stibines were prepared utilizing the reaction of alkyl or arylmagnesium halides with $SbCl_3$ (51). The preparation of O-phenylenebis(dimethylstibine) (52) as well as that of some p-chlorophenyldialkylstibines (53) was published.

Antimony V: Several pentaalkyl antimony compounds including Bu_3SbMe_2, Bu_2SbMe_3, $BuSbMe_4$, Me_5Sb, and Bu_5Sb were prepared using alkylmagnesium halides as reagents (54). Triorganoantimony (V) compounds R_3SbL containing potentially tridentate ligands were prepared (55). Geol and co-workers (56) prepared pentacoordinate cationic antimony complexes such as $[R_3SbL_2][ClO_4]_2$ [R=Me, Ph; L-MNSO, etc. and compounds of the type R_3SbX_2 (R=Me, Ph; $X=N_3$, NCO) (57). Organostibonium salts with iron-antimony σ bonds $[RR'Sb(FeCp(CO)_2]_2\ BPh_4$ have been prepared (58). Antimonytetrachlorideacetate-and formate have been prepared (59). The five coordinate compound $Ph_4SbC_6H_4Me\text{-}p$ (60) was reported and its' solvolytic reactions in alcohols were studied.

BISMUTH

Complexes of Bi (III) with 1-nitroso-2-napthol have been prepared and studied (61). The ^{19}F N.M.R. of compounds of the type p-FC_6H_4Bi-$(pC_6H_4NMe_2)_2$ was studied (62) and reported.

ARSENIC, ANTIMONY AND BISMUTH AS ACCEPTORS

As, Sb, Bi (III) acceptors: Adducts of $AsBr_3$, $SbBr_3$, and $BiBr_3$ with dioxane (63) were prepared and studied. Bipyridyl and terpyridyl complexes of As, Sb, and Bi halides were the subject of investigations by Brodie and Wilkins (64). A N.M.R. study indicated that $AsCl_3$ forms complexes with benzene, toluene, and similar compounds (65). The nature of the complexes formed when SeO_3 interacts with As_2O_3 and Sb_2O_3 was investigated (66). The reaction of tartaric acid with antimony (III) oxide yields species of composition $H[C_4H_2O_6Sb(H_2O)]$, $H[C_4H_2O_6Sb]$, and $M^{2+}[C_4H_2O_6Sb(H_2O)_2]$ (67). The antimony trichloride-tributyl phosphate system has been studied (68). Both acetonitrile and benzonetrile form complexes with $SbCl_3$ and $SbBr_3$ (69). Bromoantimonate (III) (70), chlorofluoroantimonate (III) (71), pentachloroantimonate (III) (72), pentafluoroantimonate (III) (73), and tetrafluoroantimonate (III) (74) species have been prepared.

As, Sb, Bi (V) acceptors: Arsenic and antimony pentahalides and their derivatives have been extensively utilized as Lewis type acceptors. Reports concerning the preparation and study of the following adducts have appeared during the period surveyed (XIV):

(XIV)
$NiF_3 \cdot 2KAsF_6$ (75)
$Hg_4(ASF_6)_2$ (76)
$(ClF_2SO)X$; $X=AsF_6$, SbF_6, Sb_2F_{11} (77)(78)
$(BrF_6)(AsF_6)$ (79)
$(KrF)X$; $X=AsF_6$, SbF_6 (80)
$(ClO_2SO_3F)X$ X-AsF_5, SbF_5 (81)
$Rb_2(As_2F_{10}O) \cdot H_2O$ (82)
$S_8^{++}(X)_2$; $Se^{++}(X)_2$ $X=As_2F_{11}$; Sb_2F_{11} (83)
$(C_6F_6)(AsF_6)$ (84)
$(TeF_3)(Sb_2F_{11})$ (85)
$(I_2)(Sb_2F_{11})$ (86)
$(ClO_2)(Sb_2F_{11})$ (87)
$Rb_2[Sb_2F_{10}O]$ (88)

Carbamoyl azides react with antimony pentachloride forming adducts (89). The mixed chlorohalo-and chloropseudohalodiphenylantimonates (ie) $[Me_4N][Ph_2SbCl_3X][X=Cl,Br,N_3]$ have been prepared (90). Complexes of the type $Sb(OR)Cl_4 \cdot nL$ (R=Me,Et,C_2H_4Cl; n=1,2; L=phosphoryl or amine oxide donor) have also been reported (91). The $SbF_5 \cdot Me_2O$ and $SbF_5 \cdot Me_3N$ systems have been studied; both 1:1 and 1:2 complexes are formed (92).

REFERENCES

1. G. O. Doak and L. D. Freedman, J. ORGANOMETAL. CHEM., 68, 295 (1974).
2. L. D. Freedman and G. O. Doak, J. ORGANOMETAL. CHEM., 68, 371 (1974).
3. L. D. Freedman and G. O. Doak, J. ORGANOMETAL. CHEM., 68, 411 (1974).
4. N. Logan, D. B. Sowerby, INORG. CHEM. MAIN-GROUP ELEM., 1, 231 (1973).
5. J. L. Wardell, ORGANOMETAL. CHEM., 2, 215 (1973).
6. C. A. McAuliffe, ASPECTS OF INORGANIC CHEMISTRY, MacMillan, (1973).
7. I. N. Azerbaev, A. A. Abramova, and Yu. G. Bosyakov, USP. KHIM., 43, 1384 (1974).
8. Yu. F. Gatilov, L. B. Ionov and A. P. Kukorina, ZH. OBSHCH. KHIM., 43, 2677 (1973).
9. R. D. Gigauri, G. N. Chachaua, B. D. Chernokal'skii, and M. M. Ugulava, ZH. OBSHCH. KHIM., 44, 1720 (1974)
10. Yu. F. Gatilov, F. D. Yambushev, and N. Kh. Tenisheva, ZH. OBSHCH. KHIM., 43, 2405 (1973).
11. R. D. Gigauri, M. A. Indzhiya, B. D. Chernokal'skii, M. M. Ugulava, and Ts. A. Sidamashuili, ZH. OBSHCH. KHIM., 44, 1541 (1974).
12. R. B. King and Avi Efraty, J. CHEM. SOC., PERKIN TRANS. 1371, (1974)
13. R. B. King and P. R. Heckeley, PHOSPHOROUS, 3, 209 (1974).
14. W. R. Cullen and J. T. Price, CAN. J. CHEM., 57, 1108 (1974).
15. P. Jutzi and P. Manfred, CHEM. BER., 107, 1228 (1974).
16. J. Ellermann, H. Schoessner, A. Haag, and H. Schoedel, J. ORGANOMETAL. CHEM., 65, 33 (1974).
17. Yu. F. Gatilov, B. E. Abalonin, and Z. M. Izmailova, DOKL. AKAD. NAUK. SSSR, 216, 800 (1974).
18. M. Eisenhut and R. Schmutzler, PHOSPHORUS, 4, 73 (1974).
19. L. Horner and S. Samaan, PHOSPHORUS, 4, 1 (1974).
20. G. Maerkl and C. Martin, TETRAHEDRON LETT., 4503 (1973).
21. G. Maerkl, H. Baier, and C. Martin, TETRAHEDRON LETT., 1977 (1974).
22. G. Maerkl and F. Kneidl, ANGEW. CHEM., 85, 990 (1973).
23. G. Maerkl, J. Advena, and H. Hauptmann, TETRAHEDRON LETT., 203 (1974)
24. G. Maerkl, J. Advena, and H. Hauptmann, TETRAHEDRON LETT., 303 (1974).
25. L. I. Zakharkin and N. F. Shemyakin, IZV. AKAD. NAUK. SSSR KHIM., 940 (1974).
26. H. D. Smith and M. F. Hawthorne, INORG. CHEM., 13, 2312 (1974).
27. A. R. Siedle and L. J. Todd, J. CHEM. SOC., CHEM. COMMUN., 914 (1973).
28. J. L. Little, S. S. Pao, and K. K. Sugathan, INORG. CHEM., 13, 1752 (1974).
29. J. E. Byrne and C. R. Russ, J. INORG. NUCL. CHEM., 36, 35 (1974).
30. R. Demuth, Z. NATURFORSCH. TEIL B, 29, 42 (1974).
31. H. Preiss and D. Hass, Z. ANORG. ALLG. CHEM., 404, 190 (1974).
32. O. Adler and F. Kober, J. ORGANOMETAL. CHEM., 72, 351 (1974).
33. F. Kober, Z. ANORG. ALLG. CHEM., 401, 243 (1973).
34. F. Kober and W. J. Ruehl, Z. ANORG. ALLG. CHEM., 406, 52 (1974).
35. U. Wannagat, M. Schlingmann, and H. Autzen, CHEM. ZTG., 98, 372 (1974).
36. R. Demuth and J. Grobe, Z. NATURFORSCH. TEIL B, 29, 35 (1974).
37. W. Wolfsberger, Z. NATURFORSCH. TEIL B, 29, 35 (1974).
38. F. Kober and W. J. Ruehl, J. FLUORINE CHEM., 4, 65 (1974).
39. J. Kaufmann and F. Kober, J. ORGANOMETAL. CHEM., 71, 49 (1974).
40. R. Engler, Z. ANORG. ALLG. CHEM., 406, 74 (1974)

41. R. Engler, Z. ANORG. ALLG. CHEM., 407, 35 (1974).
42. D. Breitinger and G. P. Arnold, INORG. NUCL. CHEM. LETT., 10, 517 (1974).
43. G. O. Doak and L. D. Freedman, SYNTHESIS, 5, 328 (1974).
44. P. S. Kendurkar and R. S. Tewari, J. ORGANOMETAL. CHEM., 60, 247 (1973).
45. J. I. G. Cadogan and I. Gosney, J. CHEM. SOC. PERKIN TRANS., 460 (1974).
46. J. I. G. Cadogan and I. Gosney, J. CHEM. SOC. PERKIN TRANS., 466 (1974).
47. R. Bravo, M. Durand, and J. P. Laurent, ORG. MAGN. RESONANCE, 5, 357 (1973).
48. K. H. Mitschke and H. Schmidbaur, CHEM. BER., 106, 3645 (1973).
49. G. L. Kok, SPECTRO. CHIM. ACTA. PART A, 30A, 961 (1974).
50. B. D. Chernokal'skii, L. A. Vorob'eva, R. R. Shagidullin, I. A. Lamanova, and T. V. Zykova, ZH. OBSHCH. KHIM., 43, 264 (1974).
51. M. Benmalik, H. Chermette, C. Martelet, D. Sandino, and J. Towset, J. ORGANOMETAL. CHEM., 67, 53 (1974).
52. E. Skewchuk and S. B. Wild, J. ORGANOMETAL. CHEM. 71, C1-C2 (1974).
53. Y. Kato, S. Yachigo, and I. Hirao, MEM. KYUSHA. INST. TECHNOL. ENG., 4, 25 (1974).
54. A. N. Nesmeyanov, A. E. Borisov, N. V. Novikova, E. L. Fedin, and P. V. Petrovskii, IZV. AKAD. SSSR. SER. KHIM., 8, 1833 (1973).
55. F. DiBianca, E. Rivarola, A. L. Spek, H. A. Meimema, and J. G. Noltes, J. ORGANOMETAL. CHEM., 63, 293 (1973).
56. R. G. Goel, and H. S. Pradsad, J. ORGANOMETAL. CHEM., 59, 253 (1973).
57. G. A. Olah, P. Shilling, and I. Gross, J. AMER. CHEM., 96, 876 (1974).
58. Y. Matsumura, M. Harkawa, and R. Okawara, J. ORGANOMETAL. CHEM., 71, 403 (1974).
59. R. A. Laber and A. Schmidt, Z. ANORG. ALLG. CHEM., 407, 237 (1974).
60. W. E. McEwen and C. T. Lin, PHOSPHOROUS, 3, 229 (1974).
61. S. V. Patel and J. R. Raju, INDIAN J. CHEM., 12, 770 (1970)
62. D. D. Kravtson, B. D. Kason, S. I. Pombrik, and E. J. Fedin, IZV. AKAD. NAUK. SSSR. SER. KHIM., 1, 234 (1974).
63. R. C. Maheshwari, S. K. Suri, and V. Ramakrishna, INDIAN J. CHEM., 11, 1196 (1973).
64. A. M. Brodie and C. J. Wilkins, INORG. CHEM. ACTA., 8, 13 (1974).
65. M. I. Vakhrin and V. S. Grechnishkin, ZH. FIZ. KHIM., 48, 1515 (1974).
66. R. C. Paul, R. D. Sharma, S. Singh, and K. C. Malkotra, INDIAN J. CHEM., 11, 1174 (1973).
67. M. U. Usukakunov, S. U. Bleshinskii, N. I. Sibichenkova, M. S. Mironov, and L. I. Lukina, IZV. AKAD. NAUK. KIRG. SSSR SER. KHIM., 46, 15 (1973).
68. T. N. Sumarkova, T. D. Ibraeva, and Yu. A. Nevskaya, IZV. AKAD. NAUK. KAZ. SSSR SER. KHIM., 23, 15 (1973).
69. T. N. Sumarkova, R. A. Slavinskaya, and N. Yu. Vykhrest, IZV. AKAD. NAUK. KAZ. SSSR SER. KHIM., 23, 354 (1973).
70. N. K. Jha and S. S. A. Rizvi, J. INORG. NUCL. CHEM., 36, 1479 (1974).
71. R. L. Davidovich, L. A. Zemnukhova, and N. I. Sigula, ZH. NEORG. KHIM., 19, 2529 (1974).
72. R. K. Wismer and R. A. Jacobson, INORG. CHEM., 13, 1678 (1974).

73. N. Habibi, B. Ducourant, R. Fourcade, and G. Mascherpa, BULL. SOC. CHIM. FR., 21 (1974).
74. M. Mehrain, B. Ducourant, R. Fourcade, and G. Mascherpa, BULL. SOC. CHIM. FR., 757 (1974).
75. T. L. Court and M. F. A. Dove, J. CHEM. SOC., DALTON TRANS., 1995 (1973).
76. B. D. Cutforth, R. J. Gillespie, and P. R. Ireland, J. CHEM. SOC., CHEM. COMMUN., 723 (1973).
77. C. Lau and J. Passmore, J. CHEM. SOC., DALTON TRANS. 2528 (1973).
78. R. F. Dunphy, C. Lau, H. Lynton, and J. Passmore, J. CHEM. SOC. DALTON TRANS., 2533 (1973).
79. R. J. Gillespie and G. J. Schrobilgen, INORG. CHEM., 13, 1230 (1974).
80. R. J. Gillespie and G. J. Schrobilgen, J. CHEM. SOC., CHEM. COMMUN., 90 (1974).
81. W. R. Cullen and J. T. Price, CAN. J. CHEM., 57, 1108 (1974).
82. W. Haase, ACTA CRYSTALLOGR. SECT. B, 30, PT. 70, 1722 (1974).
83. P. A. W. Dean, R. J. Gillespie, and P. K. Ummat, INORG. SYNTH., 15, 213 (1974).
84. D. M. Byler and D. F. Shriver, INORG. CHEM., 13, 2697 (1974).
85. A. J. Edwards and P. Taylor, J. CHEM. SOC., DALTON TRANS., 2150 (1973).
86. G. G. Davis, R. J. Gillespie, P. R. Ireland, and J. M. Sowa, CAN. J. CHEM., 52, 2048 (1974).
87. A. J. Edwards and R. J. C. Sells, J. CHEM. SOC., DALTON TRANS., 1726 (1974).
88. W. Haase, ACTA, CRYSTALLOGR., SECT. B, 30B, 2508 (1974).
89. W. Buder and A. Schmidt, CHEM. BER., 106, 2877 (1973).
90. N. Bertazzi, L. Pellerito, and G. C. Stocco, INORG. NUCL. CHEM. LETT., 10, 855 (1974).
91. R. C. Paul, H. Madan, and S. L. Chadha, J. INORG. NUCL. CHEM., 36, 737 (1974).
92. S. Brownstein and M. Farrall, CAN. J. CHEM., 52, 1958 (1974).

CHALCOGENS

John S. Thayer
Department of Chemistry
University of Cincinnati
Cincinnati, Ohio 45221

OXYGEN

A monograph has been published on activation mechanisms of oxygen (1).

Ozone will react with various hypohalites to form chloro-oxygen compounds (2):

$$ClOSO_2F + 2\ O_3 \longrightarrow ClO_2SO_3F + 2\ O_2 \qquad (1)$$

Ozone has also been used to prepare singlet oxygen (3):

$$C_6H_5C(:O)H + O_3 \longrightarrow C_6H_5CO_2H + O_2 \qquad (2)$$

Chemonuclear reactors have been used to prepare ozone (4).

Nitrous oxide will yield O^- upon an irradiated magnesium oxide surface (5); this ion further reacts with CO or O_2 to give CO_2^- and ozonide anion. All three species were studied by ESR spectroscopy.

The compound O_2SbF_6 may be used to remove Xe, Rn and radioactive Po or Bi from the atmosphere (6).

Interest in the cleavage and subsequent reaction of the O-O bond in peroxides continues strong. Decomposition of $[(C_6H_5)_3SbBr]_2O_2$ gives monatomic oxygen as a product (7). The thermal rearrangement of triphenylgermyltrialkylsilylperoxides gives Si-O-C bonds exclusively (8):

$$(C_6H_5)_3GeOOSiR_3 \longrightarrow (C_6H_5)_3GeOSiR_2OR \qquad (R = C_6H_5,\ alkyl) \qquad (3)$$

The reaction of t-butylhydroperoxide with diphenyldisulfide is inhibited by the presence of diphenylselenoxide (9). Treatment of trifluoromethylperoxides with oxides of nitrogen give peroxynitrates (10):

$$CF_3OOH + N_2O_5 \longrightarrow CF_3OONO_2 \qquad (4)$$

$$CF_3OOF + N_2O_4 \longrightarrow CF_3OONO_2 \qquad (5)$$

Reviews on the chemistry of peroxides with fluoro substituents (11) and of the stable radical (12)

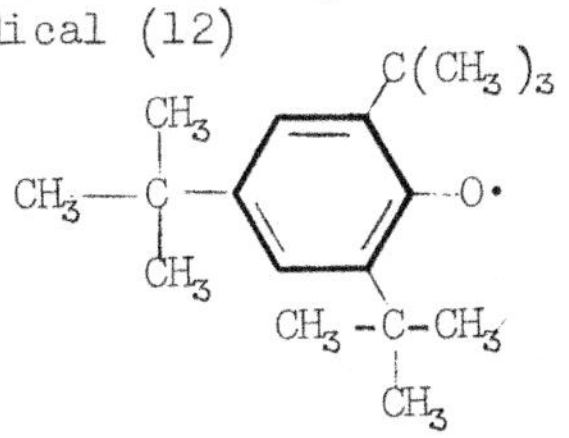

have appeared.

SULFUR

Various classes of sulfur compounds were reviewed during this past year (13-15).

Polysulfur Compounds

The reaction of $S_{10}Cl_2$ with hydrogen polysulfides yields macrocyclic allotropes (10) of sulfur (16):

$$S_{10}Cl_2 + H_2S \longrightarrow S_{10+x} + 2\ HCl \qquad (x = 8,10) \qquad (6)$$

The species S_{20} has two C_2 axes and four six-membered rings in a chair configuration, with one end open for further bonding. Amorphous sulfur reacts with cycloheptatriene in a sulfolanepyridine solvent to form the tricyclic compound $C_8H_7S_3$ (17):

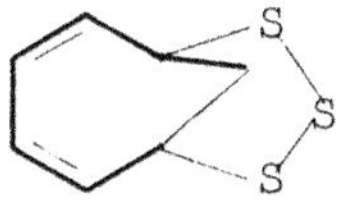

Mixtures of propane and butane will react with S_8 at 700-1000° and one atmosphere pressure to give CS_2; the reaction goes more smoothly in an electric discharge system (18).

A study of polysulfide formation in aqueous solution has appeared (19). Oxidation of aqueous sodium sulfide gives thiosulfate in good yields with no gas emission (20). A method for the quantitative determination of S_2Cl_2 has been proposed (21):

$$S_2Cl_2 + 10\ ICl + 8\ H_2O \longrightarrow 2\ SO_4^{2-} + 5\ I_2 + 16\ H^+ + 12\ Cl^- \qquad (7)$$

Sulfur Halides

Thionyl chloride or S_2Cl_2 will react with hydrogen iodide at -78° to give the corresponding iodides (22). A mixture of SCl_2 and S_2Cl_2 will react with fluorine in the presence of PCl_3 or $SbCl_3$ to give SF_4 in 95%

purity (23). Sulfur hexafluoride will react with various nonmetal oxides to form the corresponding fluorides (24):

$$SF_6 + SiO_2 \longrightarrow SO_2F_2 + SiF_4 \quad (8)$$

Chlorocarbon derivatives of fluorosulfonic acid have been prepared (25):

$$CHCl_3 + F_2SO_3 \longrightarrow Cl_2CHOSO_2F + ClF \quad (9)$$

$$Cl_2C{=}CHCl + F_2SO_3 \longrightarrow ClFCHCCl_2OSO_2F \quad (10)$$

Sulfur-Nitrogen Compounds

Research activity continues strong in this area of sulfur chemistry. Tetrasulfur tetranitride will react with organotin nitrides (26):

$$(CH_3)_3Sn\ _2NX + S_4N_4 \longrightarrow (CH_3)_3SnN{=}S{=}NX \quad (11)$$

$$(X = (CH_3)_3Sn^- \ , \ (CH_3)_3Si-)$$

The compound $[(CH_3)_3SnN{=}]_2S$ will react with methyltrichlorosilane to give a bicyclic species (27):

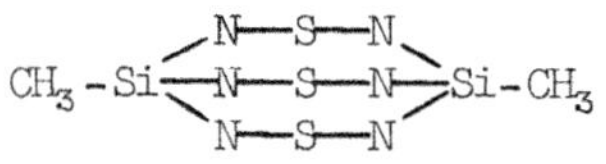

The related compound $[(CH_3)_3SiN{=}]_2S$ will react with chlorine at -70^0 to give the N-chloro derivative

$$(CH_3)_3SiN{=}S{=}NSi(CH_3)_3 + Cl_2 \longrightarrow (CH_3)_3SiN{=}S{=}NCl \quad (12)$$

which then reacts with SCl_2 to give either of two products (28):

$$(CH_3)_3SiN{=}S{=}NCl + SCl_2 \longrightarrow [(CH_3)_3SiN{=}S{=}N-]_2S \quad (13)$$

$$(CH_3)_3SiN{=}S{=}NCl + 2\ SCl_2 \longrightarrow S_4N_4 \quad (14)$$

The N-chloro compound will also react with trifluoromethylacetyl chloride to give a cyclic species:

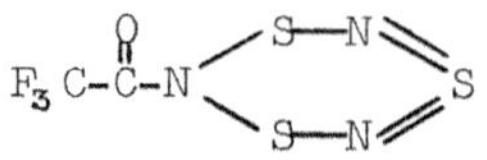

N-sodiohexamethyldisilazane will react with OSF_4 to give a compound containing hexavalent sulfur (28):

$$[(CH_3)_3Si]_2NNa + OSF_4 \longrightarrow [(CH_3)_3SiN{=}]_3S \quad (15)$$

The synthesis of S_3N_2Cl and $S_3N_2SO_3Cl$ have been reported (29). Two interesting fluorosulfur-nitrogen compounds have been prepared (30):

$$[(CH_3)_3SiN=]_2SF_2 \quad + \quad OSF_4 \longrightarrow N\equiv \overset{F}{\underset{F}{S}}-N=\overset{F}{\underset{F}{S}}=O \qquad (16)$$

$$N\equiv \overset{F}{\underset{F}{S}}-N=\overset{F}{\underset{F}{S}}=O \quad + \quad (C_6H_5)_4As^+ \xrightarrow{H_2O} (C_6H_5)_4As^+[HN=\overset{F}{\underset{F}{S}}=N-SO_2F]^- \qquad (17)$$

An anionic nitrogen-sulfur species has been reported (31):

$$(C_4H_9)_4N^+\ OH^- \quad + S_7NH \longrightarrow (C_4H_9)_4N^+\ NS_4^- \qquad (18)$$

Miscellaneous

Dialkylsulfides or dialkylsulfoxides can be converted to the corresponding sulfones by reaction with organic peroxides in the presence of base (32). Reaction of $C_5H_5S^+\ ClO_4$ with phenyllithium gives oligomeric species instead of the hoped-for monomeric 1-thiabiphenyl (33).

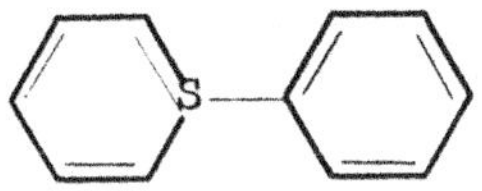

Treatment of carbon monosulfide with halogens in an electric discharge gives halogenation of the sulfur atom (34)

$$CS \quad + \quad Cl_2 \longrightarrow CSCl_4 \quad + \quad Cl_3CSCl \qquad (19)$$

$$3\ CS \quad + \quad 3\ HCl \longrightarrow (HClC=S)_3 \qquad (20)$$

SELENIUM

Two monographs on selenium derivatives have been published recently. One discusses the chemistry and biology of organic selenium compounds (35), while the other reviews selenium chemistry generally (36).

Pentafluoroselenic acid serves as the starting point for a number of oxyselenium fluorides (37):

$$HOSeF_5 \quad + \quad [(CH_3)_3Si]_3N \longrightarrow (CH_3)_3SiOSeF_5 \qquad (21)$$

$$(CH_3)_3SiOSeF_5 \quad + \quad NaOSi(CH_3)_3 \longrightarrow NaOSeF_5 \qquad (22)$$

$$NaOSeF_5 \xrightarrow{\Delta} OSeF_4 \quad + \quad Se_2O_2F_8 \qquad (23)$$

Ammonium selenite may be converted to ammonium selenate by electrolysis (38). Reaction between $SeCl_4$ and SeO_3 yields chloroselenates (39). Wynne has reported that the selenium atom of selenocyanate ion can be halogenated (40):

$$SeCN^- + SOCl_2 \longrightarrow Cl_2SeCN^- \qquad (24)$$

$$SeCN^- + Br_2 \longrightarrow Br_2SeCN^- \qquad (25)$$

$$SeCN^- + SO_2Cl_2 \longrightarrow Cl_3Se^- \qquad (26)$$

Hydrogen selenide reacts with methylmercuric nitrate (41):

$$H_2Se + CH_2HgNO_3 \longrightarrow (CH_3Hg)_3Se^+ \; NO_3^- \qquad (27)$$

However, methylmercuric bromide simply gives $(CH_3Hg)_2Se$.

Diphenylselenium dichloride reacts with ammonia and its derivatives (42). Ammonia, methylamine and dimethylamine reduce this compound to diphenylselenide. Trimethylsilyl amines give other reactions:

$$(C_6H_5)_2SeCl_2 + (CH_3)_3SiN(CH_3)_2 \longrightarrow (C_6H_5)_2\underset{Cl}{Se}N(CH_3)_2 \qquad (28)$$

$$(C_6H_5)_2SeCl_2 + [(CH_3)_3Si)_2NH \longrightarrow [(C_6H_5)_2Se{=}]_2N^+Cl^- \qquad (29)$$

Para-tolylselenyl chloride reacts with ethylene (43):

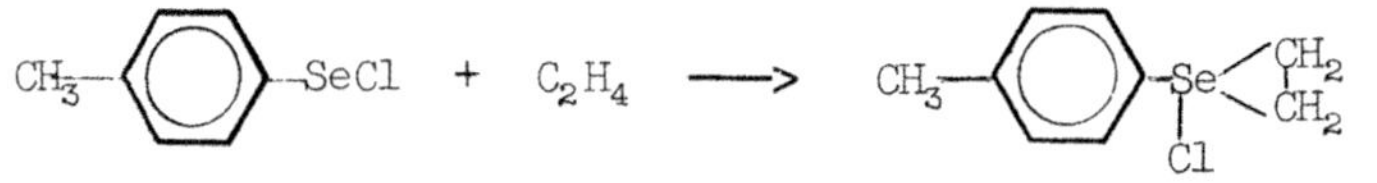

(30)

While the selenacyclopropane can be isolated, it rearranges readily to p-$CH_3C_6H_4SeCH_2CH_2Cl$.

TELLURIUM

A monograph on the organic chemistry of tellurium has recently been published (44). Biological effects of organotellurium compounds have been discussed in a review article (45).

Tellurium hexafluoride reacts with RbF and CsF in C_6F_6 (46):

$$CsF + TeF_6 \longrightarrow CsTeF_7 \qquad (31)$$

$$2\,RbF + TeF_6 \longrightarrow Rb_2TeF_8 \qquad (32)$$

These compounds are stable in the solid state but not in HF solution. Chlorine fluoride reacts with TeF_4, $TeCl_4$, or TeO_2 to give $ClTeF_5$ (47). Solid state studies of the Te-TeI_4 system show TeI, but not TeI_2, as an intermediate phase (48). Pyrolysis of $NaOTeF_5$ gives $Na_2TeO_2F_4$, while $LiOTeF_5$ gives $Te_2O_2F_8$ (37). Hydrogen telluride will react with methylmercuric bromide to give $(CH_3Hg)_2Te$ (41).

Reaction of diarylditellurides with iodine in the presence of tetra-alkylphosphonium iodides give salts (49):

$$Ar_2Te_2 \ + \ 2\ I_2 \xrightarrow{R_4P^+} 2\ ArTeI_2^- \qquad (33)$$

Various synthetic approaches to organotellurium iodides have been reported (50), of which the following is of particular interest:

$$R_2TeI_2 \ + \ R'I \xrightarrow[Na_2CO_3]{Na_2SO_3} R_2R'Te^+ \ I^- \qquad (34)$$

Spectroscopic studies of dialkyltellurium tetraiodides show them to be solid-state adducts of formula $R_2TeI_2 \cdot I_2$, which dissociate in solution (51).

REFERENCES

1. O. Hayaishi, ed., MOLECULAR MECHANISMS OF OXYGEN ACTIVATION, Academic Press, New York, 1974.
2. C. J. Schack and K. O. Christe, INORG. CHEM., 13, 2378 (1974).
3. F. E. Stary, D. E. Emige, and R. W. Murray, J. AMER. CHEM. SOC., 96, 5671 (1974).
4. M. Steinberg, M. Beller, and J. R. Powell, CHEM. ABS., 81, 111041 (1974).
5. C. Naccache and M. Che, CHEM. ABS., 80, 119905 (1974).
6. L. Stein, CHEM. ABS., 81, 175598 (1974).
7. J. Dahlmann and K. Winsel, Z. CHEM., 14, 232 (1974).
8. V. A. Yablokov, A. P. Tarabarina, and N. V. Yablokova, SYN. REACTIV. INORG. METAL-ORG. CHEM., 4, 339 (1974).
9. P. Southwell-Keely, I. L. Johnstone, and E. R. Cole, CHEM. ABS., 81, 151733 (1974).
10. F. A. Hohorst and D. D. DesMarteau, INORG. CHEM., 13, 715 (1974).
11. R. A. DeMarco and J. M. Shreeve, ADV. INORG. CHEM. RADIOCHEM., 16, 109 (1974).
12. E. Mueller, CHEM. ABS., 81, 119733 (1974).
13. F. Seel, ADV. INORG. CHEM. RADIOCHEM., 16, 297 (1974).
14. A. W. Jache, ADV. INORG. CHEM. RADIOCHEM., 16, 177 (1974).
15. H. W. Roesky, CHEM. ZEIT., 98, 121 (1974).
16. M. Schmidt, E. Wilhelm, T. Debaerdemaeker, E. Hellner, and A. Kutoglu, Z. ANORG. ALLGEM. CHEM., 405, 153 (1974).
17. H. Fritz and C. D. Weis, TET. LETTERS, 1974, 1659.
18. V. A. Borodulya, L. M. Vinograd, N. W. Grebenko, and A. A. Peliks, CHEM. ABS., 81, 138163 (1974).
19. W. F. Giggenbach, INORG. CHEM., 13, 1724 (1974).
20. H. Kashiwase and K. Takeuchi, CHEM. ABS., 81, 5159 (1974).
21. P. N. K. Nambisan, CURR. SCI., 43, 581 (1974).
22. D. K. Padma, IND. J. CHEM., 12, 320 (1974).
23. W. Becher and J. Massonne, CHEM. ZEIT., 98, 117 (1974).
24. A. P. Hagen, D. J Jones, and S. R. Ruttman, J. INORG. NUCL. CHEM., 36, 1217 (1974).
25. L. F. R. Cafferata and J. E. Sicre, INORG. CHEM., 13, 242 (1974).

26. H. W. Roesky and H. Wiezer, CHEM. BER., 107, 3186 (1974).
27. H. W Roesky and H. Wiezer, ANGEW. CHEM. 86, 130 (1974).
28. W. Lidy, W. Sundermeyer, and W. Verbeek, Z. ANORG. ALLGEM. CHEM., 406, 228 (1974).
29. R. C. Paul, R. P. Sharma, and R. D. Verma, IND. J. CHEM., 12, 418 (1974).
30. O. Glemser and R. Hoefer, Z. NATURFORSCH., 29B, 121 (1974).
31. T. Chivers and I. Drummond, INORG. CHEM., 13, 1222 (1974).
32. S. Suyama and Y. Ogata, CHEM. ABS., 81, 13110 (1974).
33. G. H. Senkler, J. Stackhouse, B. E. Maryanoff, and K. Mislow, J. AMER. CHEM. SOC., 96, 5648 (1974).
34. K. J. Klabunde, C. M. White, and H. F. Efner, INORG. CHEM., 13, 1778 (1974).
35. D. L. Klayman and W.H.H. Günther, ed., ORGANIC SELENIUM COMPOUNDS, Wiley-Interscience, New York, 1973.
36. R. A. Zingaro and W. C. Cooper, ed., SELENIUM, Van Nostrand-Reinhold, New York, 1974.
37. K. Seppelt, Z. ANORG. ALLGEM. CHEM., 406, 287 (1974).
38. A. Jokuziene and I. V. Yanitskii, CHEM. ABS., 81, 9076 (1974).
39. R. C. Paul, R. D Sharma, and K. C. Malhotra, IND. J. CHEM., 12, 320 (1974).
40. K. J. Wynne and J. Golen, INORG. CHEM., 13, 185 (1974).
41. D. Breitinger and W. Morrell, INORG. NUCL. CHEM. LETT., 10, 409 (1974).
42. V. Horn and R. Paetzold, Z. ANORG. ALLGEM. CHEM., 404, 213 (1974).
43. D. G. Dennis and G. H. Schmid, CAN. J. CHEM., 52, 1027 (1974).
44. K. J. Irgolic, ORGANIC CHEMISTRY OF TELLURIUM, Gordon and Breach, New York, 1974.
45. J. S. Thayer, J. ORGANOMETAL. CHEM., 76, 265 (1974).
46. H. Selig, S. Sarig, and A. Abramowitz, INORG. CHEM. 13, 1508 (1974).
47. C. Lau and J. Passmore, INORG. CHEM., 13, 2278 (1974).
48. R. Kniep, A. Rabenau, and H. Rau, J. LESS COMM. METALS, 35, 325 (1974).
49. N. Petragnani, L. Torres, K. J. Wynne, and D. J Williams, J. ORGANOMETAL. CHEM., 76, 241 (1974).
50. J. S. Thayer and K. V Smith, SYNTH. INORG. METALORG. CHEM., 3, 101 (1973).
51. K. V. Smith ad J. S. Thayer, INORG. CHEM., 13, 3021 (1974).

HALOGENS AND PSEUDOHALOGENS

M. J. Steffel

Department of Chemistry
The Ohio State University Marion Campus
Marion, Ohio 43302

HALOGENS

Interhalogen Compounds and Ions. The hexafluorobromine(VII) cation results from oxidation of BrF_5 by $Kr_2F_3^+$ or KrF^+ (1). The new compounds $BrF_6^+AsF_6^-$ and $BrF_6^+Sb_2F_{11}^-$ form in BrF_5 solutions of $Kr_2F_3^+AsF_6^-$, $KrF^+AsF_6^-$, or $Kr_2F_3^+SbF_6^-$ in yields not exceeding 20%. Attempts to prepare BrF_7 were unsuccessful.

Octafluoroperiodate salts have been reported (2). Reaction of IF_7 and nitrosyl fluoride produces $NO^+IF_8^-$. If cesium fluoride is added to the reaction mixture $CsIF_8$ forms.

Use of $NaICl_2$ as a mild iodinating agent has been recommended (3). Reaction periods can be cut to half the time required for iodination by ICl.

A description of syntheses of interhalogen fluorosulfates IX_2SO_3F and I_2XSO_3F (X = Cl, Br) by oxidative addition of a halogen or interhalogen to iodine(I) fluorosulfate has been published (4). These compounds contain the interhalogen cations IX_2^+ and I_2X^+. This is apparently the first isolation of a compound containing I_2Br^+.

Pure BrCl can be prepared by the irradiation of a CCl_2F_2 solution of bromine and excess chlorine at -79° (5). Probably there is *in situ* formation of BrCl when mixtures of $SbCl_5$ and bromine or lithium bromide, and of $SbCl_3$ and bromine are used for chlorobromination (6).

Oxides and Oxo Compounds. A new bromine oxide, Br_2O_3, can be prepared by the controlled thermal decomposition of Br_2O_4 *in vacuo* (7).

There is infrared spectroscopic evidence for the existence of nitrogen hypofluorite (8). This, as well as its more stable isomer nitrosyl fluoride, formed when fluorine and nitric oxide were codeposited onto a cesium iodide window at $8^{O}K$.

A systematic study of adducts of chlorine oxide trifluoride with pentafluorides of Group VA and VB elements has been reported (9). These are $ClOF_2^+MF_6^-$ (M = P, As, Sb, Bi, V, Nb, Ta).

The reactions of chloryl fluorosulfate with AsF_5 and with SbF_5 produce $ClO_2SO_3F \cdot AsF_5$ and $ClO_2Sb_2F_{11}$ (10). Infrared and Raman spectra support the formulation of these as ionic compounds containing ClO_2^+.

The new bis(perchlorato)bromate(I) anion has been isolated in $Cs[Br(OClO_3)_2]$ (11). This compound forms slowly when a mixture of cesium bromide and $ClOClO_3$ is kept at -45°.

An examination of reactions of ozone with covalent hypohalites showed that oxidative oxygenation of the terminal halogen occurs (12). Interaction of $BrOClO_3$ and ozone in CF_3Cl at -45° produced the new compound $O_2BrOClO_3$.

The simplest perfluoroalkyl perchlorate, trifluoromethyl perchlorate, has been prepared by the action of chlorine perchlorate on trifluoromethyl iodide (13). Iodine perchlorate is a postulated by-product, equation (1).

$$CF_3I + 2ClOClO_3 \longrightarrow CF_3OClO_3 + Cl_2 + IClO_4 \qquad (1).$$

A structural study of CF_3OClO_3 has been published (14).

Adducts, $R_4NClO_4 \cdot BCl_3$, form when R_4NClO_4 (R = CH_3, C_2H_5, C_4H_9) and BCl_3 interact in chloroform (15). On the basis of infrared spectra these are formulated as perchloratoborates, $[R_4N][Cl_3BOClO_3]$.

Fluorine. Recent reviews describe fluorinated peroxides (16) and fluorosulfuric acid and its salts and derivatives (17).

In an investigation of complex fluoride salts (18) it was learned that some salts previously thought to contain $H_3F_4^-$ (19) were probably hexafluorosilicates. This work led to synthesis of HF_2^- salts, anilinium hydrogen difluoride and *p*-toluidinium hydrogen difluoride, in greater than 99% yield by interaction of liquid hydrogen fluoride and the base at room temperature.

Poly(carbon monofluoride) produced by a flow method with careful temperature control has a limiting stoichiometry of $CF_{1.12 \pm 0.03}$ (20). There are also new static syntheses of $(CF_x)_n$ (21). Investigation of the structure of this material is underway (22).

The hexafluorobenzene cation can be prepared by oxidation of C_6F_6 using the dioxygenyl cation (23). The C_6F_6 and WF_6 are condensed onto O_2AsF_6, O_2SbF_6, or $O_2Sb_2F_{11}$ at -196° and the mixture is warmed to 0° producing $C_6F_6^+$ salts in at least 40% yield.

A new type of reaction for fluorine fluorosulfate is its interaction with hydrogenated haloalkanes (24), equation (2). With hydrogenated

$$CHCl_3 + F_2SO_3 \xrightarrow{-78^\circ} CHCl_2OSO_2F + ClF \qquad (2)$$

haloalkenes addition occurs, equation (3). Both fluorosulfate

$$CHCl{=}CCl_2 + F_2SO_3 \xrightarrow{-78^\circ} CHClFCCl_2OSO_2F \qquad (3)$$

products shown in these equations are new compounds.

Chlorine fluorosulfate will add to gaseous haloalkenes (25). Some products are $ClCF_2CF_2OSO_2F$, $FCCl_2CFClOSO_2F$, $ClCF_2CFClOSO_2F$, $FCCl_2CF_2OSO_2F$, $CF_3CFClCF_2OSO_2F$, and $(CF_3)_2CClCF_2OSO_2F$.

Selenium tetrafluoride is reported to be a convenient fluorinating agent (26). It is prepared by reacting ClF_3 with selenium.

Transition metal pentafluorides and tetrafluorides can be prepared in high yields by reductive fluoride elimination (27). The reducing agent can be either hydrogen or silicon, equations (4), (5), and (6), in hydrogen

$$2MoF_6 + H_2 \longrightarrow 2MoF_5 + 2HF \qquad (4)$$

$$4MoF_6 + Si \longrightarrow 4MoF_5 + SiF_4 \qquad (5)$$

$$4\,MoF_5 + Si \longrightarrow 4MoF_4 + SiF_4 \qquad (6)$$

fluoride at room temperature. The method is reported to be a general one.

New fluorometal carbonyl derivatives have been reported. Both $[Cr(CO)_5F]^-$ and $[W(CO)_5F]^-$ can be prepared by interaction of $[M_2(CO)_{10}]^{2-}$ and silver fluoride in tetrahydrofuran and by ultraviolet irradiation of a dichloromethane solution containing $M(CO)_6$ and F^- (28). Reactions also occur with the analogous molybdenum compounds but products were not isolated. Attempts at synthesis of $[Ir(CO)(AsPh_3)_2F]$ produced both the desired compound and $[Ir(AsPh_3)_3F]$ (29).

Bromine. Controlled pyrolysis of thiotriazyl bromide produces thiazyl bromide (30), equation (7).

$$2S_4N_3Br \longrightarrow S_4N_4 + 2BrSN + 2S \qquad (7)$$

Iodine. Methoxoiodine(V) fluorides are the products of reactions between IF_5 and $(CH_3)_2Si(OCH_3)_2$ (31). These are $IF_{5-n}(OCH_3)n$ (n = 1 to 4). Also, $I(OCH_3)_5$ was thought to be a product but was too unstable for analyses to be obtained.

Methyliodine(II) fluoride, CH_3IF, has been observed, but not isolated, as the product in the crossed molecular beam reaction between fluorine and methyl iodide (32).

Synthesis of trifluoromethyliodine(III) difluoride, CF_3IF_2, is effected by interaction of CF_3I and fluorine at -78° (33).

Trifluoromethyliodine(V) tetrafluoride, CF_3IF_4, is the main product of the reaction between CF_3I and ClF_3 in perfluoro-n-hexane at -78° (34). There is evidence that CF_3IF_2 also forms. Interactions of CF_3IF_4 and methylmethoxosilanes produce $CF_3IF_{4-n}(OCH_3)_n$ (n = 1 to 4).

Pentafluorophenyliodine(V) tetrafluoride, $C_6F_5IF_4$, results when C_6F_5I is oxidized by ClF_3 below room temperature (35).

Products of oxidations of other iodine(I) compounds by ClF_3 (36) are shown in (8) and (9).

$$F_5SCF_2CF_2I + ClF_3 \longrightarrow F_5SCF_2CF_2IF_2 \text{ and } F_5SCF_2CF_2IF_4 \qquad (8)$$

$$F_5S(CF_2)_4I + ClF_3 \longrightarrow F_5S(CF_2)_4IF_4 \qquad (9)$$

Iodine tris(trifluoromethanesulfonate), $(CF_3SO_3)_3I$, can be prepared by reaction of $(CF_3CO_2)_3I$ and CF_3SO_3H (37).

PSEUDOHALOGEN COMPOUNDS OF MAIN GROUP ELEMENTS

Group II. New thiocyanate compounds are $Be(CH_3CN)_2(NCS)Cl$ and $Be(CH_3CN)_2(NCS)_2$ (38). These weak electrolytes (1:1) are prepared by reaction of $BeCl_2 \cdot 2CH_3CN$ with thiocyanate ion in acetonitrile.

Beryllium chloride reacts with liquid chlorocyanogen at 0° forming $BeCl_2 \cdot 2NCCl$ (39). Chlorocyanogen also forms complexes with other divalent metals: $M(NCCl)_6X_2$ (M = Mg, Co, Ni; X = $SbCl_6^-$, $InCl_4^-$) (40).

Group III. Eight new pseudohalo complexes containing thallium(III),

L_2TlX (L = indenyl, cyclopentadienyl; X = N_3, CN, NCO, NCS), have been prepared from the analogous chloro complexes (41).

Group IV. Chlorocyanogen is cleaved in its gas phase reaction with $HSiCl_3$ in the presence of triethylamine; Cl_3SiCN results (42).

The new silanes $HSi(NCO)_3$ and $HSi(NCS)_3$ form in the reaction of a trihalosilane with the appropriate silver pseudohalide (43). During the reaction $Si(NCX)_4$ (X = O, S) also forms. Mixed halo/pseudohalo silanes can be prepared by exchange reactions.

There are new azido and mixed azido/thiocyanato complexes of tin(IV) (44) and lead(IV) (45): $[(C_6H_5)_3M(N_3)_2]^-$ and $[(C_6H_5)_3M(N_3)(NCS)]^-$ for Sn and Pb, $[(CH_3)_2Sn(N_3)_4]^{2-}$, $[(C_6H_5)_2Sn(N_3)_2(NCS)_2]^{2-}$, $[(C_6H_5)_2Pb(N_3)_4]^{2-}$, and $[(C_6H_5)_2Pb(N_3)(NCS)_2]^-$. All were isolated as tetramethylammonium or tetraphenylarsonium salts.

Group V. Pure $[R_3Sb(N_3)_2]$ and $[R_3Sb(NCO)_2]$ (R = CH_3, C_6H_5), uncontaminated with hydrolysis products, have been prepared (46). The azides form in a benzene solution of R_3SbCl_2, hydrazoic acid, and sodium azide. Interaction of R_3SbCl_2 and silver cyanate in ether produces the isocyanates.

New 6-coordinate organoantimony(V) complexes are $[(C_6H_5)_2SbCl_3X]^-$ (X = N_3, NCS) (47). They were prepared from $(C_6H_5)_2SbCl_3 \cdot 2H_2O$ and were isolated as tetramethylammonium or tetraphenylarsonium salts.

The bismuth(III) complexes $M[BiL(SCN)_5]$ (M = Mg, Ca, Sr, Zn; L = thiorea) are prepared from bismuthyl carbonate, thiocyanic acid, thiourea, and the metal carbonate (48). Both thiocyanate and thiourea are coordinated through sulfur.

Group VI. A new reaction, that of trifluoromethanesulfonamide with chlorosulfonylisocyanate, produces CF_3SO_2NCO (49). Phosphorus pentasulfide reacts with this isocyanate converting it to the new compound CF_3SO_2NCS. Reaction with phosphorus pentachloride converts CF_3SO_2NCO to $CF_3SO_2NCCl_2$.

Selenocyanate ion reacts with thionyl chloride and with bromine (1:1 molar ratio) forming dihalocyanoselenate(II) ions (50). Compounds that have been isolated are $[(C_3H_7)_4N][SeBr_2CN]$ and $[R_4N][SeCl_2CN]$ (R = CH_3, C_3H_7).

PSEUDOHALO COMPLEXES OF TRANSITION METALS

Coverage is not exhaustive. Studies have been included if they describe several complexes or if the bonding of a pseudohalide is of particular interest.

A review describes preparation, structures, spectra, and chemical properties of mixed phosphine cyano complexes (51).

Titanium, Zirconium, Hafnium. By varying the amount of ammonium thiocyanate that is allowed to react with $TiCl_4$ in tetrahydrofuran or acetonitrile, the complexes $[TiL_2Cl_{4-n}(NCS)_n]$ (n = 1, 2, 3; L = THF or CH_3CN) can be isolated (52).

It is reported that cyanate is oxygen-bonded in $[ML(OCN)_4]$ (M = Zr, Hf; L = 2, 2'-bipyridine or 1, 10-phenanthroline), $[ML_2(OCN)_4]$ (M = Zr, Hf; L = pyridine or dimethylsulfoxide), and $Zr(OH)(OCN)_3 \cdot$dioxane (53).

Niobium and Tantalum. The complexes $[MCl_4(N_3)]_2$ (M = Nb, Ta) are products of reactions of ClN_3 or sodium azide with MCl_5 (54). They are dimeric with the two units linked via α-N atoms of azide. The anionic complex $[NbCl_5(N_3)]^-$ results when a chloroform solution of $NbCl_5$ and tetraethylammonium azide is warmed from -78° to room temperature (55).

Chromium, Molybdenum, Tungsten. Ansolvous substitution reactions of K_3 $[Cr(NCS)_6]$ with tertiary aromatic monoamines gave 39 complexes containing $[CrL_2(NCS)_4]^-$ (56).

Pseudohalo complexes of molybdenum(VI) and tungsten(VI) have been reported (57). The ions $[MO_2(NCS)_4]^{2-}$ precipitate as tetraalkylammonium salts from freshly acidified solutions of ammonium paramolybdate and of $Na_2WO_4 \cdot 2H_2O$ in the presence of a large excess of thiocyanate ion. The neutral complexes Cp_2MOX_2 (M = Mo, W; X = N_3, CN, NCO, SCN) are prepared from the analogous chlorides (58).

Manganese and Rhenium. A new cyanomanganate(III) complex, oxo-bridged $K_7[(CN)_5MnOMn(CN)_5]$ CN, is the final product resulting from reaction of potassium permanganate and potassium cyanide in a saturated aqueous solution (59).

Transition metal complexes of cyanate, thiocyanate, and selenocyanate react with benzoylhydrazine in acetone and ethanol or in water forming $ML_2(NCO)_2$ (M = Mn, Co), $CdL(NCO)_2$, $ML_2(NCS)_2$ (M = Mn, Co), $CdL(NCS)_2$, $ML_3(NCSe)_2$ (M = Mn, Zn), and $ML_2(NCSe)_2$ (M = Co, Ni) (solvation not shown) (60). Similarly, o-, m- , and p-$HOC_6H_4CONHNH_2$ and o-, m-, and p-$CH_3OC_6H_4CONHNH_2$ react in water with $M(NCS)_2$ (M = Mn, Co, Ni, Cd) giving $ML_2(NCS)_2 \cdot H_2O$ (61). Reaction of acetylhydrazine with metal pseudohalides produces $ML_2(NCS)_2 \cdot H_2O$ (M = Mn, Fe, Co, Ni, Zn, Cd) and $ML_2(NCSe)_2 \cdot H_2O$ (M = Mn, Co, Ni, Zn, Cd) (62).

Hexamethylphosphoramide complexes $[ML_2(NCS)_2]$ of Mn, Co, Zn or Cd contain M-NCS groups whereas there is bridging thiocyanate in the analogous Ni complex (63).

By varying the metal pseudohalide/ligand ratio a series of complexes containing pyridine N-oxide (P) or quinoline N-oxide (Q) was prepared (64). These are shown in Table 1 (solvation not shown).

Table I

Some Pseudohalo Complexes

Complex	Metal	Ligand	X
ML_4X_2	Co	P	S
ML_3X_2	Mn	P	S, Se
		Q	Se
	Co	P, Q	S, Se
	Ni	P	S, Se
	Zn	P	S, Se
ML_2X_2	Mn	Q	S
	Co	P	S, Se
	Ni	P	S, Se
		Q	S, Se
MLX_2	Co	P	S

Other new manganese(II) complexes are $[MnL_2(NCSe)_2]$ (L = pyridine; β-, γ- picoline; 3, 5-lutidine; 4-ethylpyridine; isoquinoline) (65).

Reaction of $[Re(CO)_2(NO)(PPh_3)_2]$ and an aroyl azide, $RCON_3$ (R = C_6H_5, p-$CH_3C_6H_4$), yields $[Re(NCO)(NO)(PPh_3)_2(CONHCOR)]$ (66) in protic solvents. In benzene the product is $[Re(CO)(NO)(PPh_3)_2(RCONCO)]$.

To prepare $[Re(NCS)_6]^{2-}$, ReO_4^- was reduced by tin(II) in hydrochloric acid in the presence of ammonium thiocyanate; it was isolated as its tetraphenylarsonium salt (67).

Iron and Ruthenium. Ferrous ferricyanide can be prepared in aqueous solution by adding a ferrous salt to a ferricyanide compound (68). The most important reaction condition is maintenance of reducing conditions by using sucrose, potassium iodide, or excess $K_4[Fe(CN)_6]$.

In an investigation of the pH and temperature dependence of reactions of $[Fe_2L_4O]^{4-}$ (L = 1, 10-phenanthroline or 2, 2'-bipyridine), it was found that at pH = 1.2-1.5 and at room temperature cyanide ion immediately effects oxo-bridge rupture and reduction of iron(III) to iron(II) giving $[FeL_2(CN)_2]$ (69). In contrast, at pH = 3.8-4.0 and a temperature of 45-50° iron(III) persists and $[FeL_2(CN)_2]CN$ forms.

Interaction of $FeCl_2 \cdot 4H_2O$, a pseudohalide salt, and 2-(2-pyridyl)-imidazole produces cis-$[FeL_2X_2]$ (X = N_3, CN, NCS) (70).

The complexes $[Fe(CO)(NO)L_2X]$ (L = PPh_3 and X = N_3, CN, NCO, SCN, SeCN; L = $PMePh_2$ and X = N_3, SCN) form via interaction of the sodium salt of the pseudohalide and $[Fe(CO)_2(NO)L_2]PF_6$ (71).

Reactions involving $RuCl_3$, ammonium thiocyanate, carbon monoxide, and a tetraalkylammonium chloride give rise to $[R_4N]_2[Ru(CO)_2(NCS)_4]$ (R = CH_3, C_2H_5) and $[(C_2H_5)_4N]_2[Ru_2(CO)_4(NCS)_6]$ (72).

Cobalt, Rhodium, Iridium. Salts of $[CoL_2X_2]^{2-}$ and $[CoLX_3]^{2-}$ (L = $N(CN)_2$; X = OCN, SCN) are prepared from $Co[N(CN)_2]_2$ and the appropriate pseudohalide salt in acetone (73).

There is reported to be N-bonding of the pseudohalogen in trans-$[CoL(DMG)_2(NCX)]$ (L = p-aminobenzenesulfonamide; p-aminobenzenesulfocyanamide; 2-(p-sulfanil)pyridine; X = O, S, Se,) (74, 75). In contrast, when L = nicotinamide, isolation of both trans-$[CoL(DMG)_2(NCS)]$ and trans-$[CoL(DMG)_2(SCN)]$ was reported (76).

When a diastereomeric mixture of $[CpCo(C_3F_7)(CNCHCH_3C_6H_5)I]$ reacts with silver perchlorate and ammonium thiocyanate the four N- and S-bonded diastereomers of $[CpCo(C_3F_7)(CNCHCH_3C_6H_5)(CNS)]$ are formed (77). Separation gives a pure N-bonded complex that is stable in tetrahydrofuran and as a solid but isomerizes in benzene and toluene.

Trans-$[Co(acac)_2py(SCN)]$ precipitates from a mixture of potassium thiocyanate, pyridine, $[Co_2(acac)_4(OH)_2]$ and glacial acetic acid in methanol at 0° (78). In chloroform it undergoes linkage isomerization to trans-$[Co(acac)_2py(NCS)]$. This undergoes geometric isomerization in refluxing chloroform to cis-$[Co(acac)_2py(NCS)]$. Linkage and geometric isomerization also occur in the solid state.

Polymeric complexes, [M(4,4'-bipyridyl)$(SCN)_2$] (M = Co, Ni, Zn, Cd, Hg), result when an excess of the ligand is added to hot solutions of the metal thiocyanate in ethanol (79).

Isolation of both [$Rh(PPh_3)_3(NCO)$] and [$Rh(PPh_3)_3(OCN)$] provides the first example of solid state linkage isomers of cyanate (80). Reaction of [$Rh(PPh_3)_3Cl$] and tetraphenylarsonium cyanate in acetonitrile produces the -NCO form. When ethanol is added to the reaction mixture the -OCN isomer is obtained.

Replacement of SO_3F or SO_3CF_3 from [$Ir(CO)(PPh_3)_2(CH_3)ClX$] (X = SO_3F, SO_3CF_3) provides a synthetic route to complexes that cannot be obtained by oxidative addition (81). Pseudohalo complexes prepared in this manner are [$Ir(CO)(PPh_3)_2(CH_3)ClX$] (X = N_3, NCO, NCS).

Nickel, Palladium, Platinum. Ability of cyanate to bond via oxygen to a transition metal has been verified by an X-ray structural study of [$Ni_2L_2(OCN)_2$] [$B(C_6H_5)_4$]$_2$ (L = 2,2', 2"-triaminotriethylamine) (82). Cyanate groups bridge nickel atoms in an end-to-end fashion. Analogous thiocyanato and selenocyanato complexes have also been studied (83).

Reaction of piperidine (L) and [$Ni(NH_3)_4(NCO)_2$] gives the 5-coordinate complex [$NiL_3(NCO)_2$] (84). In contrast, when piperidine reacts with $Ni(NCX)_2$ (X = S, Se), 6-coordinate [$NiL_2(NCX)_4$] forms. With 2, 6-dimethylpiperidine, 4-coordinate [$NiL_2(NCX)_2$] (X = O, S) results.

Five-coordinate complexes form when $Ni(NCS)(ClO_4)$ or $Ni(CN)_2 \cdot 4H_2O$ reacts with bisdiphenylphosphinomethane (L); these are [$NiL_2(NCS)$] ClO_4 and [$NiL_2(CN)$] CN (85). With nickel thiocyanate [$NiL_2(NCS)_2$] forms; in dichloromethane this dimerizes to [$NiL(NCS)_2$]$_2$.

By varying the amount of imidazole (L) allowed to react with nickel thiocyanate [NiL_6] $(NCS)_2$, trans- [$NiL_4(NCS)_2$], [$NiL_2(NCS)_2$], and [$NiL(NCS)_2$] can be isolated (86). The latter two complexes apparently have end-to-end bridging of thiocyanate.

It is reported that only steric factors affect the mode of thiocyanate attachment in the series [$PdL(SCN)_2$] (L = $Ph_2PCH_2PPh_2$) (87), [(PdL'(NCS)(SCN)] (L' = $Ph_2PCH_2CH_2PPh_2$) (88), and [PdL"$(NCS)_2$] (L" = $Ph_2PCH_2CH_2CH_2PPh_2$) (87). However, other workers claim this conclusion is overstated (89).

Isomerization of [PdL(CNS)] [$B(C_6H_5)_4$] (L = 4-methyl-1,1, 7, 7, -tetraethyldiethylenetriamine) is also attributed to steric factors (90). In solution there is conversion from Pd-SCN to Pd-NCS; this reverses in the solid state. In contrast, neither $[PtL(SCN)][B(C_6H_5)_4]$ nor $[PdL'(NCS)][B(C_6H_5)_4]$ (L' = tripyridine) isomerizes.

Apparently the nature of other ligands also influences thiocyanate bonding in $[PdL_2(SCN)_2]$ where L = $(EtO)_3P$ and in $[PdL_2(NCS)_2]$ where L = $(i\text{-}PrO)_3P$ (91).

Thiocyanate displacement of chloride from cis- $[Pt(Ph_2PC{\equiv}CPR_2)_2Cl_2]$ (R = C_6H_5, C_2H_5, $\underline{i}\text{-}C_3H_7$, $\underline{t}\text{-}C_4H_9$) produces the mixed thiocyanato/isothiocyanato complexes cis-$[Pt(Ph_2PC{\equiv}CPR_2)_2(NCS)(SCN)]$ (92).

A new azido platinum complex, $[Pt(CH_3)_3(N_3)]$, is prepared by reaction of sodium azide with $[Pt(CH_3)_3]_2SO_4$ (from $[Pt(CH_3)_3Cl]$ and silver sulfate) (93). It is reported to be the first trimethylplatinum compound with a covalent Pt-N bond. Although the complex is monomeric in benzene the base peak in the mass spectrum indicates a tetramer.

Addition of nitromethane to $[Pt(PPh_3)_4]$ in a protic solvent is reported to be a safe method for preparing trans-$[Pt(PPh_3)_2(CNO)_2]$ (94). This fulminato complex rearranges to the isomeric isocyanato complex cis-$[Pt(PPh_3)_2(NCO)_2]$ in the presence of organic carbonyl compounds (95). Also, trans-$[Pt(PPh_3)_2(CNO)_2]$ reacts with organic thiocarbonyl compounds giving cis-$[Pt(PPh_3)_2(NCS)_2]$ and is reduced by phosphines to trans-$[Pt(PPh_3)_2(CN)_2]$ (95). Reaction of $[Pt(PPh_3)_4]$ and formoldoxime hydrochloride produces trans-$[Pt(PPh_3)_2(CN)Cl]$ (94).

Copper and Silver. Several new cyano complexes of copper(II) have recently been synthesized and characterized (96, 97, 98).

Interaction of copper nitrate, an amine, and NH_4NCX (X = O, S) produces $[CuL_2(NCO)_2]$ and $[CuL_2(SCN)_2]$ (L = propylenediamine, diethylenediamine) (99). In contrast, selenocyanate gives 6-coordinate complexes, $[(CuL_2(SeCN)]NO_3$, in which there are SeCN bridges.

Synthesis of $KCu(SeCN)_3$ from copper nitrate and potassium selenocyanate has been reported (100).

The mixed complexes $K[Ag(CN)(SCN)]$ and $K_2[Ag(CN)(SCN)_2]$ have been isolated (101).

REFERENCES

1. R. J. Gillespie and G. J. Schrobilgen, INORG. CHEM., 13, 1230 (1974).
2. C. J. Adams, INORG. NUCL. CHEM. LETT., 10, 831 (1974).
3. J. Hebky, Czech. Patent 151719 (1974); CHEM. ABSTR., 80, 133003b (1974).
4. W. W. Wilson and F. Aubke, INORG. CHEM., 13, 326 (1974).
5. M. Schmeisser and K. H. Tytko, Z. ANORG. ALLG. CHEM., 403, 231 (1974).
6. S. Uemura, A. Onoe, and M. Okano, BULL. CHEM. SOC. JAP., 47, 143 (1974).
7. J.-L. Pascal, A. C. Pavia, J. Potier, and A. Potier, C. R. ACAD. SCI., SER. C, 279, 43 (1974).
8. R. R. Smardzewski and W. B. Fox, J. AMER. CHEM. SOC., 96, 304 (1974).
9. R. Bougon, T. Bui Huy, A. Cadet, P. Charpin, and R. Rousson, INORG. CHEM., 13, 690 (1974).
10. P. A. Yeats and F. Aubke, J. FLUORINE CHEM., 4, 243 (1974).
11. K. O. Christe and C. J. Schack, INORG. CHEM., 13, 1452 (1974).
12. C. J. Schack and K. O. Christe, INORG. CHEM., 13, 2378 (1974).
13. C. J. Schack, D. Pilipovich, and K. O. Christe, INORG. NUCL. CHEM. LETT., 10, 449 (1974).
14. C. J. Schack and K. O. Christe, INORG. CHEM., 13, 2374 (1974).
15. K. V. Titova and V. Ya. Rosolovskii, IZV. AKAD. NAUK SSSR, SER. KHIM., 2174 (1974); CHEM. ABSTR., 82, 25309p (1975).
16. R. A. De Marco and J. M. Shreeve, ADVAN. INORG. CHEM. RADIO-CHEM., 16, 109 (1974).
17. A. W. Jache, ADVAN. INORG. CHEM. RADIOCHEM., 16, 177 (1974).
18. K. M. Harmon, S. L. Madeira, and R. W. Carling, INORG. CHEM., 13, 1260 (1974)
19. J. F. T. Berliner and R. M. Hann, J. PHYS. CHEM., 32, 1142 (1928).
20. R. J. Lagow, R. B. Badachhape, J. L. Wood, and J. L. Margrave, J. AMER. CHEM. SOC., 96, 2628 (1974).
21. R. J. Lagow, R. B. Badachhape, J. L. Wood, and J. L. Margrave, J. CHEM. SOC., DALTON TRANS., 1268 (1974).
22. L. B. Ebert, J. I. Brauman, and R. A. Huggins, J. AMER. CHEM. SOC., 96, 7841 (1974).
23. T. J. Richardson and N. Bartlett, J. CHEM. SOC., CHEM. COMMUN., 427 (1974).
24. L. F. R. Cafferata and J. E. Sicre, INORG. CHEM., 13, 242 (1974).
25. A. V. Fokin, Yu. N. Studnev, L. D. Kuznetsova, and V. L. Rud, IZV. AKAD. NAUK SSSR, SER. KHIM., 471 (1974); CHEM. ABSTR., 81, 24970q (1974).

26. G. A. Olah, M. Nojima, and I. Kerekes, J. AMER. CHEM. SOC., 96, 925 (1974).
27. R. T. Paine and L. B. Asprey, INORG. CHEM., 13, 1529 (1974).
28. W. Douglas and J. K. Ruff, J. ORGANOMETAL. CHEM., 65, 65 (1974).
29. C. A. McAuliffe and R. Pollock, J. ORGANOMETAL. CHEM., 69, C13 (1974).
30. S. C. Peake and A. J. Downs, J. CHEM. SOC., DALTON TRANS., 859, (1974).
31. G. Oates, J. M. Winfield, and O. R. Chambers, J. CHEM. SOC., DALTON TRANS., 1380 (1974).
32. J. M. Farrar and Y. T. Lee, J. AMER. CHEM. SOC., 96, 7570 (1974).
33. J. Baumanns, L. Deneken, D. Naumann, and M. Schmeisser, J. FLUORINE CHEM., 3, 323 (1973).
34. G. Oates and J. M. Winfield, J. CHEM. SOC., DALTON TRANS., 119 (1974).
35. J. A. Berry, G. Oates, and J. M. Winfield, J. CHEM. SOC., DALTON TRANS., 509 (1974).
36. G. Oates and J. M. Winfield, J. FLUORINE CHEM., 4, 235 (1974).
37. M. Schmeisser, P. Sartori, and B. Lippsmeier, Z. NATURFORSCH., TEIL B, 28, 573 (1973).
38. H. Boehland and W. Hanay, Z. CHEM., 14, 286 (1974).
39. J. MacCordick, C. R. ACAD. SCI., SER. C, 278, 1177 (1974).
40. A. P. Zuur, P. L. A. Eversteijn, B. T. de Haas, A. Willemsen, and W. L. Groeneveld, INORG. NUCL. CHEM. LETT., 10, 213 (1974).
41. N. Kumar, B. L. Kalsotra, and R. K. Multani, J. INORG. NUCL. CHEM., 36, 1157 (1974).
42. V. G. Noskov, L. N. Kalinina, A. A. Kirpichnikova, and M. A. Englin, ZH. OBSHCH. KHIM., 43, 2419 (1973); CHEM. ABSTR., 80, 59993f (1974).
43. F. Hoefler, G. Jaegerhuber, and W. Veigl, MONATSH. CHEM., 105, 539 (1974).
44. R. Barbieri, N. Bertazzi, C. Tomarchio, and R. H. Herber, J. ORGANOMETAL. CHEM., 84, 39 (1975).
45. N. Bertazzi, G. Alonzo, G. C. Stocco, and L. Pellerito, J. ORGANOMETAL. CHEM., 80, 225 (1974).
46. R. G. Goel and D. R. Ridley, INORG. CHEM., 13, 1252 (1974).
47. N. Bertazzi, L. Pellerito, and G. C. Stocco, INORG. NUCL. CHEM. LETT., 10, 855 (1974).
48. H. Ladzińska-Kulińska, ROCZ. CHEM., 48, 333 (1974).
49. E. Behrend and A. Haas, J. FLUORINE CHEM., 4, 83 (1974).
50. K. J. Wynne and J. Golen, INORG. CHEM., 13, 185 (1974).
51. P. Rigo and A. Turco, COORD. CHEM. REV., 13, 133 (1974).
52. H. Boehland and H. Steinecke, Z. CHEM., 14, 65 (1974).
53. A. M. Golub, T. P. Lishko, and V. P. Chistoplyasova, ZH. NEORG. KHIM., 19, 1509 (1974).

54. J. Straehle, Z. ANORG. ALLG. CHEM., 405, 139 (1974).
55. M. Kasper and R. D. Bereman, INORG. NUCL. CHEM. LETT., 10, 443 (1974).
56. I. Gănescu, D. Oprescu, and Cs. Várhelyi, MONATSH. CHEM., 105, 525 (1974).
57. B. J. Brisdon and D. A. Edwards, INORG. NUCL. CHEM., LETT., 10, 301 (1974).
58. S. P. Anand, J. INORG. NUCL. CHEM., 36, 925 (1974).
59. R. F. Ziolo, R. H. Stanford, G. R. Rossman, and H. B. Gray, J. AMER. CHEM. SOC., 96, 7910 (1974).
60. G. V. Tsintsadze, P. V. Gogorishvili, L. K. Nagornaya, and T. N. Tsivtsivadze, TR. GRUS. POLITEKH. INST., No. 5 (1972); REF. ZH., KHIM. 1973, Abstr. No. 14V228; CHEM. ABSTR., 80, 66253t (1974).
61. Yu. Ya. Kharitonov, R. I. Machkhoshvili, N. B. Generalova, and R. N. Shchelokov, ZH. NEORG. KHIM., 19, 1124 (1974).
62. G. V. Tsintsadze, P. V. Gogorishvili, L. K. Nagornaya, T. N. Tsivtsivadze, and R. I. Machkhoshvili, ISSLED. OBL. KHIM. KOMPLEKSN. PROSTYKH SOEDIN. NEK. PEREKHODNYKH REDK. MET., 2, 194 (1974); CHEM. ABSTR., 82, 25295f (1975).
63. V. V. Skopenko, A. S. Tryashin, and V. Ya. Zub, UKR. KHIM. ZH. (RUSS. ED.), 40, 792 (1974); CHEM. ABSTR., 81, 130259z (1974).
64. G. B. Aitken and G. P. McQuillan, J. CHEM. SOC., DALTON TRANS., 2637 (1973).
65. A. K. Das and D. V. Ramana Rao, INDIAN J. CHEM., 12, 898 (1974).
66. G. La Monica. S. Cenini, and M. Freni, J. ORGANOMETAL. CHEM., 76, 355 (1974).
67. T. Okubo and H. Koshimua, NIPPON KAGAKU KAISHI, 1649 (1974); CHEM. ABSTR., 81, 144882j (1974).
68. R. Robinette and R. L. Collins, J. COORD. CHEM., 3, 333 (1974).
69. B. J. C. Lima, F. David, and P. G. David, J. CHEM. SOC., DALTON TRANS., 680 (1974).
70. Y. Sasaki and T. Shigematsu, BULL. CHEM. SOC. JAP., 47, 109 (1974).
71. G. Dolcetti, L. Busetto, and A. Palazzi, INORG. CHEM., 13, 222 (1974).
72. A. R. M. Abbadi, M. A. Shehadeh, and J. V. Kingston, J. INORG. NUCL. CHEM., 36, 1173 (1974).
73. H. Koehler, T. P. Lischko, H. Hartung, and A. M. Golub, Z. ANORG. ALLG. CHEM., 403, 35 (1973).
74. V. N. Shafranskii and I. L. Fusu, ZH. NEORG. KHIM., 19, 776 (1974).
75. V. N. Shafranskii and I. L. Fusu, ZH. NEORG. KHIM., 19, 2195 (1974).
76. D. G. Batyr, M. P. Starysh, V. N. Shafranskii, and Yu. Ya. Kharitonov, ZH. NEORG. KHIM., 19, 2779 (1974).
77. H. Brunner and W. Rambold, Z. NATURFORSCH., TEIL B, 29, 367 (1974).

78. L. J. Boucher, D. R. Herrington, and C. G. Coe, INORG. CHEM., 13, 2290 (1974).
79. I. S. Ahuja and R. Singh, INDIAN J. CHEM., 12, 107 (1974).
80. S. J. Anderson and A. H. Norbury, J. CHEM. SOC., CHEM. COMMUN., 37 (1974).
81. D. Strope and D. F. Shriver, INORG. CHEM., 13, 2652 (1974).
82. D. M. Duggan and D. N. Hendrickson, INORG. CHEM., 13, 2056 (1974).
83. D. M. Duggan and D. N. Hendrickson, INORG. CHEM., 13, 2929 (1974).
84. R. Eggli and W. Ludwig, INORG. CHIM. ACTA, 7, 697 (1973).
85. K. K. Chow and C. A. McAuliffe, INORG. CHIM. ACTA, 10, 197 (1974).
86. J. C. Jansen and J. Reedijk, Z. NATURFORSCH., TEIL B, 29, 527 (1974).
87. G. J. Palenik, W. L. Steffen, M. Mathew, M. Li, and D. W. Meek, INORG. NUCL. CHEM. LETT., 10, 125 (1974).
88. G. Beran and G. J. Palenik, J. CHEM. SOC., CHEM. COMMUN., 1354 (1970).
89. J. E. Huheey and S. O. Grim, INORG. NUCL. CHEM. LETT., 10, 973 (1974).
90. J. L. Burmeister, R. L. Hassel, K. A. Johnson, and J. C. Lim, INORG. CHIM. ACTA, 9, 23 (1974).
91. L. F. Gogolyukhina, A. D. Troitskaya, and G. A. Levshina, ZH. OBSHCH. KHIM., 44, 223 (1974).
92. Y. S. Wong, S. Jacobson, P. C. Chieh, and A. J. Carty, INORG. CHEM., 13, 284 (1974).
93. K.-H. Von Dahlen and J. Lorberth, J. ORGANOMETAL. CHEM., 65, 267 (1974).
94. K. Schorpp and W. Beck, CHEM. BER., 107, 1371 (1974).
95. W. Beck, K. Schorpp, and C. Oetker, CHEM. BER., 107, 1380 (1974).
96. M. Wicholas and T. Wolford, INORG. CHEM., 13, 316 (1974).
97. D. M. Duggan, R. G. Jungst, K. R. Mann, G. D. Stucky, and D. N. Hendrickson, J. AMER. CHEM. SOC., 96, 3443 (1974).
98. D. M. Duggan and D. N. Hendrickson, INORG. CHEM., 13, 1911 (1974).
99. R. N. Dash and D. V. Ramana Rao, J. INDIAN CHEM. SOC., 51, 211 (1974).
100. E. Söderbäck, ACTA CHEM. SCAND., 28, 116 (1974).
101. A. N. Sergeeva, D. I. Semenishin, and A. V. Mazepa, ZH. NEORG. KHIM., 18, 2956 (1973).

SCANDIUM, YTTRIUM, LANTHANIDES AND ACTINIDES

David R. Lorenz and John R. Wasson

Department of Chemistry
University of Kentucky
Lexington, Kentucky 40506

SCANDIUM

Pale pink trichlorotris(tetrahydrofuran)scandium(III) can be obtained by heating anhydrous scandium chloride with excess tetrahydrofuran (1). Hydrated scandium nitrate, prepared by evaporation of a solution of Sc_2O_3 and dilute nitric acid, reacts with a methanol solution of tropolone (trop) to yield pale yellow HSc(trop)$_4$ and red-brown Sc(trop)$_3$ (2). The interaction of scandium chloride and tropolone in ethanol can be employed to prepare the acid dimer [HSc(trop)$_4$]$_2$ (3,4). Scandium(III) complexes with acetylacetonate, tropolonate, oxalate, pyridine-2,6-dicarboxylic acid and the mixed oxalate phenanthroline complex have been prepared (5) and their X-ray photoelectron spectra recorded. The preparation of a number of basic scandium sulfates has been described (6). The preparations of basic and acidic scandium selenites have been described (7). Scandium-45 nuclear magnetic resonance has been employed (8) to characterize aqueous solutions of scandium salts as well as ScF_6^{3-}.

YTTRIUM AND LANTHANIDES

Values of E°[M^{3+}/M^{2+}] for the lanthanides have been estimated (9). Lanthanum nickel hydride, $LaNi_{5.3}H_6$, can be obtained (10) by exposing the intermetallic compound to hydrogen. Brown tricyclopentadienyl cerium(IV) hydride and amide can be obtained (11) from the reaction of tricyclopentadienyl cerium(IV) chloride with sodium hydride and sodamide, respectively, in tetrahydrofuran. Carbonyl compounds of Pr, Gd and Ho can be prepared by codeposition of the metal atoms with carbon monoxide in argon matrices (12).

By reaction of lanthanum and gadolinium metals with $NaNH_2$ and NH_3 in a high temperature autoclave the compounds $Na_3[La(NH_2)_6]$, $Na_3[Gd(NH_2)_6]$ and $Na[Gd(NH_2)_4]$ have been prepared (13). Single crystals of $K_3La(NH_2)_6$ have been prepared by the reaction of potassium and lanthanum metals at 200°C and 4000 atm. NH_3 pressure for 5 days (15). The reaction of hydrated rare earth oxalates with hydrazine vapor produces unstable $Ln_2(C_2O_4)_3 \cdot 4N_2H_4(1\text{-}2)H_2O$ (Ln = Tb, Dy, Ho, Er or Y) and $Ln_2(C_2O_4)_3 \cdot 3N_2H_4(2\text{-}4)H_2O$ (Ln = Tm, Yb, or Lu) (15). Erbium and ytterbium complexes with 1,2-cyclohexanediamine - N,N,N',N'-tetraacetic acid can be prepared from the metal nitrates. The rupture of metal-ligand bonds due to β decay has been studied (16). Lanthanide metals form complexes with acetoacetanilide (HAAA) with the formula $[Ln(AAA)_2(OH)H_2O)]$, Ln = lanthanide metal, which are reported (17) to be unstable in solution, but have been characterized in the solid state. The interaction of amino acids and lanthanide ions in aqueous solution has been investigated (18).

Rare earth carbonates can be treated with L-aspartic acid in stoichiometric amounts to yield the tris-chelates with the general formula $Ln(Asp\text{-}H)_3 \cdot 3H_2O$ in aqueous solutions (19) which may then be evaporated to isolate the complexes. Complex formation in water-urea-Gd, Dy and Yt sulfate systems has been characterized (20). Stability constants have been reported for La, Eu, Gd, Tb and Lu complexes with benzoic and m- and p-hydroxybenzoic acids but solids were not isolated (21). The reduction of Ce(IV) with malic acid and the first hydrolysis constant of Ce(IV) have been described (22). Aspartic acid chelates of Pr(III) and Gd(III) have been investigated in solution (23) but solids were not isolated. The extraction of Nd and Er into chloroform with 1,10-phenanthroline and 2,3-hydroxynaphthoic acid has been described (24). Citrates of Nd, La and Lu have been characterized by PMR and electronic spectroscopy but solids were not isolated (25). Substituted phenylacetic acids react with ceric ammonium nitrate in aqueous acetonitrile to yield benzyl radicals as intermediates (26). Stability constants for La, Th and Sc complexes with hexamethylenediaminetetraacetic acid and iminodiacetic acid have been described (27) but solids were not isolated. Aqueous solutions of $[M(EDTA)]^-$ (M = La, Pr, Eu, Gd) complexes, prepared by mixing stoichiometric quantities of MCl_3 and EDTA solutions, have been employed as NMR shift reagents for the study of conformations of adenosine 5'-monophosphate and cytidine 5'-monophosphate in solution (28). Complexes (1:1:1) of EDTA, benzoylacetone and Nd, Ho and Er have been characterized in solution but solids were not isolated (29). Formation constants for yttrium and all of the rare earth thenoyltrifluoroacetonates, $R(TTA)_3$, have been reported (30). Mono-, bis- and tris-benzoylacetonato complexes of Eu(III) have been characterized in aqueous methanol solutions but solids were not isolated (31). The syntheses of several chiral Eu(III) and Pr(III) tris-β-diketonate NMR shift reagents have been described. Sodium methoxide and the β-diketone in methanol can be added to a methanol solution of the hydrated metal chloride in order to prepare the β-diketonate chelate. Distillation, sublimation and recrystallization can be employed to purify the metal chelate (32). Hydrated tris(2,4-pentanedionato)europium(III), when refluxed with tetraphenylporphine (H_2TPP) in 1,2,4-trichlorobenzene, yields Eu(TTP)(acac). The reaction is a general one for lanthanide elements and can be employed to prepare NMR shift reagents (33).

Bivalent holmium has been prepared by gamma irradiation of Ho_2O_3. This has been verified by polarography, chronopotentiometry and titrations (34). Pure monoclinic crystals of TbOOH (terbium oxide hydroxide) can be prepared (35) by heating Tb_4O_7 in platinum boats in a quartz tube with hydrogen at 800°C for 20 hours. Europium tungsten bronzes can be obtained by heating Eu_2O_3-WO_3-W or WO_3-Eu mixtures in sealed containers

(36). Crystals of the series $Y_{3-c}Pr_cScFe_4O_{12}$ can be grown from a flux of the oxides (37). The growth of laser quality neodymium pentaphosphate, NdP_5O_{14}, crystals from mixtures of Nd_2O_3 and H_3PO_4 has been described (38). The Tb^{3+}-activated $In_2Si_2O_7$ phosphor can be prepared (39) by firing a mixture of In_2O_3, SiO_2 and Tb_4O_7 at 1200-1500°C in air. Pr^{3+}-activated CaF_2, YF_3, α-$NaYF_4$ and LaF_3 can be prepared (40) by firing mixtures of the dry fluorides. The formation of $LaVO_3$ involves heating La_2O_3 and V_2O_3 for 24 hours (41). Rare earth hexamolybdates, $Ln_2O_3 \cdot 6MoO_3$, can be prepared from the appropriate melts of the metal oxides (42). The cerium(III) sulfate-water system has been studied (43) and several hydrates characterized. The conditions of formation and composition of hydrated lanthanide, scandium and yttrium hydroxides obtained by action of solutions of ammonia and alkalis on the crystalline hydrated chlorides, nitrates and sulfates have been determined (44). The preparative reactions and phase equilibria of solid lanthanide hydroxide nitrates in aqueous media at 400° and 1200 atm. have been investigated(45). Ammonium samarium sulfate tetrahydrate crystals can be prepared by letting aqueous samarium sulfate-ammonium sulfate (1:4) solutions stand in air (46). Black tetragonal scheelite-type compounds of stoichiometry $Eu_x^{II}MoO_4$ ($0.67 \leq x \leq 1.00$) and $Ln^{III}MoO_4$ (Ln = Y, Gd-Lu) can be prepared (47) by heating mixtures of the metal oxides in sealed tubes. Glasses ($3Ln_2O_3 \cdot 5Ga_2O_3$) of lanthanide oxides (Ln_2O_3-Ln = Pr, Nd, Sm, Eu, Gd) with Ga_2O_3 have been prepared by reaction of the Ln_2O_3 with Ga_2O_3 under high pressure conditions (48). The oxide glass $3Gd_2O_3 \cdot 5Fe_2O_3$ was prepared (49) from Fe_2O_3 and Gd_2O_3 by use of a laser furnace and high pressure. Adducts of the lanthanide metals with acetylurea have been formed (50) and found to have the general formula $Ln(ClO_4)_3 \cdot 4AU$ (Ln = lanthanide ions and Y, and AU = acetylurea). They were prepared by addition of acetylurea in chloroform to an ethylacetate solution of the hydrated metal perchlorate. The lanthanide metals form adducts with dipropionamide (DP) to yield compounds of the general formula $Ln(ClO_4)_3 \cdot 4DP$. The ligands are in the trans-trans planar form with coordination occuring through the carbonyl group (51). The adducts of the lanthanide metal perchlorates with N,N-dimethylacetoacetamide (DMAA) have been prepared (52) in ethanol and precipitated from solution by addition of triethylorthoformate. They are hygroscopic and have been studied spectroscopically, and found to have the general formula $Ln(ClO_4)_3 \cdot 4DMAA$ (Ln = lanthanide ions and Y). Adducts between lanthanide perchlorates and thioxane-oxide (TSO) can be prepared (53) by interaction of metal salts and ligand in ethanol corresponding to the compositions $Ln(ClO_4)_3 \cdot 9TSO$ (Ln = La-Nd), $Sm(ClO_4)_3 \cdot 8.5TSO$, $Ln(ClO_4)_3 \cdot 8TSO$ (Ln = Eu-Er) and $Ln(ClO_4)_3 \cdot 7TSO$ (Ln = Tm-Lu, Y).

Hydrated lanthanide(III) chlorides react with excess sodium diethyldithiophosphate in hot ethanol to yield compounds of the type $(C_6H_5)_4AsLn(S_2P\{OC_2H_5\}_2)_4$ after addition of tetraphenylarsonium chloride and extraction into hot benzene (54). Samarium(II) selenide and ytterbium(II) selenide can be obtained by reducing the sesquiselenides with the respective gaseous metals in a high vacuum (55). Rare earth (La, Ce, Pr, Nd, and Sm) mono- and poly-phosphides have been prepared (56) by the reaction of the powdered pure elements in a sealed silica tube in vacuo at 500-700°C.

$Ca_8Ln_5F_{31}$ (Ln = Y, Ho–Lu) phases can be obtained (57) by solid phase reactions of CaF_2 with LnF_3 compounds. The interaction of lanthanum and cerium trifluorides with calcium and lanthanum oxides in molten NaF–LiF eutectic mixture at 800°C has been studied - the presence LaOF and CeOF are indicated (58). Anhydrous $EuCl_3$ in tributylphosphate forms a tris-adduct (59). Samarium diiodide can be prepared by direct reaction of iodine vapor with the metal in a tungsten crucible or by the reaction of samarium with mercuric iodide. Thallium triiodide can be obtained by reaction of iodine vapor with the etal in a tungsten or tantalum crucible and purified by sublimation (60). Lanthanide chloride and isothiocyanate adducts of tetramethylmalonamide (TMMA) of compositions $LnCl_3 \cdot 2TMMA$ (Ln = La–Lu, Y, except Pm and Tm), $Ln(NCS)_3 \cdot 3TMMA$ (Ln = La–Yb, Y, except Pm) and $Lu(NCS)_3 \cdot 2TMMA$ can be obtained (61) by interaction of the appropriate hydrated metal salt with TMMA in ethanol and triethylorthoformate. The fused salt systems $ACl–CeCl_3$, where A = Li, Na, K, Rb or Cs, have been studied (62) and from the thermodynamics of the systems, the formation of the compounds $CeCl_4^-$ and $CeCl_6^{3-}$ was indicated. Ammonium nonafluorodi-lanthanate(III) monohydrates, $(NH_4)_3Ln_2F_9 \cdot H_2O$ (Ln = Y, La, Pr, Sm, Er) have been prepared by the reaction of the metal nitrates with ammonium fluoride in aqueous solution (63). Ammonium undecafluorolanthanate(III) monohydrates, $(NH_4)_2Ln_3F_{11} \cdot H_2O$ (Ln = Y, Dy, Ho, Er) can be obtained (63) by the reaction of the lanthanide bromide with ammonium fluoride in methanol.

ACTINIDES

Fission yields for the reaction of uranium with 14MeV neutrons (64) and 170MeV protons (65) have been reported as has the fission with mono-energetic neutrons with 130–1700 keV energy (66). Neptunium trihalides (chloride and bromide) can be prepared in vacuo by reacting NpO_2 with CCl_4 and hydrogen or by reacting NpO_2 with excess molten $AlBr_3$ (67), pure $NpBr_3$ was then sublimed from the reaction mixture. The comparative solution chemistry of nobelium has been described (68). Bis(cyclooctatetra-enyl)protactinium(IV) can be prepared <u>via</u> the reaction:

$$PaCl_4 + 2K_2C_8H_8 \xrightarrow{THF} Pa(C_8H_8)_2 + 4KCl \qquad (69).$$

Bis(tetraethylammonium)hexafluoroprotactinate(IV) can be obtained by heating PaF_4 with the ammonium fluoride in oxygen-free propylene carbonate. The analogous U(IV) compound can be similarly prepared (70). The preparation of triscyclopentadienyl plutonium(III) has been accomplished (71) by reaction of dicesium hexachloroplutonate(IV) or anhydrous trichloroplutonium(III) with magnesium biscyclopentadienide in THF under nitrogen:

$$(1)\ Cs_2PuCl_6 + MgCp_2 \xrightarrow{THF} Cp_3Pu + MgCl_2 + CsCl$$

$$(2)\ 2PuCl_3 + 3MgCp_2 \xrightarrow{THF} 2Cp_3Pu + 3MgCl_2 + \cdots$$

The crystal structure of californium metal has been determined by electron diffraction (72) of a thin film of the metal prepared by reducing Cf_2O_3 with lanthanum metal and condensing the volatile californium metal on electron microscopy grids.

$ThN(NH_2)$ can be prepared (73) by reaction of ThNI with alkali amide in liquid (20°C) or supercritical ammonia (300°C, 4000 atm). Similarly, $Th_3N_2(NH)_3$ can be obtained. $Th(NH)_2$ can be prepared by the reaction of the metal with ammonia (325°C, 6000 atm) in the presence of some alkali amide (73). Tris[bis-trimethylsilylamido]monochlorothorium(IV) can be obtained by reacting anhydrous thorium tetrachloride with lithium bis-trimethylsilylamide at room temperature (74). Thorium has been found to form a blue complex with very high molar absorptivity ($\sim 10^5$) with cetyltrimethylammonium bromide and chromeazurol 5 in aqueous solution (75). The insertion of CX_2 (X = O, S, Se) and of COS into the metal-nitrogen bonds of thorium(IV) and uranium(IV) dialkylamides yields the corresponding carbamates (76). Complexes of thorium(IV) and uranium(IV) perchlorates with the phosphine oxides - trimethylphosphine oxide (tmpo), triphenylphosphine oxide (tppo), tri-n-propylphosphine oxide (tprpo), hexamethylphosphoramide (hmpa) and octamethylphosphoramide (ompa) were prepared (77) by reaction of the metal salts in alcoholic solutions with the appropriate phosphine oxide, to give the compounds - $Th(ClO_4)_4 \cdot 5tppo$, $Th(ClO_4)_4 \cdot 6hmpa$, $Th(ClO_4)_4 \cdot 4tprpo$, $U(ClO_4)_4 \cdot 6tmpo$, $U(ClO_4)_4 \cdot 6hmpa$, $U(ClO_4)_4 \cdot 4ompa$, $U(ClO_4)_4 \cdot tprpo$, $UO_2(ClO_4)_2 \cdot 4tmpo$, $UO_2(ClO_4)_2 \cdot 4tprpo$, $UO_2(ClO_4)_2 \cdot 4tppo$, $UO_2(ClO_4)_2 \cdot 5hmpa$. The kinetics and mechanism of the disproportionation of U(V) has been investigated (78) and should be of relevance to synthetic work. The formation of gas phase $PbThCl_6$ from $PbCl_2$ and $ThCl_4$ has been described (79). The gas phase molecules $AgThCl_5$, $CuTh_2Cl_9$, Tl_2ThCl_6, Tl_2UCl_6, TlU_2Cl_9 and $ThUCl_6$ have been characterized (80) by mass spectroscopy.

Unstable yellow-green uranyl chloride bis triethylammonium chloride can be obtained (81) by mixing solutions of triethylamine in HCl and UO_2Cl_2 in HCl. The complex contains the $UO_2Cl_4^{2-}$ anion. Complexes of the composition $UO_2(SO_4)L_2$ and $UO_2SO_4L'_3$ (L = dimethylformamide, dimethyl sulfoxide, tetramethylurea; L' = antipyrine, glycine) can be obtained (82) by interaction of the ligands with aqueous solutions of uranyl sulfate. The reaction of uranyl tris(quinolinoate) with oxalic acid in water yields (oxine H)$_2$ $[(UO_2)_2(C_2O_4)_2(oxine)_2(H_2O)_2]$ (83). Polymeric compounds result upon treating uranyl chromate with various neutral ligands (84). An NMR study of the dimeric 1:1 complex of UO_2^{2+} with malic acid has been reported (85) but the solid dimer was not isolated. A number of uranyl-macrocyclic ligand complexes have been prepared (86) <u>via</u> template syntheses:

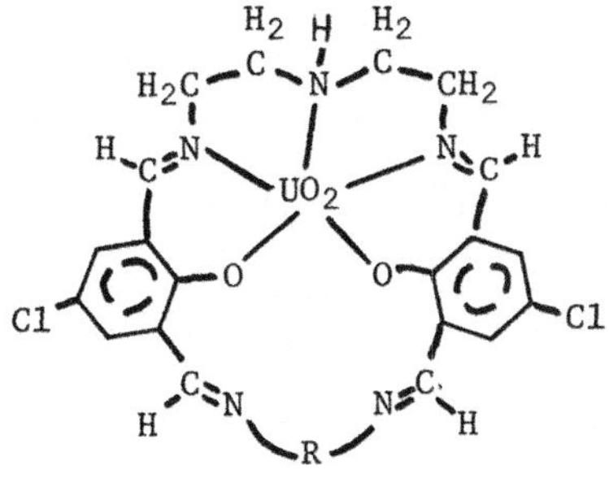

R
$-CH_2-CH_2-$
$-CH_2-CH_2-CH_2-$
$-CH_2-CH(CH_3)-$

A variety of phosphine oxide complexes of Th(IV), U(IV), Np(IV) and UO_2^{2+} nitrates have been described (87). Reddish orange dioxo(methanol)-bis(tropolonato)uranium(VI) results upon interaction of methanol solutions of tropolone and uranyl nitrate hexahydrate. The complex can be heated to remove methanol and reacted with a variety of ligands which replace methanol (88). The crystal structure of the pyridine adduct was determined. The compound $(Ph_4As)_2[UO_2(PDC)_2]\cdot 6H_2O$ has been prepared (89) by reacting aqueous solutions of uranyl nitrate, pyridine-2,6-dicarboxylic acid (H_2PDC) and tetraphenylarsonium chloride at 80°C. The product was isolated as well shaped greenish-yellow crystals, and the infrared, thermal analysis and crystal data were obtained. Adducts of uranyl o-hydroxyarylcarbonyl complexes (UO_2L_2B) where B = H_2O, pyridine, pyridine N-oxide, have been prepared (90) by dissolving the carbonyl compound in methanol, adding it to uranyl nitrate hexahydrate in methanol, stirring and then adding aqueous NaOH, pyridine or pyridine N-oxide and allowing the adduct to precipitate out. The photochemical reduction of uranyl ion in citric acid solutions (91) and the inhibition of Tl(I) ion on the photoreduction of uranyl ion in lactic acid solutions (92) have been described. The photoreduction of uranyl ion in sulfuric acid with laser light and ethanol has also been reported (93). The $UO_2(NCS)_6^{4-}$ ion can be extracted from aqueous solution with long chain alkylammonium compounds (94).

Thermodynamic parameters for the formation of $M(SCN)_{aq}^{2+}$ (M = Eu, Am, Cm, Bk) complexes have been determined (95) but solids were not isolated. Crystals of β-NH_4UF_5, ammonium pentafluorouranate(IV), can be obtained from dilute boiling solutions of formic acid, uranyl fluoride and ammonium fluoride, coupled with photochemical reduction (96). The compound $UZrF_7$ has been prepared (97) according to:
$UF_4 + \frac{1}{4} Zr + \frac{3}{4} ZrF_4 \longrightarrow UZrF_7$. The compound contains U(III). Uranium hexachloride can be reduced, with chloride ions to give hexachlorouranate(V), with methylene chloride (CH_2Cl_2) to give uranium pentachloride, and with chlorine azide in methylene chloride, NH_4UCl_6 is the product and in carbon tetrachloride, $[C(N_3)_3]UCl_6$ is the product (98). The interaction of nitrate ion with U(IV) in hexamethylphosphoramide has been studied and a variety of related complexes isolated (99). Tetrakis-(1,1,1-trifluoro-4-phenylbutane-2,4-dionato)uranium(IV), $U(tfba)_4$, can be prepared by extraction of U(IV) from aqueous HCl at pH$\sim$1 with a solution of Htfba in carbon tetrachloride. $U(tfba)_4$ can be employed as an NMR shift reagent (100). Mixed N,N'-ethylenebis(salicylideneiminato), salen, complexes of U(IV) of the type $U(salen)X_2$ (X = Cl, acetylacetonato, dipivaloylmethanato, N-methylsalicylideneiminato, dibenzoylmethanato) have been described (101). $U(salen)_2$ and UCl_4 undergo a ligand redistribution reaction in tetrahydrofuran (THF) to yield $U(salen)Cl_2\cdot 2THF$ (101).

Lactam complexes of thorium and uranium tetrahalides, $ThCl_4(aco)_4\cdot$ acetone, $ThI_4\cdot 8L$, $UCl_4\cdot 4L$, $UBr_4(pip)_6$, $UBr_2(pip)_6(ClO_4)_2$ and $UI_4\cdot 8L$ where L is 2-piperidone(pip), 1-aza-cyclooctan-2-one(aco) and ε-caprolactam have been isolated (102) from reactions of the appropriate metal salts and ligands in acetone. Complexes of uranium and thorium halides (chloride and bromide) with various sulphoxides may be prepared and isolated (103) to give compounds of the form - $UX_4 \cdot 2tbso$, $UCl_4 \cdot 2dibso$, $UBr_4 \cdot 4dibso$, $ThCl_4 \cdot 3dibso$, $(UCl_2 \cdot 6dmso)(ClO_4)_2$, and $(UCl_3 \cdot 5dmso)ClO_4$ where tbso- di-t-butyl sulphoxide, dibso- di-isopropylsulphoxide, dmso-dimethyl suphoxide. A new thallium(I) uranate(VI) of composition Tl_4UO_5 can be prepared by heating thallium(I) carbonate and uranium trioxide in 2:1 mole ratio at low pressure for 20-30 hours at 200°C (104). High purity ^{234}U may be separated from aged ^{238}Pu by precipitation of the uranium as $NaUO_2(Ac)_3$ from an aqueous sodium acetate solution (105). The compound $U_2O_3F_6$ has been prepared (106) by the thermal decomposition of UOF_4 in vacuum at 290°C for one hour to yield the yellow powdered product. The bonding of carbon dioxide adsorbed and chemisorbed on uranium oxides has been studied (107) by infrared spectroscopic methods. U_3O_8 can be prepared by heating uranyl nitrate in air at 800°C for 3 days (108). This can be further reacted with chlorine gas at atmospheric pressure in a quartz boat at approximately 1000°C to give the product UO_2Cl_2. The preparation and X-ray photoelectron spectra of a number of uranium oxide phases have been reported (109). Uranium and rare earth tungsten bronzes can be prepared (110) in vacuum by solid state reaction of W, WO_3 and the appropriate metal oxide according to the reaction:

$$(3x)M_sO_t + (3s-xt)WO_3 + (xt)W \longrightarrow (3s)M_xWO_3 \ .$$

The reaction of UO_2 in molten alkali nitrate melts to form solid solutions of the type $Na_{2-x}M_xU_2O_7$ (M = alkali metal) has been investigated (111). Bright yellow $UO_4 \cdot 2NH_3 \cdot 2HF$ can be prepared (112) by addition of hydrogen peroxide to a solution containing 9M ammonium fluoride and 0.002 M uranyl ion at 50°C.

REFERENCES

1. J. L. Atwood and K. D. Smith, J. C. S. DALTON TRANS., 921 (1974).
2. T. J. Anderson, M. A. Neuman and G. A. Melson, INORG. CHEM., 13, 158 (1974).
3. A. R. Davis and F. W. B. Einstein, INORG. CHEM., 13, 1880 (1974).
4. T. J. Anderson, M. A. Neuman and G. A. Melson, INORG. CHEM., 13, 1884 (1974).
5. A. Hamer, D. Tisley and R. Walton, J. INORG. NUCL. CHEM., 36, 1771 (1974).
6. V. M. Shatskii, B. I. Bashkov, L. N. Komissarova and A. M. Grevtsev, ZH. NEORGAN. KHIM., 19, 2013 (1974).
7. L. N. Komissarova and A. S. Znamenskaya, ZH. NEORGAN. KHIM., 19, 295 (1974).
8. Yu. A. Buslaev, S. P. Petrosyants, V. P. Tarasov and V. I. Chagin, ZH. NEORGAN. KHIM., 17, 1790 (1974).
9. D. A. Johnson, J. C. S. DALTON TRANS., 1671 (1974).
10. T. K. Halstead, J. SOLID STATE CHEM., 11, 114 (1974).
11. S. Kapur, B. L. Kalsotra and R. K. Multani, J. INORG. NUCL. CHEM., 36, 932 (1974).
12. J. L. Slater, T. C. De Vore and V. Calder, INORG. CHEM., 13, 1808 (1974).

13. G. Linde and R. Juza, Z. ANORG. ALLG. CHEM., 409, 191 (1974).
14. C. Hadenfeldt, B. Gieger, and H. Jacobs, Z. ANORG. ALLG. CHEM., 403, 319 (1974).
15. G. V. Bezdenezhnykh, V. A. Sharov and E. I. Krylov, ZH. NEORGAN. KHIM., 19, 2048 (1974).
16. T. Asanr, S. Okada, K. Sakamoto, S. Taniguchi and Y. Kobayashi, J. INORG. NUCL. CHEM., 36, 1433 (1974).
17. N. K. Dutt and R. De, INDIAN J. CHEM., 12, 1320 (1974).
18. R. Prados, L. G. Stadtherr, H. Donato, Jr. and R. B. Martin, J. INORG. NUCL. CHEM., 36, 689 (1974).
19. O. Farooq, A. Malik and N. Ahmad, ACTA CHIM. ACAD. SCI. HUNG., 83, 343 (1974).
20. A. A. Sopueva, K. S. Sulaimankulov and L. A. Tokmergenova, ZH. NEORGAN. KHIM., 19, 1095 (1974).
21. I. M. Batyaev and N. L. Puzankova, ZH. NEORGAN. KHIM., 19, 2023 (1974).
22. Z. Amjad and A. McAuley, J. C. S. DALTON TRANS., 2521 (1974).
23. R. C. Agarwal, J. INDIAN CHEM. SOC., 51, 772 (1974).
24. V. T. Mishchenko, E. I. Tselik and N. S. Poluektov, ZH. NEORGAN. KHIM., 19, 2028 (1974).
25. N. A. Kostromina and T. V. Ternovaya, ZH. NEORGAN. KHIM., 19, 2036 (1974).
26. W. S. Trahanovsky, J. Cramer and D. W. Brixius, J. AMER. CHEM. SOC., 96, 1077 (1974).
27. N. I. Kurkina, N. Yu. Petrova and N. A. Skorik, ZH. NEORGAN. KHIM., 19, 661 (1974).
28. C. M. Dobson, R. J. P. Williams and A. V. Xavier, J. C. S. DALTON TRANS., 1762 (1974); C. D. Barry, C. M. Dobson, R. J. P. Williams and A. V. Xavier, IBID., 1765 (1974).
29. M. A. Tishchenko, I. I. Zheltvai and N. S. Poluektov, ZH. NEORGAN. KHIM., 19, 1793 (1974).
30. J. Alstad, J. H. Augustson and L. Farbu, J. INORG. NUCL. CHEM., 36, 899 (1974).
31. A. G. Gorynshko and N. K. Davidenko, ZH. NEORGAN. KHIM., 19, 1772 (1974).
32. H. L. Goering, J. N. Eikenberry, G. S. Koermer and C. J. Latimer, J. AMER. CHEM. SOC., 96 1493 (1974); M. D. McCreary, D. W. Lewis, D. L. Wernick and G. M. Whitesides, IBID., 96, 1038 (1974).
33. C.-P. Wong, R. F. Venteicher, W. DeW. Horrocks, Jr., J. AMER. CHEM. SOC., 96, 7149 (1974).
34. D. Apers, R. Block, and P. Capron, J. INORG. NUCL. CHEM., 36, 1441 (1974).
35. A. N. Christensen and S. Quézel, J. SOLID STATE CHEM., 9, 234 (1974).
36. C. S. Dimbylow, I. J. McColm, C. M. P. Barton, N. N. Greenwood and G. E. Turner, J. SOLID STATE CHEM., 10, 128 (1974).
37. P. D. Dernier, E. M. Gyorgy and W. H. Grodkiewicz, J. SOLID STATE CHEM., 10, 122 (1974).
38. H. G. Danielmeyer, J. P. Jeser, E. Schönherr and W. Stetter, J. CRYSTAL GROWTH, 22, 298 (1974).

39. Y. Tsujimoto, Y. Fukuda, S. Sugai and M. Fukai, J. LUMINESCENCE, 9, 475 (1974).
40. J. L. Sommerdijk, A. Bril and A. W. de Jager, J. LUMINESCENCE, 9, 288 (1974).
41. P. Dougier and P. Hagenmuller, J. SOLID STATE CHEM., 11, 177 (1974).
42. L. Z. Gokhman, G. V. Lysanova, D. A. Dulin and A. V. Pashkova, ZH. NEORGAN. KHIM., 19, 2018 (1974).
43. T. P. Spacibenko, ZH. NEORGAN. KHIM., 19, 899 (1974).
44. V. A. Musorin, V. V. Sakharov and L. M. Zaitsev, ZH. NEORGAN. KHIM., 19, 1476 (1974).
45. J. M. Haschke, INORG. CHEM., 13, 1812 (1974).
46. B. Eriksson, L. O. Larsson, K. Niinisto and U. Skoglund, INORG. CHEM., 13, 290 (1974).
47. E. Banks and M. Nemiroff, INORG. CHEM., 13, 2715 (1974).
48. S. Yajima, K. Okamura and T. Shishido, CHEM. LETT., 545 (1974).
49. S. Yajima, K. Okamura and T. Shishido, CHEM. LETT., 1531 (1974).
50. C. Airoldi and Y. Gushikem, J. INORG. NUCL. CHEM., 36, 1892 (1974).
51. O. Alves, Y. Gushikem, and C. Airoldi, J. INORG. NUCL. CHEM., 36, 1079 (1974).
52. M. Perrier and G. Vincentini, J. INORG. NUCL. CHEM., 36, 1187 (1974).
53. G. Vicentini and M. Perrier, J. INORG. NUCL. CHEM., 36, 77 (1974).
54. A. A. Pinkerton, INORG. NUCL. CHEM. LETT., 10, 495 (1974).
55. T. Petzel, INORG. NUCL. CHEM. LETT., 10, 119 (1974).
56. S. Ono, K. Nomura and H. Hayakawa, J. LESS COMMON METALS, 38, 119 (1974).
57. P. P. Fedorov, O. E. Izotova, V. B. Alexandrov and B. P. Sobolev, J. SOLID STATE CHEM., 9, 368 (1974).
58. V. F. Gorbunov and G. P. Novoselov, ZH. NEORGAN. KHIM., 19, 1734 (1974).
59. V. V. Yastrebov, O. V. Galaktionova, E. N. Lebedeva and S. S. Korovin, ZH. NEORGAN. KHIM., 19, 1252 (1974).
60. C. Hirayama, P. M. Castle, R. W. Liebermann, R. J. Zollweg and F. E. Camp, INORG. CHEM., 13, 2804 (1974).
61. G. Vicentini, M. Perrier, L. B. Zinner and M. I. Amin, J. INORG. NUCL. CHEM., 36, 771 (1974).
62. G. N. Papatheodorou and O. J. Kleppa, J. PHYS. CHEM., 78, 178 (1974).
63. R. C. Russo and H. M. Haendler, J. INORG. NUCL. CHEM., 36, 763 (1974).
64. J. Blachot, L. C. Carraz, P. Cavallini, C. Chauvin, A. Ferrieu and A. Moussa, J. INORG. NUCL. CHEM., 36, 495 (1974).
65. I. Haldoersen, A. C. Pappas and M. Skarestad, J. INORG. NUCL. CHEM., 36, 245 (1974).
66. J. G. Cuninghame, J. A. B. Goodall and H. H. Willis, J. INORG. NUCL. CHEM., 36, 1453 (1974).
67. W. Carnall, H. Crosswhite, R.Pappalardo, D. Cohen, S. Fried, P. Lucas and F. Wagner, Jr., J. CHEM. PHYS., 61, 4994 (1974).
68. R. J. Silva, W. J. McDowell, O. L. Keller, Jr. and J. R. Tarrant, INORG. CHEM., 13, 2233 (1974).

69. J. Goffart, J. Fuger, D. Brown and G. Duyckaerts, INORG. NUCL. CHEM. LETT., 10, 413 (1974).
70. D. Brown, B. Whittaker and N. Edelstein, INORG. CHEM., 13, 1805 (1974).
71. L. R. Crisler and W. G. Eggerman, J. INORG. NUCL. CHEM., 36, 1428 (1974).
72. R. Haire and R. Baybarz, J. INORG. NUCL. CHEM., 36, 1295 (1974).
73. H. Blunck and R. Juza, Z. ANORG. ALLGEM. CHEM., 410, 9 (1974).
74. D. C. Bradley, J. S. Ghotra and F. A. Hart, INORG. NUCL. CHEM. LETT., 10, 209 (1974).
75. B. Evtimova, ANAL. CHIM. ACTA, 68, 222 (1974).
76. K. W. Bagnall and E. Yanir, J. INORG. NUCL. CHEM., 36, 777 (1974).
77. K. Bagnall and M. Wakerley, J. LESS COMMON METALS, 35, 267 (1974).
78. A. Ekstrom, INORG. CHEM., 13, 2237 (1974).
79. M. Binnewies and H. Schäfer, Z. ANORG. ALLGEM. CHEM., 410, 149 (1974).
80. M. Binnewies and H. Schäfer, Z. ANORG. ALLGEM. CHEM., 407, 327 (1974).
81. H. Brusset, N.-Q. Dao and F. Haffner, J. INORG. NUCL. CHEM., 36, 791 (1974).
82. R. N. Shchelokov, I. M. Orlova and G. V. Podnebesnova, ZH. NEORGAN. KHIM., 19, 744 (1974).
83. E. N. In'kova and V. P. Arsent'eva, ZH. NEORGAN. KHIM., 19, 1364 (1974).
84. R. N. Shchelokov, I. M. Orlova and G. V. Podnebesnova, ZH. NEORGAN. KHIM., 19, 1581 (1974).
85. J. D. Pedrosa and V. M. S. Gil, J. INORG. NUCL. CHEM., 36, 1803 (1974).
86. U. Casselato, M. Vidali and P. A. Vigato, INORG. NUCL. CHEM. LETT., 10, 437 (1974).
87. K. W. Bagnall and M. W. Wakerley, J. C. S. DALTON TRANS., 889 (1974).
88. S. Degetto, G. Marangoni, G. Bombieri, E. Forsellini, L. Baracco and R. Gasiani, J. C. S. DALTON TRANS., 1933 (1974).
89. G. Marangoni, S. Degetto, R. Graziani, G. Bombieri and E. Forsellini, J. INORG. NUCL. CHEM., 36, 1787 (1974).
90. J. Haigh, J. MOL. STR., 24, 281 (1974).
91. A. Ohyoshi and K. Ueno, J. INORG. NUCL. CHEM., 36, 379 (1974).
92. Y. Yokoyama, M. Moriyasu and S. Ikeda, J. INORG. NUCL. CHEM., 36, 385 (1974).
93. J. T. Bell and S. R. Buxton, J. INORG. NUCL. CHEM., 36, 1575 (1974).
94. T. Sato, S. Kotani and M. L. Good, J. INORG. NUCL. CHEM., 36, 451 (1974).
95. W. F. Kinard and G. R. Choppin, J. INORG. NUCL. CHEM., 36, 1131 (1974).
96. R. A. Penneman, R. R. Ryan and A. Rosenzweig, ACTA CRYST., B30, 1966 (1974).
97. G. Fonteneau and J. Lucas, J. INORG. NUCL. CHEM., 36, 1515 (1974).
98. W. Kolitsch and U. Müller, Z. ANORG. ALLG. CHEM., 410, 21 (1974).
99. J. G. H. du Preez and C. P. J. Van Vuuren, J. INORG. NUCL. CHEM., 36, 81 (1974).

100. G. Folcher, J. Paris, P. Plurien, P. Rigny and E. Soulie, J. C. S. CHEM. COMM., 3 (1974).
101. F. Calderazzo, M. Pasquali, T. Salvatori, J. C. S. DALTON TRANS., 1102 (1974).
102. J. G. H. du Preez, M. L. Gibson and P. J. Steenkamp, J. INORG. NUCL. CHEM., 36, 579 (1974).
103. J. G. H. du Preez and M. L. Gibson, J. INORG. NUCL. CHEM., 36, 1795 (1974).
104. A. Giridharan, M. Udupa and G. Aravamudan, Z. ANORG. ALLG. CHEM., 407, 345 (1974).
105. E. Russell, J. INORG. NUCL. CHEM., 36, 1659 (1974).
106. P. W. Wilson, J. INORG. NUCL. CHEM., 36, 1783 (1974).
107. C. Colmenares, J. PHYS. CHEM., 78, 2117 (1974).
108. E. Cordfunke and G. Prins, J. INORG. NUCL. CHEM., 36, 1291 (1974).
109. G. C. Allen, J. A. Crofts, M. T. Curtis, P. M. Tucker, D. Chadwick and P. J. Hampson, J. C. S. DALTON TRANS., 1296 (1974).
110. J. F. Thomas and M. J. Sienko, J. CHEM. PHYS., 61, 3920 (1974).
111. C. Toussaint and A. Avogadro, J. INORG. NUCL. CHEM., 36, 781 (1974).
112. H.-G. Bachmann, H. Z. Dokuzoguz and H. M. Muller, J. INORG. NUCL. CHEM., 36, 795 (1974).

TITANIUM, ZIRCONIUM, AND HAFNIUM

John J. Alexander

Department of Chemistry
University of Cincinnati
Cincinnati, Ohio 45221

GENERAL

This review has been organized according to the nature of groups bonded to Ti, Zr or Hf starting with the halogens and working back to hydrogen. Since many of the newly-synthesized compounds contain more than one family of elements bonded to one of the Group IVB metals, this classification scheme has an element of arbitrariness. In general, some choice has had to be made about which bonded group is of major interest. So, for example, a complex formed from $TiCl_4$ and N-donor ligands is placed under Group V. Compounds containing bonded elements of interest from more than one periodic group are placed under the first group discussed. Thus, a complex containing both N- and O-donor groups will appear under Group VI compounds. In this spirit several organometallic compounds are mentioned, but usually only when they involve reactions forming bonds other than M-C. The organometallic aspects of Ti, Zr and Hf compounds are treated in a complementary series of reviews (1).

An interesting account of the discovery of Zr has appeared (2).

Two reviews were published of general interest: one concerning the properties of transition metal oxides including Ti, Zr and Hf (3) and a second concerning the structures of Ti, Zr and Hf compounds determined by x-ray diffraction (4).

Optically active Ti complexes were prepared during 1974. Benzene solutions of $(\eta^5\text{-Cp})(\eta^5\text{-Cp}')Ti(OR)Cl$ ($Cp=C_5H_5$; $Cp'=C_5H_4CH(CH_3)_2$, $R=C_6H_5$, $C_6H_3(CH_3)$-3-$(CH(CH_3)_2)$-6; $Cp'=C_5H_4CH(C_6H_5)_2$, $R=C_6H_4(CH_3)$-2 were treated with optically active mandelic acid to afford products with a center of asymmetry at Ti. Either enantiomer could be obtained depending on the absolute configuration of the mandelic acid employed for the asymmetric destruction (5). By starting with compounds $(\eta^5\text{-Cp})(\eta^5\text{-Cp}^*)Ti(OR)Cl$ containing an optically active center at the Cp^* (Cp^*=

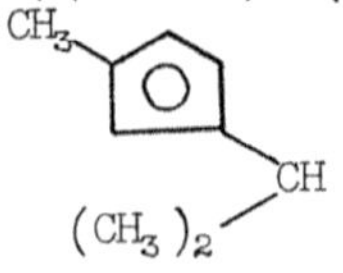

R = 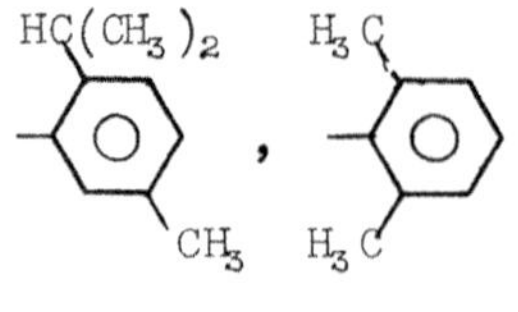

) diasteromers of

higher specific rotation were formed (5). Several other Ti diasteromers of this type have been prepared by these French workers (6,7,8).

A report of the use of $TiCl_4$ as a shift reagent for pmr and cmr has appeared (9). In $CDCl_3$ protons close to the site of coordination to this Lewis acid are shifted in alcohols, ethers, ketones, esters, acids, amides and nitriles. Workers using TiH_2 should be aware of a report (10) that this compound has shown some deleterious physiological effects in rats.

HALIDES AND PSEUDO HALIDES

A new preparative method for the simple halides of Zr(III) has appeared (11). It involves reduction of $Zr(IV)X_4$ (X=Cl,Br,I) by Al metal in a solution of AlX_3 at high temperature in a sealed ampoule. Because Hf(IV) reduction is slower under these conditions, Hf impurities are separated.

Two reports of mixed fluoride oxides have been published. From a solution of K_2TiF_6, $BaCl_2$ and K_2CO_3 or $(NH_4)_2CO_3$ at pH 2.7-4.0 a precipitate of $BaTiOF_4 \cdot 0.5\ H_2O$ was obtained (12). Reaction 1 affords a fluoride oxide containing niobium which is said (13) to have a rutile structure:

$$Ti_2O_3 + NbO_2F \xrightarrow{900^\circ} Ti_2NbO_5F \qquad (1)$$

Several fluoride sulfates of Zr have resulted from reactions 2 and 3 carried out in $H_2SO_4(aq)$ (14).

$$Zr(SO_4)_2 \cdot 4H_2O + 2MF \longrightarrow M_2Zr(SO_4)_2F_2 \cdot 3H_2O + H_2O \qquad (2)$$

$$ZrF_4 \cdot 3H_2O + M_2SO_4 \longrightarrow M_2Zr(SO_4)F_4 + 3\ H_2O \qquad (3)$$

$(M=K, Rb, Cs, NH_4)$

When NH_2Cl is allowed to react with $TiCl_4$, $ZrCl_4$ or $HfCl_4$ in CCl_4, the complex chlorides $(NH_4)_2MCl_6$ (M=Ti, Zr, Hf) are produced (15). These compounds decompose to the nitrides on pyrolysis.

Pseudo halides of the type $(\eta^5\text{-}C_5H_5)_2MX_2$ (M=Ti, Zr; X=$N(CN)_2$, $C(CN)_3$, $ONC(CN)_2$) were prepared (16) by treating the dichlorides or dibromides with stoichiometric quantities of silver dicyanamide, silver tricyanomethide, or silver nitrosodicyanamide, respectively, in tetrahydrofuran (THF) or acetonitrile. The orange-red Ti compounds and colorless Zr compounds decompose in water but are soluble in polar organic solvents.

Red-black hexacoordinate Ti thiocyanato complexes of the type $[TiCl_{4-n}(NCS)_n(THF)_2] \cdot 3THF$ (n=1-3) result from the reaction of $TiCl_4 \cdot 2THF$ with NH_4NCS in 1:1, 1:2 and 1:3 molar ratios in THF (17). In acetonitrile the THF molecules in the first coordination sphere are replaced to give $[TiCl_{4-n}(NCS)_n(CH_3CN)_2]$ (n=1-3). The thiocyante is N-bonded to Ti. Work on the known gray-green compound $K_5Ti(CN)_8$ prepared from $TiBr_3$ and KCN in $NH_3(\ell)$ rules out either octahedral six-coordination or eight-coordination for Ti. The complex ion seems to be $Ti(CN)_7^{4-}$ having a distorted structure like $Mo(CN)_7^{4-}$ in $K_4Mo(CN)_7 \cdot 2H_2O$ rather than a pentagonal bipyramidal structure like $V(CN)_7^{4-}$ (18). Treatment of TiBr and KCN in liquid NH_3 gives black pyrophoric $K_4Ti(CN)_4$ (18).

GROUP VI COMPOUNDS

Simple Compounds with Bonds to O, S, Se: A cubic modification of TiO can be made by shock wave compression of an equimolar mixture of Ti and TiO_2 in steel ampoules (19). The more usual method of preparation of this oxide by heating Ti and TiO_2 gives a mixture of cubic and monoclinic phases on spontaneous cooling.

The spinel Co_2TiO_4 is the product when $Ti(O\text{-}i\text{-}C_3H_7)_4$ is hydrolyzed at 75° C in basic solution in the presence of 0.8 mole percent Co(II) and the product heated to 800° C (20). By heating $RbO_{0.56}$ with TiO_2 in a 2:1 ratio to 550° C in a Ag bomb, Rb_2TiO_3 was obtained (21). This oxide exhibits 4-coordination around Ti.

Reduction of ZrO_2 or HfO_2 with CS_2 at temperatures of 1300° C and 1400° C, respectively, gave pure samples of ZrS_2 or HfS_2 (22). Reaction of the elements in an evacuated SiO_2 tube in the presence of I afforded sulfides of the composition Ln_2MS_5 (Ln= any lanthanide except Eu, Yb; M= Zr, Hf). The following selenides could be similarly prepared: La_2ZrSe_5, Ce_2ZrSe_5, Pr_2ZrSe_5, Nd_2ZrSe_5, Pm_2ZrSe_5, Sm_2ZrSe_5, Gd_2ZrSe_5, Tb_2ZrSe_5, La_2HfSe_5 and Ce_2HfSe_5 (23). These compounds have the U_3S_5 structure.

Several intercalation compounds of TiS_2 have been reported. When TiS_2 is placed in an alkali halide melt at 900-1000° in a H_2S/CS_2 stream, hexagonal plates of $M_x(TiS_2)$ (M= alkali metal; x= 0.3-0.5) are obtained having TiS_2^{x-} layers with xM^+ ions between. These react with water to give hydrates; polar organic solvents will replace water (24). Starting with $(NH_2NH_2)TiS_2$, intercalation compounds of the types $(RNH_2)_xTiS_2$ (R= C_4H_9 - C_8H_{17}; x= 0.95-1.02) (25) and $(R_2NH)_xTiS_2$ (R= C_3H_7 - C_8H_{17}; x= 0.95-1.02) (26) can be prepared by heating the hydrazine compound with the desired amine for several days. Replacement of octylamine and dioctylamine will afford similar compounds of nonyl and decyl mono- and diamines.

Basic salts of Hf have been made by dissolving HfO_2 in boiling H_2SO_4, evaporating to dryness, re-dissolving the residue in 2M H_2SO_4 and maintaining the solution at 325° for 3 weeks in a gold-lined autoclave. The products are $Hf(OH)_2SO_4$ and the monohydrate (27).

Several Ti and Zr alkoxides were prepared from the chlorides and alcohols (28). N-methylaminoalkoxides could be made by refluxing $TiCl_4$ with the alcohols in benzene (29). Prepared in this way were $Ti(OR)_{4-n}$ - $(OCHR'CH_2NR''R''')_n$ (R= C_2H_5, i-C_3H_7; n= 1-4; R'= R''= H, R'''= CH_3; R'= H, R''= R'''= CH_3; R'= R''= R'''= CH_3). The tetrakis compounds react with phenyl isocyanate to give insertion products:

$$Ti(OCHR'CH_2NR''R''')_4 + 4C_6H_5NCO \rightarrow Ti[N(C_6H_5)\overset{\displaystyle O}{\overset{\|}{C}}OCHR'CH_2NR''R''']_4 \qquad (4)$$

$$(R'=R''=H, R'''=CH_3;\ R'=H, R''=R'''=CH_3;\ R'=R''=R'''=CH_3).$$

Alkoxides resulted from the reactions of $TiCl_4$ and $(\eta^5-C_5H_5)_2TiCl_2$ with one mole of $LiOCH_2CH_2CH{=}CH_2$. The products, $TiCl_3(OCH_2CH_2CH{=}CH_2)$ and $(\eta^5-C_5H_5)_2TiCl(OCH_2-CH_2CH{=}CH_2)$, did not exhibit olefin-Ti coordination (30). $TiCl_3(OCH_2CH_2CH{=}CH_2)$ gives a 1:1 complex with N,N,N',N'-tetramethylethylenediamine and a 2:1 complex with dimethyl ether. Acid salts result from reacting $TiCl_4$ with various acids. CH_3SO_3H gives $TiCl_{4-n}(SO_3CH_3)_n$ (n=1-4); these compounds give six-coordinate 1:2 adducts with amines (31). Similarly, $TiCl_4$ will react with slight excess of strong monobasic acids in SO_2Cl_2 to give yellow precipitates of $TiCl_2(SO_3CH_3)_2$, $TiCl_2(SO_3F)_2$, $TiCl_2(PO_2F_2)_2$ and $TiCl_2(CO_2CF_3)_2$. $TiCl_3(SO_3CH_3)$ could be made by using excess $TiCl_4$ while other monosubstituted compounds were the products of ligand redistribution reactions (32).

A compound $\{(C_6H_5CH_2)_3Ti\}_2O$ with a Ti-O-Ti bridge may be made by refluxing $Ti(CH_2C_6H_5)_4$ in ether (33). Tetrakis complexes $Ti(SC_6F_5)_4$ and $Hf(SC_6F_5)_4$ result from refluxing Ti or Hf tetrahalides with the thiol (34).

<u>Complexes with O-Donors</u>: The carbonate complexes $K_4M(CO_3)_4$ and $K_6M_2(OH)_2(CO_3)_6$ (M=Zr, Hf) are reported (35) to be eight coordinate with bidentate $CO_3^{=}$ groups. $TiCl_4$ gives 1:1 presumably six-coordinate complexes with the ligands (36).

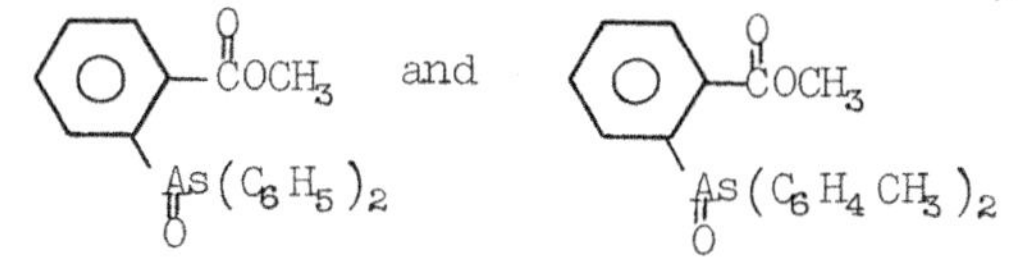

An attempt to incorporate Zr into a boron cage structure via the reaction of $Zr(acac)_4$ (acac=acetylacetonate) with $Na_2(1,2-B_9C_2H_{11})$ gave a compound with the proposed structure shown in Figure I:

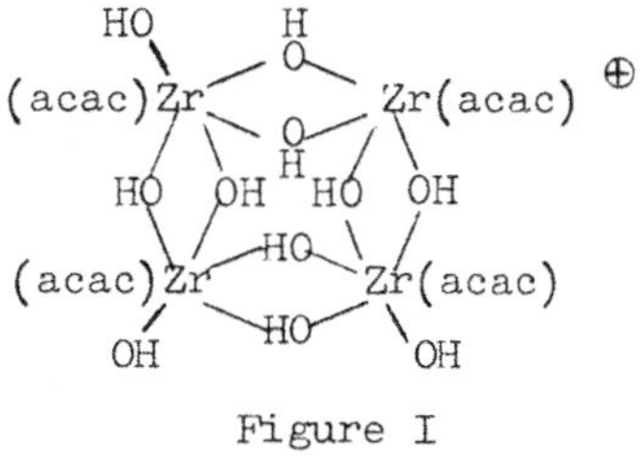

$[1,2-B_9C_2H_{12}]^{\ominus}$

Figure I

Analogous compounds were obtained from $Na_2(1,7-B_9C_2H_{11})$ or by employing $Hf(acac)_4$ (37). Many new β-diketonates were reported last year. By refluxing the diketone in CH_2Cl_2 with TiF_4, Fay and Lowry (38) prepared $TiF_2(diket)_2$ where diket is acetylacetonate, benzoylacetonate, dibenzoylmethanate, dipivaloylmethanate, trifluoroacetylacetonate, benzoyltrifluoroacetonate, thenoyltrifluoroacetonate and hexafluoroacetylacetonate. $Ti(OC_2H_5)_4$ gave $Ti(OC_2H_5)_2(diket)_2$ where diket is acetylacetonate, trifluoroacetylacetonate and hexafluoroacetylacetonate on similar treatment (38). Ligand exchange was studied in these compounds by nmr. Other β-diketonates prepared by similar means include numerous Ti(IV) complexes of β-diketo esters (39) and some of Zr(III) and Hf(III) (40) as well as of Zr(IV) and Hf(IV) (41).

Chelates involving dicarboxylic acids have also appeared in the recent literature. $TiCl_4$ on treatment with excess oxalic acid followed by $In(NO_3)_3$ gives $In[Ti(OH)_2(C_2O_4)_2]_3 \cdot 3H_2O$ (42). A hydroxo complex $[In(OH)_2][O_2TiO(OH)_2] \cdot 2H_2O$ results if NH_3(aq) is employed in place of $H_2C_2O_4$. By adding $MOCl_2$ (M=Zr, Hf) to a two-fold excess of potassium glutarate at pH 10, precipitates of the composition $M_4(C_5H_6O_4)_3(OH)_{10} \cdot 8H_2O$ (M=Zr, Hf) were obtained. Infrared results indicate the presence of mono- and bidentate carboxylate groups (43). With ammonium succinate precipitates of the composition $M_4(C_6H_8O_4)_5(OH)_6 \cdot 12H_2O$ (M=Zr, Hf) result (44). These colorless crystals have tetrameric cyclic $(MO)_4$ groups.

$TiCl_3$ is said (45) to give a 1:1 complex with 5,5'-thiodisalicylic acid presumably containing coordinated O and S.

Complexes with O- and N-Donors: $TiCl_3$ reacts according to equation 4 to produce a compound having a Ti-O-Ti linkage (46) in 98-99% yield.

$$4TiCl_3 + O_2 + 8C_5H_5N \xrightarrow{py} 2[Ti_2Cl_6O\ 4py] \qquad (4)$$

Luminescent complexes of the formulas $Zr(LH)Cl_2$, ZrL_2Cl_2 and $Zr(HL)_2(L)_2Cl_2$ are obtained by treating $ZrCl_4$ in ethanol with 1:1, 1:2, or 1:3 ratios, respectively, of HL =

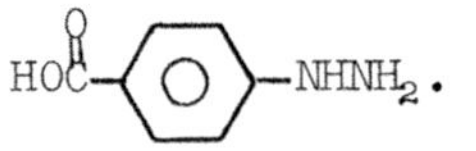

Possibly N and O are coordinated to Zr in one or more of these compounds (47). Other complexes potentially containing both N- and O-donors are those of 4-anilino-3-penten-2-one (48), salicylaldehyde salicylhydrazone (49), and 2-(pyridylazo)resorcinol (50).

Reaction of $TiCl_3 \cdot 3THF$ with a potentially quadrindentate Schiff base (bis-salicylidine)ethylenediimine gives a red compound of the structure shown in Figure II in which Ti(III) has been oxidized to Ti(IV).

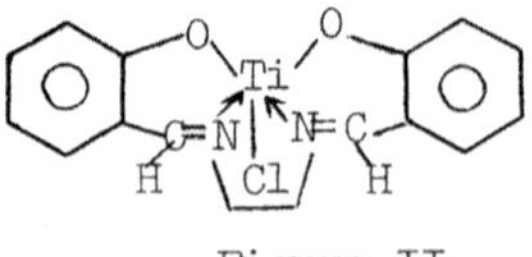

Figure II

With bis(salicylidine)phenylenediamine two compounds are obtained as shown in Figures III and IV.

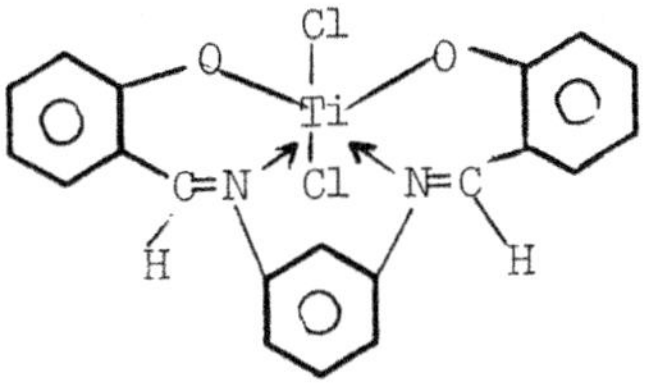

Figure III

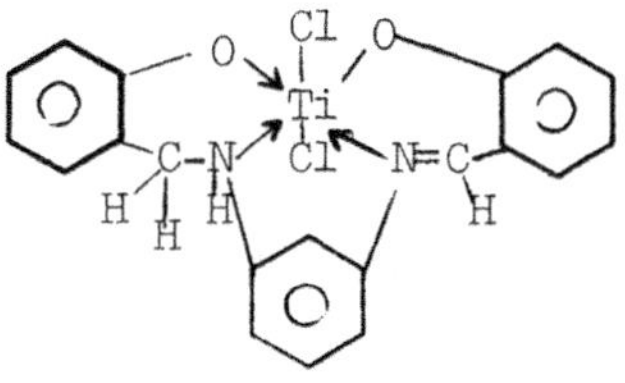

Figure IV

The compound in Figure IV results from a surprising reduction of one of the C=N groups (51). Several Zr(IV) complexes of Schiff bases of the composition $Zr(OC_3H_7)_2(SB)_2$ and $Zr(SB)_4$ (SB=Schiff bases) are obtained (52) in quantitative yield on treatment of $Zr(OC_3H_7)_4 \cdot C_3H_7OH$ with the stoichiometric quantity of Schiff base. These complexes are said to be eight-coordinate around Zr.

Complexes with S-, Se- and Te-Donors: $(\eta^5\text{-}C_5H_5)_2Ti^{II}(CO)_2$ gives oxidative addition products of the composition $[(\eta^5\text{-}C_5H_5)_2Ti(SR)]_2$ (R=CH_3, C_6H_5, $CH_2C_6H_5$, p-$CH_2C_6H_4CH_3$) on treatment with the disulfides. The red-violet products are easily hydrolyzed. These dimers will undergo bridge-splitting on refluxing in toluene with $(\eta^5\text{-}C_5H_5)_2Ti(CO)_2$ to afford $[(\eta^5\text{-}C_5H_5)_2Ti(\mu\text{-}SR)_2Ti(\eta^5\text{-}C_5H_5)_2]$ (53).

$(\eta^5\text{-}C_5H_5)Ti^{III}Cl_2$ will undergo oxidative addition reactions with disulphides in THF (but not in benzene) to afford $(\eta^5\text{-}C_5H_5)Ti^{IV}Cl_2(SR)$ (R=CH_3, C_6H_5). Thiuram disulfides add oxidatively to give $(\eta^5\text{-}C_5H_5)X_2Ti(\langle^S_S\rangle CNR_2)$ (X=Cl, Br; R=CH_3, C_2H_5) (53a).

Similar compounds are prepared starting with $(\eta^5\text{-}C_5H_5)_2TiCl_2$. On treatment with $LiCH_2SC_6H_5$ at -20° yellow orange $(\eta^5\text{-}C_5H_5)_2Ti(CH_2SC_6H_5)_2$ is obtained which gives $(\eta^5\text{-}C_5H_5)_2Ti(SC_6H_5)_2$ and $(\eta^5\text{-}C_5H_5)TiSC_6H_5$ on heating in benzene (54).

Analogous Se and Te compounds may be made in ~80% yield by reactions 5 and 6 (55):

$$(\eta^5\text{-}C_5H_5)_2TiCl_2 + 2REMgBr \xrightarrow{THF} (\eta^5\text{-}C_5H_5)_2Ti(ER)_2 + 2MgBrCl \quad (5)$$

E = Se; R=CH_3, $CH_2C_6H_5$, p-$C_6H_4CH_3$
E = Te; R=C_6H_5, p-$C_6H_4CH_3$

$$(\eta^5\text{-}C_5H_5)_2ZrCl_2 + 2Li\ ER_2 \xrightarrow{THF} (\eta^5\text{-}C_5H_5)_2Zr(ER)_2 + 2LiCl \quad (6)$$

E = Se; R=C_6H_5, p-$C_6H_4CH_3$)
E = Te; R=C_6H_5)

Titanocenophane derivatives containing chelating polysulfides may be prepared via reaction 7 (56):

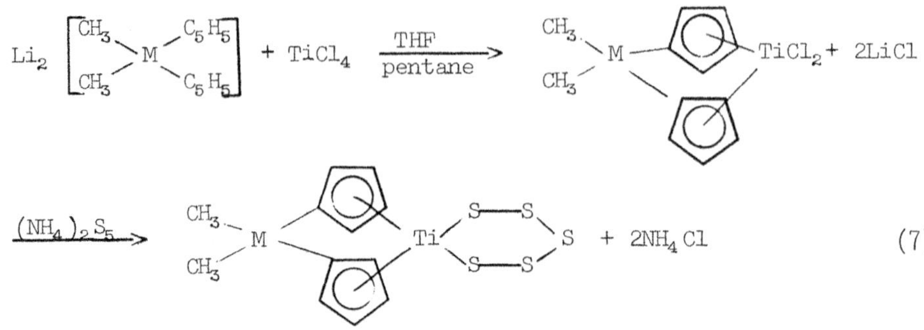

(7)

(M=Si, Ge)

Dithiocarbamates give chelates with $TiCl_4$ of composition $Ti(S_2CNR_2)_nCl_{4-n}$ ($n=2$-4, $R=CH_3$, i-C_3H_7, i-C_4H_9; $n=3$, $R=C_2H_5$) when $TiCl_4$ and anhydrous sodium dithiocarbamates are refluxed in benzene or dichloromethane (57). When $n=3$, the complex is a seven-coordinate pentagonal bipyramid having Cl in axial position (58).

Tris(benzenedithiolato) complexes of Ti, Zr, and Hf may be prepared as tetraalkylammonium salts in nearly 70% yield by reaction 8 (59).

$$M(NR_2)_4 + H_2S_2C_6H_4 + 2HS_2C_6H_4^- \longrightarrow M(S_2C_6H_4)_3^{2-} + 4NHR_2 \qquad (8)$$

(M=Ti, Zr, Hf)

GROUP V COMPOUNDS

Nitrides and Amides: Sodium amide will react with ZrNI in liquid ammonia to give green moisture-sensitive $ZrN(NH_2)$ in 96% yield (60). Amides containing cyclic amide groups $Ti[N(CH_2)_n]_4$ ($n=2$-6) can be obtained by refluxing $Ti[N(CH_3)_2]_4$ with the appropriate amine under N_2 in dry benzene (61). The compounds are yellow crystals or oils.

Complexes: Active research continues into nitrogen fixation with compounds of Ti. Dark blue-black $[(\eta^5\text{-}C_5(CH_3)_5)_2Ti]_2N_2$ may be isolated on treating a toluene solution of $[\eta^5\text{-}C_5(CH_3)_5]_2TiH_2$ with N_2 and cooling to 0^0 (62). At -80^0 $[\eta^5\text{-}C_5(CH_3)_5]_2TiN_2$ may be precipitated as a blue-purple solid. Treatment of either of these with HCl gives some N_2H_4. A similar green complex of Zr formulated as $[(\eta^5\text{-}C_5(CH_3)_5)_2Zr]_2(N_2)_3$ has also been prepared. It dissolves to give paramagnetic red solutions and also gives N_2H_4 on treatment with HCl (63). Russian workers have isolated $[(\eta^5\text{-}C_5H_5)_2Ti]_2N_2$ (64). This compound gives hydrazine on treatment with methanolic HCl or ammonia on reaction with phenyl lithium followed by hydrolysis.

The N_2 group in $ReCl(N_2)[P(CH_3)_2(C_6H_5)]_4$ coordinates to $TiCl_4$ to give $[ReCl(P(CH_3)_2(C_6H_5))_4N_2]_2TiCl_4$ solvates in THF, benzene or CH_2Cl_2. The compound contains two bridging dinitrogen groups (65). These compounds were previously formulated as derivatives of Ti(III). In $(C_2H_5)_2O$, red-brown $[ReCl(P(CH_3)_2(C_6H_5))_4N_2][Ti_2Cl_6O][(C_2H_5)_2O]$ is formed. The structure of this species is not known.

Octamethylcyclotetrasilazane (OMT) reacts with $TiCl_4$ in benzene to give a yellow compound whose formula is $(TiCl_4)_2OMT$. The complex is very air sensitive and has the structure shown in Figure V.

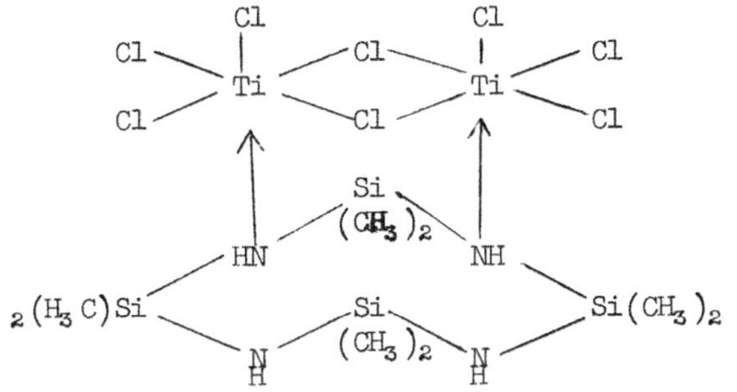

Figure V

Hexamethylcyclotrisilazane gives a similar species while with $TiCl_3$ OMT gives $(TiCl_3)_2OMT\cdot 2THF$ where THF molecules serve to complete the octahedral coordination around Ti(66).

Both 1:1 and 1:2 complexes of $TiCl_4$ with imines $C_6H_5CH{=}NR$ ($R{=}C_6H_5, C_4H_9$) were isolated (67). ZrX_4 (X=Cl,Br,I) forms complexes of composition $ZrX_4(3,5\text{-}L)_2$, $ZrX_4(2,4\text{-}L)_2$ and $ZrX_4(2,6\text{-}L)$ where L is lutidine on mixing stoichiometric amounts of the reactants in benzene (68). $ZrCl_3$ gives complexes of the type $ZrCl_3$ $(3,4\text{-}L)_2$ and $ZrCl_3$ $(2,4\text{-}L)_{1.2}$. The latter stoichiometry may result from insertion of lutidines into chains of $ZrCl_6$ octahedra in the solid (68).

A number of complexes with N-donor ligands have been reported among them complexes of $Zr(OCN)_4$ and $Hf(OCN)_4$ with 2,2'-bipy and 1,10-phen (69), of C_6H_5 Zr $Cl_3\cdot 3THF$ with py, benzonitrile, 2,2'-bipyridine, 4,4'-bipyridine and other ligands (70) and of $Zr(CH_2C_6H_5)_4$ with several mono- and bidentate ligands (71).

GROUP IV COMPOUNDS

Besides their important role as model systems for N fixation, low valent Ti compounds have been reported to fix CO (72). Solutions of $TiCl_3\cdot 3THF$ in THF were treated with Mg and CO. On evaporation of solvent, a black product was obtained which gave acetylene on hydrolysis.

The isocyanide complexes $(\eta^5\text{-}C_5H_5)_2TiCl(CNC_6H_{11})$ and $[(\eta^5\text{-}C_5H_5)TiCl_2\text{-}(CNC_6H_{11})]_2$ may be prepared by reaction of the ligand with $[(\eta^5\text{-}C_5H_5)_2\text{-}$

$TiCl]_2$ or $[(\eta^5-C_5H_5)TiCl_2]_2$ in a bridge-splitting reaction. The isocyanide does not insert into the Ti-C bond of $(\eta^5-C_5H_5)Ti(CH_2C_6H_5)Cl$ but instead simply replaces the benzyl group (73).

Two rather unexpected reactions leading to organometallic compounds may be mentioned briefly. First, $Ti(N(CH_3)_2)_4$ will react with $Mo(CO)_6$, $W(CO)_6$, $Fe(CO)_5$ or $Ni(CO)_4$ to give a metal carbene complex containing the group $M{-}C\begin{smallmatrix}\nearrow OTi(N(CH_3)_2)_3 \\ \searrow N(CH_3)_2\end{smallmatrix}$ (M=Mo, W, Fe, Ni) (74). Second, on reaction of $(\eta^5-C_5H_5)_2TiCl_2$ with $LiN[Si(CH_3)_3]_2$, a deprotonation of a CH_3 group occurs to afford the organometallic complex depicted in Figure VI (75).

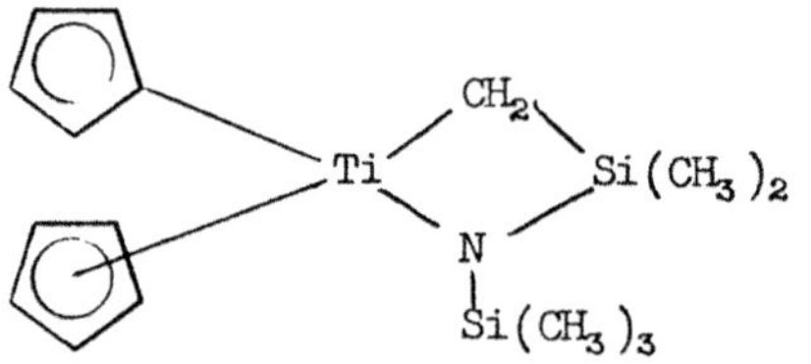

Figure VI

GROUP III COMPOUNDS

Preparative conditions for MB_2 (M=Ti, Zr, Hf) from reaction of metal oxides with B have been given (76).

HYDRIDES

The intermetallic compound FeTi is reported to react with H_2 at ~65 atm pressure and temperatures from 450° to 600° to give two hydrides of composition FeTiH and $FeTiH_2$. Both have dissociation pressures > 1 atm at 0°C (77).

REFERENCES

1. P. C. Wailes, J. ORGANOMETAL. CHEM. 79, 201 (1974).
2. J. T. Allsop, EDUC. CHEM. 10, 222 (1973).
3. C.N.R. Rao, in SOLID STATE CHEM. 1974, C.N.R. Rao, ed., M. Dekker, New York, 1974. p. 455.
4. M. B. Hursthouse, MOL. STRUCT. DIFFR. METHODS 1, 443 (1973).
5. A. Dormond, J. Tirouflet and F. LeMoige, J. ORGANOMETAL. CHEM. 69, C7 (1974).
6. A. Dormond, Ou-Khan and J. Tirouflet, C. R. ACAD. SCI. SER. C 278, 1207 (1974).
7. J. C. Leblanc, C. Moise and T. Bounthakna, C. R. ACAD. SCI. SER. C 278, 973 (1974).
8. C. Moise, J. C. Leblanc and J. Tirouflet, TET. LETT. 1974, 1723.
9. A. K. Bose, P. R. Srinivasan and G. Trainer, J. AMER. CHEM. SOC. 96, 3670 (1974).

10. G. A. Shkurko and I. T. Brakhova, POROSH. MET. 13, 100 (1973); CA 80:56188m (1974).
11. E. M. Larsen, J. W. Moyer, F. Gil-Arnao and M. J. Camp, INORG. CHEM. 13, 574 (1974).
12. A. M. Golub, I. I. Voitko and S. P. Rozhenko, IZV. AKAD. NAUK. SSSR, NEORG. MAT. 10, 287 (1974); CA 81:42095n (1974).
13. J. Senegas and J. Galy, C. R. ACAD. SCI., SER. C. 277, 1243 (1973).
14. R. L. Davidovich, M. A. Medkov and S. B. Ivanov, IZV. AKAD. SCI. SSSR, SER. KHIM. 1974, 1222.
15. H. B. Fritz and W. Treptow, Z. NATURFORSCH. 28b, 584 (1973).
16. K. Issleib, H. Köhler and G. Wille, Z. CHEM. 13, 347 (1973).
17. H. Böhland and H. Steinecke, Z. CHEM. 14, 65 (1974).
18. D. Nicholls, T. A. Ryan and K. R. Seddon, J. CHEM. SOC., CHEM. COMM. 1974, 635.
19. E. D. Ruchkin, A. E. Strelyaev, V. A. Litvinova, E. V. Lazareva and S. S. Batsanov, ZH. NEORG. KHIM. 18, 2855 (1973).
20. M. Inagaki, Y. Masuda, C. Shibata and S. Naka, J. INORG. NUCL. CHEM. 36, 2623 (1974).
21. W. Schartau and R. Hoppe, Z. ANORG. ALLGEM. CHEM. 408, 60 (1974).
22. B. T. Kaminskii, G. N. Prokof'eva, A. S. Plygunov and P. A. Galitskii, POROSH, MET. 1973, 6; CA 80: 5305q (1974).
23. P. C. Donohue and W. Jeitschko, MATER. RES. BULL. 9, 1333 (1974).
24. R. Schöllhorn and A. Weiss, Z. NATURFORSCH, 28b, 711 (1973).
25. A. Weiss and R. Ruthardt, Z. NATURFORSCH. 28b, 249 (1973).
26. A. Weiss and R. Ruthardt, Z. NATURFORSCH. 28b, 522 (1973).
27. M. Hansson, ACTA CHEM. SCAND. 27, 2455 (1973).
28. P. T. Joseph, INDIAN CHEM. J., ANNU. 1972, 135.
29. P. C. Bharara, V. D. Gupta and R. C. Mehrota, Z. ANORG. ALLGEM. CHEM. 403, 337 (1974).
30. R. J. H. Clark and M. A. Coles, J. CHEM. SOC., DALTON TRANS. 1974, 1462.
31. R. C. Paul, V. P. Kapila and S. K. Sharma, J. INORG. NUCL. CHEM. 36, 1933 (1974).
32. J. R. Dalziel, R. D. Kett, P. A. Yeats and F. Aubke, CAN. J. CHEM. 52, 231 (1974).
33. A. Jacot and D. Roulet, CHIMIA 28, 15 (1974).
34. R.J.H. Clark and D. Kaminaris, INORG. CHIM. ACTA 11, L7 (1974).
35. J. Dervin, J. Faucherre and H. Pruszek, REV. CHIM. MINER. 11, 372 (1974).
36. S. S. Sandhu and H. Singh, J. INDIAN CHEM. SOC. 50, 373 (1973).
37. A. R. Seidle, J. INORG. NUCL. CHEM. 35, 3429 (1973).
38. R. C. Fay and R. N. Lowry, INORG. CHEM. 13, 1309 (1974).
39. P. D. Podolesov, V. B. Jordanovska, M. R. Korunoski and D. N. Tošev, J. INORG. NUCL. CHEM. 36, 1495 (1974).
40. M. Kh. Minacheva, E. M. Brainina and L. I. Terekhova, IZV. AKAD. NAUK. SSSR, SER. KHIM, 1973, 2359.
41. E. N. Brainina and M. Minacheeva, IZV. AKAD. NAUK SSSR, SER. KHIM 1974, 2306.
42. E. Macarovici, E. Perte, M. Birou and M. Strajescu, STUD. UNIV. BABES-BOLYAI, SER. CHEM. 18, 123 (1973); CA 81:32704W (1974).

43. L. A. Tsurik and V. V. Serebrennikov, TR. TOMSK. GOS. UNIV. 240, 115 (1973); CA 80:70253k (1974).
44. L. A. Tsurik and V. V. Serebrenikov, TR. TOMSK. GOS. UNIV. 240, 124 (1973); CA 80:140719t (1974).
45. P. C. Srivastava, K. B. Pandeya and H. L. Nigam, J. INORG. NUCL. CHEM. 35, 3613 (1973).
46. C. D. Schmulbach, C. C. Hinckley, C. Kolich, T. A. Ballentine and P. J. Nassiff, INORG. CHEM. 13, 2026 (1974).
47. V. R. Timofeeva and M. S. Novakovskii, ZH. OBSHCH. KHIM. 44, 971 (1974).
48. A. K. Srivastava, V. B. Rana and M. Mohan, INORG. NUCL. CHEM. LETT. 10, 791 (1974).
49. K. K. Narang and A. Aggarwal, INORG. CHIM. ACTA 9, 137 (1974).
50. F. I. Lobanov, H. V. Quy and I. M. Gibalo, ZH. NEORG. KHIM. 19, 2376 (1974).
51. F. L. Bowden and D. Ferguson, J. CHEM. SOC., DALTON TRANS. 1974, 460.
52. S. R. Gupta and J. P. Tandon, MONATSH. FÜR CHEM. 104, 1283 (1973).
53. G. Fachinetti and C. Floriani, J. CHEM. SOC., DALTON TRANS. 1974, 2433.
53a. R.S.P. Coutts and P. C. Wailes, J. ORGANOMETAL. CHEM. 73, C5 (1974).
54. R. Taube and D. Steinborn, J. ORGANOMETAL. CHEM. 65, C9 (1974).
55. M. Sato and T. Yoshida, J. ORGANOMETAL CHEM. 67, 395 (1974).
56. H. Köpf and W. Kahl, J. ORGANOMETAL. CHEM. 64, C37 (1973).
57. A. N. Bhat, R. C. Fay, D. F. Lewis, A. F. Lindmark and S. H. Strauss, INORG. CHEM. 13, 886 (1974).
58. D. F. Lewis and R. C. Fay, J. AMER. CHEM. SOC. 96, 3843 (1974).
59. M. J. Bennett, M. Cowie, J. L. Martin and J. Takatas, J. AMER. CHEM. SOC. 95, 7504 (1973).
60. H. Blunck and R. Zuga, Z. ANORG. ALLGEM. CHEM. 406, 145 (1974).
61. H. Bürger and U. Dämmgen, Z. ANORG. ALLGEM. CHEM. 407, 201 (1974).
62. J. E. Bercaw, J. AMER. CHEM. SOC. 96, 5087 (1974).
63. J. M. Manriquez and J. E. Bercaw, J. AMER. CHEM. SOC. 96, 6229 (1974).
64. Yu. G. Borod'ko, I. N. Iveleva, L. M. Kachapina, S. S. Salienko, A. K. Shilova and A. E. Shilov, ZH. STRUKT. KHIM. 14, 1112 (1973).
65. R. Robson, INORG. CHEM. 13, 475 (1974).
66. J. Hughes and G. R. Willey, J. AMER. CHEM. SOC. 95, 8758 (1975).
67. V. A. Kogan, A. S. Egorov, O. A. Osipov and V. P. Sokolov, ZH. OBSHCH. KHIM. 43, 2719 (1973); CA 80:77780p (1974).
68. E. M. Larsen and T. E. Henzler, INORG. CHEM. 13, 581 (1974).
69. A. M. Golub, T. P. Lishko and V. P. Chistoplyasova, ZH. NEORG. KHIM. 19, 1509 (1974).
70. J. F. Clarke, G.W.A. Fowles and D A. Rice, J. ORGANOMETAL. CHEM. 76, 349 (1974).
71. J. F. Clark, G.W.A. Fowles and D.A. Rice, J. ORGANOMETAL. CHEM. 74, 417 (1974).
72. B. Jezowska-Trzebiatowska and P. Sobota, J. ORGANOMETAL. CHEM. 61, C43 (1973).
73. C. Floriani and G. Fachinetti, J. CHEM. SOC., Dalton Trans. 1973, 1954.
74. W. Petz, J. ORGANOMETAL. CHEM. 72, 369 (1974).

75. C. R. Bennett and D. C. Bradley, J. CHEM. SOC. CHEM. COMMUN. 1973, 29.
76. T. I. Serebryakova and E. V. Marek, METALLOTHERM. PROTESSY KHIM. MET., MATER. KONF. 1971 (pub. 1971), 78; CA 80:77705t (1974).
77. J. J. Reilly and R. H. Wiswall, Jr., INORG. CHEM. 13, 218 (1974).

VANADIUM, NIOBIUM AND TANTALUM

Fred W. Moore and George A. Tsigdinos

Research Laboratory
Climax Molybdenum Company of Michigan
A Subsidiary of AMAX INC
Ann Arbor, Michigan 48105

GENERAL

An annual review for 1973 covering the organometallic chemistry of these elements has been published (1) and the chemistry of the pentafluorides of these elements has been reviewed (2). The synthesis of inorganic compounds of vanadium (3a), niobium and tantalum (3b), as well as the synthesis of heteropoly compounds containing vanadium or niobium (3c) have been critically reviewed. The synthesis of carbonyl (3d) and cyclopentadienyl complexes (3e) of these elements are also included in this review.

VANADIUM

Oxides, Hydroxides, and Oxygen-Containing Compounds: The solid state chemistry of vanadium oxides has been reviewed (4) and a book on vanadates has appeared (5). A review of oxides of the type V_nO_{2n-1} has been published (6). Reduction of ammonium metavanadate with hydrogen at 500° yields pure V_2O_3 (7). Duttonite, $VO(OH)_2$, was prepared under hydrothermal conditions from $VO_2 \cdot H_2O$ which was formed from the reaction of V_2O_5 with SO_2 (8). Vanadium bronzes have been prepared containing NH_4^+ ion (9), cesium or rubidium (10,11), and alkaline earth elements (12).

The compound $KNa(VO_3)_2$ was found to form in the KVO_3-$NaVO_3$-H_2O system (13). The compound $K_2BeV_6O_{11} \cdot 5H_2O$ can be precipitated from a solution of $Be(NO_3)_2$ and K_3VO_4 at pH 4 (14). A study of the MgO-V_2O_5 system revealed the formation of a metavanadate and a divanadate (15). The divanadates of Mg (16,17), Cu (17,18), Co, Ni, and Zn (17) have been prepared. The reaction of $CaCl_2$ with V_2O_5 at 700-1300° yields $Ca_5V_2O_{10}$ as orthorhombic crystals (19) and several complex compounds were obtained by reacting the alkaline earth oxides with mixtures of VO_2 and V_2O_5 (20). The compound $CdVO_3$ was prepared from CdO and VO_2 under high pressure at 1200° (21). The vanadates of Al (22-24), Ga (24), In (22,24) and Tl (24,25) have been examined in detail. The mixed vanadate, $CaZr(VO_4)_2$, has been prepared and characterized (26). The new perovskites, Ba_2NbVO_6 and Sr_2TaVO_6, were prepared at high temperatures from the oxides (27). Compounds containing Se^{4+} with V(IV) or V(V) have been prepared from the appropriate oxides

(28,29). A study of the MoO_3-MVO_3 (M = Li or Na) reveals the formation of $M_2V_2Mo_3O_{15}$ (30), whereas the V_2O_5-Cs_2MoO_4 system yields several complex phases (31). The reaction of uranyl acetate with ammonium vanadate at 70-80° yields a uranyl trivanadate (32). The various polymorphic forms of ThV_2O_7 have been prepared by both aqueous and solid state methods (33). A vanadyl perrhenate has been isolated by an aqueous reaction (34).

The preparation of YVO_4 by high temperature reactions of Y_2O_3 with $NaVO_3$ was examined in detail (35) and the mixed crystal Y(PO_4, VO_4) has been prepared by high temperature reactions (36). Rare earth vanadites containing V(III) have been isolated from the solid state reaction of V_2O_3 with the rare earth oxide at 1400° (37).

Several mixed heteropoly salts containing V have been prepared. The following salts have been described: $Cs_5(PW_{11}VO_{40})$ (38), $K_7(MnV_{13}O_{38})$ (39), $K_7(NiV_{13}O_{38})$ (39), $M_4(V_2W_4O_{19})$ (M = Li, K, Rb) (40), $M_6(V_4W_8O_{36})$ (M = Li, K, Rb) (40), $Na_5V_2W_4O_{38}$ (41), $Na_5V_2W_{10}O_{38}$ (41), $K_7H_2W_{11}VO_{40}$ (42). Salts containing the ions $HV_4O_{12}^{3-}$ (43) and $V_5O_{14}^{3-}$ have also been isolated (44).

<u>Halogen Containing Compounds</u>: The pyrazole and imidazole complexes of VCl_2 of the type $VCl_2 \cdot L_4$ and $VCl_2 \cdot L_6$ have been isolated (45) and a wide variety of complexes were prepared with heterocyclic amines and VX_2 (X = Cl, Br, I) (46). The solid state reaction of VF_3 with several monovalent metal fluorides yields a variety of compounds mainly of the type M_3VF_6 (47). The reaction of VCl_3 with Et_4NN_3 yields the compound $(Et_4N)(VNCl_3)$ which in turn forms an unstable complex with CH_3CN that decomposes to $(Et_4N)(VN_3Cl_3)$ (48). The reaction of chloramine with VCl_4 yields the complex $(NH_4)_2VCl_6$ in good yield (49). Vanadium tetrachloride is reduced by $B_2[N(CH_3)_2]_4$ to the compound $VCl_3 \cdot [N(CH_3)_2]CH_2$ which affords $VCl_3 \cdot 3py$ on reaction with pyridine (50). The etherates, $H(VCl_4) \cdot 5Et_2O$, $H_2(VCl_5) \cdot 7Et_2O$, and $H(VCl_3Br) \cdot 5Et_2O$, were isolated from the reaction of VCl_3 with the corresponding hydrohalic acid (51).

The compound Cs_2VF_6 was prepared and its structure was determined (52). Vanadium oxodichloride was prepared by refluxing $VOCl_3$ with sulfur (53) and complexes of $VOCl_2$ have been prepared with aromatic amines, heterocyclic compounds (53), and urea (54). The compound VOL_2I, where HL is 8-quinolinol, was isolated by reacting the ligand with V(V) and KI (55). Red brown $(NH_4)_3V(O_2)_2F_4$ was prepared from the reaction of H_2O_2 with a HF solution of V_2O_5 (56) and salts of $[VO_2(C_2O_4)F]^{2-}$ are isolated from HF solutions of $H_2C_2O_4$, F^-, and VO_3^- (57). Complexes of the type $VOF_3(AA)$, $VO_2F(AA)$, and $VOF_3 \cdot L$, where AA is o-phen. or bipy and L is pyridine oxide, have been isolated and characterized (58). A 1:1 adduct is formed between VF_5 and $ClOF_3$ by direct reaction (59). The reactions of a β-diketone (HL) with $(Et_4N)(VOF_4)$ yields compounds of the type $(Et_4N)(VOLF_3)$ (60). The compounds, $Cl_2VO(O_2PCl_2)$ and $Cl_2VO(O_2PCl_2)_2$, result from the reaction of $P_2O_3Cl_4$ with $VOCl_3$ or VCl_4, respectively (61). The heating of alkali metal salts of $VOCl_4^{3-}$ in a chlorine stream yields complexes of the type $A_9V_3O_3Cl_{14}$ (62). Complexes of the formulation VOL_2Cl result from the reaction of $VOCl_3$ with a variety of β-diketones (HL) in cyclohexane (63). Vanadium oxotribromide is obtained in good yield by the reaction of $VOCl_3$ with HBr at 0° and $VOBr_2$

is formed by the thermal decomposition of $VOBr_3$ (64). The oxodibromide reacts with liquid NH_3 to yield $VO(NH_3)_5Br_2$, whereas $VOBr_3$ undergoes ammonolysis (64).

Refractory and Related Compounds: Extremely high purity vanadium has been prepared by electron beam melting (65). The compound VH_2 was prepared by reacting activated V powder with H_2 at low temperatures (66). Vanadium diboride is formed by heating tablets of VH and $Na_2B_{12}H_{12}$ in argon (67) and VC is formed from VH and carbon black at 800° (68). Crystals of V_5S_8 were grown by chemical transport (69). The nonstoichiometric phases Na_xVS_2 were prepared from the elements and characterized (70). The reaction of VOCl with liquid NH_3 yields $VO(NH_2)$ which decomposes at 160° to $V(NH_2)_3$ and V_2O_3, whereas VOCl reacts at 700° with NH_3 to form VN (71). Carbonates of V(III) were formed by reacting VCl_3 with $(NH_4)_2CO_3$ and these materials can be readily converted to $VO(OH)_2$ (72). The system V_2O_5-P_2O_5-H_2O was examined and several complex compounds were observed (73), but the reaction of H_3PO_4 with V_2O_5, $VOCl_2$, and V at 220-415° yields $VO(PO_3)_3$, $V(PO_3)_3$, and $VH_2P_3O_{10}$ (74). The anhydrous methanesulfonate of V(III) was prepared from VCl_3 and the acid (75) and the reaction of $ZnMe_2$ with $OV(OPr)_3$ in pentane yields $CH_3VO(OPr)_2$ (76). The compound $K_4[V(CN)_6]\cdot 3H_2O$ was prepared from VSO_4 and KCN and on heating in argon above 480° VC was isolated (77). The salt $K_4[V(CN)_7]\cdot 2H_2O$ has been obtained from VCl_3 (78) or by electrolytic reduction (79).

Coordination Compounds: The chemistry of covalent organic derivatives of vanadium has been reviewed (80). The compound $HV(PF_3)_6$ was prepared by ultraviolet radiation of PF_3 solutions of bis(diglyme)sodium hexacarbonyl vanadate(-I) (81). A complex of 5-aminoindazole (L) of the form $VOLSO_4$ was prepared and characterized (82). The reaction of VCl_3 with sodium disubstituted dithiocarbamates in ethanol solutions yields tris(dithiocarbamate) complexes of V(III) (83). Complexes of V(III) and V(IV) were isolated with thiovanol and characterized (84). A low symmetry complex of bis(2-methyl-8-quinolinato) of oxovanadium(IV), $VO(Quin)_2$, was prepared from $VOSO_4$ and the sodium salt of the ligand (85). The quinoline oxide (L) complexes of VO^{2+} of the form VOL_2X_2 (where X is Cl^-, Br^-, ClO_4^-, NCS^-) and VOL_2X (where X is SO_4^{2-}, $C_2O_4^{2-}$) have been characterized (86). A stable vanadyl (IV) complex of $(C_6H_{11})_2PS_2$ was prepared from a methanolic solution of $VOSO_4$ and the ammonium salt of the ligand (87). A series of vanadyl(IV) diaryldithiophosphinates have been prepared by refluxing $VOSO_4$ with the acid form of the ligand in methanol solution or with the sodium salt of the ligand in aqueous solution (88, 89). A series of solid adducts of 4-methylpyridine oxide with a variety of vanadyl(IV) β-diketonates have been prepared (90). Several oxovanadium(IV) Schiff base complexes have been isolated and characterized (91-95). Vanadyl(IV) complexes of a substituted aromatic oxime (96), a salicylhydrazone (97), and caprohyhroxamic acid (98) have been characterized. A vanadium(V) complex of the type VO_2L is formed with o-hydroxynaphthohydroxamic acid (HL) (99). The reaction of $VOCl_3$ with hexamethyl disilazane yields the imine $(Me_3SiO)_3V{=}N{-}Si(Me)_3$ (100). Vanadium(V) complexes of form VO(OH)L, where HL is 8-quinolinol or 2-methyl-8-quinolinol were prepared as enzyme models (101).

NIOBIUM AND TANTALUM

General: The synthesis and properties of niobium and tantalum compounds have been reviewed (102).

Oxides and Oxygen Containing Phases: Single crystals of the T form of Nb_2O_5 have been obtained by slow cooling under 20 kilobars of pressure (103). The NbO_2-Nb_2O_5 system was studied at 1000-1100° and the new compounds $NbO_{2.42}$ and $NbO_{2.47}$ were found (104). The solid solutions $Nb_xTa_yO_2$, where $0.8 < x + y < 1.0$, are formed by the effect of shock waves on a mixture of Nb_2O_5 with Ta_2O_5 (105).

The formation of compounds between Nb_2O_5 and other metal oxides has been reviewed (106). The first oxoniobate(III), $LiNbO_2$, was obtained by heating Li_2O, NbO_2 and NbO (107). The organic tantalate, $C_6H_{11}Cl_3TaO_4$, was prepared from bis (2-methoxy)ethoxide and its structure was determined (108). Several niobates were observed in the Li_2CO_3-Nb_2O_5 system (109) and several niobates were found in the Na_2O-Nb_2O_5 system (110). Calcium metatantalates were obtained by fusion of CaO-Ta_2O_5 at 1400° or by sintering $CaCO_3$ with Ta_2O_5 (111). Compounds of the type $A_nB_nO_{3n+2}$, where A = Ca, Ca and Ni, B = Nb, n = 4, 5, 6, 7, were formed by heating $Ca_2Nb_2O_7$ with $LiNbO_3$ (112). Strontium orthoniobate was prepared by the reaction of NbO_2 with SrO or $SrCO_3$ at 1000-1200° (113). An amorphous phase of niobium strontium barium oxide hydroxide resulted from the reaction of niobium hydroxide oxide with $Ba(OH)_2$ and $Sr(OH)_2$ (114). The perovskites, Ba_2NbVO_6 and Sr_2TaVO_6, were isolated from the component oxides at 1100-1200° (27). Calcination of MgO and NbO_2 at 1000-1500° results in $MgNbO_3$ with an ilmenite type structure (115). The titanium niobates, $TiNb_2O_7$ and $Ti_2Nb_{10}O_{29}$, were prepared from the oxides and their structures were determined (116). The structure of the phase $(Nb,Zr)O_{2.46}$ in the Nb_2O_5-ZrO_2 system was also determined (117). Some sodium and potassium polyniobates were isolated from the reaction of the alkali metal hydroxides with $(NbO)_2HAsO_4$ (118). The compound $SbNbO_4$ was formed by heating the hydroxides at 700° (119). Phases of the type $(V,Nb)_2O_3$, (V,Nb)O, $(V,Ta)_2O_3$, $(V,Ta)O_2$, $(Ti,Ta)O_2$, and (Ti,Ta)O have been prepared and characterized (120). Conditions for preparing bismuth niobate and tantalate from methanolic solutions of the pentachloride were examined (121). Iron orthoniobate was prepared from the oxides at 1200° (122) and ferric hexatantalate has been prepared by an aqueous preparation (123). Cobalt hexaniobate was also prepared by a precipitation technique and it decomposed to $CoNb_2O_6$ at 690° (124). Fergusonite, $YNbO_4$, was prepared by reaction of the oxides (125) and $NdNbO_4$ was prepared from the hydroxides (126). Europium orthoniobate is reduced by hydrogen to $EuNbO_3$ at 1100°, but other rare earth niobates are not reduced (127). The Eu(II) perovskites, $EuTaCrO_6$, $EuTaAlO_6$, $EuMgTa_2O_9$, and $Eu_3MnTa_2O_9$, were prepared from the oxides (128). The similar compounds, $EuTiNbO_6$ and $SmTiNbO_6$, were isolated from the oxides at 960° and the compound Pb_2CrTaO_6 was prepared at 800° under high pressure (130). A study of the system K_2O-Nb_2O_5-WO_3 revealed the formation of $K_2O \cdot Nb_2O_5 \cdot 2WO_3$ as a definite phase (131). The new pyrochlore type compounds, $NbWO_{5.5}$ and $TaWO_{5.5}$, were observed during the thermal decomposition of the mixed hydroxides (132). The peroxy salt, $Cs_4Nb_2W_4O_{19}(O_2) \cdot 3H_2O$ was prepared by acidification of the lithium salts (133) and several salts of $[Nb(C_2O_4)_4]^{3-}$ have been characterized (134).

Halides and Halogen Containing Compounds: The niobium and tantalum halides $M_6X_{14}\cdot 8H_2O$ (X = Cl, Br) are synthesized in high yields from $A_4M_6X_{18}$ (A = alkali metal) by a disproportionation reaction (135). The cubic compound K_3NbF_7 was synthesized by reduction of molten KF-NbF_5 at 850° with Nb metal (136). The thiochloride $NbSCl_2$ was formed by heating $NbCl_4$ with Sb or As sulfides at 200° (137) and salts of the type $ATaSCl_4$, where A is an alkali metal, are formed from the hexachlorotantalate and Sb_2S_3 at 300° (138). Four new oxofluorotantalates, K_3TaOF_6, $K_2Ta_4O_9F_4$, $K_3TaO_2F_4$, and $K_2Ta_2O_3F_6$, were prepared by heating the calculated quantities of Ta_2O_5, KF, $KTaO_3$, CaF_2, and K_2TaF_7 at 700-900° (139). The optimum conditions for chlorinating Nb_2O_5 and Ta_2O_5 to the pentachlorides were determined (140). A series of peroxyfluorotantalates of the type $Et_4N[TaO_2F(diket)]$, where diket is β-diketone, were prepared by reacting the diketone with the new compound $Et_4N[TaO_2F_4(H_2O)]$ (141) and salts of the type $(NH_4)_3M(O_2)F_4$ (M = Nb, Ta) were obtained from H_2O_2 containing solutions of the metal pentoxide in 30% HF (56). Oxochloro(diketonato)niobates of the formulation Et_4N $[NbOCl_3\cdot diketonate]$ have been prepared and characterized (142). The oxygenyl fluorometallates, O_2NbF_6, $O_2Nb_2F_{11}$, and $O_2Ta_2F_{11}$, have been synthesized by heating the metal at 300-500° in a mixture of O_2 and F_2 (143). Amide adducts of the type $MX_5\cdot nL$, where M = Nb, X = Cl, n = 1,3,4; X = Br, n = 1,3; M = Ta, X = Br, n = 1,3, were isolated from several amides (144) and similar complexes have been isolated with hexamethylphosphoramide as the ligand (145). Complexes of TaF_5 with $(CH_3)_2O$ and $(CH_3)_3N$ have also been synthesized (146) and aromatic amine complexes of $NbCl_5$ have been examined (147). The fluoroniobates, $(AH)(NbOF_4)$, where A is a variety of organic base cations, have been characterized (148) and polymeric $TaOCl_3(NR_3)$, where R = octyl has been prepared from the oxochloride (149). The reaction of $NbCl_5$ with $(Et)_4N_3$ in $CHCl_3$ yields the compound $Et_4NNbN_3Cl_5$ (48) and the reaction of NH_2Cl with MCl_5 (M = Nb, Ta) yields NH_4MCl_6 salts (49). The azides MCL_4N_3 (M = Nb, Ta) can be obtained from the reaction of the corresponding pentahloride with ClN_3 or NaN_3 (150). A 1:1 complex between $ClOF_3$ with either NbF_5 or TaF_5 is formed when they are reacted in a vacuum line (59). The 1:2 adduct $NbOCl_3\cdot 2OSMe_2$ was prepared from $NbOCl_3$ and its dissociation in CH_3CN was studied (151). A range of methyl niobium(V) and methyl tantalum (V) oxohalide complexes of the formula $MeMOX_2\cdot 2L$ have been synthesized with a variety of ligands (152).

Refractory and Related Compounds: High purity Nb metal was obtained by hydrogen reduction of $NbCl_5$ (153). The formation of a hydride in the Nb-H system at low temperatures was observed (154) and NbH_2 was synthesized from activated Nb at -20 to -60° (66). The boride Nb_3B_4 was formed from the metal and $Na_2B_{12}H_{12}$ (67) and the conditions for boride formation from BX_3 and MX_5 (X = Br, Cl; M = Nb, Ta) by thermal decomposition were studied (155). The nitrides of Nb and Ta have been examined by several workers (156-160) and the oxide-nitride of niobium was formed by heating the chloroimido derivatives formed from $NbOCl_3$ and NH_3 at 600° (161). The reaction of hydrogen with mixtures of $SiCl_4$ and $NbCl_5$ or $TaCl_5$ yields Nb_5Si_3 and Ta_5Si_3 respectively (162). Several nonstoichiometric intercalation compounds of TaS_2 (163) and Ta_2S_2C (164) have been characterized. The formation of the compounds Nb_3Ge, Nb_3Sn, Nb_3Pb has been investigated under high pressures and temperatures (165). Niobium phosphide was prepared by electrolysis of a mixture of metaphosphate, alkali metal halide, and Nb_2O_5 at 700-800° (166). The Nb-Ga system has been examined and the compounds $NbGa_3$, Nb_5Ga_4, Nb_3Ga, and Nb_3Ga_2 have been identified (167, 168). A review of the preparation and properties of $NbSe_2$ has appeared (169) and the mixed compounds $TaS_{2-x}Se_4$ (x = 0-2) have been characterized (170). A new sulfocarbide of tantalum was formed by the reaction of Ta with CS_2 at 800-1000° (171) and the compounds $Cu_3MS_xSe_{4-x}$ (M = Nb, Ta) have been isolated from the reaction of the elements (172). A high pressure form of Nb_3Te has been characterized (173).

Coordination Compounds: The preparation of colorless dimethyl sulfoxide complexes of the composition $M[NbO(C_2O_4)(dmso)_2]$ (M = NH_4, K, Cs, Rb) were prepared and characterized (174). Niobium pentaethoxide forms a variety of complexes with oximes (175) and similar complexes are formed between $Nb(OCH_3)_5$ and several heterocyclic bases (176). The reaction of $Nb(OR)_5$ with PhNCO yields insertion complexes of the type $(RO)_{5-n}Nb[NPh(CO_2R)]n$ (n = 1-5) in benzene solution (177). Compounds of the formulation L_4NbCl_5 (L = benzimidazole and 2-substituted imidazoles) were prepared and they were found to be 8 coordinate (178). Niobium(V) Schiff base complexes of the type NbL_2Cl_3 (179, 180), $NbCl_3L$ (180), and $Nb(Oi\text{-}Pr)_{5-x}L_x$ (x = 1-4) (181) have been isolated.

REFERENCES

1. P. C. Wailes, J. ORGANOMET. CHEM. 75, 325 (1974).
2. R. D. Peacock, ADVAN. FLUORINE CHEM. 7, 113 (1973).
3. "METHODIUM CHIMICUM, VOL. 8, TRANSITION ELEMENTS AND THEIR COMPOUNDS." Ed. by K. Niedenzu and H. Zimmer, Georg Thieme Verlag, Stuttgart, 1974.

   a. R. J. H. Clark, VANADIUM, pg. 111.
   b. L. Kolditz, NIOBIUM AND TANTALUM, pg. 123
   c. G. A. Tsigdinos, HETEROPOLY COMPOUNDS, pg. 497.
   d. R. B. King, CARBONYLS OF TRANSITION METALS, pg. 393
   e. H. Rosenberg, SANDWICH COMPOUNDS, pg. 457.
4. C. N. R. Rao, SOLID STATE CHEM. 1974, 455.
5. M. V. Moskhosoev, A. P. Nakhodnova, and B. V. Slobodin, "CHEMISTRY OF RARE ELEMENTS, PT. 2, VANADATES," Donestk. Gos. Univ., Donestk, U.S.S.R., 1972.
6. Y. Kachi, NIPPON BUTSARI GAKKAISHI, 27, 910 (1972).
7. Y. Abe and H. Yomaguchi, JAPAN KOKAI 74-47, 294, May, 1974.
8. R. Bak, DAEHAN HWAHAK HWOEJEE, 17, 424 (1973)
9. T. Palanisamy, J. Gopolakrishnan, and M. V. C. Sastri, J. SOLID STATE CHEM. 9, 273 (1974).
10. V. L. Volkov, L. L. Surat, and A. A. Fotiev, IZV. AKAD. NAUK SSSR, NEORG. MATER. 10, 181 (1974).
11. L. L. Surat, V. L. Volkov, and A. A. Fotiev, ZH. NEORG. KHIM. 19, 2875 (1974).
12. V. L. Volkov and A. A. Fotiev, IZV. AKAD. NAUK SSSR, NEORG. MATER. 10, 182 (1974).

13. A. P. Yatsenko, A. A. Ivakin, M. P. Glazyrin, and N. S. Matseeva, ZH. NEORG. KHIM. 19, 3171 (1974).
14. A. P. Nakhodnova, L. N. Kovgan, M. V. Mokhosoev, ZH. NEORG. KHIM. 19, 67 (1974).
15. E. Pollert, SILIKATY, 17, 337 (1973).
16. R. Gopal and C. Calvo, ACTA CRYSTALLOGR. 30B, 2491 (1974).
17. J. C. Pedregsa, E. J. Baran, and P. J. Aymonino, Z. ANORG. ALLGEM. CHEM. 404, 308 (1974).
18. D. Mercurio-Lavaud and B. Frit, C. R. ACAD. SCI., SER. C, 277, 1104 (1973).
19. A. A. Fotiev and V. A. Markorov, ZH. NEORG. KHIM. 19, 1116 (1974).
20. J. C. Bouloux, J. Galy, and P. Hagenmuller, REV. CHIM. MINER. 11, 48 (1974).
21. B. L. Chamberland and P. S. Danielson, J. SOLID STATE CHEM. 10, 249 (1974).
22. A. P. Nakhodnova, L. F. Chernysh, and M. V. Mokhosoev, U.S.S.R. Pat. 414,196, Feb. 5, 1974.
23. L. F. Chernysh, A. P. Nokhodnova, and M. V. Mokhosoev, IZV. AKAD. NAUK SSSR, NEORG. MATER. 10, 640 (1974).
24. B. G. Golovkin and A. A. Fotiev, RUSS. J. INORG. CHEM. 18, 1362 (1973).
25. M. Touboul, M. Ganne, C. Cucke, and M. Touanoux, Z. ANORG. ALLGEM. CHEM. 410, 1 (1974).
26. V. P. Balko, G. M. Rylov, and V. V. Bokakin, ZH. STRUKT. KHIM. 15, 155 (1974).
27. J. C. Bernier, C. Chauvel, and O. Kahn, J. SOLID STATE CHEM. 11, 265 (1974).
28. G. Meunier, M. Bertaud, and J. Galy, ACTA CRYSTALLOGR. 30B, 2834 (1974).
29. A. A. Kuznetsova, V. V. Kovalev, and V. T. Kalinnikov, ZH. NEORG. KHIM. 19, 2274 (1974).
30. B. N. Slobodin, M. V. Mokhosoev, and N. G. Kabanova, RUSS. J. INORG. HCEM. 18, 1181 (1973).
31. B. V. Slobodin, M. V. Mokhosoev, and P. T. Zvoleiko, ZH. NEORG. KHIM. 19, 388 (1974).
32. K. Stadnicka, ROCZ. CHEM. 47, 202 (1973).
33. E. J. Baran, L. A. Gentil, J. C. Pedregosa, and P. J. Aymonino, Z. ANORG. ALLGEM. CHEM. 410, 301 (1974).
34. Yu. A. Busulaev, A. A. Kuznetsova, V. V. Kovalev, and V. T. Kalinnikov, ZH. NEORG. KHIM. 19, 3398 (1974).
35. A. A. Fotiev and M. Ya. Khodov, TR. INST. KHIM. URAL. NAUCH. TSENTR., AKAD. NAUK SSSR 25, 53 (1973).
36. B. Hajek and A. Much, J. LESS-COMMON METALS 35, 339 (1974).
37. T. Shinike, G. Adachi, and J. Shiokawa, NIPPON KAGAKU KAISHI 1974, 244 and 1038.
38. I. M. Novoselova, V. F. Barkovskii, Z. I. Shishkina, and I. A. Leont'eva, ZH. NEORG.KHIM. 19, 3078 (1974).
39. G. B. Kauffman, R. Fuller, J. Felser, C. M. Flynn, Jr., and M. T. Pope, INORG. SYNTH. 15, 103 (1974).
40. D. F. Takezhanova, D. U. Begalieva, A. K. Il'yasova, and A. B. Bekturov, DOKL. AKAD. NAUK SSSR 216, 335 (1974).
41. D. F. Takezhanova, D. U. Begalieva, A. K. Il'yasova, and A. B. Bektirov, DOKL. AKAD. NAUK SSSR 215, 1139 (1974).

42. C. M. Flynn, Jr., M. T. Pope, and S. O'Donnell, INORG. CHEM. 13, 831 (1974).
43. J. Fuchs, Z. NATURFORSCH., PT. B, 28, 389 (1973).
44. L. Zurkova, V. Sucha, C. Vargova, and J. Corba, COLLECT. CZECH. CHEM. COMMUN. 39, 387 (1974).
45. M. Issogoni, N. Katsaros, E. Vrachnou-Astra, and E. Olympios, INORG. CHIM. ACTA 9, 131 (1974).
46. M. Khamar, L. F. Larkworthy, K. C. Patel, D. J. Phillips, and G. Beech, AUST. J. CHEM. 27, 41 (1974).
47. J. C. Cretenet, REV. CHIM. MINER. 10, 399 (1973).
48. M. Kasper and R. D. Bereman, INORG. NUCL. CHEM. LETT. 10, 443 (1974).
49. H. P. Fritz and W. Treptow, Z. NATURFORSCH., PT. B, 28, 584 (1973).
50. R. F. Kiesel and E. P. Schram, INORG. CHEM. 13, 1313 (1974).
51. A. G. Galinos and E. D. Manesis, ANN. CHIM. (PARIS) 8, 369 (1973).
52. K. Walterson, K. A. Wilhelmi, A. Carpy, and J. Galy, BULL. SOC. FR. MINERAL. CRISTALLOGR. 97, 13 (1974).
53. A. K. Datta and M. A. Hand, Z. ANORG. ALLGEM. CHEM. 407, 75 (1974).
54. Z. O. Dzhamalova, O. F. Khodzhaev, and N. A. Parpiev, ZH. NEORG. KHIM. 19, 3032 (1974).
55. Y. Anjaneyulu, A. S. R. Murty, V. P. R. Rao, PROC. NAT. ACAD. SCI., INDIA, SECT. A, 43, Pt. 1-2, 84 (1973).
56. G. Pausewang, Z. NATURFORSCH., PT. B., 29, 272 (1974).
57. H. Rieskamp and R. Mattes, Z. NATURFORSCH., PT. B, 29, 271 (1974).
58. J. Sala-Pala and J. Guerchais, J. MOL. STRUCT. 20, 169 (1974).
59. R. Bougon, T. Bui Huy, A. Cadet, P. Charpin, and R. Rousson, INORG. CHEM. 13, 690 (1974).
60. J. Sala-Pala and J. Guerchais, BULL. SOC. CHIM. FR. 1974, 2913 (1973).
61. A. F. Shihada and K. Dehnicke, Z. NATURFORSCH., PT. B, 28, 268 (1973).
62. A. I. Morozov, ZH. NEORG. KHIM. 19, 1810 (1974).
63. J. P. Fackler, Jr., SYN. REACTIV. INORG. METAL-ORG. CHEM. 4, 61 (1974).
64. A. Anagnostopoulos, D. Nicholls, and M. E. Pettifer, J. CHEM. SOC., DALTON TRANS., 1974, 569.
65. J. Bressers, R. Creten, and G. Van Holsbeke, J. LESS-COMMON METALS 39, 7 (1975).
66. A. A. Chertkov, L. N. Padurets, and V. I. Mikheeva, DOKL. AKAD. NAUK SSSR 215, 610 (1974).
67. T. Hanslik and F. Mares, CZECH. PAT. 153,828, June 15, 1974.
68. G. V. Samsonov, V. Vaumanko, L. N. Okhremchuk, B. M. Rud, and A. M. Lemeshko, IZV. AKAD. NAUK SSSR, NEORG. MATER. 10, 52 (1974).
69. M. Saeki, M. Nakano, and M. Nakahira, J. CRYST. GROWTH 1974, 24, 154.
70. G. A. Wiegers, R. Van der Meer, H. Van Heiningen, H. J. Kloosterboer, and A. J. A. Alberink, MATER. RES. BULL. 9, 1261 (1974).
71. N. I. Vorob'ev, V. V. Pechkovskii, and L. V. Kobets, ZH. NEORG. KHIM. 19, 3 (1974).
72. V. P. Chayi and F. T. Yakushevskaya, MATER. MEZHOTRASL. SOVESHCH. METOD. POLUCH. ANAL. FERRITOVYKH, SEGNETO-P'EZOELEKTRICHESKIKH MATER. SYR'YA NIKH, 3RD 2, 111 (1971).
73. L. A. Pechenina, U. Sedmalis, and Z. Konstants, LATV. PSR ZINAT. AKAD. VESTIS, KIM. SER. 1974, 536 (1974).
74. A. V. Lavrov, L. S. Guzeeva, and P. M. Fedorov, IZV. AKAD. NAUK SSSR, NEORG. MATER. 10, 2180 (1974).

75. R. C. Paul, V. P. Kapila, N. Palta, and S. K. Sharma, INDIAN J. CHEM. 12, 825 (1974).
76. K. H. Thiele, B. Adler, H. Grahlert, and A. Lachowicz, Z. ANORG. ALLGEM. CHEM. 403, 279 (1974).
77. V. V. Dovgei, K. N. Mikhalevich, and A. N. Sergeeva, RUSS. J. INORG. CHEM. 19, 513 (1974).
78. R. A. Levenson, R. J. G. Dominguez, M. A. Willis, and F. R. Young, III, INORG. CHEM. 13, 2761 (1974).
79. K. Mikhalevich, A. N. Sergeeva, and V. V. Dovgei, U.S.S.R. PAT. 394,306, August 22, 1973.
80. R. N. Grimes, U. S. NAT. TECH. INFORM. SERV., AD REP. 1973, NO. 766842/9.
81. T. Kruck and H. U. Hempel, ANGEW. CHEM. 86, 233 (1974).
82. S. A. A. Zaidi and A. S. Farooqi, J. INORG. NUCL. CHEM. 35, 4320 (1973).
83. L. Que, Jr. and L. H. Pignolet, INORG. CHEM. 13, 351 (1974).
84. K. M. Kanth, K. B. Pandeya, and H. L. Nigam, INDIAN J. CHEM. 11, 1031 (1973).
85. H. J. Stoklosa, J. R. Wasson, and B. J. McCormick, INORG. CHEM. 13, 592 (1974).
86. G. Durgaprasad and C. C. Patel, INDIAN J. CHEM. 11, 1300 (1973).
87. H.J. Stoklosa, G. L. Seebach, and J. R. Wasson, J. PHYS. CHEM. 78, 962 (1974).
88. R. N. Mukherjee, M. S. Venkateshan, and M. D. Zingde, PROC. CHEM. SYMP. 2, 69 (1972).
89. R. N. Mukherjee, S. V. Shanbhag, M. S. Venkateshan, and M. D. Zingde, INDIAN J. CHEM. 11, 1066 91973).
90. N. S. Al-Niaimi, A. R. Al-Karaghouli, S. M. Aliwi, and M. G. Jalhoom, J. INORG. NUCL. CHEM. 36, 283 (1974).
91. H. Okawa, I. Ando, and S. Kida, BULL. CHEM. SOC. JPN., 47, 3041 (1974).
92. R. L. Dutta and G. P. Sengupta, J.INDIAN CHEM. SOC. 50, 640 (1973).
93. G. M. Klesova, L. V. Mosina, V. V. Zelentsov, Yu. V. Yablokov, and V. I. Spitsyn, ZH. NEORG. KHIM. 19, 1155 (1974).
94. A. Pasini and M. Gullotti, J. COORD. CHEM. 3, 319 (1974).
95. G. Durgaprasad and C. C. Patel, INDIAN J. CHEM. 12, 840 (1974).
96. R. N. Saksena and K. K. Panday, J. INDIAN CHEM. SOC. 50, 609 (1973).
97. K. K. Narang and A. Aggarwal, INORG. CHIM. ACTA 9, 137 (1974).
98. V. F. Popova and A. L. Markman, TR. TASHKENT. POLITEKH. INST. 74, 13 (1971).
99. B. G. Ostrobrod and A. L. Markman, TR. TASHKENT POLITEKH. INST. 74, 16 (1971).
100. A. F. Shihada, Z. ANORG. ALLGEM. CHEM. 408, 9 (1974).
101. L. W. Amos and D. T. Sawyer, INORG. CHEM. 13, 78 (1974).
102. A. K. Sharov, A. A. Fotiev, and E. I. Krylov, "SYNTHESIS AND PROPERTIES OF NIOBIUM, TANTALUM, AND TITANIUM COMPOUNDS," Nauka, Moskow, 1974.
103. S. Tamura, K. Kato, and M. Goto, Z. ANORG. ALLGEM. CHEM. 410, 313 (1974).
104. J. F. Marucco, J. SOLID STATE CHEM. 10, 211 (1974)
105. A. V. Anan'in, O. N. Breusov, A. N. Dremin, V. N. Drobyshev, and S. V. Pershin, ZH. NEORG. KHIM. 19, 710 (1974).

106. G. Tilloca, REV. INT. HAUTES TEMP. REFRACT. 10, 183 (1973).
107. G. Meyer and R. Hoppe, ANGEW. CHEM. 86, 819 (1974).
108. M. G. B. Drew and J. D. Wilkins, INORG. NUCL. CHEM. LETT. 10, 549 (1974).
109. E. K. Belyaev and V. F. Annopol'skii, IZV. AKAD. NAUK SSSR, NEORG. MATER. 10, 1507 (1974).
110. P. Appendino, ANN. CHIM. (ROME), 63, 547 (1973).
111. L. V. Favorskaya, G. A. Nikiforova, E. N. Pankova, and I. V. Kirichenko, TEKHNOL. MINER. SYR'YA 1972, 58.
112. M. Nanot, F. Queyroux, J. Giles, A. Carpy, and J. Galy, J. SOLID STATE CHEM. 11, 272 (1974).
113. G. Kasimov, E. G. Vovkotrub, and E. I. Krylov, ZH. NEORG. KHIM. 19, 60 (1974).
114. V. I. Spitsyn, V. M. Ermolaev, V. V. Sakharov, and M. A. Zakharov, ZH. NEORG. KHIM. 19, 3237 (1974).
115. G. G. Kasimov, E. M. Elovskikh, V. G. Teplov, V. I. Dvoinin, and B. P. Filin, ZH. OBSHCH. KHIM. 44, 2581 (1974).
116. R. B. Von Dreele and A. K. Cheetham, PROC. ROY. SOC. LONDON 338, 311 (1974).
117. B. Nolander and R. Norin, ACTA CHEM. SCAND. 28A, 252 (1974).
118. Z. M. Titov and A. K. Sharov, TR. INST. KHIM., URAL NAUCH. TSENTR. AKAD. NAUK SSSR 25, 145 (1973).
119. A. M. Sych, V. I. Popolitov, and L. A. Eremenko, ZH. NEORG. KHIM. 19, 2391 (1974).
120. G. P. Shveikin, S. I. Alyamovskii. V. A. Perelyaev, and Uy. G. Zainulin, TR. INST. KHIM., URAL NAUCH. TSENTR., AKAD. NAUK SSSR 25, 3 (1973).
121. A. M. Sych and L. A. Eremenko, U.S.S.R. PAT. 393,214, August 10, 1973.
122. G. G. Kasimov, E. G. Vovkotrub, I. G. Razanov, and A. U. Ospanov. IZV. AKAD. NAUK SSSR, NEORG. MATER. 10, 934 (1974).
123. Z. E. Batura., V. I. Bezrukov, and N.G. Kisel, ZH. NEORG. KHIM. 19, 2200 (1974).
124. M. S. Sinakina, V. I. Bezrukov, and N.G. Kisel, ZH. NEORG.KHIM. 19, 2945 (1974).
125. V. S. Darshne and V. V. Deshpande, PROC. CHEM. SYMP. 2, 325 (1972).
126. A. M. Sych, L. A. Eremenko, L. A. Zastavker, and M. M. Nekrasov, IZV. AKAD. NAUK SSSR, NEORG. MATER. 10, 496 (1974).
127. E. I. Krylov, E. M. Elovkikh, G. G. Kasimov, and B. P. Filin, ZH.
128. H. Parant, J. C. Bernier, and P. Poix, C. R. ACAD. SCI., SER. C, 278, 49 (1974).
129. E. I. Krylov, A. E. Borosov, Yu. N. Makurin, and G. G. Kasimov, IZV. AKAD. NAUK SSSR, NEORG. MATER. 10, 74 (1974).
130. L. G. Sevast'yanova, E. V. Zubova, K. P. Burdina, and Yu. N. Venevtsev, IZV. AKAD. NAUK SSSR, NEORG. MATER. 10, 1359 (1974).
131. T. Ikeda and S. Shimizu, JAP. J. APPL. PHYS. 13, 256 (1974).
132. D. Groult, C. Michel, and B. Raveau, J. INORG. NUCL. CHEM. 36, 61 (1974).
133. M. Dabbabi, M. Boyer, and P. Souchay, C. R. ACAD. SCI., SER. C, 278, 1047 (1974).
134. E. N. Traggeim, R. N. Shchelokov, and M. A. Michnik, KHIM. KHIM. TEKHNOL., TR. YUBILEINOI CONF., POSVYASHCH. 70-LETIYU INST. 1970, 294.

135. F. W. Koknat, J. A. Parsons, and A. Vongvusharintra, INORG. CHEM. 13, 1699 (1974).
136. L. O. Gilpatrick and L. M. Toth, INORG. CHEM. 13, 2242 (1974).
137. I. S. Morozov and N. P. Dergacheva, U.S.S. R. PAT. 430,064, May 30, 1974.
138. A. I. Morozov, U.S.S.R. PAT. 430,058, May 30, 1974.
139. J. Chaminade, M. Vlasse, M. Pouchard, and P. Hagenmuller, BULL. SOC. CHIM. FR. 1974, 1791.
140. I. A. Glukhov, L. M. Shalukhina, and A. Sharipov, IZV. AKAD. NAUK TADZH. SSR, OTD. FIZ.-MAT. GEOL-KHIM. NAUK 1973, 48 (1973).
141. J. Y. Calves and J. E. Guerchais, J. FLUORINE CHEM. 1974, 47 (1974).
142. R. Kergoat and J. E. Guerchais, J. LESS-COMMON METALS 39, 91 (1975).
143. A. J. Edwards, W. E. Falconer, J. E. Griffiths, W. A. Sunder, and M. J. Vasile, J. CHEM. SOC., DALTON TRANS. 1974, 1129.
144. F. J. R. Masaguer and M. R. Bermejo, AN. QUIM. 69, 1099 (1973).
145. F. J. R. Masaguer and J. Sordo, An. Quim. 69, 1263 (1973).
146. S. Brownstein and M. J. Farrall, CAN. J. CHEM. 52, 1958 (1974).
147. L. A. Ugulava, N. I. Pirtskhalava, and A. F. Peresun'ko, SOOBSHCH. AKAD. NAUK GRUZ. SSR 75, 593 (1974).
148. A. K. Sengupta and N. G. Mandal, J. INDIAN CHEM. SOC. 51, 579 (1974).
149. S. Sinitsyna, V. G. Khlebodarov, and N. A. Bukhtereva, ZH. NEORG. KHIM. 19, 1532 (1974).
150. J. Strähle, Z. ANORG. ALLGEM. CHEM. 405, 139 (1974).
151. J. G. Riess, R. C. Muller, and M. Postel, INORG. CHEM. 13, 1802 (1974).
152. C. Santini-Scampucci and J. G. Riess, J. CHEM. SOC., DALTON TRANS. 1974, 1433.
153. I. V. Petruevich and L. A. Nisel'son, NAUCHN. TR. NAUCHNO-ISSLED. PROEKTN. INST. REDKOMET. PROMSTI. 1972, No. 42, 37.
154. G. R. Holmquist and H. K. Birnbaum, U. S. TECH. INFORM. SERV., AD REPT NO. 763480, 1973.
155. B. Armas, REV. INT. HAUTES TEMP. REFRACT. 10, 231 (1973).
156. G. Brauer and E. Mohr-Rosenbaum, Z. ANORG. ALLGEM. CHEM. 405, 225 (1974).
157. J. Spitz, J. Chevallier, and A. Aubert, J. LESS-COMMON METALS 35, 181 (1974).
158. A. Deschanvres, A. Maisseu, G. Nouet, and J. Vicens, J. LESS-COMMON METALS 35, 237 (1974).
159. J. Strähle, Z. ANORG. ALLGEM. CHEM. 402, 47 (1973).
160. I. P. Borovinskaya and A. G. Merzhanov. METALLOTERM. PROTSESSY KHIM. MET., MATER. KONF. 1971, 58.
161. N. I. Vorob'ev and L. V. Kobets, RUSS. J. INORG. CHEM. 18, 1581 (1973).
162. F. N. Kozlov, I. V. Petrusevich, A. E. Donskoi, and L. A. Niel'son, NAUCHN. TR. NAUCHNO-ISSLED. PROEKTN. INST. REDKOMET. PROMSTI. 1972, No. 42, 61.
163. M. S. Whittingham, J. CHEM. SOC., CHEM. COMMUN. 1974, 328.
164. R. Schoellhorn and A. Weiss, Z. NATURFORSCH., PT. B, 28, 716 (1973)
165. J. M. Leger, J. LESS COMMON METALS 34, 17 (1974).
166. A. M. Golub, F. F. Grigorenko, N. S. Slobodyanik, P. G. Nagornyi, and T. I. Fedoruk, U.S.S.R. PAT. 436,018, July 15, 1974.
167. P. Feschotte and E. L. Spitz, J. LESS-COMMON METALS 37, 233 (1974).
168. M. Drys, R. Kubiak, and K. Lukaszewicz, BULL. ACAD. POL. SCI., SER. CHIM. 21, 901 (1973).

169. Y. Ishizawa, KOTAI BUTSURI, 9, 209 (1974).
170. J. F. Revelli and W. A. Phillips, J. SOLID STATE CHEM. 9, 176 (1974).
171. M. Caillet, A. Galerie, and J. Besson, J. LESS-COMMON METALS 37, 111 (1974).
172. A. Muller and W. Sievert, Z. ANORG. ALLGEM. CHEM. 406, 80 (1974)
173. J. F. Cannon, D. L. Robertson, H. T. Hall, and A. C. Lawson, J. PHYS. CHEM. SOLIDS 35, 1181 (1974).
174. N. Brnicevic and C. Djordjevic, J. CHEM. SOC., DALTON TRANS. 1974, 165.
175. R. Bohra, A. K.Rai, and R. C. Mehrotra, INDIAN J. CHEM. 12, 855 (1974)
176. L. G. Hubert-Pfalzgraf and J. G. Riess, J. CHEM. SOC., DALTON TRANS. 1974, 585.
177. R. C. Mehrotra, A. K. Rai, and R. Bohra, J. INORG. NUCL. CHEM. 36, 1887 (1974).
178. N. S. Biradar, T. R. Goudar, and V. H. Kulkarni, J. INORG. NUCL. CHEM. 36, 1181 (1974).
179. N. S. Biradar and T. R. Goudar, CURR. SCI. 43, 740 (1974).
180. N. S. Biradar, T. R. Goudar, V. B. Mahale, and V. H. Kulkarni, J. KARNATAK UNIV. 18, 19 (1973).
181. S. R. Gupta and J. P. Tandon, ANN. SOC. SCI. BRUXELLES, SER. 1, 88, 251 (1974).

CHROMIUM, MOLYBDENUM AND TUNGSTEN

George A. Tsigdinos and Fred W. Moore

Research Laboratory
Climax Molybdenum Company of Michigan
A Subsidiary of AMAX INC
Ann Arbor, Michigan 48105

GENERAL

A number of reviews and papers on all general aspects of the chemistry of chromium, molybdenum and tungsten have appeared during 1974. An annual survey on the organometallic aspects of chromium, molybdenum and tungsten containing 385 references has been published (1). Sixty-three papers presented at the First International Symposium on Molybdenum Chemistry held at the University of Reading in 1973 have been published in a single volume of the Journal of Less-Common Metals (2). A review with 237 references on heteropoly tungstates and heteropolymolybdates has also appeared (3). Similarly, a review with 560 references on transition-metal oxides including tungsten has also been published (4). The molecular structure (by X-ray diffraction) of chromium, molybdenum and tungsten compounds has been reviewed (5). The chemistry of trivalent chromium has been summarized in a review with 53 references (6). A book on the Crystal Chemistry and Stereochemistry of Molybdenum Compounds has been published (7). A critical review of preparative procedures for compounds of chromium, molybdenum and tungsten and for heteropoly compounds has been compiled (8).

CHROMIUM

Oxides and Hydroxides: The decomposition of chromyl chloride in the presence of oxygen at 380-400° gave pure CrO_2 at pressures over 60 atm (9). $Cr(OH)_3$ gels undergo hydrothermal reaction at 200-50° to give orthorhombic CrOOH (10). Phase diagram studies for the systems $M_2SO_4 \cdot 3MoO_3$-Cr_2O_3, where M = K, Rb and Cs, have been carried out at 450-400°. In all cases the compounds $M_2SO_4 \cdot 3MoO_3 \cdot Cr_2O_3$ were identified (11). Contrary to previous reports, CrO_3 and graphite do not react directly to form intercalation compounds. Only in the presence of glacial acetic acid a compound is obtained (12).

Chromates and Chromites: Finely dispersed chromium borates of type $Cr_2O_3 \cdot B_2O_3$ and $2CrO_3 \cdot 3B_2O_3 \cdot nH_2O$, where n = 13 - 18, were obtained from sodium or potassium nitrate solutions. The compounds were characterized by X-ray (13). The pure perovskite of composition Pb_2CrTaO_6 was prepared at 800° and at 30-40 kbar. Dialectric studies showed a weak anomaly at

-50° (14). The thermal decomposition of various zinc chromates was studied by various instrumental techniques and the products obtained were identified (15). A double salt of composition $(NH_4)_2Cr_2O_7 \cdot (NH_4)_2MoO_4$ was prepared by heating at 50° a solution of $(NH_4)_2CrO_4$ and MoO_3. The product decomposes without melting (16). Single crystals of $SrCr_2O_4$ were prepared and investigated by X-ray. The coordination about Cr^{+3} is octahedral and about Sr^{+2} is trigonal prismatic (17).

Halogen-Containing Compounds of Chromium: A novel preparation of chromium(II) chloride was reported. Metal pellets react with HCl in THF under nitrogen to give pure $CrCl_2 \cdot$THF (18). The direct fluorination of CrO_3 in a static thermal system has yielded CrO_2F_2 and $CrOF_4$, the product obtained depending on the reaction temperature (19). The composition of the vapor phase in the cesium chloride-chromium(II) chloride system indicated the presence of $CsCrCl_3$ (20). An intercalation compound of chromium trichloride-graphite forms by the direct reaction of the materials. Between room and liquid nitrogen temperatures, the magnetic properties of the inserted chloride are the same as that of the free chloride (21). Gaseous complexes of $CrCl_3$ are formed during chemical transport reactions involving $AlCl_3$, $GaCl_3$, $FeCl_3$ and $NbCl_5$ as the transport agents (22). The anion $CrMoCl_9^{-3}$ was isolated as the brown salt $[(n\text{-}C_4H_9)_4N]CrMoCl_9$ from ethylene dichloride solutions containing $[(n\text{-}C_4H_9)_4N]_2MoCl_6$ and $[(n\text{-}C_4H_9)_4N]Cl$. The complex contains both divalent chromium and molybdenum (23). Complexes of quininine with chromium(II) chloride of type $L.3CrCl_3 \cdot 6H_2O$ were prepared from the metal halide and the ligand and characterized by spectra and magnetic moments (24). Thermally stable (up to 360°) μ-halobis (μ-phosphinato) chromium(III) polymers of type $[CrX(OPRR^1O)_2]_n$, where R = R^1 = Ph, C_8H_{17}; R = Me, R^1 = Ph; X = Cl, Br, I, were prepared by the reaction of anhydrous $CrCl_3$ and the appropriate phosphinic acids (25).

Refractory and Related Compounds: Thallium thiochromate, $TlCr_3S_5$, the first thiochromate(III) with honeycomb structure was prepared by the thermal decomposition of $TlCrS_2$ at 800°. The compound is a metallic conductor (26). Phase relations between spinel and defect NiAs structures in the system $M_{1-x}Ni_xCr_2S_4$ (where M = Mn, Fe, Co) were investigated. The spinel structure is stable between x=0 and x=0.3 (27). Single crystals of $Cr_2S_{3-x}Se_x$, where x=0.75-2.7, were prepared by chemical transport with $AlCl_3$ or $CrCl_3$ above 980°(28). Cr_2S_3 was prepared by the reaction of tris(dipyridyl)chromium with sulfur, by reaction of Ph_3Cr with H_2S, or by reaction of $CrCl_3 \cdot 3$THF with Li and H_2S (29). The new compounds RCr_3S_6 (R = Y, Gd, Dy, Ho or Er) were synthesized by high temperature reactions of $RCrS_3$ and $CrCl_3$ or I_2 (30). The compounds $(M,Co)Cr_2S_4$ with M = Mn, Zn, or Cd were formed by chemical transport involving $AlCl_3$. When $CrCl_3$ + HCl mixture was used as the transport agent, $FeCr_2S_4$ and $CoCr_2S_4$ were obtained (31). Phase studies of the ternany system U-Cr-C showed the formation of $UCrC_2$ (32).

Miscellaneous: Tris(dialkylamino)stibine complexes of type $(Me_3N)_3SbCr(CO)_5$ were prepared by the reaction of the stibine with $Cr(CO)_6$ (33). The stoichiometry of the formation of a surface complex of Cr(II) (~1% on silica) with molecular nitrogen was studied. A 1:2 Cr-N_2 complex forms at 1 atm (34). Transition metal-carbene complexes like cis-and trans-

$(CO)_4Cr[C(OMe)Me][PR_3]$, $(CO)_5Cr[C(OMe)Me]$, $(CO)_5Cr(PR_3)$ and trans-$(CO)_4Cr(PR_3)_2$ were prepared by the thermal decomposition of $(CO)_5Cr[C(OMe)(PR_3)Me]$ (35). The preparation of pentaaquo(3-chloropyridine)-chromium(III) and pentaaquo (3-cyanopyridine)chromium(III) perchlorates were prepared by the reduction of the respective pyridine adducts of diperoxychromium(VI) species with acidic ferrous perchlorate solution (36). The preparation and redox behavior of oxocentered complex ion $[Cr_3O(MeCO_2)_6L_3]^+$, where L = py, β-picoline, has also been reported (37). A method for preparing salts of the dithionatopentamminechromium(III) ion has been described. The IR and UV spectra of the compounds prepared have been given (38). $Me_4M(SnCl_3)$ has been reported to react with $Cr(CO)_6$ or $Cr(C_6H_6)(CO)_3$ in refluxing THF to give $Me_4N[Cr(CO)_5SnCl_3]$ or $(Me_4N)_3[Cr(CO)_3(SnCl_3)_3]$(39). The preparation and characterization of the dioctylphosphinate $Cr(OP(C_8H_{17})_2O)_2$ containing divalent chromium has been reported (40). Chromium hexacarbonyl reacts with 1,4-diselenan to form yellow-green crystals of 1,4-diselenantetracarbonylmetal (41). $[CrFe_2O(RCO_2)_6(H_2O)_3]NO_3.nH_2O$ (R = CH_2Cl,Et,Ph) and $[Cr_2FeO(RCO_2)_6(H_2O)_3]NO_3.nH_2O$ were prepared by the reaction of $Cr(NO_3)_3$ and $Fe(NO_3)_3$ with $(RCO)_2O$ in the presence of absolute ethanol (42). The reaction of o-tetrachlorobenzoquinone with $Cr(CO)_6$ yielded neutral tris (o-benzoquinone) metal complex, $Cr(O_2C_6Cl_4)_3$ (43). Chromium(0) 1,2-dithionates like $[Cr(S_2C_2Ph_2)_3]$ have been prepared and characterized (44). When $Cr_2(SO_4)_3$ was treated with ethylenediamine and barium sulfamate, $[Cr(en)_3](SO_3NH_2)_3$ was precipitated (45). The preparation of several polynuclear compounds of chromium(III) with alkyldinitrilotetracetic acids has been reported (46). The reaction of N_2O_5 with potassium trisoxalatochromate(III) yielded $K_3[Cr(NO_3)_6]$ with a magnetic moment of 3.70 BM (47). Chromium hexacarbonyl was prepared by the autoclave reaction of CO with powdered Cr metal obtained by the reduction of $CrCl_3.3THF$ with potassium (48). Tricarbonylchromium complexes of structurally isomeric borazine derivatives were prepared from $Cr(CO)_3(MeCN)_3$ and the corresponding borazine in dioxane or without solvent (49). The heteropoly compound $K_5[SiCrW_{11}O_{39}(H_2O)].10H_2O$ and the corresponding acid have been prepared and characterized (50). The preparation of triple sulfates containing sodium, magnesium and chromium have been prepared and their dehydration behavior studied (51).

MOLYBDENUM

Oxides and their Hydrates: Five species of hydrous MoO_3 were prepared under various conditions and their structure and chemical behavior ascertained. These three were the yellow mono and dihydrates and a white mono-hydrate (52). The acid properties of MoO_3 with respect to the O_2^- ion in a molten equimolar KCl-NaCl mixture was investigated at 700° (53). The crystal structure of the white alpha molybdic acid, $MoO_3.H_2O$, was reported. It is related to the layer structure of MoO_3 which can be thought of as a condensation product of $MoO_3.H_2O$ chains (54). The kinetics of the formation of MoO_{3-x} from a mixture of MoO_3 and α-Al_2O_3 was studied. The species formed at 600-700° were identified as $MoO_{2.88}$, $MoO_{2.76}$ and $MoO_{2.75}$ (55). The B_2O_3-MoO_3 binary oxide system was studied by thermal, optical and X-ray examination but no binary solid phases were observed (56). Study of the ternary system lead oxide-boron oxide-molybdenum trioxide showed the phases $2PbO.MoO_3$, $PbO.MoO_3$, $4PbO.B_2O_3$, $2PbO.B_2O_3$, $5PbO.4B_2O_3$ and $PbO.2B_2O_3$ and the ternary phase

$15PbO.2B_2O_3.2MoO_3$, congruently melting at 772° (57). $UMo_{10}O_{32}$ was prepared by heating $UMo_{10.5}O_{32}$ at 260°. Alpha, beta, and gamma $U_3Mo_{10}O_{32}$ were also prepared (58). An examination of the potassium metaphosphate-tungsten trioxide-molybdenum trioxide system has been reported (59). The compound $M_2SO_4.3MoO_3.Cr_2O_3$ was identified in the system M_2SO_4-MoO_3-Cr_2O_3 where M = K, Rb and Cs (11).

Simple Molybdates: The salt $(NH_4)_2Cr_2O_7.(NH_4)_2MoO_4$ was prepared from aqueous solutions of $(NH_4)_2Cr_2O_7$ and MoO_3 at 50° (16). In the $ThMo_2O_8$-Cs_2MoO_4 system, 2:1, 1:1, and 1:3 compounds have been reported to form (60). The H_2O-WO_3-MoO_3 system at 25° was studied. Solubility curves for MoO_3, $ZnMo_3O_{10}.5H_2O$ and $ZnMo_2O_7.5H_2O$ were established (61). The compounds Sb_2MoO_6 and $Sb_2(MoO_4)_3$ were established in the Sb_2O_3-MoO_3 system. Sb_2MoO_6 decomposes in air at 400° (62). Double molybdates of alkali metals and bismuth, $M^+ Bi(MoO_4)_2$ where M = Li, Na, K, Rb or Cs have been reported (63). The kinetics and mechanism of the synthesis of cadmium molybdate from $CdCO_3$ + MoO_3 or CdO + MoO_3 have been studied (64). The formation of $MnMoO_4$ by the solid state reaction of manganese oxides and MoO_3 has also been studied (65). The polymorphic transformation of $CoMoO_4$ from 25-1200° has been studied. Two new modifications found that are responsible for the explosion-wise transition that is observed in this molybdate (66). The determination of the structure of $UMo_{10}O_{32}$ indicates the presence of hexagonal bipyramids of $(UO_2)_{96}$ and MoO_6 octahedra (67). A study of the crystal structure of α-UMo_2O_8 indicates that it is different from that of α-$ThMo_2O_8$ (68). The new compounds with the scheelite structure, $Bi_3(FeO_4)(MoO_4)_2$ and $Bi_3(GaO_4)(MoO_4)_2$ have been prepared (69). Several double molybdates of scandium with sodium or indium with sodium have been studied (70). New tetragonal scheelite-type compounds of stoichiometry $Eu_x^{+2}MoO_4 (0.67 \leq X \leq 1.00)$ and $Ln^{+3}MoO_4$ (Ln = Y, Gd, Lu) have been prepared by heating reaction mixtures in sealed tubes. These are the first compounds reported in which all the Mo is present as MoO_4 tetrahedra (71). The conditions for the formation and several of the properties of double molybdates of the rare earth elements of the yttrium subgroup and the alkali metals has been reviewed (72). The potassium molybdate-lanthanum molybdate system was studied by DTA and X-ray from 300-1100°. The compounds $KLa(MoO_4)_2$ and $K_5La(MoO_4)_4$ were observed (73). The molybdates $Zr(MoO_4)_2$, $Hf(MoO_4)_2$, Y_6MoO_{12}, Y_2MoO_6, and $Y_2(MoO_4)_3$ were synthesized and their optical properties determined (74). The equilibria observed in the Cs_2MoO_4-MoO_3 system have been reported in detail (75). A review with 82 references has been published dealing with the preparation, structure and properties of $R^{+3}M^+(MoO_4)_2$, where R is the rare earth, M is the alkali metal (76).

Isopoly Compounds: Rare earth hexamolybdates of type $Ln_2O_3.6MoO_3.nH_2O$ (Ln = La, Pr, Nd, Sm, Eu, Gd) were produced by the reaction of Ln_2O_3 with MoO_3 at 660 - 720° (77). New compounds like $NaMoO_4$, Na_3MoO_4 and $K_6Mo_2O_4$ were prepared by the reaction of sodium or potassium metal vapors with MoO_2 and Na_2MoO_4 at 300 - 600° at vapor pressures 0.01 - 0.1 torr. The chemical and physical properties of the new materials is described (78). Thallium(I) trimolybdate, $Tl_2Mo_3O_{10}$, was prepared and examined by X-ray powder patterns. The space group is P1 or P$\bar{1}$ the density being 4.90 (78). Equilibrium phase diagram studies on the system Cs_2MoO_4-V_2O_5 showed the presence of the compounds $Cs_4Mo_2V_6O_{23}$, $Cs_4Mo_2V_{10}O_{33}$ and $Cs_2MoV_{10}O_{25.75}$. The latter is a cesium oxygen vanadium molybdenum bronze (79). The solubility of ammonium paramolybdate has been increased by the addition of a water soluble aliphatic amine. Solutions thus prepared were used for the impregnation of catalyst supports (80). Acidification of sodium molybdate solutions were studied thermometrically. The major species formed at 3M ionic strength were $HMo_2O_8^{-3}$, $Mo_7O_{24}^{-6}$, $Mo_8O_{26}^{-4}$ and MoO_3 (81).

Heteropoly Compounds: Saturated solutions of 12-molybdosilicic acid were prepared by ion exchange without the preliminary synthesis of the corresponding heteropoly salts (82). The beta form of 12-molybdosilicic acid was prepared and isolated as a yellow hygroscopic solid from 1:1 mixtures of water-dioxane (83). The neptunium heteropoly acid $H_8[NpMo_{12}O_{42}].18H_2O$ was prepared from ammonium paramolybdate and $(NH_4)_2Np(NO_3)_6$ solution at 65 - 70°. It is isomorphous with the corresponding $H_8[UMo_{12}O_{42}].18H_2O$ (84). Methods for preparing 12-molybdophosphoric and 12-molybdosilicic acids and their metal salts of Mn, Co, Ni, Cu, La, and Ag have been described. The acids and salts thus prepared were examined for their thermal behavior, solubility, and hydrolytic stability in aqueous and mixed solvents (85). The thermal decomposition of the crystalline isomers of 12-molybdosilicic acid has been studied and the nature of the phases obtained was established (86). The thermal decomposition of reduced 12-molybdosilicic acid was studied. It was found that the stability of these acids increases with the degree of reduction (87). The reactivity of the molybdenum oxides obtained from such decomposition was studied in the oxidation of carbon monoxide (88). The surface area of the oxides obtained from such composition is ~10 m^2/g (89). Spectroscopic studies of heteropoly compounds of the type $[ZMo_{12}O_{42}]^{-8}$ where Z = Th^{+4}, U^{+4} and Ce^{+4} showed that they have the same structure as solids as in solution. $[CeM_2Mo_{12}O_{42}]^{-4}$ ions are formed with divalent metals M^{+2}. Th^{+4} reacts with $[CeMo_{12}O_{42}]^{-8}$ and $[UMo_{12}O_{42}]^{-8}$ to form complexes like $[ThCeMo_{12}O_{42}]^{-4}$ (90). Nanomolybdophosphate, $[PMo_9O_{31}(OH)_3]^{-6}$, a new heteropoly compound, was obtained as hexagonal $Na_3H_3[PMo_9O_3(OH)_3].10H_2O$ by evaporation of solutions containing ~1g $Na_2MoO_4.2H_2O$ and 0.5 g $NaH_2PO_4.2H_2O$ and its structure determined (91). A novel structure has been reported in $(NH_4)_5[(MoO_3)_5(PO_4)(HPO_4)].3H_2O$. The anion consists of a ring of five distorted MoO_6 octahedra, the crown thus formed is capped on one face by PO_4^{-3} and on the other by HPO_4^{-2} (92). Spectral studies have indicated that reduced molybdoantimolylphosphoric acid has the stoichiometry $PSb_2Mo_{10}O_{40}$ (93). A solution study of the arsenate-molybdate-H^+ system indicated that no molybdoarsenate is stable over an extended pH range. The species $H_nAs_2Mo_{18}O_{62}^{n-6}$ is stable at 0.3>pH. The tetrameric $H_8[As_4Mo_{12}O_{50}]$ has the same stability range (94). The latter was prepared by dissolving MoO_3 in H_3AsO_4 solution followed by acidification (95). The molecular structure of the hexamolybdotellurate ion in the crystal complex with telluric acid, $(NH_4)_6[TeMo_6O_{24}].Te(OH)_6.7H_2O$ has been reported (96). The compound $LiH_2[AlMo_6O_{24}H_6].8H_2O$ was prepared and characterized (97). The replacement of W by Mo in $(NH_4)_6[P_2W_{18}O_{62}]$ to give $(NH_4)_6[P_2Mo_4W_{14}O_{62}]$ increases the sensitivity of the reaction toward mercapto compounds and other organic ligands (98). The basicity of $H_5[PW_6Mo_6O_{40}]$ was found to be three in glacial acetic acid (99). The dilatometric titration of MoO_4^{-2} with H_5IO_6 resulted in the formation of

$[IMo_6O_{24}]^{-5}$, with the anion $[HIMo_4O_{20}]^{-8}$ being the intermediate (100).

Thiomolybdates and Chalcogenides: Structure studies have shown that 40-50 wt % substitution of Nb and Ta atoms for Mo in MoS_2 semiconductors can be accomplished without changing the structure. However, a large amount of stacking faults results in the structure (101). The synthesis of molybdenum disulfide from Mo powder and S was studied at 640-1020°. Above 720° the process is diffusion controlled. The preparation of MoS_2 from $CaMoO_4$ in molten Na_2CO_3 or K_2CO_3 in 92-95% yields was developed. The lubricating power of these MoS_2 samples was similar to that of naturally occurring MoS_2 (102). $MoSe_2$ was prepared from the elements in a halide reaction medium (103). Germanium reacted with MoO_3 at 950-1050° to give $MoGe_2$ of microhardness 930-40 Kg/cm^2 (104). Layer intercalation compounds of MoS_2 with alkali and alkaline earth metals show cation exchange behavior after decomposition with water or aqueous ammonia (105). The compounds $CoMo_2S_4$ and $FeMo_2S_4$ are isostructural, their structures consist of octahedral chains of MoS_6 and CoS_6 extending along the a and b axis of the monoclinic cell (106). Thermodynamic calculations and experimental data have shown that Mo monocarbides are stabilized at high pressures. At 25 kbars the decomposition of the hexagonal η-carbide occurred at $\leq$1380° (107). The compound Mo_4Si_3C was synthesized in a rotating plasma reactor (108). Ternary phase studies of U, Mo and C showed the formation of $UMoC_2$ (32). The preparation of ternary complex borides like $ThMoB_4$ has also been reported (109). Hydrazinium tetrathiomolybdate, $N_2H_6MoS_4$, has been prepared by dissolving MoO_3 in 40% N_2H_4 and saturating with H_2S (110). The unusually stable sulfur-bridged ion, $Mo_2S_2O^{+2}$, was separated from a $Na_2Mo_2O_2S_2$(cysteine)$_2$ solution by addition of HCl. The anion is stable in 10M HCl and evolved no H_2S after several weeks (111). The novel salt, $K_3(MoOS_3)Cl$ was prepared from Na_2MoO_4-H_2S solutions by precipitation with an organic solvent (112). The Ph_4P^+ salts of $M(MoOS_3)_2^{-2}$ (M = Zn and Ni) and $M(MoO_2S_2)_2^{-2}$(M = Ni and Co) were prepared by the reaction of Ph_4PCl with aqueous solutions of Cs_2MoOS_3 or $(NH_4)_2MoO_2S_2$ containing the appropriate transition metal salt (113).

Halogen-Containing Compounds: The interaction of graphite with MoF_6 has been studied thermogravimetrically. The lamellar products obtained have the composition C_5MoF_6, C_7MoF_6, and C_8MoF_6 (114). Reaction of MoO_3 with organic bases (L) in 40% HF gave salts $(LH)_2MoO_2F_4$ where L = γ-picoline, 2,6-lutidine, quinoline, etc. (115) The preparation of hydroxyfluoromolybdates(V) of the type $Mo(OH)F_6^{-2}$ has been reported (116). Compounds extracted from fluoride solutions containing tri-n-octylamine have the composition $(R_3NH)_2MoOF_4$ (117). Novel methods for preparing MoF_5 and MoF_4 were devised. Reduction of MoF_6 with silicon metal yielded MoF_5 and further reduction yielded MoF_4 (118). Molybdenum trichloride has been synthesized by the reaction of $MoCl_5$ with anhydrous $SnCl_2$ or red P at 210-220° (119). The reduction of $(NH_4)_2MoO_4$ by a mixture of HI and HCl or by HBr has been shown a convenient method for preparing $(NH_4)_2MoOCl_5$ and $(NH_4)_2MoOBr_5$ (120). The KCl-$MoCl_3$ system was investigated by differential thermal analysis. The compound K_3MoCl_6 was found with a congruent m.p. at 865° and a polymorphic transition at 460° (121). Polynuclear metal halides like $(H_3O)_2[(Mo_6X_8)X_6]\cdot 6H_2O$, where X = Cl, Br, were prepared by the reduction of MoX_5 with Al (122). Double-bridged di-μ-oxo-bis[oxomolybdates(V)], μ-oxo-μ-(sulfato, perchlorato, oxalato)-bis[oxomolybdates(V)], and tetrameric di-μ-oxo-bis[di-oxo-oxodi-molybdates(V)] have been obtained by appropriate synthesis routes (123). The oxidation of molybdenum dichloride with octachlorocyclopentene at 270° has yielded $[Mo_6Cl_{12}]Cl_3$ in which the oxidation state of Mo is +2.5. The salt $(Et_4N)_3[(Mo_6Cl_{12})Cl_6]$ was also readily obtained (124). Routes for preparing the anions $Mo_2Cl_9^{-3}$ have been devised (125). The direct synthesis of the anion $CrMoCl_4^{-3}$ has been reported (23). Gaseous complexes of $MoCl_3$ have been shown to form in chemical transport reactions involving $AlCl_3$, $GaCl_3$, etc. (22). $MoOCl_4$ reacts with anisole to give $MoOCl_2(C_6H_4OOMe)_2$ (126). The hydrolysis of cis-$[MoOX_3L]$, where X = Cl, Br and L = 2,2-bipyridyl, 1,10-phenanthroline, gave the dioxobridged compounds $Mo_2O_4X_2L_2$ (127). Tetrakis(dithiocarboxylato)molybdenum(IV) has been synthesized and its properties determined (128). The reaction of $MoCl_6^{-3}$ or $MoCl_6^-$ with $Mo(CO)_4Cl_3^-$ in CH_2Cl_2 resulted in the formation of $Mo_2Cl_9^{-3}$ and $MoCl_9^{-2}$, respectively (129). The reaction of $(NH_4)_2Mo_2Cl_9 \cdot H_2O$ with Et_2S, 2,5-dithiahexane and other ligands has yeilded complexes of formula $Mo_2Cl_4L_4$ and $Mo_2Cl_4(LL)_2$ (130). The preparation of the compound $Mo_2(en)_4Cl_4$ has been described (131). Halides of molybdenum with sulfur or selenium like $MoSCl_3$, $MoSeCl_3$, $MoSCl_2$, $MoSeCl_2$ and MoS_2Cl_2 have been described (132). The oxidation of $MoCl_4(PPrPh_2)_2$ by p-$MeC_6H_4SO_3N_3$ gave $MoCl_4(NPPrPh_2)(OPPrPh_2)$ (133). The reaction of $Mo_2Cl_8^{-4}$ with pyridine, 2,2'-pyridine and 1,10-phenanthroline yielded new complexes of type $Mo_2Cl_4L_2$ (134). Cationic molybdenum(IV) oxohaloisocyanide complexes have been synthesized and their properties determined (135). Et_4NBr reacts with $MoOCl_4$ to form $(Et_4N)_2$ $MoOCl_4Br$ (136).

<u>Miscellaneous</u>: Isothiocyanate complexes of Mo(VI), cis-$(R_4N)_2[MoO_2(NCS)_4]$ have been prepared and found to behave as 2:1 electrolytes in acetonitrile (137). Zerovalent complexes of molybdenum were obtained by the reaction of $Mo(CO)_6$ with benzil bis(phenylimines) (138). The preparation of Li_4 $[Mo_2Me_8]$ $\cdot 4Et_2O$ was accomplished by the reaction of LiMe with $Mo_2(O_2CMe)_4 \cdot Et_2O$ (139). The coordination compounds $MoOCl_3 \cdot 2(CH_3)_2NCOH$ and $[(C_2H_5)_4N]_2MoO(NCSe)_5$ have been prepared and characterized (140). Cationic carbonylnitrosyl complexes like $[Mo(CO)(NO)L_2]PF_6$ have been prepared from $Mo(CO)_4L$, $(Ph_2PCH_2)_2$=L, and $NOPF_6$ (141). The preparation of π-ethylene complexes of Mo(0) and hydrido-(acetylacetonato) complex of Mo(II) with diphosphine ligands has been reported (142). $ZrOCl_2$ reacts with $K_3[MoO(OH)(CN)_4]$ to form $(ZrO)_3[MoO(OH)(CN)_4]_2$ (143). Photolysis of $Mo(CO)_5L$ (L = NH_3, n-$PrNH_2$ and piperidine) in the presence of excess L yielded $Mo(CO)_4L_2$ complexes (144). Hexakis(dimethyl-aminato)- and hexakis(tert-butoxy)dimolybdenum(III) have been prepared from $MoCl_3$ and $LiNMe_2$ or LiOBu (145). Complexes of Mo(VI or V) with 8-quinolinol have been prepared and the implications of the weak Mo-N bonding for the Mo-flavin interaction in metalloflavoenzymes has been discussed (146). Pentacarbonyl(tetraphosphorus trisulfide)molybdenum, $Mo(CO)_5P_4S_3$, has been prepared and its structure determined (147). $[Mo_2(EtOAc)_4](CF_3SO_3)_4$ has been obtained by dissolving $Mo_2(OAc)_4$ in anhydrous EtOAc containing CR_3SO_3H (148). The first solid flavin complexes of molybdenum(IV) have been isolated and characterized thermally and spectrally. Their general formulation is $MoOCl_3$(H-flavin) (149). When MeLi was added to $Li_2[MoH_2Ph_2Me_2Ph_2] \cdot 2THF$ in Et_2O at -20° the diamagnetic complex $Li_4[MoH_2Me_2Ph_2]$ precipitated (150). The preparation of the complex $(C_6H_6)Mo[P(cy)_2Me]_2(N_2)$ where cy = cycloxexyl

and of the μ-dinitrogen-bis[1,2-bis(dimethylphosphino)ethane]hydrido μ-[(1,3,5-trimethylbenzene)]molybdenum has been described (151). Tetrakis-diethyldithiocarbamatomolybdenum(IV) was prepared from $Mo(CO)_6$ and Et_2NCS_2H in refluxing acetone (152). The reaction of o-tetrachlorobenzoquinone with $Mo(CO)_6$ yielded the neutral $Mo(O_2C_6Cl_4)_3$ complex (43). The preparation of 1,4-diselenantetracarbonyl complexes of molybdenum has been reported (41). $Me_4N[Mo(CO)_5SnCl_3]$ was obtained by the reaction of $Mo(CO)_6$ with $Me_4N(SnCl_3)$ (39). The complex $(Me_2N)_3SbMo(CO)_5$ formed when $Mo(CO)_6$ was reacted with $(Me_3N)_3Sb$ (33). $K_4[MoO(OH)(CN)_4]2H_2O$ and $K_4[MoO_2(CN)_4]\cdot 8H_2O$ have been prepared by irradiating basic solutions of $K_4[Mo(CN)_8]\cdot 2H_2O$ for 24 and 12 hr, respectively (153).

TUNGSTEN

Oxides: The ternary system lead(II) oxide-boron oxide-tungsten trioxide was investigated. Six binary and one ternary phase, were identified (154). An investigation of the ternary system KPO_3-WO_3-MoO_3 and of the binary system KPO_3-WO_3 was also published (59). The system Bi_2O_3-WO_3 was studied at 500-1100°. Four intermediate phases, $7Bi_2O_3\cdot WO_3$, $7Bi_2O_3\cdot 2WO_3$, $Bi_2\cdot WO_3$, and $Bi_2O_3\cdot 2WO_3$, were found (155). The ordered perovskites $Ba_6Sm_2W_3O_{18}$ and $Ba_6Sm_2U_3O_{18}$ form a complete series of $Ba_2Sm_{0.67}U_{1-x}W_xO_6$ (x = 0.1, 0.25, 0.4, 0.5, 0.6, 0.75, 0.9, 1.0) solid solutions. These are prepared by heating the oxides in the desired ratio at 1300° (156). The thermodynamics of the reactions of Y_2O_3 and La_2O_3 with W in vacuum was analyzed. The reaction $M_2O_3 + 16W \rightarrow 16(3M_2O_3\cdot WO_3) + MO_{(g)}$ is thermodynamically most probable (157). The congruently melting compound $ZnWO_4$ (m 1230°) was observed in the phase diagram of the ZnO-WO_3 system (158).

Tungsten Bronzes and Simple Tungstates: The structure of the high pressure form of $BaWO_4$-II was found to have resemblance to either scheelite or wolframite (159). Double tungstates of alkali metals and bismuth of type $MBi(WO_4)_2$ have been prepared and characterized (63). Polyanions containing W and Nb like the peroxy salt $Cs_4Nb_2W_4O_{19}(O_2)\cdot 3H_2O$, have been prepared and characterized (160). Reduction of the peroxy complex gave $Cs_3HNb_2W_4O_{19}\cdot 10H_2O$ (160). Pure perovskite of the composition $Pb(Ga_{2/3}W_{1/3})O_3$ was prepared at 40 kbar and 800°. This modification was found to be metastable under atmospheric conditions (14). The crystal structure of Fe_2WO_6 was investigated. It consists of zigzag chains of FeO_6 and WO_6 octahedra sharing edges (161). A new series of tungsten compounds of the eschynite type $LnVWO_6$, where Ln is a rare earth, has been synthesized (162). The crystalline basic zirconium tungstate, isomorphorus with $ZrMo_2O_7(OH)_2\cdot 2H_2O$ was prepared and characterized (163). Binary systems of lanthanide tungstates and tungsten pentoxide have been reported (164). $FeWO_4$ was prepared by the solid-phase reaction of FeO and WO_3 at 800° (165). The compound $Ba(HC_2O_4)_2\cdot 2H_2O$ precipitates out of solutions of $BaWO_4$ and oxalic acid (166). Tungstates of Zr, Th, U, V, Sn, and Ti were prepared and their ion exchange properties determined (167). The preparation and characterization of several tungstates of rare earths has been reported (70,76,168-174). Zirconium tungsten bronzes of perovskite structure Zr_xWO_3, $0<x<0.075$ were prepared and their structure studied (175). A subsolidus phase study of the K_2O-Nb_2O_5-WO_3 was made. The ternary compound $K_2O\cdot Nb_2O_5\cdot 2WO_3$ of the pyrochlore structure was found

(176). A short review on sodium tungsten bronzes has appeared (177).

Isopoly and Heteropoly Tungstates: Aqueous solutions of Na_2WO_4 in 3M $NaNO_3$ acidified with nitric acid always yielded a paratungstate of composition $5Na_2O \cdot 12WO_3 \cdot 28H_2O$ at 25° after 160 days (178). Equilibria studies on acidified tungstate solutions have been reported (179). The potassium paratungstate-ammonium paratungstate-water system was studied in detail at 25° (180). The crystal structure of disodium octaammonium paratungstate, $Na_2(NH_4)_8[H_2W_{12}O_{42}] \cdot 12H_2O$ has been determined. The anion $[H_2W_{12}O_{42}]^{-10}$ is present in the structure (181). $NaWO_2$, Na_3WO_4 and $K_6Mo_2O_4$ were prepared by the reaction of Na or K vapors with WO_2 and $NaWO_4$ or KWO_4 at 300-600° and pressure 0.01-0.1 torr (78). The solubility of ammonium paratungstate was increased 25-fold by the reaction of an aliphatic primary amine in water (80). A basicity of three was obtained in acetic acid for the heteropoly compounds $1/2B_2O_3 \cdot WO_3 \cdot nH_2O$ and $1/2B_2O_3 \cdot 10WO_3 \cdot V_2O_5 \cdot nH_2O$ (99). A 6-tungstochromate(III) anion was produced by the reaction of chromium nitrate with paratungstate ion (182). Two tungstoborate anions were prepared. Tetragonal $[BW_{12}O_{40}]^{-5}$ corresponding to alpha $[SiMo_{12}O_{40}]^{-4}$ and hexagonal $[BW_{12}O_{40}]^{-5}$ not corresponding to beta $[SiMo_{12}O_{40}]^{-4}$ (183). The acids and potassium salts of new heteropoly compounds containing fluorine, $H_4[W_{12}O_{38}F_2]$ and $H_5[H_2W_{12}O_{39}F]$ have been prepared and characterized (184). The formation of the heteropoly anions $[SiW_{11}VO_{40}]^{-6}$ and $[PW_{11}VO_{40}]^{-5}$ has been reported (185). The isomerism and properties of 9-tungstoheteropoly compounds containing W and Mo has been discussed (186). Heteropoly tungstates containing phosphorus +3 and +5 have been prepared. The sodium and potassium salts of $[SbW_9O_{33}]^{-9}$ were isolated (187). The compounds $K_5[SbW_{11}CrO_{40}H_2] \cdot 10H_2O$ and $H_5[SiW_{11}CrO_{40}H_2] \cdot 12H_2O$ have been prepared and characterized (50). Alkali salts of tungstovanadate anions have also been prepared (188). The yellow heteropoly anion $[H_2W_{11}VO_{40}]^{-7}$, isolated as the potassium salt, was obtained by the chlorine oxidation of $[H_2W_{11}VO_{48}]^{-8}$ (189). The crystalline blue $H_5[W_{10}VO_{34}OH] \cdot nH_2O$ was obtained by the reduction of the corresponding oxidized form with $SnCl_2$ or $TiCl_3$ (190).

Halogen-Containing Compounds: WSF_4 reacts with HF in acetonitrile to form the anions $W_2S_2F_9^-$ and $WSF_{5-x}Cl_x^-$ where x = 1-4 (191). Oxo and peroxofluoro tungstate anions were prepared by the reaction of H_2WO_4 with HF and H_2O_2 (192). Salts like $(R_3NH)_2WOF_4$ are obtained in extracting tungsten from fluoride solutions by tri-n-octylamine (117). Aliphatic alcohols react with WOF_4 to form complexes like $WOF_3(OR) \cdot HOR$ (193). Polynuclear compounds like $(H_3O)_2(W_6X_8)X_6 \cdot 6H_2O$ (x = Cl or Br) were prepared by the reduction of WCl_6 with Al (122). $PhWCl_3$ has been prepared by the reaction of WCl_6 with Ph_4Sn (194).

Miscellaneous: The reduction of $W(N_2)_2(Ph_2CH_2CH_2PPH_2)_2$ gave a monohaptohydrido(2-) species which was characterized by simple-crystal X-ray (195). Hydrazinium tetrathiotungstate, $N_2H_6WS_4$, was prepared from WO_3 and 40% N_2H_4 at 50-60° (104). $[Ni(WS_4)(en)_2]$ was prepared by mixing aqueous solutions of $Nien_2Cl_2$ and $(NH_4)_2WS_4$ (196). $(Ph_4A)_2M(WO_2S_2)_2$, A = As, P and M = Ni, Co, were prepared by adding an aqueous solution to $Ni(NO_3)_2$ or $CoSO_4$ and Ph_4ACl to a freshly prepared solution of $(NH_4)_2WO_2S_2$ (197). The crystal structure of $(NH_4)_2WO_2S_2$ has been reported. The anion is tetrahedral

(198). $(Me_2N)_3Sb$ reacts with $W(CO)_6$ to give $(Me_2N)_3SbW(CO)_5$ (33). The preparation of (trichlorostannyl)tungsten(0) complex has been reported (39). 1,4-Diselenantetracarbonyl complex of tungsten, $C_8H_8O_4WSe_2$, has been prepared and characterized (41). The reaction of o-tetrachlorobenzoquinone with $W(CO)_6$ has yielded the neutral complex, $W(O_2C_6Cl_4)_3$ (42). RNH_2 and $W(CO)_6$ react in xylene to form $W(CO)_3(RNH_2)_3$ where R = Bu, cyclohexyl, $PhCH_2$ (199). Cis-$(R_4N)_2[WO_2(NCS)_4]$, where R = Me, Et, was precipitated from acidified solutions of tungstate in the presence of thiocyanate (137). The complex boride $ThWB_4$ has been reported (130). In the U-C-W system the compounds UWC_2 and UWC_{2-x} (x = 0.12) were shown to form (32). $K_4W(CN)_8 \cdot 2H_2O$ was prepared by the reduction of WO_4^{-2} by KBH_4 in potassium cyanide (200). Thermolysis of $W(CO)_6$ at 400° gave $WO_{0.39}C_{0.43}$. In air it is oxidized to WO_3 (201).

REFERENCES

1. D. J. Darensbourg and M. Y. Darensbourg, J. ORGANOMET. CHEM, 83, 309-425 (1974).
2. J. LESS COMMON METALS, 36, 1-533 (1974).
3. T. J. R. Weakly, STRUCT. BONDING (BERLIN), 18, 131-76 (1974)
4. C.N.R. Rao, SOLID STATE CHEM., 455-563 (1974)
5. Lj. Manojlovic-Muir, MOL. STRUCT. DIFFR. METHODS, 1, 459-79 (1973).
6. H.M.N.H. Irving, J. SOC. LEATHER TECHNOL. CHEM., 58, 51-62 (1974).
7. M. A. Porai-Koshits and L. O. Atovmyan, Nauka: Moscow, USSR, 1974, 231 p.
8. K. Niedenzu and H. Zimmer (Editors), "Methodicum Chimicum," Vol. 8, Georg Thieme Verlag, Stuttgart, 1974.
   J. R. Wasson, "Chromium," Chapter 14
   F. W. Moore, "Molybdenum," Chapter 15
   D. G. Tsisley and R.A. Walton, "Tungsten," Chapter 16
   G. A. Tsigdinos, "Heteropoly Compounds," Chapter 32
9. A. Funke and P. Kleinert, Z.ANORG. ALLG. CHEM., 403, 156 (1974).
10. M. A. Alario-Franco, J. Fenerty and K. S. W. Sing, REACT. SOLIDS, PROC. INT. SYMP., 7th, 327 (1972).
11. V. L. Butukhanov, E. I. Get'man and M. V. Mokhosoev, ZH. NEORG. KHIM., 19, 1979 (1974).
12. L. B. Ebert, R. A. Huggins and J. I. Brauman, CARBON, 12, 199 (1974).
13. E. P. Gladkaya and O. N. Chudnovskaya, OBSHCH. PRIKL. KHIM., 6, 84 (1974).
14. L. G. Sevast'yanova et. al., IZV. AKAD. NAUK SSSR, NEORG. MATER., 10, 1359 (1974).
15. R. P. Clarl and F. W. Reinhardt, THERMOCHIMICA ACTA, 8, 185 (1974).
16. N. S. Kozlov et. al., DOKL. AKAD. NAUK BELORUSS. SSR, 18, 344 (1974).
17. H. Pausch and H. Müller-Buschbaum, Z. ANORG. ALLG. CHEM., 405, 1 (1974).
18. L. F. Larkworthy and M. H. O. Nelson-Richardson, CHEM. IND. (LONDON), (4), 164 (1974).
19. A. J. Edwards, W. E. Falconer and W. A. Sunder, J. CHEM. SOC., DALTON TRANS., 541 (1974).
20. I. A. Rat'kovskii et. al., ZH. FIZ. KHIM., 48, 231 (1974).
21. R. Vangelisti et. al., C.R. ACAD. SCI. SER. C, 278, 869 (1974).
22. W. Domke and J. Karbinski, Z. ANORG. ALLG. CHEM., 403, 116 (1974).

23. M. S. Matson and R. A. D. Wentworth, J. AM. CHEM. SOC., 96, 7837 (1974).
24. J. Tsangoris and G. T. Baxevanidis, Z. NATURFORSCH., TEIL 13, 29, 532 (1974).
25. H. D. Gillman, P. Nannelli and B. P. Block, J. INORG. NUCL. CHEM. 35, 4053 (1973).
26. C. Platte and H. Sabrowsky, NATURWISSENSCHAFTEN, 60, 474 (1974).
27. M. Robbins et. al., J. SOLID STATE CHEM., 9, 170 (1974).
28. H. D. Lutz et. al., Z. NATURFORSCH., TEIL B, 28, 685 (1973).
29. H. D. Lutz, H. Haenseler and W. Becker, Z. NATURFORSCH., TEIL B, 29, 385 (1974).
30. T. Takahashi, K. Ametani and O. Yamada, J. CRYST. GROWTH, 24-25, 151 (1974).
31. H. D. Lutz, et. al., MONATSCH. CHEM., 105, 849 (1974).
32. O. S. Ivanov and Z. M. Alekseeva, FIZ-KHIM. ANAL. SPLANOV URANA, TORIYA TSIRKONIYA, 120 (1974).
33. A. Kiennemann and R. Kieffer, C.R. ACAD. SCI., SER. C, 279, 355 (1974).
34. H. L. Krauss and B. Rebenstorf, Z. ANORG. ALLG. CHEM., 402, 113 (1973).
35. H. Fischer, E. O. Fischer and F. R. Kreissl, J. ORGANOMETAL. CHEM., 64, C41, 1974.
36. A. Bakac, R. Marcec, and M. Orchanovic, INORG. CHEM., 13, 57 (1974).
37. S. Uemura, A. Spencer and G. Wilkinson, J. CHEM. SOC. DALTON TRANS., 2565 (1973).
38. J. M. Coronas and J. Casabo, ANN. QUIM., 70, 330 (1974).
39. T. Kruck and H. Breuer, CHEM. BER., 107, 263 (1974).
40. H. D. Gillman, U.S. NAT. TECH. INFORM. SERV. AD REP. No. 770156/8GA, 1973.
41. G. Hunter and R. C. Massey, J. CHEM. SOC. DALTON TRANS., 1872 (1974).
42. V. V. Zelentsov et. al., DOKL. AKAD. NAUK SSSR, 216, 844 (1974).
43. C. G. Pierpont, H. H. Downs and T. G. Rukavina, J. AM. CHEM. SOC., 96, 5573 (1974).
44. Y. Masuda, Y. Shiomi and S. Misumi, NIPPON KAGAKU KAISHI, (9), 1645 (1974).
45. K. A. Kotyaeva and I. I. Kalinichenko, ZH. NEORG. KHIM., 19, 2112 (1974).
46. A. Richard and P. Souchay, C. R. ACAD. SCI., SER. C., 227, 973 (1973).
47. R. J. Fereday, MALAYSIAN J. SCI., 2, 95 (1973).
48. R. D. Rieke, J. ORGANOMET. CHEM., 76, C19 (1974).
49. M. Scotti and H. Werner, HELV. CHIM. ACTA., 57, 1234 (1974).
50. V. I. Spitsyn, T. I. Bogatyreva and T. I. Boyatyreva, ZH. NEORG. KHIM., 19, 1898 (1974).
51. R. Perret et. al., C. R. ACAD. SCI., SER. C., 278, 951 (1974).
52. N. Sotani, NIPPON KAGAKU KAISHI, (4), 673 (1974).
53. V. I. Shapoval, A. S. Avaliani and N. A. Gasviani, SOOBSHCH. AKAD. NAUK GRUZ. SSR, 72, 105 (1973).
54. I. Boeschen and B. Krebs, ACTA CRYSTALLOGR., SECT. 13, 30, 1795 (1974).
55. Y. Kotera and A. Ueno, BULL. CHEM. SOC. JAP., 46, 3431 (1973).
56. V. T. Mal'tsev, P. M. Chobanyan, and V. L. Volkov, RUSS. J. INORG. CHEM., 18, 1068 (1973).
57. V. T. Mal'tsev. P. Chobanyan and V. L. Volkov, ZH. NEORG. KHIM., 19, 497 (1974).

58. V. N. Serezhkin et al., ZH. NEORG. KHIM., 19, 1036 (1974).
59. L. V. Semenyakova and I. G. Kokarovtseva, ZH. NEORG. KHIM., 18, 3068 (1973)
60. N. N. Bushuev and V. K. Trunov, DOKL. AKAD. NAUK SSSR, 217, 827 (1974).
61. J. Meullemeestre and E. Penigrault, BULL. SOC. CHIM. FR., 9 (1974).
62. M. Parmentier, A. Courtois and C. Gleitzer, BULL. SOC. CHIM. FR., 75 (1974).
63. P. V. Kletsov, V. A. Vinokurov and R. F. Klevtsova, KRISTALLOGRAFIYA, 18, 1192 (1973).
64. E. V. Tkachenko and V. M. Zhukovskii, ZH. PRIKL. KHIM. (LENINGRAD), 46, 2129 (1973).
65. J. Ziolkowski and P. Courtine, ANN CHIM. (PARIS), 8, 303 (1973).
66. J. Chojnacki, P. Kozlowski and J. Haber, J. SOLID STATE CHEM., 11, 106 (1974).
67. V. N. Serezhkin, L. M. Kovba and V. K. Trunov, KRISTALLOGRAFIYA, 19, 379 (1974).
68. V. N. Serezhkin, L. M. Kovba and V. K. Trunov, RADIOKHIMIYA, 16, 231 (1974).
69. A. W. Sleight and W. Teitschko, MATER. RES. BULL., 9, 951 (1974).
70. Y. A. Velikodnyi and V. K. Trunov, IZV. AKAD. NAUK SSSR, NEORG. MATER., 10, 1290 (1974).
71. E. Banks and M. Nemiroff, INORG. CHEM., 13, 2715 (1974).
72. V. E. Plyushchev et al., KHIM. KHIM. TEKHNOL., TR. YUBILEINOI KONF., POSVYASHCH. 70-LETIYU INST. (MOSK. TONKOI KHIM, TEKHNOL.) 353, 1970 (Pub. 1972).
73. T. P. Rybakova and V. K. Trunov, ZH. NEORG. KHIM., 19, 1070 (1974).
74. N. V. Gavrish et al., IZV. AKAD. NAUK SSSR, NEORG. MATER. 9, 1766 (1973).
75. H. R. Hoekstra, INORG. NUCL. CHEM. LETT., 9, 1291 (1973).
76. A. A. Provotorov and V. A. Balashov, USP. KHIM., 42, 1788 (1973).
77. L. Z. Gokhman et al., ZH. NEORG. KHIM., 19, 2018 (1974).
78. M. Toubol, P. Toledano and G. Perez, C. R. ACAD. SCI., SER. C, 278, 417 (1974).
79. B. V. Slobodin, M. V. Mokhosoev and P. T. Zvoleiko, RUSS. J. INORG. CHEM., 19, 211 (1974).
80. Y. Hirose and H. Murata, Japan 73 42,359, August 31, 1970.
81. N. D. Jespersen, J. INORG. NUCL. CHEM., 35, 3873 (1973).
82. F. Y. Kul'ba, E. G. Zlotnikov and Y. Makasov, ZH. NEORG. KHIM., 19, 1536 (1974).
83. G. A. Tsigdinos and C. J. Hallada, J. LESS-COMMON METALS, 36, 79 (1974).
84. E. A. Torchenkova et al., DOKL. AKAD. NAUK SSSR, 216, 1073 (1974).
85. G. A. Tsigdinos, IND. ENG. CHEM., PROD. RES. DEV., 13, 267 (1974).
86. P. Rabette, D. Olivier and H. Pezerat, REV. CHIM. MINER., 11, 157 (1974).
87. P. Rabette, D. Olivier, J. LESS-COMMON METALS, 36, 299 (1974).
88. P. Rabette and D. Olivier, REV. CHIM. MINER., 11, 185 (1974).
89. P. Rabette and D. Olivier, C. R. ACAD. SCI., SER. C, 277, 559 (1973).
90. V. I. Spitsyn et al., Z. CHEM., 14, 1 (1974).
91. H. D'Amour and R. Allmann, NATURWISSENSCHAFTEN , 61, 31 (1974).
92. J. Fischer, L. Ricard and P. Toledano, J. CHEM. SOC., DALTON TRANS., 944 (1974).

93. J. E. Going and S. J. Eisenreich, ANAL. CHIM. ACTA., 70, 95 (1974).
94. R. Contant, BULL. SOC. CHIM. FR., 3277 (1973).
95. P. Souchay and R. Contant, BULL. SOC. CHIM. FR., 3287 (1973).
96. H. T. Evans, Jr., ACTA CRYSTALLOGR., SECT. B, B30, 2095 (1974).
97. Q. S. Holguin, R. Cruz and C. A. Campero, REV. INST. MEX. PETROL., 5, 57 (1973).
98. L. M. Karina and N. A. Tsvetkov, TR. MUSK. INST. TONKOI KHIM. TEKHNOL., (2), 26 (1972).
99. I. K. Latichevskii and N. A. Polotebnova, ZH. NEORG. KHIM., 18, 3298 (1973).
100. G. Wiese and D. Boese, Z. NATURFORSCH., TEIL B, 29 (1974).
101. V. L. Kalikhman and E. P. Gladchenko, TR. VSES. NAUCH.-ISSLED. INST. ELECKTRO-MEKH., (38), 390 (1972).
102. A. N. Zelikman et al., KITAL'KOGENIDY, (2), 47 (1970).
103. W. Kwestroo, PREP. METHODS SOLID STATE CHEM., 563 (1974).
104. V. N. Bondarev, METALLOTERM. PROTSESSY KHIM. MET.,MATER. KONF., 290 (1971)
105. R. Schoellhorn and A. Armin, J. LESS-COMMON METALS, 36, 229 (1974).
106. J. Guillevic, J-Y. Le Marouille and D. Grandjean, ACTA CRYSTALLOGR., B30, 111 (1974).
107. A. Y. Shinyaev et al., DOKL. AKAD. NAUK SSSR, 212, 1336 (1973).
108. C. Cevales and L. Angelin, DTSCH. KERAM. GES., 51, 258 (1974).
109. P. Rogl and H. Nowotny, MONATSCH. CHEM., 105, 1082 (1974).
110. M. R. Udupa and G. Aravamudan, PROC. CHEM. SYMP., 2, 127 (1972).
111. B. Spivac and Z. Dori, J. CHEM. SOC., CHEM. COMMUN., (23), 909 (1973).
112. A. Müller, et al., Z. ANORG. ALLGEM. CHEM., 403, 310 (1974).
113. E. Koeniger-Ahlborn and A. Müller, ANGEW. CHEM., 86, 709 (1974).
114. A. A. Opalovskii et al., ZH. NEORG. KHIM., 19, 2071 (1974).
115. M. C. Chakravorti and S. C. Pandit, J. INORG. NUCL. CHEM. 35, 3644 (1973).
116. M. C. Chakravorti and S. C. Pandit, J. INDIAN CHEM. SOC., 50, 618 (1973).
117. Y. S. Tseryuta et al., ZH. NEORG. KHIM., 19, 200 (1974).
118. R. T. Paine and L. B. Asprey, INORG. CHEM., 13, 1529 (1974).
119. D. V. Drobot and E. A. Sapranova, ZH. NEORG. KHIM., 19, 228 (1974).
120. H. K. Saha and A. K. Banerjee, INORG. SYNTH., 15, 400 (1974).
121. D. V. Drobot and E. A. Sapranova, RUSS. J. INORG. CHEM., 18, 1067 (1973).
122. W. C. Dorman and R. E. McCarley, INORG. CHEM., 13, 491 (1974).
123. B. Jezowska-Trzebiatowska et al., INORG. CHEM., 13, 617 (1974).
124. D. L. Kepert, R. E. Marshall and D. Taylor, J. CHEM. SOC., DALTON TRANS., 506 (1974).
125. W. H. Delphin and R. A. D. Wentworth, INORG. CHEM., 13, 2037 (1974).
126. I. A. Glukhov et al., ZH. NEORG. KHIM., 19, 963 (1974).
127. H. K. Saha, M. Bagchi and S. Chakrovorty, J. INORG. NUCL. CHEM., 36, 455 (1974).
128. O. Piovesana and L. Sestili, INORG. CHEM., 13, 2745 (1974).
129. W. H. Delphin, R. A. D. Wentworth and M. S. Matson, INORG. CHEM., 13, 2552 (1974).
130. J. San Filippo, H. J. Sniadoch and R. L. Grayson, INORG. CHEM., 13, 2121 (1974).
131. A. R. Bowen and H. Taube, INORG. CHEM., 13, 2245 (1974).

132. D. Britnell, G. W. A. Fowles and D. A. Rice, J. CHEM. SOC., DALTON TRANS., 219 (1974).
133. D. Scott and A. G. Wedd, J. CHEM. SOC., CHEM. COMMUN., (14), 527 (1974).
134. J. V. Dobcnik and P. Segedin, MONATSH. CHEM., 105, 142 (1974).
135. M. Novotny and S. J. Lippard, INORG. CHEM., 13, 828 (1974).
136. J. P. Brunette and M. J. F. Leroy, J. INORG. NUCL. CHEM., 36, 289 (1974).
137. B. J. Brisdon and D. A. Edwards, INORG. NUCL. CHEM. LETT., 10, 301 (1974).
138. D. Walther, Z. ANORG. ALLG. CHEM., 405, 8 (1974).
139. F. A. Cotton et al., J. AM. CHEM. SOC., 96, 3824 (1974).
140. A. M. Golub et al., RUSS. J. INORG. CHEM., 18, 1120 (1973).
141. N. G. Connelly, J. CHEM. SOC., DALTON TRANS., 2183 (1973).
142. K. Takashi et al., J. CHEM. SOC., CHEM. COMMUN., 136 (1974).
143. Kabir-Ud-Din, A. A. Khan and M. A. Beg, J. INORG. NUCL. CHEM., 36, 470 (1974).
144. M. Wrighton, INORG. CHEM., 13, 905 (1974).
145. M. H. Chisholm and W. Reichert, J.AM. CHEM. SOC., 96, 1249 (1974).
146. L. W. Amos and D. T. Sawyer, INORG. CHEM., 13, 78 (1974).
147. A. W. Cordes, et al., INORG. CHEM., 13, 132 (1974).
148. E. H. Abbott, F. Schoenewolf and T. Backstrom, J. COORD. CHEM., 3, 255 (1974).
149. J. Selbin, J. Sherrill and C. H. Bigger, INORG. CHEM., 13, 2544 (1974).
150. B. Heyn and H. Schreer, Z. CHEM., 14, 28 (1974).
151. M. L. H. Green and W. E. Silverthorn, J. CHEM. SOC., DALTON TRANS., 2164 (1974).
152. Z. B.Varadi and A. Niewpoort, INORG. NUCL. CHEM. LETT., 10, 801 (1974).
153. R. L. Marks and S. E. Popielski, INORG. NUCL. CHEM. LETT., 10, 885 (1974).
154. B. T. Mal'tsev et al., ZH. NEORG. KHIM., 18, 3314 (1973).
155. S. N. Hoda and L. L. Y. Chang, J. AM. CERAM. SOC., 57, 323 (1974).
156. H. J. Schittenhelm and S. Kemmler-Sack, Z. ANORG. ALLE. CHEM., 407, 181 (1974).
157. V. A. Levitskii, V. N. Chentsov and Y.Y. Skolis, ZH. FIZ. KHIM., 48, 566 (1974).
158. I. P. Kislyakov, I. N. Smirnova and G. I. Boguslavskaya, IZV. VYSSH. UCHEB. ZAVED., KHIM. KHIM. TEKHNOL., 16, 1440 (1973)
159. I. Kawada, K. Kato and T. Fujita, ACTA CRYSTALLOGR., SECT. B, B30, 2069 (1974).
160. M. Dabbari, M. Boyer and P. Souchay, C. R. ACAD. SCI., SER. C, 278, 1047 (1974).
161. J. Senegus and J. Galy, SOLID STATE CHEM., 10, 5 (1974).
162. R. Salmon et al., REV. CHIM. MINER., 11, 71 (1974).
163. A. Clearfield and R. H. Blessing, J. INORG. NUCL. CHEM., 36, 1174 (1974).
164. I. N. Belyaev and C. A. Voropaeva, TEZISY DOKL. MEZHOTRASL. SOVESHCH. METOD. POLUCH. ANAL. FERRITOVYKH, SEGNETO-, P'EZOELEKTRICHESKIKH KONDENS. MATER. SYR'YA NIKH, 4th, 40 (1971).
165. G. M. Zhiteteva, T. N. Komarova and L. A. Anfilogova, FIZ.-KHIM. ISSLED. VZAIMODEISTVIYA SOLEI SHCHELCHNYKH METAL. RASPLAVAKH PROD.

DESTRUKTSII SAPROPELITOV, 176 (1972).
166. Y. M. Potashnikov and O. I. Kovalevskii, ZH. OBSHCH. KHIM., 43, 1869 (1973).
167. M. Qureshi, J. P. Gupta and V. Sharma, TALANTA, 21, 102 (1974).
168. A. A. Evdokimov and V. K. Trunov, ZH. NEORG. KHIM., 19, 1412 (1974).
169. A. A. Evdokimov and V. K. Trunov, ZH. NEORG. KHIM., 18, 3072 (1973).
170. V. K. Trunov and V. K. Rybakov, ZH. NEORG. KHIM., 19, 347 (1974).
171. A. A. Evdokimov and V.K. Trunov, ZH. NEORG. KHIM., 18, 3157 (1973).
172. V. K. Rybakov and V. K. Trurov, ZH. NEORG. KHIM., 18, 3152 (1973).
173. V. K. Trunov and V. K. Rybakov, ZH. NEORG. KHIM., 19, 636 (1974).
174. A. A. Evdokimov and V.K. Trurov, ZH. NEORG. KHIM., 19, 232 (1974).
175. T. Ekstroem and R. J. D. Tilley, MATER. RES. BULL., 9, 999 (1974).
176. T. Ikeda and S. Shimizu, JAP. J. APPL. PHYS., 13, 256 (1974).
177. J-P. Randin, J. CHEM. ED., 51, 32 (1974).
178. J. Chojnacka and L. Czepirski, ROCZ. CHEM., 47, 1831 (1973).
179. R. Arnek and Y. Susaki, ACTA CHEM. SCAND., 28, 20 (1974).
180. I. N. Belyaev, R. I. Baylei and K. G. Burtseva, ZH. NEORG. KHIM., 19, 2261 (1974).
181. H. D'Amaur and R. Allmann, Z. KRISTALLOGR., KRISTALLGEOM., KRISTALLPHYS., KRISTALLCHEM., 138, 5 (1973).
182. V. I. Spitsyn et al., RUSS. J. INORG. CHEM., 18, 1150 (1973).
183. G. Herve and A. Terez, C. R. ACAD. SCI., SER. C, 278, 1417 (1974).
184. F. Chauveau and P. Souchay, J. INORG. NUCL. HCEM., 36, 1761 (1974).
185. G. Salamon-Bertho, REV. CHIM. MINER., 9, 683 (1972)
186. R. Contant, J. M. Fruchart and G. Herve, C. R. ACAD. SCI., SER. C, 278, 199 (1974).
187. C. Tourne et al., C. R. ACAD. SCI., SER. C, 277, 643 (1974).
188. D. F. Takezhanova et al., DOKL. AKAD. NAUK SSSR, 216, 335 (1974).
189. C. M. Flynn, M. T. Pope and S. O'Donnell, INORG. CHEM., 13, 831 (1974).
190. E. F. Tkach and N. A. Polotebnova, ZH. NEORG. KHIM., 19, 155 (1974).
191. Y. A. Buslaev, Y. V. Kokunov and Y. D. Chubar, DOKL. AKAD. NAUK SSSR, 213, 1083 (1973).
192. R. G. Beiles, V. S. Samorodova and Z. E. Rozmanova, ZH. NEORG. KHIM., 18, 2978 (1973).
193. Y. A. Buslaev et al., ZH. NEORG. KHIM., 19, 1196 (1974).
194. W. Grahlert, K. Milowski and U. Longbein, Z. CHEM., 14, 287 (1974).
195. G. A. Heath, R. Mason and K. M. Thomas, J. AM. CHEM. SOC., 96, 259 (1974).
196. M. C. Chakravorti and A. Mueller, INORG. NUCL. CHEM. LETT., 10, 63 (1974).
197. A. Mueller, H. H. Heinsen and G. Vandrish, INORG. CHEM., 13, 1001 (1974).
198. W. Gonschorek, T. Hahn and A. Mueller, KRISTALLOGR., KRISTALLGEOM., KRISTALLPHYS., KRISTALCHEM., 138, 380 (1973).
199. S. C. Tripathi et al., J. INDIAN CHEM. SOC., 51, 220 (1974).
200. J. G. Leipoldt, L. D. Bok and P. J. Cilliers, Z. ANORG. ALLG. CHEM., 407, 350 (1974).
201. O. D. Krichevskaya et al., ZH. PRIKL. KHIM. (LENINGRAD), 46, 2593 (1973).

MANGANESE, TECHNETIUM AND RHENIUM

Geoffrey Davies

Department of Chemistry
Northeastern University
Boston, Massachusetts 02115

GENERAL

Discussions of manganese chemistry concerned with the spectroscopy of spinels (1), oxides (2), halides (3), Shiff-base complexes (4), and the preparation and characterization of a wide range of manganese(III) and -(IV) alkyldithiocarbamates (5) supplemented the appearance of Part C, Section 1 of Gmelin's "Manganese" (6). Rhenium chemistry (7) and analysis (8), rhenium oxides (9), the preparation of highly pure Re by electrodialysis (10) and the preparation of an important series of thiuram disulfide derivatives of lower oxidation state rhenium (11) have also received attention.

General reviews of manganese metalloenzymes (12), the chemistry of elements of the first transition period (13) and of the organometallic chemistry of the elements (14) have been published.

MANGANESE

New developments in higher oxidation states were a notable feature in 1974 (5).

Lower Oxidation States: Inoue et al (15) have given a comprehensive account of the preparation and magnetic behavior of $Mn(bipy)_3 \cdot 2THF$ and $LiMn(bipy)_3 \cdot 4THF$.

Manganese (II): A selection of the simpler compounds reported in 1974 is collected in Table I. Of particular note are the use of molten salts e.g. $Mn(py)_4Cl_2$, as thermal storage devices (16), the preparation of 8-coordinate complexes with 1,10 phenanthroline, o-phenylenediamine and 8-aminoquinoline (17) and the identification of another manganese complex carrying a protonated ligand (18).

Kallinikov et al (30) report a useful synthetic route to anhydrous carboxylate salts through the thermal or photochemical reactions of (Cp) $Mn(CO)_3$ in xylene; 60-80% yields of $Mn(OOCR)_2$ are reported for R ranging from methyl to 2-quinolinyl (30). Chasan et al (31) have investigated the synthetic advantages of zeolites in the preparation of anhydrous complexes.

The template synthesis approach has yielded interesting 7-coordinated species from the reaction of pyridine 2,6 dicarboxaldehyde and primary amines in the presence of manganese (II) in methanol (32). Tetraphenyl porphyrin (H_2L) forms species MnL and MnLX (X = Cl^-, AcO^-, CN^-). The former species gives an isolable NO complex (33).

Among the useful solvates are the hydrated phosphites, whose physical properties have been described (34). A selection of other solvates are collected in Table II.

Higher Oxidation States: Development of the interesting chemistry of manganese (III) complexes continues (see, e.g., the excellent work of ref 5).

The electrochemistry of $Mn(H_2O)_6^{3+}$ (44) and the characterization of $Mn(TEA)_2^{3+}$, $3OH^-$ (TEA = triethanolamine) (45) have been investigated. Although the latter complex is stable in strong alkali, it is readily reduced to "$Mn(TEA)^{2+}$" at dropping mercury or Pt electrodes. A longstanding part of manganese(III)-CN chemistry has been discounted by a crystallographic proof that the major product of the reaction of $KMnO_4$ with KCN is gold-brown, diamagnetic $K_7[(CN)_5MnOMn(CN)_5]CN$ when the reaction is run under controlled conditions (46). Cueto (47) reported a series of studies of the bright red product from the supernatant solution of the $MnCO_3$/KCN/H_2O_2 reaction. The neutral complex is formulated as $Mn(CN)_2(H_2O)_3OH$ and is converted by acid to Mn^{2+}_{aq} (47). Thermal decomposition of manganese(III)

TABLE I

ANHYDROUS COMPLEXES OF MANGANESE (II)

COMPLEX	COMMENTS	REF.
*MnA	$Mn(OAc)_2 + H_2A \xrightarrow[\text{reflux}]{\text{EtOH}}$	19
*MnF_2B_2	$Mn(BF_4)_2$ + B $\xrightarrow{\text{aq EtOH}}$	20
$Mn[O_2P(EtO)_2]_2$	$MnCl_2$ + triethylphosphate	21
$Mn[O_2PR_2]$	R = octyl	22
*$Mn(OEPPh_3)_2X_2$	X = Cl, Br, I corrects previous report	23
$TlMnF_3$	$MnF_2 + TlF \xrightarrow{\text{dilHF}}$	24
*$MnG_2(NCSe)_2$	$MnNO_3 + KSeCN \xrightarrow{Me_2CO}$	25
*$MnH_4(NCSe)_2$	N-bonded	26
*<u>cis</u>-$Mn(J)_2Cl_2$	X-ray	27
*<u>trans</u>-MnK_nX_2	X = Br, I, NCS n = 4-10	28
*MnP_4Cl_2	$MnCl_2 + L(1:5) \xrightarrow{\text{solid}}$ N-bonded	29

*Footnotes:

A = N(2-hydroxyphenyl)salicylaldimine; B = 3,5 dimethyl-pyrazole

E = P or As; G = hexamethylphosphoramide

H = pyridine, substituted pyridine, quinoline;

J = 1,2 ethanediol; K = urea; P = acrylamide

TABLE II

SOME SOLVATES OF MANGANESE (II)

SUBSTANCE	COMMENTS	REF.
*$MnCl_2 \cdot 4A$		35
*$MnI_2 \cdot nB$	n = 1, 2	36
$Mn(NO_3)_2 \cdot 4H_2O$	X-ray	37
$Mn(OAc)_2 \cdot 4H_2O$	X-ray	38
*$MnC_2 \cdot 2H_2O$		39
$MnF_2 \cdot n\ N_2H_4$	n = 1, 2 n = 1 more stable	40
*$MnE_2Cl_2 \cdot nH_2O$	$MnCl_2 + E \xrightarrow[H_2O]{1:2}$	41
$MnI_2(N_2H_4)_2 \cdot H_2O$ $Mn(OAc)_2(N_2H_4) \cdot 2H_2O$	$N_2H_4 \cdot 2HI + MnI_2 \longrightarrow$	42
*$MnX_2G_\chi \cdot nS$	X = NCS, NCSe, NCO χ = 2, 3, n = 1.5-3.0	43

*Footnotes:
A = $SeC(NH_2)_2$; B = hexamethylenetetramine;
H_2C = malonic acid; E = o-hydroxybenzoylhydrazine
G = benzoylhydrazine; S = Me_2CO, EtOH, etc.

dialkyldithiocarbamates in DMF is first-order, and the rate decreases with the length of the alkyl chain (48).

The preparation and properties of $Mn(OL)_3Cl_3$ (L = py, Ph_3P and Ph_3As) (49) and the thermal conversions

$$(H_2phen)MnCl_5 \xrightarrow[\text{vacuum}]{N_2} Mn(phen)Cl_3 + Mn(phen)Cl_2 \quad (1)$$

$$(H_2bipy)MnCl_5 \xrightarrow[\text{vacuum}]{N_2} Mn(bipy)Cl_3 + Mn(phen)Cl_2 \quad (2)$$

which give easily isolable manganese (III) products (50), both result in potentially useful species. $Mn(OAc)_3$, one of the cornerstones of manganese (III) chemistry, has been shown to exhibit Jahn-Teller distortion in the solid state (51) and tris-(tropolinato)manganese (III) evidently contains 2 differently distorted molecules in the unit cell, a highly unusual situation (52). Complexes of the type Mn(acac)XL (X = NCS, N_3, Cl, Br, L = py, MeOH) lose L on mild heating which is analogous to the behavior of some porphyrins (53). $Mn(acac)_3$ reacts with carboxylic acids RCO_2H to give $Mn(acac)_2O_2CR$, which are reduced on heating to $Mn(acac)O_2CR$ (54). A reaction of $Mn(OAc)_3$ with HL

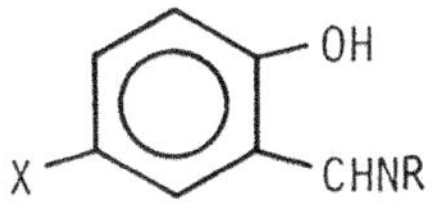

(3)

(R = Ph, m-MeC_6H_4, p-ClC_6H_4; X = H, Br) in methanol to give high spin derivatives MnL_3 has also been reported (55). The high intensities of the absorption bands of tris-(dipyrromethenato)manganese (III) suggests considerable covalency in the bonding (56). X-ray structural studies of $MnL_3{\cdot}MeOH$ (HL = oxine) have shown it to be the mer-isomer (57).

Schiff base complexes of manganese continue to attract attention.

Boucher and Herrington (58) have studied the electronic and CD spectral properties of the ligand obtained by condensing salicylaldehyde or o-hydroxyacetophenone with -1,2-propanediamine. The complexes are of the type MnLX, analogous to square-pyramidal $Mn(sal\text{-}NR)_2OAc$ (R = m-, p-$NO_2C_6H_9$) (59). $Mn(sal\text{-}NPh)_3$ is distorted from octahedral symmetry (59). An interesting "one dimensional" system is catena-μ-acetato (NN′ ethylene bis-salicylaldiminato)manganese (III), the structure of which has been reported (60).

A comprehensive account of the synthesis of 13-vanadomanganates(IV) has been published by Pope and his coworkers (61). A reexamination of the properties of the oxalato manganese(IV) species $K_2[Mn(OH)_2(C_2O_4)_2].2H_2O$ has suggested a basic $[Mn_2O_2(C_2O_4)_4]^{4-}$ structure (62).

Reduction of $KMnO_4$ or $NaMnO_4$ with $MnCl_2$ in acetic acid gives δ-MnO_2, which is regarded as a highly disordered, layered alkali - Mn(II, III) manganate(IV) (63). This substance is converted to α-MnO_2 and other products on prolonged heating at 100° in the presence of K^+ and NH_4^+ (63).

TECHNETIUM

Radiopharmaceuticals: Research interest in the radiopharmaceutical chemistry of "$^{99}Tc^m$" continued at a high level in 1974. A review of the radiochemistry of technetium by Murakami (64) emphasises preparative and purification methodology. Excess Sn(II) (commonly used in the preparation of radiopharmaceuticals for reduction of technetium(VII)) causes extensive liver damage in rats (65), and deleleterious accumulations of NO_2^- (produced by radiolysis of NO_3^- in some commercial generators) have also been noted (66). Alternative reducing agents which have been proposed include Zr/dilute HCl and Al/HF (67). A technique for acetone elution of $^{99}Tc^m$ from ^{99}Mo on silica gel has been patented (68): 95% yields of high purity product are claimed and acetone is noted as a good solvent for subsequent biochemical labelling (68). Weakly chelated reduced Tc species interact with Sephadex (69) but the effect is much less marked on Biogel resin (70). These effects are undoubtedly associated with the acid-dissociation equilibria of the more weakly chelated lower oxidation state complexes. Very encouraging results have been obtained in studies of the direct labelling of erythrocytes (71) and hemoglobin (72). The latter investigations have shown that Tc is associated with the β-globin chain (72). Albumin labelling also has been reviewed (73) and several recommended procedures for obtaining stable, uniform microaggregates have been published (74).

Bone scanning studies have included an investigation of the effect of varying ratios of $SnCl_2$/1-hydroxyethane 1,1-diphosphonate on the uptake of $^{99}Tc^{m}$ in synthetic hydroxyapatite (75). Several authors agree that pyrophosphate is the best of polyphosphates as a label carrier (76). Synthetic work on labelled tetracycline for kidney and gall-bladder scanning has been reported (77). Preferential accumulation in necrotic tissue is evident (78). Liver-specific chelate complexes with thiols such as the primary mercaptoalcohols and p-mercaptobenzoic acid have been prepared by reduction of $^{99}TcO_4^-$ with Zn(II), Cu(I), Fe(II) and Sn(II) (79).

OTHER COMPLEXES

Variable temperature pmr studies of (monohydrogen mesoporphyrin dimethylesterato)tricarbonyltechnetium(I) in $Cl_2CHCHCl_2$ have shown that this species is fluxional (as are two porphyrin derivatives of rhenium(I) (80)). This raises the question as to whether the same situation obtains in other low oxidation state metalloporphyrins. A distorted octahedral co-ordination environment has been found in $Tc[PPh(OEt)_2]_4Cl_2$ (81)

Refluxing NH_4TcO_4 with HX (X = Cl, Br) and $PhMe_2P$ in a ratio of 1:3:15 produces mer-$TcX_3(PMe_2Ph)_3$ (82). Reduction of the phosphine ratio to 5 under similar conditions provides trans-$Tc(PMe_2Ph)_2X_4$ (82). Cyclic voltametric studies of TcO_4^- in the pH range 0.5-14 have resulted in the suggestion that the TcO_2 and Tc_2O_3 products may be useful as synthetic precursors (83). The $^{99}Mo \longrightarrow {}^{99}Tc^m$ process in dilute aqueous H_2SO_4 results in the formation of stable chelates of technetium(V) on extraction with chloroform containing 8-hydroxyquinoline and α- benzoin oxime(84).

Technetates(VI) may be obtained by controlled-potential cathodic reduction of pertechnetate salts. For example, olive green crystals of $(Me_4N)_2TcO_4$ were obtained by reduction at a platinum sheet, and $KTcO_4$ in DMSO was reduced at a mercury electrode to K_2TcO_4 (85).

RHENIUM

1974 was a relatively quiet year in rhenium chemistry, with attention being focussed largely on oxidation states(V)-(VII) (but see ref. 11).

Rhenium (I): $(Ph_3P)_2Cl_2ReON_2CPh$ reacts with arphos (arphos = $Ph_2AsCH_2CH_2PPh_2$) in methanol to give the slightly soluble $ReCl(N_2)(arphos)_2$ complex in >50% yield. This complex reacts with $InCl_3$ to give $ReCl(N_2)(arphos)_2Cl$ in the absence of oxygen (86). $Re_2(CO)_{10}$ reacts with monhydrogenmeso-tetraphenylporphyrin in refluxing decalin to give (monhydrogenmeso-tetraphenylporphyrinato)$Re(CO)_3$, with three nitrogen atoms of the porphyrin ring attached to Re (87).

Two previously reported products of the reaction of $[ReCl(N_2)(PPhMe_2)_4]$ with $TiCl_3$ have been shown to be derivatives of titanium(IV). The intensly-colored adduct $[ReCl(N_2)(PPhMe_2)_4]_2 \cdot TiCl_4$ is trinuclear, with two bridging nitrogen atoms (88).

Rhenium (II): $ReCl_3$ and $Re_2Cl_8^{2-}$ react with PEt_3, PPr_3 and PEt_2Ph to give derivatives of the type $Re_2X_4(PEt_3)_4$. The chlorocomplex has a quadruple Re-Re bond and reacts with CCl_4 to give $(Et_3PCl)_2Re_2X_4Cl_4$ (89).

Rhenium (III): Protonation of $ReCl_2(NH_3)(N_2Ph)(PMe_2Ph)$ gives rise to multiple bonding between Re and the protonated ligand in $ReCl_2(NH_3)(N_2HPh)(PMe_2Ph)_2^+$ (90). The crystal and molecular structure of cis-dichloropentane-2,4-dionato-trans-bis(triphenylphosphine) rhenium(III) shows a roughly octahedral arrangement of ligand atoms (91).

Rhenium (IV): An improved method of preparing ReS_2 by reaction between the elements at 735° has been published (92). ReO_4^- is reduced by Sn(II) in the presence of NH_4SCN to the $Re(NCS)_6^{2-}$ ion, which is N-bonded (93).

Rhenium (V): Unidentate, cis-nitrato groups have been identified in the salt $[Ph_4As]$ $[Re(CO)_4(NO_3)_2]$ which is obtained in the reaction between $Re(CO)_5NO_3$ and Ph_4AsNO_3 (94). An X-ray study of trans-$[Re(CH_3NH_2)_4$-$(CH_3N)Cl]ClO_4$ shows the Re atom to be out- of-plane (95). Structural studies of the racemic $ReO_2(en)_2^+$ ion show it to be trans (96).

$ReCl_5$ reacts with NO in benzene or carbon tetrachloride to give $ReCl_4NO{\cdot}C_6H_6$ or $ReCl_5NO$, respectively (97). The latter complex undergoes the following reactions;

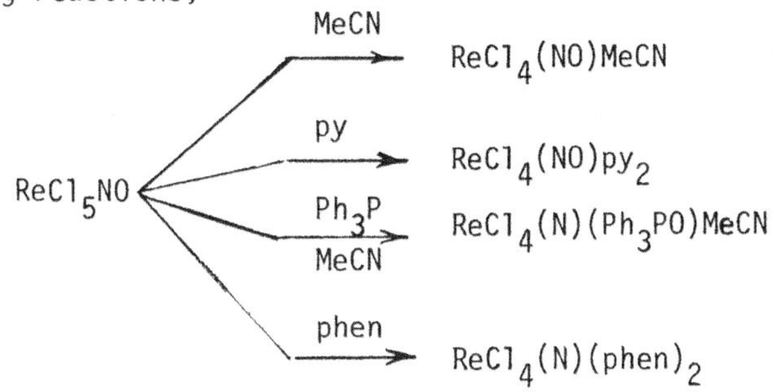

The first of these products is oxidized to $ReOCl_3(NO)MeCN$ under aerobic conditions (97).

Equilibria between $ReOCl_5^{2-}$ and thiourea and N, N´ ethylenethiourea involve the species $ReOCl_{5-n}(tu)_4$ (n=1 -4) in 6M HCl (98). The disproportionation of H_2ReOCl_5 has also been investigated in weakly acid solution. K_2ReOCl_5 reacts with $EtSCH_2CH_2SEt$ (L) in 20% v/v HCl to give diamagnetic, neutral $ReOCl_3L$ (100). An attempted recrystallization from acetone gave $ReO(OH)Cl_2L$, which was reconverted to the starting material by treatment with HCl (78).

Chakravorti and Chaudhuri (101) have reported the following reactions of $[ReO_2(py)_4]F$

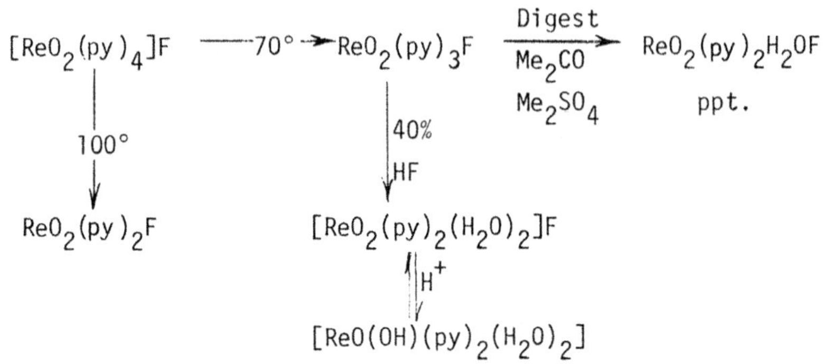

$ReO_2(py)_2(H_2O)F$ is reported to be weakly paramagnetic (101).

Pope and his coworkers (102) have taken advantage of the solubility of Keggin anions $XMZ_{11}O_{40}^{n-}$, (where X = central co-ordinated atom, Z = W or Mo and M is a second heteroatom), in non-aqueous solvents to prepare these species where M is Re(V). The red-purple anion is oxidized by Br_2 to dark green $PRe^{VI}W_{11}O_{40}^{3-}$ and pale yellow $PRe^{VII}W_{11}O_{40}^{2-}$. The X-band esr spectrum of the Re(VI) anion shows extremely large hyperfine parameters (102).

Rhenium (VI): $ReOF_4$ has C_{4v} symmetry in the gas phase (103). Wilkinson and coworkers (104) have obtained air-sensitive, crystalline samples of $ReO(Me)_4$ by the reaction of MeLi with $ReOCl_3(PPh_3)_2$ and $ReOCl_4$. $ReO(Me)_4$ is reported to be thermally stable in the absence of oxygen (104).

Rhenium (VII): The rapid extraction and separation of ReO_4^- from 3M HCl or H_2SO_4 with 4-methyl-2-pentanol has been described (105). The preparation, separation and properties of ReO_2Cl_3, obtained by heating stoichiometric mixtures of ReO_3Cl with $WOCl_4$, $MoCl_4$ or $ReOCl_4$ at <100°, have been given in detail (106). ReO_3Cl reacts with Me_4NCl in an inert solvent to give $[Me_4N]_2[ReO_3Cl_3]$. The anion also has C_{3v} symmetry if Et_4NCl is used in the reaction (107).

REFERENCES

1. P.D. Bhalerao, D.K. Kulkarni and V.G. Kher, PRAMANA 1, 230 (1973).
2. S. Huefner and G.K. Wertheim, PHYS. REV., B8, 4857 (1973).
3. Y. Sakisaka, T. Ishii and T. Sagawa, J. PHYS. SOC. JAPAN, 36, 1372 (1974); V.G. Nemoshkalenko, V.G. Aleshin and A.I. Sehkevich, METALLO-FIZIKA 52, 31 (1974).
4. J.H. Burness, J.G. Dillard and L.T. Taylor, INORG. NUCL. CHEM. LETT., 10, 387, 1137 (1974).
5. A.R. Hendrickson, R.L. Martin and N.M. Rohde, INORG. CHEM., 13, 1933 (1974); see also K.L. Brown, CRYST. STR. COMMUN., 3, 493 (1974); AUSTR. J. CHEM., 27, 2075, 2083, 2089 (1974); R.Y. Saleh and D.K. Straub, INORG. CHEM., 13, 3017 (1974).
6. "Manganese, Part C, Section 1", Gmelin Institute, (Verlag Chemie, Weinheim), 1973.
7. G. Rouschais, CHEM.REV., 74, 531 (1974).
8. L.V. Borisova and A.N. Ermakov, Nauka, Moscow, 1974, 318 pp.
9. C.N.R. Rao, SOLID STATE CHEM., 215 (1974).
10. O.E. Krein, L.V. Borisova, V.A. Zarinskii, O.D. Prasalova and V.I. Levedev, ZHUR. PRIKLAD KHIM (LENINGRAD)., 47, 1072 (1974).
11. J.F. Rowbottom and G. Wilkinson, DALTON TRANS., 684 (1974).
12. M.C. Scrutton, G.H. Reed and A.S. Mildvan, ADV. EXP. MED. BIOL., 13, 383 (1974).
13. R. Davis, INORG. CHEM., TRANS. ELEM., 2, 179 (1973); 3, 185 (1974).
14. P.M. Treichel, J. ORGANOMETALL. CHEM., 83, 427 (1974)
15. M. Inoue, K. Hara, T. Horiba and M. Kubo, BULL CHEM. SOC. JAPAN, 49, 2137 (1974).

16. J.F. Bald, US 3, 811, 948 (1974).

17. B. Chiswell and E.J. O'Reilly, INORG. CHIM. ACTA, 7, 704 (1973).

18. G. Brun and G. Jourdan, COMPT. REND., C279, 129 (1974).

19. F. Maggio, T. Pizzini and V. Romano, INORG. NUCL. CHEM. LETT., 10, 1005 (1974).

20. M.A. Guichelaar, J.A. Van Hest and J. Reedjik, ibid, 10, 999 (1974).

21. R.C. Paul, V.P. Kapila, R.S. Battu and S.K. Sharma, IND. J. CHEM., 12, 827 (1974).

22. H.D. Gillmann, INORG. CHEM., 13, 1921 (1974).

23. S. Casey, W. Levason and C.A. McCauliffe, DALTON TRANS, 886 (1974).

24. G.S. Rao and S.K. Gupta, IND. J. CHEM., 11, 956 (1973); see also H.J. Seifert, T. Krimmel and W. Heinemann, J. THERMAL ANAL., 6, 175 (1974).

25. A.S. Tryashin, V.V. Skopenko and D.A. Stakhov, ZHUR. NEORG. KHIM, 18 2658 (1973); M.W.G. DeBolster, B. Nieuwenhuijse and J. Reedjik, Z. NATURFORSCH., B28, 104 (1973).

26. A.K. Das and D.V.R. Rao, IND. J. CHEM., 12, 898 (1974).

27. B.M. Antti, ACTA CHEM. SCAND, 27, 3513 (1973).

28. G.V. Tsintsadze et al, ZHUR. STRUKT. KHIM, 15, 306 (1973); see also J.P. Barbier and R. Hugel, INORG. CHIM. ACTA, 10, 93(1973); Z Durski et al., ROCZ. CHEM., 48, 1615 (1974).

29. A.E. Shvelashvili, L.P. Sarashvili and R.M. Vashikidze, ZHUR. NEORG. KHIM., 19, 568 (1974)

30. V.T. Kallinikov et al, DOKL. AKAD. NAUK. SSR., 213, 844 (1973).

31. D.E. Chasan et al, INORG. NUCL. CHEM. LETT., 11, 41 (1975).

32. N.W. Alcock, D.C. Liles, M. McPartlin and D.A. Tasker, CHEM. COMM., 727 (1974).

33. B.B. Wayland and L.W. Olsen, INORG. CHIM. ACTA, 11, L23 (1974).

34. M. Ebert and J. Eysseltova, MONATSCH. CHEM., 105,1030 (1974).

35. N.D. Grekova, K. Nogoev and K. Sulaimankov, ZHUR. NEORG. KHIM., 18, 3107 (1973).

36. D.T. Altybaeva, D.B. Sarukaeva, B.I. Imanakunov and P.T. Yun, ISVEST. AKAD. NAUK. KIRG. SSR., 51 (1973).

37. D. Popov, R. Herak, B. Preleshik and B. Ribar, ZEIT. KRIST., 137, 280, 318 (1973).

38. E.F. Bertaut et al, ACTA CRYST. B30, 2234 (1974).

39. L. Walter-Levy, J. Perrotey and J.W. Visser, BULL. SOC. CHIM.FRANCE, 2596 (1973).

40. P. Glavic, A. Bole and J. Slivnik, J. INORG. NUCL. CHEM., 35, 3979 (1973).

41. J. Mach, CHEM. ABS. 81, 44787p (1974).

42. R.Y. Aliev, ZHUR. OBSHCH. KHIM, 43, 2572 (1973).

43. G.V. Tsintsadze et al, CHEM. ABS, 80, 66253t, 66272Z (1974).

44. D.R. Rossemsky and R.J. Hill, FARADAYTRANS. I., 70, 1140 (1974).

45. J. Klepetar and K. Stulik, J. ELECTROCHEM., INTERFAC. ELECTROCHEM., 55, 255 (1974).

46. R.F. Ziolo, R.H. Stanford, G.R. Rossman and H.B. Gray, J. AMER. CHEM. SOC., 96, 7910 (1974).

47. G.L. Cueto, CHEM ABS., 80, 90513y (1974); 81, 85391y (1974)

48. R.I. Novoselov and I.P. Sokolovskaya, ISV. SIB. OTD. AKAD. NAUK SSSR, 60 (1974).

49. E. Contreras, V. Riera, R. Uson, CHEM. ABS., 80, 43582k (1974).

50. K. Akabori, CHEM. LETT., 1481 (1974).

51. J.P. Fackler and A. Avdeef, INORG. CHEM., 13, 1864 (1974).

52. A. Avdeef and J.P. Fackler, ibid, 13, 1854 (1974); A. Avdeef, DISS. ABS., 35, 111 (1974).

53. R.R. Stults, V.W. Day, E.L. Tassett and R.S. Marianelli, INORG. NUCL. CHEM. LETT., 9, 1259 (19743); J. AMER. CHEM. SOC., 96, 2650 (1974).

54. A.F. Nikolaev et al, ZHUR PRIKLAD. KHIM. (LENINGRAD) 46, 2718 (1973).

55. M.M. Patel and R.P. Patel, J. IND. CHEM. SOC.,51, 770 (1974).

56. Y. Murakami, K. Sakata, K. Kenji and Y. Matsuda, BULL. CHEM. SOC. JAP. 47, 3021 (1974).

57. R. Hems, T.J. Cardwell and M.F. Mackay, AUSTR. J. CHEM., 28, 443 (1975).

58. L.J. Boucher and D.R. Herrington, INORG. CHEM., 13, 1105 (1974); J. INORG. NUCL. CHEM., 36, 531 (1974).

59. M.M. Patel, R.P. Patel, K.A. Patel and N.L. Patel, IND. J. CHEM., 11, 1177 (1973); see also K. Dey and K.C. Ray, J. IND. CHEM. SOC., 50, 631 (1973); C.L. White, A.H. White, C.A. Willis and K.S. Murray, DALTON TRANS., 1793 (1974); B.C. Shama and C.C. Patel, IND. J. CHEM., 11, 941 (1973).

60. J.E. Davies, B.M. Gatehouse and K.S. Murray, DALTON TRANS, 2523 (1973)

61. G.B. Kauffman et al, INORG. SYNTH., 15, 103 (1974).

62. Y. Yoshino, I. Taminaga, M. Masakazu and T. Uehiro, BULL. CHEM. SOC. JAPAN, 47, 2787 (1974).

63. K. Hochstrasser and W. Feitknecht, THERM. ANAL. PROC. INT. CONF. 3rd, 1971, 2, 79 (1972).

64. Y. Murakami, RADIOISOTOPES, 23, 238 (1974).

65. J. McRae, R.M. Sugar, B. Shipley and G.R. Hook, J. NUCL. MED., 15, 151 (1974).

66. A.G. Dumortier, O.M. Jeghers, P. Decostre and C. Fallais, INT. J. APPL. RADIAT. ISOTOPES, 25, 189 (1974).

67. J. Steigman, W.C. Eckelman, G. Meinkin, H.S. Isaacs and P. Richards, J. NUCL. MED., 15, 75 (1974).

68. Y. Yukio and Y. Maki, JAPAN KOKAI 74 14,900; see also H. Streckler and K.H. Tetzlaff, GER. OFFEN, 2, 236, 565 (1974); M.W. Billinghurst, M. Groothedde and R. Palser, J. NUCL. MED., 15, 266 (1974); J. Steigman and P. Richards, SEM. NUCL. MED., 4, 269 (1974).

69. J. Steigman, H.P. Williams, P. Richards, E. Lebowitz and G. Meinkin, J. NUCL. MED., 15, 318 (1974).

70. M.W. Billinghurst and R.F. Palser, J.NUCL. MED., 15, 722 (1974).

71. R.F. Barth and O. Singla, J. IMMUNOL., 112, 756 (1974); see also W.C. Eckelman, T.D. Smith and P. Richards, BROOKHAVEN NATL. LAB REPORT 18497 (1973).

72. M.K. Dewanjee, J. NUCL. MED., 15, 703 (1974).

73. B.A. Rhodes, SEM. NUCL. MED., 4, 281 (1974).

74. L.A. Monroe, W.L. Thompson, N.S. Anderton and J.A. Burdine, J. NUCL. MED., 15, 192 (1974); D.M. Lyster, J.R. Scott, E.K. Mincey and R.T. Morrison, J. NUCL. MED., 15, 198 (1974); H.C. Wieland, G.M. Grames, C. Jansen and T. Davidson, J. NUCL. MED., 15, 808 (1974).

75. A.J. Tofe and M.D. Francis, J. NUCL. MED., 15, 69 (1974).

76. I. Ikeda, O. Inoue, J. Uchida, Y. Sugisawa, K. Yoshihiko, K. Kurata and S. Kato, RADIOISOTOPES, 23 1 (1974); L.O.M.J. Smithuis and J.B.F. Spijkers, PHARM. WEEKBLAD., 109, 185 (1974); B.M. Bowen and E.S. Garnett, J. NUCL. MED., 15, 652 (1974); N. Ader and L.L. Capin, GER. OFFEN., 2, 344, 802 (1974); U. Buell, D. Balser and K.W. Frey, FORTSCHR. GEB. ROENTGENSTR. NUKLEARMED, 120, 481 (1974); K.D. Schwartz, M. Berger, H. Meyer and D. Ausborn, RADIOBIOL. RADIOTHER. 15, 63 (1974); R.L. Bell, J. TENN. MED. ASS., 67, 121 (1974).

77. C.P. Fliegel, M.K. Dewanjee, L.B. Holman, M.A. Davis and S. Treves, RADIOLOGY, 110, 407 (1974); J. NUCL. MED., 15, 176 (1974).

78. M.K. Dewanjee and E.W. Prince, J. NUCL. MED., 15, 577 (1974).

79. R.A. Jackson and T.F. Bolles, GER. OFFEN., 2, 343, 607 (1974).

80. M. Tsutsui and C.P. Hrung, J. AMER. CHEM. SOC., 96, 2638 (1974).

81. G. Bandoli, D.A. Clemente, U. Mazzi and E. Tondello, CRYST. STRUCT. COMM., 3, 293 (1974).

82. U. Mazzi, G. Depaoli, G. Rizzardi and L. Magon, INORG. CHIM. ACTA, 10, L2 (1974).

83. G.A. Mazzocchin, F. Magno, U. Mazzi and R. Portanova, INORG. CHIM. ACTA, 9, 263 (1974).

84. V. Mikulaj, P. Řajec and V. Faberova, CHEM. ZVEST., 28, 37 (1974);' RADIOCHEM. RADIOANALYT. LETT., 17, 375 (1974).

85. K. Schwochan, L. Astheimer, J. Hauck and H.J. Schenk, ANGEW. CHEM., 86, 350 (1974).

86. D.J. Darensbourg and D. Madrid, INORG. CHEM., 13, 1532 (1974).

87. T.S. Strivastava, C.P. Hrung and M. Tsutsui, CHEM. COMM., 447 (1974); M. Tsutsui and C.P. Hrung, J. AMER. CHEM. SOC., 96, 2638 (1974).

88. R. Robson, INORG. CHEM., 13 475 (1974).

89. F.A. Cotton, B.A. Frenz, J.R. Ebner and R.A. Walton, CHEM. COMM., 4 (1974).

90. R. Mason, K.M. Thomas, J.A. Zubieta and B.L. Shaw, J. AMER. CHEM. SOC., 96, 260 (1974).

91. I.D. Brown, C.J.L. Lock and Che'ng Wan, CAN. J. CHEM., 52,1704 (1974).

92. A.A. Opalovsku, E.U. Lobkov and V.E. Federov, ZHUR PRIKLAD. KHIM. (LENINGRAD) 47, 1200 (1974).

93. T. Okubo and H. Koshimura, NIPPON KAGADU KAISHI, (9) 1649 (1974); see also M. Pavlova, N. Iordanov and D. Staikov, DOKL. BOLG. AKAD. NAUK., 27, 68 (1974).

94. C.C. Addison, R. Davis and N. Logan, DALTON TRANS., 1073 (1974).

95. R.S. Shandles, R.K. Murmann and E.O. Schlemper, INORG. CHEM., 13, 1373 (1974).

96. V.S. Sergienko, M.A. Porai-Koshits and T.S. Khodahova, ZHUR. STRUKT. KHIM., 15, 275 (1974).

97. V.N. Evreev and S.V. Murashko, ZHUR. NEORG. KHIM., 19, 753 (1974).

98. K.V. Kotegov, T.V. Zegzhda, N.V. Fadeeva and Y.N. Kukushkin, ZHUR NEORG. KHIM, 19, 737, 2118 (1974).

99. V.S.Khain and N.N. Antsygina, ZHUR. OBSHCH. KHIM., 44, 2103 (1974).

100. K.V. Kotegov, F.K. Khakimov, Y.N. Kukushkin and L.V. Konovalov, ZHUR. OBSHCH. KHIM, 44, 2237 (1974); see also M. Pavlova, J. INORG. CHEM., 36, 1623 (1974).

101. M.C. Chakravorti and M.K. Chaudhuri, J. INORG. NUCL. CHEM., 36, 757 (1974).

102. P.T. Meiklejohn, M.T. Pope and R.A. Prados, J. AMER. CHEM. SOC., 96, 6779 (1974).

103. R.T. Paine and R.S. McDowell, INORG. CHEM., 13, 2366 (1974).

104. K. Mertis, J.F. Gibson and G. Wilkinson, CHEM. COMM., 93 (1974).

105. S.P. Patil and V.M. Shide, SEPAR. SCI., 9, 259 (1974).

106. I.A. Glukhov, N.A. El'manova, S.S. Eliseev and M.T. Temurova, ZHUR. NEORG. KHIM., 19, 309 (1974).

107. U. Gerlach and C. Ringel, ZEIT. ANORG. ALLGEM. CHEM., 408, 180 (1974).

IRON

J. R. Wasson and H. J. Stoklosa*

Department of Chemistry
University of Kentucky
Lexington, Kentucky 40506

*Present address: E. I. DuPont de Nemours and Co.
Experimental Station
Wilmington, Delaware 19898

GENERAL

Oxo-bridged iron(III) complexes (1), isocyanide (2) and isonitrosoketone (3) complexes, the chemistry of iron(IV), -(V) and -(VI) compounds (4), Alfred Werner's research on optically active iron compounds (5) and iron metalloboranes and metal-boron bonding (6) have been reviewed. Reviews of geometrical considerations for the marcasite (FeS_2) type structure (7), association of Fe(III) and perchlorate ions in solution (8), changes of redox properties by coordination (9), and organoiron compounds as reagents in organic synthesis (10) have been presented. The structure, bonding and reactivity of iron and other metal nitrosyl complexes (11) and the reaction of nitrosonium salts with iron compounds (12) have been reviewed. Mössbauer spectra for a number of dinitrosyl iron compounds have been reported (13). Mössbauer studies of polymeric Fe(III) methoxides and halomethoxides (14) and iron compounds with sulfur-containing Schiff-base ligands (15) have been described.

The synthesis and ion binding of synthetic multidentate macrocyclic compounds have been discussed (16). Magnetic resonance studies of iron metalloenzymes have been reviewed (17). PMR and Gouy techniques for measuring the mass susceptibility of haemoproteins have been compared (18). The magnetic susceptibility anisotropies in low-spin iron(III) heme proteins have been evaluated theoretically (19). Interactions of histidine and other imidazole derivatives with transition metal ions in chemical and biological systems have been reviewed (20) as have the oxidation states and reversible redox reactions of metalloporphyrins (21). The conformational state of iron proteins has been reviewed (22). MO calculations for the carbon monoxide adduct of haemoglobin (23) and monomeric oxygen adducts of iron porphyrins (24) have been reported. The properties of rubredoxins have been extensively considered on a theoretical level (25-28) as have the Mössbauer spectra of the myoglobin-CO complex (29) and pentacyanoiron complexes (30). Use of hemoglobin as a system for undergraduate laboratory study has been described (31).

COMPOUNDS WITH CARBON DONORS

Potassium hexacyanoferrate(III) reacts with molten sulfur at 200°C to yield, ultimately, yellow $K_3Fe(CN)_4(CNS)_2$ which is believed to contain the thiofulminate ion (32). Reactions of cyanide ion with aqueous solutions of oxo-bridged binuclear complexes of iron(III) and 1,10-phenanthroline and 2,2'-bipyridine, $[L_2(H_2O)Fe\text{-}O\text{-}FeL_2(H_2O)]^{4+}$ have been investigated (33). Single crystals of $Cs_2MgFe(CN)_6$ were grown by slow diffusion of Mg^{2+} and $Fe(CN)_6^{4-}$ ions through a silica gel containing CsCl (34). A number of $M_2[Fe(CN)_6]$ and $M_3[Fe(CN)_6]_2$ salts (M = Cu(II), Co(II), etc.) have been prepared (35) by addition of the appropriate metal chloride solution to a solution of potassium ferri- or ferrocyanide. The related compounds $Fe_4[Fe(CN)_6]_3$, $Fe_4[Ru(CN)_6]_3$ and $Fe_4[Os(CN)_6]_3$ can be similarly prepared (36). The addition compounds $K_4[Fe(CNSbX_3)_6]$ (X = F, Cl) and $K_4[Fe(CNSbX_3)_4(CN)_2]$ (X = Cl, Br) are formed by the reaction of $K_4[Fe(CN)_6]$ with SbX_3 in the melt (X = Cl or Br) or in SO_2 solution (X = F) (37). The reaction of ferro- and ferricyanide ions with hydrogen atoms has been investigated (38). Aquopentacyanoferrate(II) was identified as a final product. Products of one-, two- and four-electron reduction of the nitroprusside ion have been prepared (39) electrolytically and studied by electrochemical and spectrophotometric methods. The preparation of $Na_2[Fe(CN)_5enH]\cdot 6H_2O$ (enH = monoprotonated ethylenediamine) has been reported (40). The preparation of high purity crystals of disodium pentacyanoammineferrate(III) dihydrate, $Na_2[Fe(CN)_5NH_3]\cdot 2H_2O$, has been described (41). Solutions from which the $[Fe(CO)_4CN]^-$ ion can be precipitated can be prepared by dissolving $Fe_2(CO)_9$ and KCN in acetonitrile (42). Treating $Fe(phen)_2(CN)_2$ (phen = 1,10-phenanthroline) with dimethyl sulfate produces $Fe(phen)_2(CNCH_3)_2^{2+}$. Addition of simple amines to $Fe(CNCH_3)_6^{2+}$ can produce either monodentate carbene ligands or chelating ligands with two carbene donor centers (43).

COMPOUNDS WITH NITROGEN DONORS

The reaction of 3,3',4-trimethyldipyrromethene hydrochloride with iron(III) chloride hexahydrate in methanol yields dark green tris(3,3',4-trimethyldipyrromethenato)iron(III). The reactions of 3,3',5,5'-tetramethyldipyrromethene and 3,3',4,4',5,5'-hexamethyldipyrromethene with hydrated Fe(II) chloride yield green bis-ligand Fe(II) chelates (44). Tetraphenylporphyriniron(III) chloride, FeTTPCl, reacts with nitric oxide to yield diamagnetic FeTTPCl(NO). In methanol FeTTPCl reacts with nitric oxide to yield FeTTP(NO) which can reversibly bind another nitric oxide molecule. Bonding models have been presented (45). Octaethylhemin(I) reacts with dinitrogen tetroxide in dichloromethane to produce meso-tetranitromethyloctaethylhemin(II), a high-spin iron(III) complex with a porphyrin ligand fully substituted in all β- and meso-positions (46). Autoxidation of olefins and other active methylene compounds by iron(III) meso-tetraphenylporphin has been investigated (47). Siderochromes and model compounds have been characterized (48). The reactions of dilute solutions of octaethylporphyrin and its Fe(II) and Fe(III) complexes with methyl, 2-cyanopropyl, t-butoxy, and benzoyloxy radicals have been described (49). Meso-tetrakis(2,4,6-trimethoxyphenyl)porphine, $T(MeO)_3PPH_2$, yields the anion, T(MeO)$_3$PP, which forms a variety of iron complexes. The reduction of [FeCl(T(MeO)$_3$PP)] yields [Fe(T(MeO)$_3$PP)] which reacts readily with pyridine to form a diadduct. The compounds [Fe(T(MeO)$_3$PP)CO] and [Fe(T(MeO)$_3$PP)CO(py)] can also be obtained (50).

Completely conjugated 14-membered ring hexaaza ligand complexes have been synthesized by the template condensation (51) of 2,6-diacetylpyridine with hydrazine on iron(II). Low-spin six-coordinate complexes of this ligand, I, are obtained with axial ligands such as halides, pseudo-halide ions, acetonitrile, nitrite and carbon monoxide. The normally planar macrocyclic ligand can be forced into a folded conformation with strong bidentate chelating ligands such as 1,10-phenanthroline or 2,2'-bipyridine. The reaction of the bis-α-diimineiron(II) complex, I, with carbon monoxide and methyl hydrazine in trifluoroethanol results (52) in the formation of a complex, [Fe(C$_{10}$H$_{20}$N$_8$)CO(CH$_3$NHNH$_2$)]$^{2+}$ containing weakly bound carbon monoxide. Dissolution of the complex in acetonitrile results in rapid loss of CO. However, if the complex is dissolved in acetonitrile under 1 atm of CO pressure, followed by the addition of KOC(CH$_3$)$_3$ with a slight excess of CH$_3$NHNH$_2$ results in the formation of [Fe(C$_{10}$H$_{19}$N$_8$)CO(CH$_3$)] III. The compound is stable in wet or dry dimethylsulfoxide solutions when under a CO atmosphere. A paramagnetic dioxygen Fe(II) complex derived from a "picket fence" porphyrin (hemoprotein model) has been described (53). An improved preparation of Fe(II) tetraphenylporphin has been reported (54).

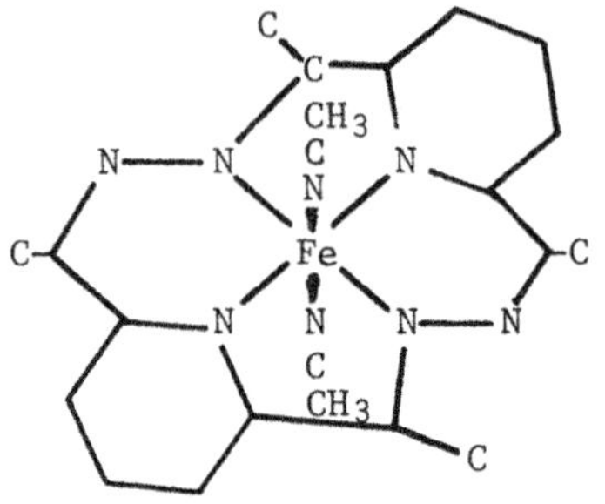

I

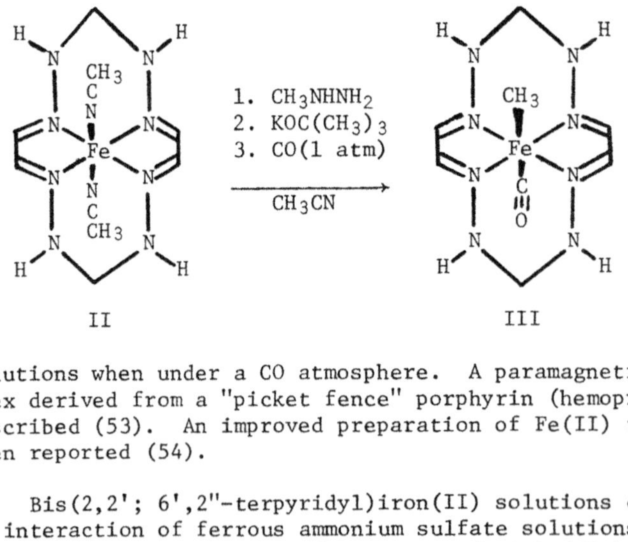

II III

Bis(2,2'; 6',2"-terpyridyl)iron(II) solutions can be prepared (55,56) by interaction of ferrous ammonium sulfate solutions with a slight excess of the ligand. Compounds of the type M(N$_2$H$_5$)$_2$(SO$_4$)$_2$ (M = Fe^{2+}, Cu^{2+}, etc.) can be prepared (57) by adding aqueous solutions of MSO$_4$ to an aqueous solution of hydrazinium sulfate. Salting in effects of metal chelates, e.g., Fe(2,2'-bipyridyl)$_3^{2+}$, and related species have been investigated (58). Tris(glyoxalbis(methylimine)iron(II) perchlorate can be prepared by adding methylamine to an aqueous mixture of glyoxal and FeCl$_2$ followed by addition of sodium perchlorate (59). Bis[2-(2-pyridyl)-imidazole]iron(II) complexes, Fe(PI)$_2$X$_2$ (X = Cl, Br, NCS, N$_3$ and CN) were prepared by mixing solutions of the ligand and the appropriate iron salt and by metathetical reactions (60). Reddish-brown carbonylbis(dimethylglyoximato)(substituted pyridine)iron(II) complexes can be obtained (61) by adding aqueous ferrous ammonium sulfate solution to a deaerated water-dioxan solution of dimethylglyoxime and the substituted pyridine which has been saturated with carbon monoxide. The liquid-liquid extraction of thiocyanotoferrate(III) complexes with tris(3,5,5-trimethylhexyl)amine has been investigated (62) but solids were not isolated. Infrared spectra of tetrahedral and octahedral Fe(III) thiocyanato complexes in the solid state and in solution have been described (63).

Tris 1-nitroso-2-naphtholate Fe(II) and Fe(III) complexes can be obtained by interaction of the potassium salt of 1-nitroso-2-naphthol with solutions of iron salts (64). Red-brown $[Fe(alanine)_2H_2O]_3O(ClO_4)_7$ results upon slow evaporation of 1:2 $Fe(ClO_4)_3$ - alanine solution in water (65). The compounds $[Fe(bipyNO)_3](ClO_4)_3 \cdot 2H_2O$ and $[Fe(phenNO)_3](ClO_4)_3$ (bipyNO and phenNO are the mono-N-oxides of 2,2'-bipyridine and 1,10-phenanthroline) are obtained by interaction of the metal perchlorate with the ligand in an ethanol-triethylorthoformate mixture (66). Ethyl picolinate (EP) forms $Fe(EP)(NCS)_2$ upon reaction with iron(II) thiocyanate (67). A number of FeL_4Cl_2 and FeL_2Cl_2 (L = substituted pyridine) complexes have been prepared by interaction of the pyridine ligand with aqueous Fe(II) chloride in a CO_2 or N_2 atmosphere (68). Bis(pyridine)-dithiocyanatoiron(II), $Fe(py)_2(NCS)_2$, can be prepared by heating $Fe(py)_4(NCS)_2$ at 117° under vacuum (69). The compounds $Fe(qnqn)Cl_2$ (70) and $Fe(mqn)Cl_2$ (71) (qnqn = trans-2-(2'-quinolyl)methylene-3-quinuclidine; mqn = 2-(N-morpholinylmethyl)-3-quinuclidine) have been prepared by interaction of the appropriate ligand with methanolic $FeCl_2 \cdot 4H_2O$ in a nitrogen atmosphere. The reaction of iron(II) perchlorate with 2-diethylphosphinomethylpyridine (Et_2PMP) yields the diamagnetic tris-chelate $[Fe(Et_2PMP)_3](ClO_4)_2$ (72). $Fe(NO)_2(PF_3)_2$ reacts with alkoxides and phenoxide by a complete F/OR exchange to yield the corresponding phosphite complexes. Only one or two fluorines exchange with OH^- to yield difluorophosphito complexes (73). The compounds $[FeX(np_3)]B(C_6H_5)_4$ (np_3 = tris(2-diphenylphosphinoethyl)amine) X = Cl, Br, I result (74) upon addition of sodium tetraphenylborate to a methylene chloride-ethanol solution of the appropriate ferrous halide and the ligand. The hydrido species $[FeHN_2(np_3)]B(C_6H_5)_4$ is obtained upon treatment of the halogeno compounds with sodium borohydride in a nitrogen atmosphere while $[FeH(CO)(np_3)]^+$ $B(C_6H_5)_4^-$ is obtained (74) similarly but in an atmosphere of carbon monoxide. Related tris(2-diphenylphosphinoethyl)phosphine compounds can be obtained in a similar fashion. A number of Fe(II) and Fe(III) compounds with di-(2-pyridyl)ketone as a ligand have been obtained. In some instances the carbonyl group undergoes metal-promoted reactions with water or alcohols (75). Iron(III) tris chelates with 4-amino-3-pentene-2-one and 4-methylamino-3-pentene-2-one are prepared (76) by heating stoichiometric quantities of the ligands and iron(III) chloride in ethanol on a water-bath for one-half hour. Yields are on the order of seventy percent. Dark brown $Fe(DAen)Cl \cdot H_2O$ can be prepared (77) by refluxing ferric chloride with the sodium salt of the ligand (DAen = Schiff base formed by condensing 2,5-dihydroxyacetophenone with ethylenediamine) in acetate buffer (pH 6-7) solutions for 2 hours. The compound can be washed with hot water and recrystallized from N,N-dimethylformamide. Iron(III) complexes with N-(3-hydroxypropyl)salicylaldimine, $[Fe(SALPA)Cl]_2$, and N-(2-phenylethyl)salicylaldimine, $Fe(SANE)_2Cl$, have been prepared and their structures determined (78). The reaction of $[Fe(SALPA)Cl]_2$ with sodium peroxide leads to the formation of dimeric $[Fe_2(SALPA)_2(SALPAH)_2]$ for which the crystal structure was also determined (79). Iron(III) complexes with pentadentate ligands derived from salicylaldehyde and 2,2'-bis(aminoethyl)sulfide, diethylenetriamine and 3,3'-bis(aminopropyl)amine have been prepared (80). Black chloro(bis{salicylideneiminephenyl}disulfido)-iron(III) is formed by interaction of the Schiff-base with iron(III) chloride in methanol (81). Salicylaldehyde salicylhydrazone forms a bis-Schiff base Fe(II) complex (82).

COMPOUNDS WITH OXYGEN DONORS

Red monoclinic $K_6[Fe_2O_6]$ can be obtained by heating $KO_{0.56}$ and "FeO" to 600°C in a closed iron tube (83). Formation of Fe_3O_4 by air oxidation of $Fe(OH)_2$ suspensions has been described (84). The free energy of formation of $LaFeO_3$ at 1204°C has been obtained (85). The $ZnO-Fe_2O_3$ catalyst can be prepared by adding aqueous ammonia to a solution of iron(III) and zinc nitrates and calcining the resultant precipitate (86). Hydrolysis reactions leading to the formation of α-, β- and γ-FeOOH have been described (87). Zinc ferrite, $ZnFe_2O_4$, can be prepared by reacting ZnO with α-Fe_2O_3 and γ-Fe_2O_3 (88). Red-brown iron(III) antimonate, $FeSbO_4$, can be obtained (89) by heating Fe_2O_3 and Sb_2O_3 for 24 hours at 1300°C. The preparation of FeO_4^{2-} solutions has been described (90). The oxide glass $3Gd_2O_3 \cdot 5Fe_2O_3$ was obtained (91) using the piston and anvil technique incorporated into a laser melting furnace. Ferric phosphate, $FePO_4$, can be synthesized by heating ferric oxide and ammonium dihydrogen phosphate. Heating $FePO_4$ iron at 800°C results in the formation of ferrous phosphate, $Fe_3(PO_4)_2$ (92). Iron(II) and -(III) doped lithium phosphate glasses can be prepared (93) by heating the appropriate metal oxide with Li_3PO_4 at 1000°C for several hours. Various hydroxides of Fe(III) were formed by addition of $NaHCO_3$ to solutions of iron(III) salts (94). A variety of complex anionic carbonates of Fe(III), e.g., $K[Co(NH_3)_6]_2[Fe_3(OH)_4(CO_3)_6]$, have been described (95).

The total synthesis of ferrichrome has been described (96). A variety of iron(III) β-diketonates have been obtained by adding the β-diketone to an aqueous methanol solution of iron(III) nitrate and then adding aqueous sodium hydroxide dropwise with vigorous stirring (97). Supplementary characterization studies of a number of phosphonate and phosphite ester complexes of Fe(III) as well as Fe(triphenylphosphine oxide)$_4$$(ClO_4)_3$ have been reported (98). Brown $[Fe(LNO)SO_4)]_2$ (LNO = 2,6-lutidine N-oxide) and pale yellow $[Fe(TPPO)_2SO_4]_x$ (TPPO = triphenylphosphine oxide) result (99) upon interaction of the ligands with methanol solutions of the metal sulfate. Brick-red $Na[Fe(EDDHA)] \cdot 4H_2O$ (H_4EDDHA = ethylenediamine di(o-hydroxyphenylacetic acid)) is prepared by stirring equimolar quantities of $H_4(EDDHA)$, freshly precipitated hydrous ferric oxide and sodium bicarbonate in 3:1 H_2O:methanol under nitrogen (100).

Brown (2-mercaptopyrimidin-4-ol)tetrakis(2-mercaptopyrimidin-4-olato-(-1))oxodiiron(III) is prepared (101) by adding an aqueous solution of 2-thiouracil (pH = 7) to aqueous $FeSO_4$. The initially green solution air oxidizes to the product. Formation constants for Fe(II)-ascorbate complexes have been investigated but solids were not isolated (102). Stability constants for Fe(II) and -(III) citrate complexes in solution have been measured (103) but solid species were not characterized. Complex formation between Fe(III) and tartrate (L) ions have been investigated potentiometrically and the species FeL and Fe_2L_2 found but solids were not isolated (104). A variety of white Fe(III) tris-monacidic phosphate or phosphonate ester chelates can be obtained (105) by suspending the anhydrous or hydrated metal halide in excess neutral ester and gradually increasing the temperature of the mixture. Triethyl phosphate reacts with iron halides at 100-250°C to form $Fe(O_2P(OC_2H_5)_2)_2$ and $Fe(O_2P(OC_2H_5)_2)_3$ (106). Air sensitive Fe(II) chelates with a variety of 2-hydroxy-4-alkoxy-benzophenones have been obtained by standard procedures (107). The compounds are monomeric in benzene.

COMPOUNDS WITH SULFUR DONORS

Synthetic analogs of active sites of iron-sulfur proteins, i.e., compounds containing ions of the type $[Fe_4S_4(SR)_4]^{2+ \text{ or } 2-}$, have been extensively characterized and their reactivities studied (108). Interaction of pyridine with iron(III) tris-dithio ligand complexes can result in the formation of pyridine diadducts of iron(II) bis-dithio ligand compounds and organic disulfide compounds (109). Iron(III) tris-thiomorpholine and N-methylpiperazine-dithiocarbamates are readily obtained (110) by interaction of the sodium salts of the ligands with aqueous iron(II) sulfate. Pentacoordinate $FeCl[SSeCNR_2]_2$ (R = CH_3, C_2H_5) compounds can be obtained (111) by interaction of tin(IV) N,N-dialkylthioselenocarbamates with ethanol-ligroin solutions of iron(III) chloride. Additional magnetic and structural studies of bis(N,N-diethyldithiocarbamato)iron(III)X (X = Cl, Br, I, SCN) compounds have been described (112). Single crystals of $FePSe_3$ can be grown by chemical vapor transport using 75 Torr of chlorine gas as a transport agent and a temperature gradient of 650-610°C (113). The hexachalcogenohypodiphosphate compounds $Fe_2P_2S_6$ and $Fe_2P_2Se_6$ can be synthesized by reaction of stoichiometric mixtures of the elements at elevated temperatures (114). A six-coordinate Fe(IV) compound with 1,1-dicarboethoxy-2,2-ethylenedithiolate has been described (115). Reactions of $Fe(mnt)_2^-$ with various amines have been investigated (116). The preparations and structures of two cubane-like iron nitrosyl complexes, $Fe_4(NO)_4(\mu_3\text{-}S)_4$ and $Fe_4(NO)_4(\mu_3\text{-}S)(\mu_3\text{-}NC(CH_3)_3)_2$ have been reported (117). Spin-free iron(III) and iron(II) complexes of thiazolidine-2-thione (ttz) having the compositions $[Fettz_2Cl_2]Cl$, $[Fettz_4Cl_2]Cl$, $[Fettz_6]Cl_3$, $[Fettz_3Br]Br_2$, $[Fettz_6]Br_3$, $[Fettz_2I_2]\cdot 2H_2O$, $[Fettz_3I]I$, $[Fettz_4I_2]$ and $[Fettz_6]I_2$ were obtained (118) by reaction of the corresponding iron salt in a little excess of molten ligand. The reaction of trans-bis(dinitrogen)bis(bisdiphenylphosphinoethane)molybdenum with tetrakis(ethylthio-μ_3-sulfido-ferrate)(2-) in the presence of excess hydrogen chloride yields tetrachloroferrate(III) among other products (119). Pyridine-2-thiol (L) complexes of the type FeL_2X_2 (X = halide) have been described (120). High-spin bis{μ-[N,N'-dimethyl-N,N'-bis(β-mercaptoethyl)ethylenediamine]}-diiron(II) and bis{μ-[N,N'-dimethyl-N,N'-bis(β-mercaptoethyl)-1,3-propanediamine]}-diiron(II) have been described (121). Iron(II) and -(III) complexes with thiomorpholin-3-one can be prepared by reaction of the appropriate iron salt in a little excess of the molten ligand (122).

COMPOUNDS WITH HALOGEN DONORS

The compounds Cs_3FeBr_5, Tl_3FeCl_5 and $(NH_4)_3FeCl_5$ were prepared (123) by a molten salt technique. Dipyridinium iron(III) pentachloride can be prepared (124) using iron(III) chloride in concentrated hydrochloric acid and pyridine. A number of ferric fluoride hydrates have been prepared (125) by reacting hydrofluoric acid with ferric hydroxides obtained from various iron(III) salts. The free energy of formation of $NaFeF_3$ has been determined (126). Hydrazinium pentafluoroferrate(III) results (127) upon interaction of aqueous hydrazinium difluoride and Fe(III) fluoride. The compounds Cs_2TlFeF_6, Cs_2KFeF_6, Rb_2KFeF_6, Rb_2NaFeF_6 and Cs_2NaFeF_6 can be obtained by fluorination at 550°C of stoichiometric mixtures of carbonate and chloride salts with $Fe(NO_3)_3 \cdot 9H_2O$ (128). Compounds of the type $MAgFeF_6$ (M = K, Rb, etc.) can be similarly obtained (129) as can be $BaLiFeF_6$ (130). Colorless $CsMgFeF_6$ and $CsZnFeF_6$ have been prepared by fluorination of simple starting materials (131). $KFeCl_3$ and $KFeBr_3$ were prepared (132) by mixing stoichiometric amounts of the appropriate potassium and iron halides in evacuated and sealed quartz ampoules which were then heated to 600°C for several hours. Pure FeOF can be prepared from the high pressure reaction of the appropriate metal oxide and metal fluoride. A homogeneous antiferromagnetic rutile-phase $Fe_xTi_{1-x}O_{2-x}F_x$ ($0 \leq x \leq 0.75$) can be obtained (133) by solid state reaction of α-Fe_2O_3, FeF_3 and TiO_2. When reacted with an aqueous solution of ammonium fluoride, freshly precipitated iron(III) hydroxide gel is partially transformed to triammonium hexafluoroferrate and not $FeOF \cdot xH_2O$ (134). Intercalated compounds of FeOCl and pyridine can be synthesized (135) by exposing FeOCl to pyridine vapor in a temperature range from 30° to 120°C. Compounds of the formula $[(C_6H_5)_3P]_3CuFeCl_3X$ (X = Cl, Br) can be prepared (136) by reaction of halotris(triphenylphosphine)copper(I) with iron(III) chloride.

MISCELLANEOUS

Iron-titanium-hydride alloy formation has been investigated (137). Iron phosphide on alumina catalysts can be prepared by heating $Fe(CO)_5$ with alumina in the presence of phosphine (138). Graphite-$FeCl_3$ and graphite-$FeCl_3$-$AlCl_3$ intercalation compounds can be obtained (139) by heating the anhydrous metal halides with graphite in sealed glass tubes between 250° and 350°C. The cathodic reduction of iron trisacetylacetonate in the presence of dienes and triphenylphosphine gives (140) complexes of the type $(diene)_2FeP(C_6H_5)_3$. Amine oxides have been found to induce the disengagement of organic ligands (L) from $(CO)_nFeL$ complexes in good yields (141). Brown crystalline bis(cyclo-octa-1,5-diene)iron(o), a useful starting material for iron(o) chemistry, has been obtained by the novel reaction of iron vapor with a solution of cyclo-octa-1,5-diene in methylcyclohexane at -120°. A description of the apparatus has been given (142). The thermal [2+2] cycloaddition reaction of vinylferrocene with tetracyanoethylene was found to occur under very mild reaction conditions in non-polar solvents and even in the solid state (143). 3-Vinylbisfulvalenediiron has been prepared in 0.02% overall yield and homopolymerized to give an 18-31% yield of homopolymer. Related copolymers can be obtained. The polymers can be oxidized with TCNQ to yield highly conducting $[Fe^{II}Fe^{III}](TCNQ)_2^-$ polysalts (144). Bis(hexamethylbenzene)-iron(II) hexafluorophosphate reacts with a variety of aromatic compounds, e.g., benzene, furan, ferrocene, etc., to form stable 1:1 charge-transfer complexes (145).

π-Cyclopentadienyl-1,2-bis(diphenylphosphino)ethane iron bromide and magnesium in tetrahydrofuran (THF) afford $(\pi\text{-}C_5H_5)(diphos)FeMgBr\cdot 3THF$ which contains a Fe-Mg (2.59Å) bond (146). Reaction of $Fe(acac)_3$, $(CH_3)_2AlOC_2H_5$ and $(C_6H_5)_2PCH_2CH_2P(C_6H_5)_2$ yields $Fe(CH_3)_2(dpe)_2$ (147). Yellow pentakis(trimethylphosphite)iron(o), $Fe\{P(OMe)_3\}_5$, can be obtained by the reduction of $[FeCl_2\{P(OMe)_3\}_3]$ with sodium amalgam in the presence of excess sodium amalgam (148). Neutral and cationic iron trimethyl phosphite complexes, e.g., $Fe(CO)_2(P(OMe)_3)_3$, have also been described (149). Red-brown bis(1-methyl-2,4,6-triphenyl-λ^5-phosphorinyl)iron has been prepared (150) by interaction of the ligand and $FeCl_2$ in THF.

Metallocarboranes have been prepared (151) by the direct reaction of small polyhedral carboranes with organoiron reagents in the gas phase or in solution without use of a prior cage-opening step. Hexaborane (10) reacts with $Fe_2(CO)_9$ in pentane to yield $\mu\text{-}Fe(CO)_4\cdot B_6H_{10}$ in high yield (152). $[2,3\text{-}(CH_3)_2C_2B_4H_4]_2Fe^{II}H_2$ can be obtained from the reaction of $FeCl_2$ in THF with $Na[2,3\text{-}(CH_3)_2C_2B_4H_5]$ along with $(CH_3)_4C_4B_8H_8$ (153). Electron delocalization in a number of Fe carboranes has been examined experimentally (154). Orange liquid $B_4H_8Fe(CO)_3$ can be prepared by the direct reaction of pentaborane (9) with $Fe(CO)_5$ and by the reaction of tetraborane (10) with $Fe_2(CO)_9$ (155). The syntheses of $[(C_4H_9)_4N][\mu\text{-}Fe(CO)_4B_7H_{12}]$ and $Fe(CO)_4B_7H_{11}$ have been described (156).

REFERENCES

1. K. S. Murray, COORD. CHEM. REVS., 12, 1 (1974).
2. F. Bonati and G. Minghetti, INORG. CHIM. ACTA, 9, 95 (1974).
3. B. C. Haldar, J. INDIAN CHEM. SOC., 51, 224 (1974).
4. W. Levason and C. A. McAuliffe, COORD. CHEM. REVS., 12, 151 (1974).
5. G. B. Kauffman, COORD. CHEM. REVS., 12, 105 (1974).
6. N. N. Greenwood and I. M. Ward, CHEM. SOC. REVS., 3, 231 (1974).
7. A. Kjekshus and T. Rakke, STRUCTURE AND BONDING, 19, 85 (1974).
8. L. Johansson, COORD. CHEM. REVS., 12, 241 (1974).
9. V. Gutmann, STRUCTURE AND BONDING, 15, 141 (1973).
10. M. Rosenblum, ACCTS. CHEM. RES., 7, 122 (1974).
11. J. H. Enemark and R. D. Feltham, COORD. CHEM. REVS., 13, 339 (1974).
12. M. T. Mocella, M. S. Okamoto and E. K. Barefield, SYN. REACT. INORG. METAL-ORG. CHEM., 4, 69 (1974).

13. H. Mosbaek and K. G. Poulsen, ACTA CHEM. SCAND., A28, 157 (1974).
14. R. L. Martin and I. A. G. Roos, AUST. J. CHEM., 27, 1833 (1974).
15. M. Bond, R. L. Martin and I. A. G. Roos, AUST. J. CHEM., 27, 2307 (1974).
16. J. J. Christensen, D. J. Eatough and R. M. Izatt, CHEM. REVS., 74, 351 (1974).
17. P. J. Quilley and G. A. Webb, COORD. CHEM. REVS., 12, 407 (1974).
18. K. D. Bartle, G. J. Dale, D. W. Jones and S. Maricic, J. MAGN. RES., 12, 286 (1973).
19. W. DeW. Horrocks, Jr. and E. S. Greenberg, MOL. PHYS., 27, 993 (1974).
20. R. J. Sundberg and R. B. Martin, CHEM. REVS., 74, 471 (1974).
21. J.-H. Fuhrhop, STRUCTURE AND BONDING, 18, 1 (1974).
22. M. Llinas, STRUCTURE AND BONDING, 17, 135 (1973).
23. M. P. Halton, INORG. CHIM. ACTA, 8, 137 (1974).
24. M. P. Halton, INORG. CHIM. ACTA, 8, 131 (1974).
25. G. H. Loew and D. Y. Lo, THEORET. CHIM. ACTA, 32, 217 (1974).
26. G. H. Loew, M. Chadwick and D. A. Steinberg, THEORET. CHIM. ACTA, 33, 125 (1974).
27. G. H. Loew and D. Lo, THEORET. CHIM. ACTA, 33, 137 (1974).
28. G. H. Loew, M. Chadwick and D. Lo, THEORET. CHIM. ACTA, 33, 147 (1974).
29. A. Trautwein, Y. Maeda, F. E. Harris and H. Formanek, THEORET. CHIM. ACTA, 36, 67 (1974).
30. A. Trautwein, F. E. Harris and I. Dezsi, THEORET. CHIM. ACTA, 35, 231 (1974).
31. A. G. Splittgerber, J. CHEM. EDUC., 51, 402 (1974).
32. S. B. Cole and J. Kleinberg, INORG. CHIM. ACTA, 10, 157 (1974).
33. B. J. C. Lima, F. David and P. G. David, J. C. S. DALTON TRANS., 680 (1974).
34. B. I. Swanson, S. I. Hamburg and R. R. Ryan, INORG. CHEM., 13, 1685 (1974).
35. S. N. Ghosh, J. INORG. NUCL. CHEM., 36, 2465 (1974).
36. H. Anoue and S. Yanagisawa, J. INORG. NUCL. CHEM., 36, 1409 (1974).
37. H. G. Nadler, J. Pebler and K. Dehnicke, Z. ANORG. ALLGEM. CHEM., 404, 230 (1974).
38. D. Zehavi and J. Rabani, J. PHYS. CHEM., 78, 1368 (1974).
39. J. Masek and E. Maslova, COLL. CZECH. CHEM. COMM., 39, 2141 (1974).
40. J. A. Olabe and P. J. Aymonino, J. INORG. NUCL. CHEM., 36, 1221 (1974).
41. A. Tullberg and N.-G. Vannerberg, ACTA CHEM. SCAND., A28, 340 (1974).
42. S. A. Goldfield and K. N. Raymond, INORG. CHEM., 13, 770 (1974).
43. D. J. Doonan and A. L. Balch, INORG. CHEM., 13, 921 (1974).
44. Y. Murakami, Y. Matsuda, K. Sakata and K. Harada, BULL. CHEM. SOC. JAPAN, 47, 458 (1974).
45. B. B. Wayland and L. W. Olson, J. AMER. CHEM. SOC., 96, 6037 (1974); B. B. Wayland, J. V. Minkiewicz and M. E. Abd-Elmageed, *ibid.*, 96, 2795 (1974).
46. J. C. Fanning, T. L. Gray and N. Datta-Gupta, J. C. S. CHEM. COMM., 23 (1974).

47. D. R. Paulson, R. Ullman, R. G. Sloane and G. L. Closs, J. C. S. CHEM. COMM., 186 (1974).
48. J. Leony and K. N. Raymond, J. AMER. CHEM. SOC., 96, 1757, 6628 (1974).
49. C. E. Castro, C. Robertson and H. Davis, BIOINORG. CHEM., 3, 343 (1974).
50. L. Vaska, A. R. Amundsen, R. Brady, B. R. Flynn and H. Nakai, FINN. CHEM. LETT., 66 (1974).
51. V. L. Goedken, Y. Park, S.-M. Peng and J. M. Norris, J. AMER. CHEM. SOC., 96, 7693 (1974).
52. V. L. Goedken and S.-M. Peng, J. AMER. CHEM. SOC., 96, 7826 (1974).
53. J. P. Collman, R. R. Gagne and C. A. Reed, J. AMER. CHEM. SOC., 96 2629 (1974).
54. S. M. Husain and J. G. Jones, INORG. NUCL. CHEM. LETT., 10, 105 (1974).
55. J. Burgess and M. V. Twigg, J. C. S. DALTON TRANS., 2032 (1974).
56. S. Musumeci, R. Rizzarelli, S. Sammartano and R. P. Bonomo, J. INORG. NUCL. CHEM., 36, 853 (1974).
57. H. T. Witteveen and J. Reedijk, J. SOLID STATE CHEM., 10, 151 (1974).
58. E. Iwamoto, M. Yamamoto and Y. Yamamoto, INORG. NUCL. CHEM. LETT., 10, 1069 (1974).
59. H. L. Chum and P. Krumholz, INORG. CHEM., 13, 514 (1974).
60. Y. Sasaki and T. Shigematsu, BULL. CHEM. SOC. JAPAN, 47, 109 (1974).
61. G. O. Morpurgo and V. Mosini, J. C. S. DALTON TRANS., 2233 (1974).
62. A. R. Burkin, N. M. Rice and M. J. Rogers, J. C. S. DALTON TRANS., 213 (1974).
63. A. Miezis, ACTA CHEM. SCAND., A28, 415 (1974).
64. M. Takada and H. Sano, BULL. CHEM. SOC. JAPAN, 47, 2341 (1974).
65. E. M. Holt, S. L. Holt, W. F. Tucker, R. O. Asplund and K. J. Watson, J. AMER. CHEM. SOC., 96, 2621 (1974).
66. A. N. Speca, N. M. Karayannis and L. L. Pytlewski, INORG. CHIM. ACTA, 9, 87 (1974); J. INORG. NUCL. CHEM., 36, 1227 (1974).
67. R. C. Paul, R. S. Chopra, R. K. Bhambri and G. Singh, J. INORG. NUCL. CHEM., 36, 3703 (1974).
68. M. Takeda, T. Tominaga and N. Saito, J. INORG. NUCL. CHEM., 36, 2459 (1974).
69. W. M. Reiff, R. B. Frankel, B. F. Litte and G. J. Long, INORG. CHEM., 13, 2153 (1974).
70. G. J. Long and D. L. Coffen, INORG. CHEM., 13, 270 (1974).
71. R. C. Dickinson and G. J. Long, INORG. CHEM., 13 262 (1974).
72. M. Schafer and E. Uhlig, Z. ANORG. ALLGEM. CHEM., 407, 23 (1974).
73. Th. Kruck, J. Waldmann, M. Höfler, G. Birkenhäger and Ch. Odenbrett, Z. ANORG. ALLGEM. CHEM., 402, 16 (1973).
74. P. Stoppioni, F. Mani and L. Sacconi, INORG. CHIM. ACTA, 11, 227 (1974).
75. V. Rattanaphani and W. R. McWhinnie, INORG. CHIM. ACTA, 9, 239 (1974).
76. A. K. Srivastava, V. B. Rana and M. Mohan, J. INORG. NUCL. CHEM., 36, 3864 (1974).
77. L. D. Dave and B. G. S. Prasad, INDIAN J. CHEM., 12, 1002 (1974).

78. J. A. Bertrand, J. L. Breece and P. G. Eller, INORG. CHEM., 13, 125 (1974).
79. J. A. Bertrand and P. G. Eller, INORG. CHEM., 13, 927 (1974).
80. R. D. Patton and L. T. Taylor, INORG. CHIM. ACTA, 8, 191 (1974).
81. J. A. Bertrand and J. L. Breece, INORG. CHIM. ACTA, 8, 267 (1974).
82. K. K. Narang and A. Aggarwal, INORG. CHIM. ACTA, 9, 137 (1974).
83. H. Rieck and R. Hoppe, Z. ANORG. ALLGEM. CHEM., 408, 151 (1974).
84. M. Kiyama, BULL. CHEM. SOC. JAPAN, 47, 1646 (1974).
85. N. Kimizuka and T. Katsura, BULL. CHEM. SOC. JAPAN, 47, 1801 (1974).
86. T. Kotanigawa and K. Shimokawa, BULL. CHEM. SOC. JAPAN, 47, 954, 1535 (1974); T. Kotanigawa, ibid., 47, 950 (1974).
87. K. Kaneko and K. Inouye, BULL. CHEM. SOC. JAPAN, 47, 1139 (1974).
88. K. R. Krishnamurthy, J. Gopalakrishnan, G. Aravamudan and M. V. C. Sastri, J. INORG. NUCL. CHEM., 36, 569 (1974).
89. J. B. Wooten, G. G. Long and L. H. Bowen, J. INORG. NUCL. CHEM., 36, 2177 (1974).
90. D. H. Williams and J. T. Riley, INORG. CHIM. ACTA, 8, 177 (1974).
91. S. Yajima, K. Okamura and T. Shishido, CHEM. LETT., 1531 (1974).
92. E. Kostiner and J. R. Rea, INORG. CHEM., 13, 2876 (1974).
93. M. Berretz and S. L. Holt, J. INORG. NUCL. CHEM., 36, 49 (1974).
94. R. J. Knight and R. N. Sylva, J. INORG. NUCL. CHEM., 36, 591 (1974).
95. A. K. Sengupta and A. K. Nandi, Z. ANORG. ALLGEM. CHEM., 403, 327 (1974).
96. Y. Isowa, M. Ohmori and H. Kurita, BULL. CHEM. SOC. JAPAN, 47, 215 (1974).
97. I. Yoshida, H. Kobayashi and K. Ueno, BULL. CHEM. SOC. JAPAN, 47, 2203 (1974).
98. N. M. Karayannis, C. M. Mikulski, M. J. Strocko, L. L. Pytlewski and M. M. Labes, INORG. CHIM. ACTA, 8, 91 (1974).
99. N. M. Karayannis, C. M. Mikulski, L. L. Pytlewski and M. M. Labes, INORG. CHIM. ACTA, 10, 97 (1974).
100. B. P. Gaber, V. Miskowski and T. G. Spiro, J. AMER. CHEM. SOC., 96, 6868 (1974).
101. I. P. Khullar and U. Agarwala, AUST. J. CHEM., 27, 1877 (1974).
102. P. Ulmgren and O. Wahlberg, ACTA CHEM. SCAND., A28, 631 (1974).
103. T. B. Field, J. L. McCourt and W. A. E. McBryde, CAN. J. CHEM., 52, 3119 (1974).
105. C. M. Mikulski, N. M. Karayannis and L. L. Pytlewski, J. INORG. NUCL. CHEM., 36, 971 (1974).
106. C. M. Mikulski, L. L. Pytlewski and N. M. Karayannis, Z. ANORG. ALLGEM. CHEM., 403, 200 (1974).
107. E. Uhlig and U. Dinjus, Z. ANORG. ALLGEM. CHEM., 397, 297 (1973).
108. B. V. DePamphilis, B. A. Averill, T. Herskovitz, L. Que, Jr. and R. H. Holm, J. AMER. CHEM. SOC., 96, 4159 (1974); L. Que, Jr., M. A. Bobrik, J. A. Ibers and R. H. Holm, ibid., 96, 4168 (1974); M. A. Bobrik, L. Que, Jr. and R. H. Holm, ibid., 96, 287 (1974).
109. R. Y. Saleh and D. K. Straub, INORG. CHEM., 13, 1559 (1974).
110. G. Marcotrigiano, G. C. Pellacani and C. Preti, J. INORG. NUCL. CHEM., 36, 3709 (1974).
111. S. Nakajima and T. Tanaka, BULL. CHEM. SOC. JAPAN, 47, 763 (1974).

112. G. E. Chapps, S. W. McCann, H. H. Wickman and R. C. Sherwood, J. CHEM. PHYS., 60, 990 (1974).
113. B. Taylor, J. Steger, A. Wold and E. Kostiner, INORG. CHEM., 13, 2719 (1974).
114. W. Klingen, R. Ott and H. Hahn, Z. ANORG. ALLGEM. CHEM., 396, 271 (1973); W. Klingen, G. Eulenberger and H. Hahn, ibid., 401, 97 (1973).
115. F. J. Hollander, R. Pedelty and D. Coucouvanis, J. AMER. CHEM. SOC., 96, 4032 (1974).
116. I. G. Dance and T. R. Miller, INORG. CHEM., 13, 525 (1974).
117. R. S. Gall, C. T.-W. Chu and L. F. Dahl, J. AMER. CHEM. SOC., 96, 4019 (1974).
118. C. Fregni, C. Preti, G. Tosi and G. Verani, J. INORG. NUCL. CHEM., 36, 3695 (1974).
119. J. Chatt, C. M. Elson and R. L. Richards, J. C. S. CHEM. COMM., 189 (1974).
120. I. P. Evans and G. Wilkinson, J. C. S. DALTON TRANS., 946 (1974).
121. W-j. Hu and S. J. Lippard, J. AMER. CHEM. SOC., 96, 2366 (1974).
122. D. De Filippo, F. Devillanova and G. Verani, J. INORG. NUCL. CHEM., 36, 1017 (1974).
123. M. Amit, A. Horowitz, E. Ron and J. Makovsky, ISRAEL J. CHEM., 11, 749 (1973).
124. R. K. Sharma and N. Kumar, J. INORG. NUCL. CHEM., 36, 1172 (1974).
125. I. Dezsi, P. J. Ouseph and P. M. Thomas, J. INORG. NUCL. CHEM., 36, 833 (1974).
126. C. E. Bamberger, B. F. Hitch and C. F. Baes, Jr., J. INORG. NUCL. CHEM., 36, 543 (1974).
127. D. Hanzel, A. Moljk and J. Slivinik, INORG. NUCL. CHEM. LETT., 10, 1 (1974).
128. E. Alter and R. Hoppe, Z. ANORG. ALLGEM. CHEM., 407, 305 (1974).
129. B. Muller and R. Hoppe, Z. ANORG. ALLGEM. CHEM., 395, 239 (1973).
130. W. Viebahn and D. Babel, Z. ANORG. ALLGEM. CHEM., 406, 38 (1974).
131. R. Jesse and R. Hoppe, Z. ANORG. ALLGEM. CHEM., 403, 143 (1974).
132. M. Amit, A. Horowitz and J. Makovsky, ISRAEL J. CHEM., 12, 827 (1974).
133. G. Pausewang, Z. ANORG. ALLGEM. CHEM., 409, 45 (1973).
134. K. J. Gallagher and M. R. Ottaway, J. C. S. DALTON TRANS., 2347 (1974).
135. F. Kanamaru, S. Yamanaka, M. Koizumi and S. Nagoi, CHEM. LETT., 373 (1974).
136. D. J. Saturnino and P. R. Girardot, INORG. CHEM., 13, 2914 (1974).
137. J. J. Reilly and R. H. Wiswall, Jr., INORG. CHEM., 13, 218 (1974).
138. E. L. Muetterties and J. C. Sauer, J. AMER. CHEM. SOC., 96, 3410 (1974).
139. T. Tominaga, T. Sakai and T. Kimura, CHEM. LETT., 853 (1974).
140. W. Schäfer, H.-J. Kerrinnes and U. Langbein, Z. ANORG. ALLGEM. CHEM., 406, 101 (1974).
141. Y. Shvo and E. Hazum, J. C. S. CHEM. COMM., 336 (1974).
142. R. Mackenzie and P. L. Timms, J. C. S. CHEM. COMM., 651 (1974).
143. Y. Shirota, K. Yoshida, T. Nogami and H. Mikawa, CHEM. LETT., 1271 (1973).

144. C. U. Pittman, Jr. and B. Surynarayanan, J. AMER. CHEM. SOC., 96, 7916 (1974).
145. D. M. Braitsch, J. C. S. CHEM. COMM., 460 (1974).
146. H. Felkin, P. J. Knowles, B. Meunier, A. Mitscher, L. Ricard and R. Weiss, J. C. S. CHEM. COMM., 44 (1974).
147. T. Ikariya and A. Yamamoto, J. C. S. CHEM. COMM., 720 (1974).
148. E. L. Muetterties and J. W. Rathke, J. C. S. CHEM. COMM., 850 (1974).
149. E. T. Libbey and G. M. Bancroft, J. C. S. DALTON TRANS., 87 (1974).
150. G. Markl and C. Martin, ANGEW. CHEM. INT. EDN., 13, 408 (1974).
151. V. R. Miller, L. G. Sneddon, D. C. Beer and R. N. Grimes, J. AMER. CHEM. SOC., 96, 3090 (1974).
152. A. Davison, D. D. Traficante and S. S. Wreford, J. AMER. CHEM. SOC., 96, 2802 (1974).
153. W. M. Maxwell, V. R. Miller and R. N. Grimes, J. AMER. CHEM. SOC., 96, 7116 (1974).
154. R. J. Wiersema and M. F. Hawthorne, J. AMER. CHEM. SOC., 96, 761 (1974).
155. N. N. Greenwood, C. G. Savory, R. N. Grimes, L. G. Sneddon, A. Davison and S. S. Wreford, J. C. S. CHEM. COMM., 718 (1974).
156. O. Hollander, W. R. Clayton and S. G. Shore, J. C. S. CHEM. COMM., 604 (1974).

COBALT and NICKEL

P. R. Mitchell

Institut für anorganische Chemie,
Universität Basel,
Basel, Switzerland.

GENERAL

It is rare for a review to be devoted primarily to the research efforts of one man, but a recent article (1) on Werner's research on optically active coordination compounds, which includes details of unpublished work of Werner and his colleagues, will be of great value to all those interested in cobalt chemistry. The chemistry of the higher oxidation states of cobalt and nickel [cobalt(III) other than Werner-type complexes, cobalt(IV), cobalt(V), nickel(III), and nickel(IV)] has been reviewed (2). Reviews on complexes of oximes (3), isocyanides (4), dinitrogen (5), and S-N chelating ligands (6) include useful references to complexes of cobalt and nickel.

Hydrogen-containing cobalt and nickel borides have been made by the reduction of cobalt(II) and nickel(II) salts with sodium tetrahydridoborate in an ethanolic solution (7). These compounds, which were tentatively assigned the formulae $(Co_2B)_5H_3$ and $(Ni_2B)_2H_3$, are useful hydrogenation catalysts. Alumina supported cobalt or nickel phosphides, prepared by pyrolysis of $Co_2(CO)_8$ or $Ni(CO)_4$ on alumina followed by reaction with phosphine, catalyse the hydrogenation and dimerization of olefins (8).

LOW-VALENT COBALT COMPLEXES

Carbonyls. A useful synthesis of octacarbonyldicobalt from anhydrous cobalt(II) iodide, zinc or cadmium dust, and carbon monoxide at atmospheric pressure has been described (9). If a larger amount of the powdered metal is used, $Zn[Co(CO)_4]_2$ or $Cd[Co(CO)_4]_2$

is formed instead. These are also formed from $Co_2(CO)_8$ and the powdered metal (9). $Hg[Co(CO)_4]_2$ may be produced in 85% yield from $Co_2(CO)_8$ and finely divided mercury (10). The metal exchange of thallium with $Hg[Co(CO)_4]_2$ and the reaction of thallium(I) nitrate with $NaCo(CO)_4$ are useful routes to thallium(I) tetracarbonylcobaltate(-I) (11): The earlier procedure from $Co_2(CO)_8$ and thallium is however still the best (12). In all of the methods careful deoxygenation is required to avoid contamination by tris(tetracarbonylcobalt) thallium(III), which can itself be prepared by the reaction of $TlCo(CO)_4$ with an excess of $Co_2(CO)_8$ (11).

The reaction of trialkyl-silanes, -germanes, or -stannanes with $Co_2(CO)_8$ gives high yields of $R_3MCo(CO)_4$ (M = Si, Ge, or Sn: R = CH_3 or C_2H_5) (13). Chlorodimethylsilane and dichloro(methyl)silane react similarly (14). Trialkylgermanes replace the trialkylsilyl group in $R_3SiCo(CO)_4$, and trialkylstannanes replace both trialkylsilyl and trialkylgermyl groups (13). Germanium tetrafluoride displaces the trimethylsilyl group in $(CH_3)_3SiCo(CO)_4$ giving $GeF_3Co(CO)_4$ (15).

The polynuclear carbonyl carbide anions, $[Co_6(CO)_{15}C]^{2-}$ and $[Co_2Rh_4(CO)_{15}C]^{2-}$, are formed by the reaction of $(CO)_9Co_3CCl$ with $[Co(CO)_4]^-$ and $[Rh(CO)_4]^-$ respectively (16), and can be isolated as the trimethyl(benzyl)ammonium salts. $[Co(CO)_4]^-$ also reacts with $[(diphos)PtCl_2]$ and $[(diars)PtCl_2]$ to give the unusual compounds $[(diphos)PtCo_2(CO)_7]$ and $[(diars)PtCo_2(CO)_7]$ (17).

The reaction of $Co_2(CO)_8$ with $R_2ECH_2ER_2$ (E = P, As or Sb) gives a series of compounds, $Co_2(CO)_6(R_2ECH_2ER_2)$, in which two cobalt atoms are joined by two bridging carbonyls, a metal-metal bond, and a bridging $R_2ECH_2ER_2$ group (18). The syntheses of $[Co(CO)_3PR_3]SnR'_3$ and $[Co(CO)_3PR_3]_2SnX_2$, by the reactions of $[Co(CO)_3PR_3]^-$ with R'_3SnCl and SnX_4 respectively, have been reported (19). Isocyanides displace up to five carbonyl groups from $Co_4(CO)_{12}$, $(C_5H_5)NiCo_3(CO)_9$, $(CO)_9Co_3CY$ (Y = Cl, CH_3 or C_2H_5), $(CO)_9Co_3S$, and $(CO)_9Co_2FeS$ with retention of the metal cluster (20): however $Co_2(CO)_8$ gives either the ionic $[(RNC)_5Co][Co(CO)_4]$ or $[Co_4(CO)_{(12-n)}(RNC)_n]$ (20), and not $[Co_2(CO)_{(8-n)}(RNC)_n]$ as was claimed earlier (21).

Phosphine and phosphite complexes. A detailed study of the synthesis and reactions of $HCo[P(OR)_3]_4$ and its derivatives has been published (22). The hydrides were synthesized by reduction of a dimethoxyethane solution of cobalt(II) chloride and the phosphite using sodium tetrahydridoborate. Deprotonation of $HCo[P(OR)_3]_4$ requires a strong base, such as potassium hydride, and prolonged reaction times at 70°. If the phosphite is sterically hindered, no deprotonation occurs. The potassium salt, $K[Co\{P(OR)_3\}_4]$, reacts with mercury(II) chloride to give $Hg\{Co[P(OR)_3]_4\}_2$, but reacts with $(CH_3)_3GeCl$ to give only $Co_2\{P(OR)_3\}_8$, probably *via* the very unstable $(CH_3)_3GeCo\{P(OR)_3\}_4$. Protonation of $HCo[P(OR)_3]_4$ occurs with trifluoroacetic acid, but the product, $[H_2Co\{P(OR)_3\}_4]^+$, could not be isolated (22).

Trifluorophosphine readily displaces dinitrogen from $[CoH(N_2)(PPh_3)_3]$ to give $[CoH(PF_3)(PPh_3)_3]$ which slowly reacts with more trifluorophosphine to give $[CoH(PF_3)_2(PPh_3)_2]$ (23), but no further substitution occurs. When trifluorophosphine is bubbled through a suspension of $[CoCl(PPh_3)_3]$ in benzene, a low yield (14%) of $[CoH(PF_3)_3(PPh_3)]$ is formed (23).

Triethylphosphite reacts with cobalt(II) chloride in solution in aqueous ethanol to give $[\{(EtO)_3P\}_5Co]Cl$, which loses phosphite when heated *in vacuo* at 90° to form $[\{(EtO)_3P\}_3CoCl]$ (24). A new general synthesis of $[Co\{P(OR)_3\}_5]^+$ from (η-cyclo-octenyl)(η-1,5-cyclo-octadiene)cobalt has been described (25): the latter compound is said to be a versatile intermediate for the synthesis of low-valent cobalt compounds, but no experimental details of its synthesis were given (25). The disproportionation of cobalt(II) to cobalt(I) and cobalt(III), well known with phosphite ligands, also occurs with phenyldifluorophosphine and cobalt(II) chloride or bromide, but not with cobalt(II) iodide (26). The syntheses of a number of five-coordinate cobalt(I) complexes containing carbonyl, cyanide or an isocyanide, in addition to phosphine, phosphonite, or phosphite ligands, have been reported (27-29).

Addition of sodium tetraphenylborate to an ethanolic solution of $[CoCl\{P(OEt)_3\}_3]$ gives a brown solid from which $[Co\{P(OEt)_3\}_2(BPh_4)]$ can be isolated; this apparently contains a coordinated tetraphenylborate group (24). Reduction of cobalt(II) salts with sodium tetrahydridoborate in the presence of triphenylphosphine or 5-phenyl-5H-dibenzophosphole (DBP) gives several types of phosphine complexes in lower oxidation states, including hydrido complexes, BH_4^- complexes, and cobalt(0) complexes (30):

$$CoX_2.6H_2O + PPh_3 \xrightarrow{NaBH_4} [CoX(PPh_3)_3] \xrightarrow{NaBH_4} [Co(PPh_3)_3(BH_4)]$$

$$[Co(PPh_3)_3(BH_4)] + N_2 \longrightarrow [Co(N_2)(PPh_3)_3(BH_4)]$$

$$CoCl_2.6H_2O + DBP \xrightarrow{NaBH_4} [CoCl(DBP)_3] \xrightarrow{NaBH_4} [Co(DBP)_4]$$

Under different conditions $[Co(DBP)_2]$ can be obtained. Sodium cyanotrihydridoborate gives similar complexes containing the coordinated BH_3CN^- ion, e.g. $[Co(BH_3CN)(DBP)_4]$, $[Co(BH_3CN)(diphos)_2]$ (30).

Other complexes. Passage of nitric oxide through a cooled (0°) suspension of bis(thioacetylthioacetonato)cobalt(II) gives the nitrosyl $[Co(NO)(SacSac)_2]$: at 20° only $[Co(NO_2)_2(SacSac)]$ is formed (31). Secondary phosphines react with hydrated cobalt(II) in solution in propan-2-ol at room temperature to give phosphine nitrosyl complexes, $[Co(NO)(HPR_2)_3]$. Some tertiary phosphines react (32) in a similar way, but only when heated under reflux: sterically hindered phosphines, such as triphenylphosphine, do not react. The reaction of $[Co(NO)(CO)_2(PPh_3)_3]$ with $\{(CH_3)_3CN\}_2S$ in hot toluene gives a tetramer $[Co_4(NO)_4\{\mu_3\text{-}NC(CH_3)_3\}_4]$ with a cubane-like structure (33).

Controlled reduction of $[Co^{II}(salen)]$ with sodium gives $Na[Co^I(salen)]$: in solution in pyridine and in tetrahydrofuran this rapidly absorbs carbon dioxide giving $Na[(py)Co(salen)(CO_2)]$ and $Na[Co(salen)(CO_2)]$ respectively. The latter compound rapidly loses carbon dioxide *in vacuo*, reverting to $Na[Co(salen)]$, whereas the former compound does so only slowly (34).

LOW-VALENT NICKEL COMPLEXES

Carbonyls. The synthesis and crystal structure of a hexanuclear anionic nickel carbonyl, $[Ni_3(CO)_3(\mu_2\text{-}CO)_3]_2^{2-}$, has been described (35): it may be identical to one said earlier to be $[Ni_4(CO)_9]^{2-}$ (36). Reduction of nickel(II)chloride with sodium methanethiolate in the presence of a tertiary phosphine and carbon monoxide gives the substituted carbonyls $L_2Ni(CO)_2$ [L_2 = $(CH_3PPh_2)_2$, $(PPh_3)_2$, diphos, and $Ph_2PCH{=}CHPPh_2$] (37). A similar reaction takes place with sodium diethylselenocarbamate as the reducing agent (38); the intermediate $[L_2Ni\{SeC(O)NEt_2\}_2]$ can be isolated. The cyclic phosphine $(C_6H_5P)_3$ displaces only one carbon monoxide from $Ni(CO)_4$ giving $[\{(C_6H_5P)_3\}Ni(CO)_3]$ in which only one phosphorus is coordinated to the nickel (39). Tetraethylammonium trichlorogermanate(II) reacts with $Ni(CO)_4$ to give $[Ni(CO)_3GeCl_3]^-$ and $[Ni(CO)_2(GeCl_3)_2]^{2-}$ (40).

Phosphine and phosphite complexes. Several different methods for the synthesis of tetrakis(tertiary phosphine)nickel(0) complexes from nickel(II) using triethylaluminium, sodium tetrahydridoborate, or zinc dust as reductants have been described (41). Some hindered phosphines gave only the coordinatively unsaturated $[(R_3P)_3Ni]$ complexes (41). $[(Et_3P)_4Ni]$, which is best prepared from bis(cyclo-octadiene)nickel(0) and triethylphosphine (42), dissociates reversibly in solution to $[(Et_3P)_3Ni]$ (41-43), which can be isolated by heating $[(Et_3P)_4Ni]$ *in vacuo* (43). $[(Et_3P)_3Ni]$ absorbs dinitrogen giving $[(Et_3P)_3Ni(N_2)]$ (42,43), although the solid isolated (43) is always nitrogen deficient. Solutions of $[(Et_3P)_4Ni]$ undergo a number of oxidative additions (42).

Reduction of an ethereal solution of the bulky phosphite, tris-[(+)-bornan-2-yl] phosphite (tbp) and bis(acetylacetonato)nickel(II) with trimethylaluminium gives two or three coordinate nickel(0) complexes $[Ni(tbp)_2]$ and $[Ni(tbp)_3]$, depending on the ratio of the reactants: this phosphite also reacts with bis(cyclo-octadiene)-nickel(0) to give $[Ni(tbp)_3]$. These coordinatively unsaturated nickel(0) complexes readily react with small ligands such as carbon monoxide, isonitriles, and olefins (44).

Reduction of nickel(II) solutions containing triphenylphosphine or 5-phenyl-5_H_-dibenzophosphole with sodium tetrahydridoborate gives [Ni(BH_4)(PPh_3)$_3$] and [Ni(DBP)$_4$] respectively (30). Nickel(0) complexes of sterically constrained cyclic phosphites can be protonated by strong acids: the products can be isolated as fairly stable solids if HBF_4, HPF_6, or hydrogen chloride are used (45).

The reactions of tetrakis(trifluorophosphine)nickel(0) with alkoxide ions (46), primary and secondary amines (47), amides (47), hydrazine (47), diaminoethane (47), and Na[N{Si(CH_3)$_3$}$_2$] (48) give nickel complexes of substituted phosphorus ligands. Substitution is usually incomplete.

COBALT(II) AND NICKEL(II) COMPLEXES

Syntheses of cobalt(II) and nickel(II) fluorophosphates have been reported (49): these can then be used in the synthesis of complexes containing fluorophosphate coordinated as a monodentate ligand (50). Other interesting cobalt(II) and nickel(II) complexes which can be prepared directly from cobalt(II) or nickel(II) salts and the ligands include complexes of cyanamide (51), dicyanamide (52, 53), perfluoropinacol (54), and o-phenylenediamine (55). A cobalt(II) perfluoropinacol complex is a reversible oxygen carrier (54). "Hexakis(o-phenylenediamine)nickel(II)" chloride actually contains two bidentate, two monodentate, and two uncoordinated ligands (55).

Addition of _t_-butyl isocyanide to anhydrous nickel(II) chloride in ethanol under carefully controlled conditions gives an isocyanide complex [NiCl_2(_t_-C_4H_9NC)(C_2H_5OH)] (56), which is a good catalyst for the polymerization of isocyanides (57).

Complexes with heavy donor atoms. Diethylammonium diethylmonoselenocarbamate reacts with [NiCl_2(PPh_3)$_2$] to give bis(diethylmonoselenocarbamato)bis(triphenylphosphine)nickel(II), which can be used to prepare nickel(0) complexes (38). Complexes of a number of sulphur containing ligands have been synthesized directly from the ligand and the metal(II) salt. The ligands include trithiocarbonate (58), dithiosquarate (59), diaryldithiophosphinates (60), tetrathiotungstate (61), monothioacetylacetonate (62,63), N,N′—ethylenebis-(monothioacetylacetonimine) (64,65) and N,N′—ethylenebis(thiosalicylideneimine) (65). The last two are the thio analogues of "baen" and "salen" respectively: like the complexes of "baen" and "salen", the thio analogues absorb oxygen giving oxygen carriers (64,65).

Direct syntheses of nickel(II) complexes of trimethylarsine (66), trimethylstibine (67), a macrocyclic PN_3 ligand (68), and bidentate PP, AsAs, PSb, and AsSb ligands (69) have been reported.

Binuclear and oligonuclear complexes. 3-Formylsalicylic acid reacts with nickel(II) acetate to give binuclear complexes, $[Ni_2L_2(OH_2)_4]$, in which the two ligand molecules each bridge the two nickel atoms (70). The similar complexes of 1,5-disubstituted pentane-1,3,5-triones reported last year have been studied in more detail (71). Evaporation of a mixture of nickel(II) acetate and 7-azaindole gives a green solid, which when heated to 100° in toluene slowly gives red crystals of tetrakis(7-azaindolato)dinickel(II) (72). This probably has a structure in which four anionic 7-azaindole ligands bridge a pair of nickel atoms, resembling the structure suggested (73) for $[Ni_2(1,8\text{-naphthyridine})_4X_2]^+$. A dimeric cobalt(II) complex of N,N′—dimethyl-N,N′—bis(β-mercaptoethyl)diaminoethane has been prepared: it has two mercaptide bridges and reacts with nitrosonium salts to give a compound which has in addition a bridging nitrosyl group (74).

N,N′—ethylenebis(salicylideneiminato)nickel(II), [Ni(salen)], reacts with organotin(IV), organolead(IV), and organothallium(III) dichlorides to give bimetallic compounds (75,76), in which the two oxygen donors of [Ni(salen)] coordinate to the second metal: the crystal structure of a tin derivative has been reported (77). Bis(β-mercaptoethylamine)nickel(II) reacts with $(CH_3)_2PbCl_2$ and with $PhTlCl_2$ in a similar way, coordinating to the lead or thallium via two sulphur donor atoms (76).

[Di(2-pyridyl)acetylene]bis(triphenylphosphine)platinum reacts with dry cobalt(II) chloride in ethanol to give [di(2-pyridyl)-

acetylene]bis(triphenylphosphine)platinumdichlorocobalt(II), in which the two pyridine nitrogen atoms of the platinum complex are coordinated to the cobalt (78).

Oxidative and reductive syntheses. The reaction of trichlorosilane with $[Ni(PPh_3)_4]$ or $[Ni(diphos)_2]$ gives the nickel(II) silyl derivatives $[(Ph_3P)_2Ni(\mu\text{-}SiCl_2)_2Ni(PPh_3)_2]$ and $[(diphos)_2NiSiCl_3]^+(SiCl_3)^-$; in the latter case $[(diphos)_2NiH]^+(SiCl_3)^-$ is also formed (79). Trichlorosilane and dichloro(methyl)silane react with $[(CH_3)_2Ni(bipy)]$ to give $[(Cl_3Si)_2Ni(bipy)]$ and $[(CH_3Cl_2Si)_2Ni(bipy)]$ respectively (80). No cobalt silyl derivatives were isolated from the reaction of trichlorosilane with $[(Ph_3P)_3Co(N_2)]$, and $[(diphos)_2CoH]$ gave only $[(diphos)_2CoH_2]^+(SiCl_3)^-$ (79).

The cobalt(III) arsine complexes cis-$[Co(tetars)(OH_2)_2](ClO_4)_3$ and cis-$[Co(diars)_2(OH_2)_2](ClO_4)_3$ are reduced by primary alcohols to dimeric cobalt(II) arsine complexes $[Co(tetars)(OH_2)]_2(ClO_4)_4$ and $[Co(diars)_2(OH_2)]_2(ClO_4)_4$ (81).

COBALT(III) COMPLEXES

Nitric oxide reacts with a deaerated solution containing cobalt(II) chloride, lithium perchlorate, and bipyridyl to give the nitrosyl $[Co(bipy)_2Cl(NO)](ClO_4)$. If nitrite is present instead of chloride $[Co(bipy)_2(NO_2)(NO)](ClO_4)$ can be isolated; this is slowly oxidized by air to give $[Co(bipy)_2(NO_2)_2](ClO_4)$. Phenanthroline gives analogous complexes. Although these complexes are weakly paramagnetic, it was suggested that they are best considered as cobalt(III) complexes with a coordinated NO^- group (82). A solution of potassium 2-aminothiophenolate, cobalt(II) chloride, and tetra(n-butyl)ammonium chloride in ethanol is rapidly oxidized by air: cooling the solution to -10° gives tetra(n-butyl)ammonium bis(2-aminothiophenolato)cobaltate(III), a paramagnetic square-planar cobalt(III) complex (83).

The cyclic hexapeptides desferri-ferrichrome and desferri-ferrichrysin react with hydrated cobalt(III) oxide to give cobalt(III) desferri-ferrichrome and cobalt(III) desferri-ferrichrysin as green solids (84). The linear hexamine, 1,14-diamino-3,6,9,12-tetraazatetradecane, reacts with $[Co(NH_3)_5Br]Br_2$ to give the cobalt(III) complex of the hexadentate ligand: the isomers were separated on SP-Sephadex (85). $[Co_2(acac)_4(OH)_2]$ reacts with potassium thiocyanate in the presence of pyridine to give the S-bonded thiocyanate complex, trans-$[Co(acac)_2(SCN)(py)]$. The S-bonded complex isomerizes in solution in chloroform to give the N-bonded isomer (86).

Syntheses in non-aqueous solvents. $[Co(NH_3)_5\{(EtO)_3PO\}]^{3+}$, prepared *in situ* from $[Co(NH_3)_5(N_3)](ClO_4)_2$ and nitrosyl tetrafluoroborate in solution in triethylphosphate, was used to prepare complexes of aryl cyanides $[Co(NH_3)_5(NCR)](ClO_4)_3$: unless rigorously anhydrous conditions are used, hydrolysis to the aryl amide complexes occurs (87). Solutions of $[Co(NH_3)_5(dmso)]^{3+}$ may be prepared (88) under much milder conditions (45°, 24 hours) than previously (89). Crystalline trans-(dimethylsulphoxide)nitrobis(dimethylglyoximato) cobalt(III) has been prepared from the analogous aquo complex (90). Removal of the chloride ion from $[Co(NH_3)_5Cl](ClO_4)_2$ with silver perchlorate in dimethylformamide gives $[Co(NH_3)_5(dmf)](ClO_4)_3$, which can be recrystallized from water without loss of the dimethylformamide (91). The reaction of $[Co(NH_3)_5(OH_2)](ClO_4)_3$ with urea in N,N-dimethylacetamide at 135° gives $[Co(NH_3)_5(NCO)](ClO_4)_2$ (92). A solution of $(NEt_4)_3[Co(CN)_5]$ in dimethylformamide absorbs oxygen from the air to give $(NEt_4)_3[Co(CN)_5(O_2)]$ with linear Co-O-O bonding (93). Synthesis of chelates of 1,4-diaminobutane (tmd), with seven-membered chelate rings is not feasible in aqueous solution. However oxygenation of cobalt(II) nitrate and 1,4-diaminobutane in solution in dimethylsulphoxide in the presence of active carbon gives a mixture of $[Co(tmd)_3]^{3+}$ and trans-$[Co(tmd)_2Cl_2]^+$, which were separated on cation exchange Sephadex (94).

Syntheses from carbonato, sulphito, and nitro intermediates. Reaction of sodium (95) or potassium (96) triscarbonatocobaltate(III) with potassium cyanide gives a deep red solution containing cis-$[Co(CN)_2(CO_3)_2]^{3-}$. This may be isolated by precipitation by $[Co(en)_3]^{3+}$, or used directly to prepare a number of other cyano complexes including cis-$[Co(CN)_2(CO_3)(NH_3)_2]^-$, $[Co(CN)_2(CO_3)(C_2O_4)]^{3-}$, $[Co(CN)_2(C_2O_4)_2]^{3-}$, $[Co(CN)(H_2O)(mal)_2]^{2-}$, $[Co(CN)_2(dtc)_2]^-$ and $[Co(CN)_2(xan)_2]^-$ (dtc = dimethyldithiocarbamate; xan = dimethylxanthate) (95,96). Triscarbonatocobaltate(III) has also been used in the synthesis of complexes of 4,7-dimethyltriethylenetetramine (97), ethylenediaminediacetic acid (98), N,N-bis(2-aminoethyl)glycine (99) and an N_4-macrocyclic ligand (100).

The reaction of $[Co(aa)_2CO_3]$ with acids gives the binuclear $[(aa)_2Co(OH)_2Co(aa)_2]$ (aa = glycinate, L-alaninate, or L-valinate) (101). The reaction of $[Co(en)_2CO_3]^+$ with tartaric acid gives a monomeric product $[Co(en)_2(tart)]Cl$ (102), under conditions somewhat different to those which gave a well-characterized dimeric species with a bridging tartrate ion (103). Displacement of the carbonate in $[Co(edda)CO_3]^-$ by diamines (98), malonate (104), and a variety of monodentate ligands (104) was used in two studies of the isomerism of ethylenediaminediacetic acid complexes.

A sulphite ligand in tris(sulphito)cobaltate(III) is easily displaced by diaminoethane in the presence of sodium bisulphite to give $[Co(SO_3)_2(en)]^-$ (105). But coordinated sulphite ion may be bonded very firmly: $[Co(NH_3)_5SO_3]^+$, which may be prepared from $S_2O_5^{=}$ and $[Co(NH_3)_5OH_2]^{3+}$ (106) or by oxygenation of a cobalt(II)/sulphite/ammonia solution (107), decomposes in acid solution to give only $[Co(NH_3)_4(OH_2)(SO_3)]^+$ (106,107); sulphite cannot be displaced. Protonation of sulphite in $[Co(SO_3)(NH_3)_5]^{2+}$, $[Co(SO_3)Cl(NH_3)_4]$, $[Co(SO_3)Br(NH_3)_4]$, and $[Co(SO_3)_2(NH_3)_4]^-$ occurs in concentrated acids but sulphite is not lost (108). However sulphite is displaced from trans-$[Co(CN)(SO_3)(NH_3)_4]$ giving $[Co(CN)(OH_2)(NH_3)_4]Cl_2$, which may be used to prepare a range of cyano complexes $[Co(CN)_5X]^{n-}$ (X = NO_2, N_3, I, SCN, or S_2O_3) (109). Diaminoethane displaces sulphite from $[Co(CN)_2(SO_3)_2(NH_3)_2]^{3-}$ giving a triammine $[Co(CN)(NH_3)_3(en)]Cl$ (110).

Sulphite will itself displace nitro groups in $[Co(NO_2)_4(NH_3)_2]^-$ and cis-$[Co(en)_2(NO_2)_2]^+$ giving $[Co(SO_3)_2(NH_3)_2]^-$ and $[Co(NO_2)(SO_3)(en)_2]$ respectively (105). Bipyridyl and phenanthroline react with $Na_3[Co(NO_2)_6]$ giving $[Co(bipy)_2(NO_2)_2]\ NO_3$ and

$[Co(phen)_2(NO_2)_2]NO_3$ respectively (111).

Binuclear and polynuclear complexes. Sulphur dioxide reacts with $[(NH_3)_3Co(\mu\text{-}OH)_3Co(NH_3)_3]^{3+}$ to give $[(NH_3)_3Co(\mu\text{-}OH)_2(\mu\text{-}SO_3)Co(NH_3)_3]^{=}$ (112) and with a range of μ-peroxo dicobalt(III) complexes to give μ-sulphato complexes (113). μ-Peroxo dicobalt(III) complexes also react with nitric oxide to give μ-nitrito complexes and nitrite ions: no evidence for the formation of μ-nitrato complexes was obtained, even when a μ-superoxo dicobalt(III) complex was used (113).

Potassium pentacyanocobaltate(II) can be oxidized with finely divided sulphur or with potassium polysulphide to $K_6[(NC)_5CoSSCo(CN)_5]$, which can also be prepared from $Ag_2[Co(CN)_5(OH_2)]$ and potassium sulphide (114). $K_3[Co(CN)_5]$ can also be oxidized with black selenium to $K_6[(NC)_5CoSeSeCo(CN)_5]$ (114) and with toluoquinone to a toluohydroquinone bridged dimer (115). Potassium cyanide reacts with the black form of $[Co(NO)(NH_3)_5]Cl_2$ to give two compounds, both with the formula $K_6[(NC)_5Co(N_2O_2)Co(CN)_5]$ (116), said to be linkage isomers.

The uncoordinated carboxyl groups in L-asparto and L-glutamato-bis(ethylenediamine)cobalt(III) react with $[Co(NH_3)_5(OH_2)]^{3+}$ and with $[Co(en)_2(OH_2)_2]^{3+}$ to give aspartate and glutamate bridged dinuclear compounds: with $[Co(en)_2(OH_2)_2]^{3+}$ trinuclear complexes are also formed (117). Monodentate oxalate in $[Rh(NH_3)_5(C_2O_4H)]^{2+}$ reacts with $[(NH_3)_3Co(\mu\text{-}OH)_3Co(NH_3)_3]^{3+}$ to give $[(NH_3)_{11}Co_2Rh(\mu\text{-}OH)_2(\mu\text{-}C_2O_4)]^{5+}$ (118). Fac-$[Co(OCH_2CH_2NH_2)_3]$ reacts with zinc chloride to give $[\{Co(OCH_2CH_2NH_2)_3\}_3Zn]Cl_2$ (119). Poly(vinylpyridine) reacts with cis-$[Co(en)_2Cl_2]Cl$ to give a water soluble polymer in which the polymer chain contains cis-$[Co(en)_2Cl(py)]^{2+}$ substituents (120).

Complexes with heavy donor atoms. Several cis-α and cis-β complexes of a linear tetradentate arsine $[(CH_3)_2As(CH_2)_3As(C_6H_5)CH_2]_2$ (tetars) have been prepared from $[Co(tetars)Cl_2]Cl$ (121). Cis-α- or cis-β-$[Co(tetars)(OH_2)_2](ClO_4)_3$ react with hydrogen peroxide to give cis-β-$[Co(tetars)(O_2)]ClO_4$, which can also be prepared by reduction of cis-α-$[Co(tetars)Cl_2]Cl$ with sodium tetrahydridoborate followed by oxidation with air (122). These sideways bonded oxygen adducts react with sulphur dioxide to give sulphato complexes, $[Co(tetars)(SO_4)]^+$, which can also be prepared directly from $[Co(tetars)(OH_2)_2](ClO_4)_3$ and sodium sulphate (122). Oxygenation of a solution of "tetars" and cobalt(II) perchlorate in ether/acetonitrile gave a peroxo bridged complex, $[(CH_3CN)(tetars)Co\text{-}O\text{-}O\text{-}Co(tetars)(CH_3CN)](ClO_4)_4$ (122). Trans-$[Co(diars)_2Cl_2]Cl$ was used in a similar way to prepare a wide range of cis-$[Co(diars)_2XY]^{n+}$ complexes (123) including some nitrato complexes. Cis-$[Co(diars)_2(OH_2)_2]^{3+}$ was used to synthesize peroxo and sulphato complexes similar to the "tetars" complexes (122).

Three arsine ligands, $E[CH_2CH_2CH_2As(CH_3)_2]_3$ (E = P, As, or Sb), react with cobalt(II) halides to give cobalt(III) complexes (124). Difluoro(phenyl)phosphine reacts with cobalt(II) chloride or bromide to give $[CoX_3(C_6H_5PF_2)_3]$ in addition to cobalt(I) complexes (26).

Air oxidation of a cobalt(II) solution containing the ligand has been used to prepare complexes with $N_2Se_2O_2$ (125), $N_2S_2O_2$ (126), N_3S_3 (127), and O_3S_3 (62,63,128) donor atoms. The syntheses of cobalt(III) complexes of tetradentate S_4-macrocyclic ligands such as 1,4,8,11-tetrathiocyclotetradecane from cobalt(II), and interconversion of cobalt(III) complexes of these thio-ether ligands, have been reported (129).

Complexes of polydentate ligands. There have been several reports of cobalt(III) complexes of 1,3-diaminopropan-2-ol (Hdap); diethanolamine (H_2dea) is displaced from $Na[Co(dea)_2]$ giving both the mixed complex Co(dea)(dap) (130), which has been studied in detail (131), and bis(1,3-diaminopropan-2-olato)cobalt(III) (132), which can also be prepared by the reaction of the ligand with cobalt(II) salts (133) or with triscarbonatocobaltate(III) (133): in contrast to an earlier report (134), the tris complex is not obtained. However the isomeric 2,3-diaminopropan-1-ol did react with $Na_3[Co(CO_3)_3]$ to give the tris complex (133). Reaction of 1,3-diaminopropan-2-ol with trans-$[Co(en)_2Cl_2]^+$ is reported to give a binuclear complex with a bridging diaminoethane, $[Co(en)(dap)NH_2CH_2]_2Cl_4$ (135).

The reaction of diaminoethane with Na[Co(edta)] under controlled conditions is reported to give a complex of monodentate diaminoethane, $[Co(enH)(Hedta)]^+$, as well as $[Co(en)(H_2edta)]^+$ (136).

Modification of coordinated ligands. A solution of copper(II) nitrate in acetic anhydride has been used to nitrate "salen" in [Co(salen)(acac)]: if a larger amount of the nitrating agent is used the acetylacetonate ligand is also nitrated (137). Nitric oxide will oxidize the coordinated nitrosyl in $[Co(en)_2(NO)Cl]Cl$ forming cis-$[Co(en)_2(NO_2)Cl]^+$: the nitric oxide is reduced to nitrous oxide (138). "Cobaloxime nitrosyl", $[Co(Hdmg)_2(NO)]$ is also oxidized by nitric oxide (138): it is oxidized by air (139) in the presence of t-butylpyridine to a nitrato complex, $[(t\text{-Bu-py})Co(Hdmg)_2(NO_3)]$; in the absence of the pyridine a nitro-complex is formed (139).

Intramolecular attack of a coordinated ammine on the pyruvate group in penta-ammine(pyruvato)cobalt(III) gives tetra-ammine-(pyruvilidene-imine)cobalt(III) (140): this will itself condense with acetylacetone in basic solution to give a triammine derivative $[(NH_3)_3Co(C_6H_{11}N_2O_2)]^{2+}$ with a tridentate NNO-ligand (141). In neutral or basic solution cis-$[Co(en)_2X(NH_2CH_2CN)]^{2+}$ (X = Cl or Br) undergoes rapid base-catalysed ring closure to form (142) $[Co(en)\{NH_2C(NH_2)=NCH_2CH_2NH_2\}X]^{2+}$.

Separation and resolution. An easy resolution of $[Co(edta)]^-$ using L-histidine has been reported (143). A wide range of dialkyldithiocarbamates react stereoselectively with $(+)_{546}$-$[Co(edta)]^-$ to give $(+)_{546}$-tris(dialkyldithiocarbamato)cobalt(III) (144). A strain of *Pseudomonas stutzeri* stereoselectively reduces $Co(en)_2(bipy)^{3+}$ and $Co(en)_2(phen)^{3+}$ and was used to resolve these complexes (145).

Chromatography on an alumina column is reported to be an easy method of removing cobalt(II) impurities from uncharged cobalt(III) complexes (146). Thin layer chromatography on high purity silica gel is a good method of separating tris(β-diketonato)cobalt(III) complexes: in some cases mer and fac isomers are readily separated (147). A further example of the finesse of separations on cation exchange "Sephadex", especially when (+)-tartratoantimonate is used as the eluant, is the separation of isomers of $[Co\{(-)\text{-}pn\}_3]^{3+}$ (148).

HIGHER OXIDATION STATES

Apart from a review (2) of these oxidation states, there has been little reported this year.

Cobalt(IV). Oxidation of tris(dialkyldithiocarbamato)cobalt-(III) with boron trifluoride gives tris(dialkyldithiocarbamato)-cobalt(IV) (149,150). There have been further reports of complex oxides of cobalt(IV) (151).

REFERENCES

1. G. B. Kauffman, COORDINATION CHEM. REV., 12, 105 (1974).
2. W. Levason and C. A. McAuliffe, COORDINATION CHEM. REV., 12, 151 (1974).
3. A. Chakravorty, COORDINATION CHEM. REV., 13, 1 (1974).
4. F. Bonati and G. Minghetti, INORG. CHIM. ACTA, 9, 95 (1974).
5. D. Sellman, ANGEW. CHEM. INTERN. ED. ENGL., 13, 639 (1974).
6. M. A. Ali and S. E. Livingstone, COORDINATION CHEM. REV., 13, 101 (1974).
7. P. C. Maybury, R. W. Mitchell and M. F. Hawthorne, J. C. S. CHEM. COMMUN., 534 (1974).
8. E. L. Muetterties and J. C. Sauer, J. AMER. CHEM. SOC., 96, 3410 (1974).
9. P. Chini, M. C. Malatesta and A. Cavalieri, CHIM. IND. (MILAN), 55, 120 (1973).
10. D. DeMontauzon and R. Poilblanc, J. ORGANOMETAL. CHEM., 54, 291 (1973).
11. D. P. Schussler, W. R. Robinson and W. F. Edgell, INORG. CHEM., 13, 153 (1974).
12. S. E. Pedersen, W. R. Robinson and D. P. Schussler, J. ORGANOMETAL. CHEM., 43, C44 (1972).
13. G. F. Bradley and S. R. Stobart, J. CHEM. SOC. DALTON, 264 (1974).
14. A. P. Hagen, L. McAmis and M. A. Stewart, J. ORGANOMETAL. CHEM., 66, 127 (1974).
15. G. C. van der Berg, A. Oskam and K. Vrieze, J. ORGANOMETAL. CHEM., 69, 169 (1974).
16. V. G. Albano, P. Chini, S. Martinengo, M. Sansoni and D. Strumolo, J. C. S. CHEM. COMMUN, 299 (1974).
17. J. Dehand and J. F. Nennig, INORG. NUCL. CHEM. LETTERS, 10, 875 (1974).
18. T. Fukumoto, Y. Matsumura and R. Okawara, J. ORGANOMETAL. CHEM. 69, 437 (1974).
19. S. R. A. Bird, J. D. Donaldson, A. F. L. Holding, S. Cenini and B. Ratcliff, INORG. CHIM. ACTA, 8, 149 (1974).
20. J. Newman and A. R. Manning, J. CHEM. SOC. DALTON, 2549 (1974).

21. M. J. Boylan, J. Bellerby, J. Newman and A. R. Manning, J. ORGANOMETAL. CHEM., 47, C33 (1973).
22. E. L. Muetterties and F. J. Hirsekorn, J. AMER. CHEM. SOC., 96, 7920 (1974).
23. M. A. Cairns and J. F. Nixon, J. ORGANOMETAL. CHEM., 74, 263 (1974).
24. L. W. Gosser and G. W. Parshall, INORG. CHEM., 13, 1947 (1974).
25. P. Meakin and J. P. Jesson, J. AMER. CHEM. SOC., 96, 5751 (1974).
26. O. Stelzer, CHEM. BER., 107, 2329 (1974).
27. E. Bordignon, U. Croatto, U. Mazzi and A. A. Orio, INORG. CHEM., 13, 935 (1974).
28. S. Otsuka and M. Rossi, BULL. CHEM. SOC. JAPAN, 46, 3411 (1973).
29. C. A. L. Becker, SYN. REACT. INORG. METAL-ORG. CHEM., 4, 213 (1974).
30. D. G. Holah, A. N. Hughes, B. C. Hui and K. Wright, CANADIAN J. CHEM., 52, 2990 (1974).
D. G. Holah, A. N. Hughes, B. C. Hui and K. Wright, INORG. NUCL. CHEM. LETTERS, 9, 835 (1973) and 10, 427 (1974).
31. A. R. Hendrickson, R. K. Y. Ho and R. L. Martin, INORG. CHEM., 13, 1279 (1974).
32. M. Bressan and P. Rigo, J. C. S. CHEM. COMMUN., 553 (1974).
33. R. S. Gall, N. G. Connelly and L. F. Dahl, J. AMER. CHEM. SOC., 96, 4017 (1974).
34. C. Floriani and G. Fachinetti, J. C. S. CHEM. COMMUN., 615 (1974).
35. J. C. Calabrese, L. F. Dahl, A. Cavalieri, P. Chini, G. Longoni and S. Martinengo, J. AMER. CHEM. SOC., 96, 2616 (1974).
36. W. Hieber and J. Ellerman, Z. NATURFORSCH., 18B, 595 (1963).
37. K. Tanaka, Y. Kawata and T. Tanaka, CHEM. LETTERS, 831 (1974).
38. K. Tanaka and T. Tanaka, INORG. NUCL. CHEM. LETTERS, 10, 605 (1974).
39. M. Baudler and M. Bock, ANGEW. CHEM. INTERN. ED. ENGL., 13, 147 (1974).
40. T. Kruck and W. Molls, Z. NATURFORSCH., 29B, 198 (1974).
41. C. A. Tolman, W. C. Seidel and L. W. Gosser, J. AMER. CHEM. SOC., 96, 53 (1974).
42. C. S. Cundy, J. ORGANOMETAL. CHEM., 69, 305 (1974).
43. C. A. Tolman, D. H. Gerlach, J. P. Jesson and R. A. Schunn, J. ORGANOMETAL. CHEM., 65, C23 (1974).
44. S. Otsuka, K. Tani, I. Kato and O. Teranaka, J. CHEM. SOC. DALTON, 2216 (1974).
45. G. K. McEwen, C. J. Rix, M. F. Traynor and J. G. Verkade, INORG. CHEM., 13, 2800 (1974).
46. T. Kruck, M. Höfler and H. Jung, CHEM. BER., 107, 2133 (1974).
47. T. Kruck, H. Jung, M. Höfler and H. Blume, CHEM. BER., 107, 2145 (1974).
48. T. Kruck, G. Mäueler and G. Schmidgen, CHEM. BER., 107, 2421 (1974).
49. C. S. Alleyne, K. O'Sullivan-Mailer and R. C. Thompson, CANADIAN J. CHEM., 52, 336 (1974).

50. C. S. Alleyne and R. C. Thompson, CANADIAN J. CHEM., 52, 3218 (1974).
51. W. Beck, H. Bock and R. Schlodder, Z. NATURFORSCH., 29B, 75 (1974).
52. H. Kohler, T. P. Lischko, H. Hartung and A. M. Golub, Z. ANORG. ALLGEM. CHEM., 403, 35 (1974).
53. H. Kohler, H. Hartung and A. M. Golub, Z. ANORG. ALLGEM. CHEM., 403, 41 (1974).
54. C. J. Willis, J. C. S. CHEM. COMMUN., 117 (1974).
55. R. C. Elder, D. Koran and H. B. Mark, INORG. CHEM., 13, 1644 (1974).
56. R. W. Stephany, R. J. M. Nolte and W. Drenth, REC. TRAV. CHIM. PAYS BAS, 92, 275 (1973).
57. R. J. M. Nolte, R. W. Stephany and W. Drenth, REC. TRAV. CHIM. PAYS BAS, 92, 83 (1973).
58. A. Bruhn and H. Seidel, Z. NATURFORSCHUNG, 29B, 123 (1974).
59. D. Coucouvanis, F. J. Hollander, R. West and D. Eggerding, J. AMER. CHEM. SOC., 96, 3006 (1974).
60. R. N. Mukherjee, M. S. Venkateshan and M. D. Zingde, J. INORG. NUCL. CHEM., 36, 1043 (1974).
61. M. C. Chakravorti and A. Müller, INORG. NUCL. CHEM. LETTERS, 10, 63 (1974).
62. M. Das and S. E. Livingstone, AUSTRALIAN J. CHEM., 27, 2109 (1974).
63. M. Das, S. E. Livingstone, S. W. Filipczuk, J. W. Hayes and D. V. Radford, J. CHEM. SOC. DALTON, 1409 (1974).
64. P. R. Blum, R. M. C. Wei and S. C. Cummings, INORG. CHEM., 13, 450 (1974).
65. A. van den Bergen, M. F. Corrigan, K. S. Murray, R. M. Slade and B. O. West, INORG. NUCL. CHEM. LETTERS, 10, 859 (1974).
66. C. D. Hem, M. Dartiguenave and Y. Dartiguenave, INORG. NUCL. CHEM. LETTERS, 10, 1039 (1974).
67. M. F. Ludmann-Obier, M. Dartiguenave and Y. Dartiguenave, INORG. NUCL. CHEM. LETTERS, 10, 147 (1974).
68. J. Riker-Nappier and D. W. Meek, J. C. S. CHEM. COMMUN., 442 (1974).
69. W. Levason and C. A. McAuliffe, INORG. CHIM. ACTA, 11, 33 (1974).
70. M. Tanaka, H. Okawa, I. Hanoaka and S. Kida, CHEM. LETTERS, 71 (1974).
71. R. L. Lintvedt, L. L. Borer, D. P. Murtha, J. M. Kuszaj and M. D. Glick, INORG. CHEM. 13, 18 (1974).
72. R. W. Brookes and R. L. Martin, AUSTRALIAN J. CHEM., 27, 1569 (1974).
73. L. Sacconi, C. Mealli and D. Gatteschi, INORG. CHEM., 13, 1985 (1974).
74. K. D. Karlin, D. L. Lewis, H. N. Rabinowitz and S. J. Lippard, J. AMER. CHEM. SOC., 96, 6519 (1974).
75. L. Pellerito, R. Cefalù, A. Gianguzza and R. Barbieri, J. ORGANOMETAL. CHEM., 70, 303 (1974).

76. L. Pellerito, R. Cefalù and A. Gianguzza, J. ORGANOMETAL. CHEM. 70, C27 (1974).
77. M. Calligaris, L. Randaccio, R. Barbieri and L. Pellerito, J. ORGANOMETAL. CHEM., 76, C56 (1974).
78. G. R. Newkome and G. L. McClure, J. AMER. CHEM. SOC., 96, 617 (1974).
79. M. F. Lappert and G. Speier, J. ORGANOMETAL. CHEM., 80, 329 (1974).
80. Y. Kiso, K. Tamao and M. Kumada, J. ORGANOMETAL. CHEM., 76, 95 (1974).
81. B. Bosnich, W. G. Jackson and S. T. D. Lo, INORG. CHEM., 13, 2598 (1974).
82. A. Vlček and A. A. Vlček, INORG. CHIM. ACTA, 9, 165 (1974).
83. P. J. M. L. Birker, E. A. De Boer and J. J. Bour, J. COORDINATION CHEM., 3, 175 (1974).
84. J. Leong and K. N. Raymond, J. AMER. CHEM. SOC., 96, 6628 (1974).
85. Y. Yoshikawa and K. Yamasaki, BULL. CHEM. SOC. JAPAN, 46, 3448 (1973).
86. L. J. Boucher, D. R. Herrington and C. G. Coe, INORG. CHEM., 13, 2290 (1974).
87. R. J. Balahura, CANADIAN J. CHEM., 52, 1762 (1974).
R. J. Balahura, INORG. CHEM., 13, 1350 (1974).
88. W. L. Reynolds, M. Biruš and S. Ašperger, J. CHEM. SOC. DALTON, 716 (1974).
89. C. R. P. MacColl and L. Beyer, INORG. CHEM., 12, 7 (1973).
90. S. T. D. Lo and D. W. Watts, INORG. CHIM. ACTA, 9, 217 (1974).
91. D. A. Buckingham, J. M. Harrowfield and A. M. Sargeson, J. AMER. CHEM. SOC., 96, 1726 (1974).
92. D. A. Buckingham, D. J. Francis and A. M. Sargeson, INORG. CHEM., 13, 2630 (1974).
93. L. D. Brown and K. N. Raymond, J. C. S. CHEM. COMMUN., 470 (1974).
94. J. Fujita and H. Ogino, CHEM. LETTERS, 57 (1974).
95. Y. Enomoto, T. Ito and M. Shibata, CHEM. LETTERS, 423 (1974).
96. S. Fujinami and M. Shibata, BULL. CHEM. SOC. JAPAN, 46, 3443 (1973).
97. D. J. Francis and G. H. Searle, AUSTRALIAN J. CHEM., 27, 269 (1974).
98. L. J. Halloran and J. I. Legg, INORG. CHEM., 13, 2193 (1974).
99. S. Nakashima and M. Shibata, BULL. CHEM. SOC. JAPAN, 47, 2069 (1974).
100. N. F. Curtis, AUSTRALIAN J. CHEM., 27, 71 (1974).
101. R. D. Gillard, S. H. Laurie, D. C. Price, D. A. Phipps and C. F. Weick, J. CHEM. SOC. DALTON, 1385 (1974).
102. R. A. Haines, E. B. Kipp and M. Reimer, INORG. CHEM., 13, 2473 (1974).
103. R. D. Gillard and M. G. Price, J. CHEM. SOC. A, 1813 (1969).
104. P. J. Garnett and D. W. Watts, INORG. CHIM. ACTA, 8, 293 (1974).
105. R. Klement, Z. NATURFORSCH. 29B, 727 (1974).

106. H. Siebert and G. Wittke, Z. ANORG. ALLGEM. CHEM., 399, 43 (1973).
107. M. A. Thacker, K. L. Scott, M. E. Simpson, R. S. Murray and W. C. E. Higginson, J. CHEM. SOC. DALTON, 647 (1974).
108. H. Siebert and G. Wittke, Z. ANORG. ALLGEM. CHEM., 405, 63 (1974).
109. N. Maki, INORG. CHEM., 13, 2180 (1974).
110 A. R. Norris, CANADIAN J. CHEM., 52, 477 (1974).
111. M. Davidson, T. W. Faulkner, M. A. Green and E. D. McKenzie, INORG. CHIM. ACTA, 9, 231 (1974).
112. H. Siebert and G. Wittke, Z. ANORG. ALLGEM. CHEM., 406, 282 (1974).
113. C.-H. Yang, D. P. Keeton and A. G. Sykes, J. CHEM. SOC. DALTON, 1089 (1974).
114. H. Siebert and S. Thym, Z. ANORG. ALLGEM. CHEM., 399, 107 (1973).
115. J. Hanzlík and A. A. Vlček, INORG. CHIM. ACTA, 8, 247 (1974).
116. H. Okamura, E. Miki, K. Mizumachi and T. Ishimori, CHEM. LETTERS, 103 (1974).
117. T. Yasui, H. Kawaguchi, Z. Kanda and T. Ama, BULL. CHEM. SOC. JAPAN, 47, 2393 (1974).
118. K. Wieghardt, Z. NATURFORSCH., 29B, 809 (1974).
119. A. N. Gerasenkova and V. V. Udovenko, RUSS. J. INORG. CHEM., 18, 1294 (1973).
120. Y. Kurimura, I. Setine, E. Tsuchida and Y. Kaino, BULL. CHEM. SOC. JAPAN, 47, 1823 (1974).
121. B. Bosnich, W. G. Jackson and S. B. Wild, INORG. CHEM., 13, 1121 (1974).
122. B. Bosnich, W. G. Jackson, S. T. D. Lo and J. W. McLaren, INORG. CHEM., 13, 2605 (1974).
123. B. Bosnich, W. G. Jackson and J. W. McLaren, INORG. CHEM., 13, 1133 (1974).
124. G. Kordosky, G. S. Benner and D. W. Meek, INORG. CHIM. ACTA, 7, 605 (1973).
125. N. V. Gerbeleu and V. G. Bodyu, RUSS. J. INORG. CHEM., 18, 1596 (1973).
126. R. J. Magee, W. Mazurek, M. J. O'Connor and A. T. Phillip, AUSTRALIAN J. CHEM., 27, 1629 (1974).
127. G. C. Pellacani and G. Peyronel, INORG. CHIM. ACTA, 9, 189 (1974).
128. M. Das and S. E. Livingstone, AUSTRALIAN J. CHEM. 27, 1177 (1974).
129. K. Travis and D. H. Busch, INORG. CHEM., 13, 2591 (1974).
130. V. N. Evreen and S. V. Murashko, RUSS. J. INORG. CHEM., 19 408 (1974).
131. V. N. Evreen, S. V. Murashko and V. E. Petrun'kin, RUSS. J. INORG. CHEM., 19, 56 (1974).
132. V. N. Evreen and S. V. Murashko, RUSS. J. INORG. CHEM., 18, 1303 (1973).
133. M. S. Okamoto and E. K. Barefield, INORG. CHEM., 13, 2611 (1974).

134. H. F. Bauer and W. C. Drinkard, J. AMER. CHEM. SOC., 82, 5031 (1960).
135. S. C. Chan and C. K. Lee, J. INORG. NUCL. CHEM., 36, 2390 (1974).
136. M. Doh, H. Ogino, J. Fujita, K. Saito and N. Tanaka, CHEM. LETTERS, 1233 (1974).
137. K. Dey and R. L. De, J. INORG. NUCL. CHEM., 36, 1182 (1974).
138. D. Gwost and K. G. Caulton, INORG. CHEM., 13, 414 (1974).
139. W. C. Trogler and L. G. Marzilli, INORG. CHEM., 13, 1008 (1974).
140. J. M. Harrowfield and A. M. Sargeson, J. AMER. CHEM. SOC., 96, 2634 (1974).
141. B. T. Golding, J. M. Harrowfield, G. B. Robertson, A. M. Sargeson and P. O. Whimp, J. AMER. CHEM. SOC., 96, 3691 (1974).
142. K. B. Nolan and R. W. Hay, J. CHEM. SOC. DALTON, 914 (1974).
143. R. D. Gillard, P. R. Mitchell and C. F. Weick, J. CHEM. SOC. DALTON, 1635 (1974).
144. L. R. Gahan, J. G. Hughes and M. J. O'Connor, J. AMER. CHEM. SOC., 96, 2271 (1974).
145. L. S. Dollimore, R. D. Gillard and I. H. Mather, J. CHEM. SOC. DALTON, 518 (1974).
146. J. R. McKellar and B. O. West, J. CHEM. SOC. DALTON, 796 (1974).
147. K. Saitoh and N. Suzuki, J. CHROMATOGRAPHY, 92, 371 (1974).
148, M. Kojima, K. Yoshikawa and K. Yamasaki, INORG. NUCL. CHEM. LETTERS, 9, 689 (1973).
149. L. R. Gahan and M. J. O'Connor, J. C. S. CHEM. COMMUN., 68 (1974).
150. R. Y. Saleh and D. K. Straub, INORG. CHEM., 13, 3017 (1974).
151. M. Jansen and R. Hoppe, Z. ANORG. ALLGEM. CHEM., 398, 54 (1973).
M. Jansen and R. Hoppe, Z. ANORG. ALLGEM. CHEM., 408, 75 and 97 (1974).

The Platinum Metals

Part I: Ruthenium, Osmium, Rhodium and Iridium

P. D. Kaplan

University of Cincinnati, Medical Center
Cincinnati, Ohio

INTRODUCTION: Two books have been published this year surveying synthetic reactions in organotransition metal chemistry along with concomitant mechanistic considerations.[1,2] There have been several general reviews on synthetic methods[3-5] and several articles reviewing specific reactions. The carbonyls continue to be an active topic of research with one review covering their photochemistry[6] and three concerning syntheses and subsequent reactions of rhodium carbonyls.[7-9] Insertion reactions of transition metal complexes have been an area of current interest. The insertion of a variety of small molecules, as SO_2, CO_2, N_2, isonitriles, into σ-bonded complexes has been surveyed.[10-11] Updated reviews on homogeneous catalysis have appeared[12] including homogeneous hydrogenation[13] and catalytic oxidation-reduction reactions.[14]

GROUP VII DONORS: The synthesis, crystal structure and magnetic susceptibility of IrF_4 have been reported.[15] IrF_4 was prepared by the decomposition of IrF_5 at 400°C in a Ni tube. The average Ir-F bond distance is 1.96Å, in the octahedral environment. Oxidative stereospecific ligand exchange has been used for the preparation of mixed halo complexes of osmium[16] in $[OsX_6]^{2-}$, in addition to their preparation by conventional techniques.[17] Seventeen of 29 possible complexes of $[OsCl_yBr_zI_{6-y-z}]^{2-}$ (y+z=0-6) were prepared by the latter route. Direct substitution of the mixed complexes was enabled by the distinct kinetic stability of the ligand-metal bond for individual halides. The influence of Π-bonding and steric factors in hexahalo Ru and Os complexes has also been studied spectrophotometrically.[18] The hexahalo iridium salts are good starting materials for the preparation of mixed ligand complexes such as trans-$[IrX_3(CO)(PPh_3)_2]$ (X=Cl,I).[19] Mixed fluoro complexes of the type trans-$[MF(CO)(PPh_3)_2]$ (M=Rh, Ir) have been synthesized in 80-90% yields from the corresponding chloride complexes.[20] Although the Rh complex is air stable, the Ir analogue must be kept in an inert atmosphere. These complexes are also useful starting materials for the preparations of other trans-$[MX(CO)(PPh_3)_2]$ where X can be a number of small molecules such as NCO, N_3, NO_2, OH (and others).

An increasing number of species containing bridging halides have been described. Di- and tri-μ-halide complexes of ruthenium have been prepared by refluxing $[RuX_2CO(PPh_3)_2DMF]$ (X=Cl,Br) in CH_2Cl_2, or $(CH_3)_2CO$, respectively.[21] Two pentacoordinate dinuclear halo bridged carbonyl complexes of Rh have been synthesized. Treating the dinuclear $[RhCl(CO)(C_2H_4)]_2$ with a nucleophile was an excellent way of producing a series of pentacoordinate dimers providing the nucleophile was a secondary or tertiary amine.[22-3] Ligand exchange of the bridging halo has been detected for some Rh and Ir complexes.[24-5] $[Rh_2(CO)_4XY]$ type intermediates have been observed spectrophotometrically in the μ-ligand substitution reaction $[Rh(CO)_2X]_2 \longrightarrow [Rh(CO)_2Y]_2$ where X≠Y; $X=Cl^-, I^-, Ac^-$ or EtS^- and $Y=I^-, RO^-$, or RS^-, (R=alkyl aryl). A similar redistribution of the halide occurred through 5-coordinate doubly bridged intermediates at 25° for $[M(CO)X(L)_2]$, (X=Cl, Br; $L=PPh_3$, PPh_2Et, $AsPh_3$; M=Rh, Ir).

GROUP VI DONORS:

Oxygen Donors: New ternary oxides of osmium and iridium of the type $M'_2M_2O_7$ (M'=Cd, Ca; M=Ir, Os) have been reported from reactions at elevated temperatures and sealed containers of the respective lower oxides. All iridium phases contain Ir^{5+} and are the first examples of iridium in an oxidation state higher than four.[26] Similar structures containing bismuth are reported for ruthenium.[27] The crystal structure of $Na_{3-x}RuO_9$ (x≃0.9) is formed by single, double, and triple chains of $[RuO_6]$ octahedra sharing edges.[28] Physicochemical properties and reactions of oxides of the types MO_4L,[29] $[MO_2L]O_2 \cdot 6H_2O$,[30] $M_2O_6L_4$,[31] $[M_3O(L)_{8.5}]^{8-}$ [32] and $[M_3O(O_2CM_e)_6L_3]^+$,[33] where M=Os, Ir, Ru; $L=C_5H_5N$, NO_2, bipyridyl, SO_4^{2-}, MeOH have been reported. In contrast to the spectral and analytical data on $K_4[OsO_2(NO_2)_2]_2O_2 \cdot 6H_2O$, the X-ray diffraction study shows a dinuclear anion with bridging O atoms and trans-OsO_2 groups. The mixed valence, trinuclear iridium (IV, IV, III) oxosulfate anion was dimeric. Electrochemical reduction of Ru in μ_3-$Ru_3O(CO_2Me)_6(MeOH)_3$ produced a mixed valence trinuclear Ru_3(II,II,III) intermediate complex. Only a monocarbonyl, e.g., $Ru_3O(CO_2Me)_6(CO)(py)_2$ was isolated from the reactions of CO with the starting material.

Oxidative additions of carboxylic anhydrides and dimethyl phosphite to Ir(I) and RH(I) complexes have been followed.[33-4] Acyclic anhydrides, RC(O)OC(O)R, react with complexes of the type trans-$[IrX(CO)L_2]$ replacing the halide, X, to form the octahedral trans-$[IrX(C(O)R)(O_2CR)(CO)L_2]$. $(MeO)_2POH$ add oxidatively to Ir(I) or Rh(I) complexes to give hydrido Ir(III) or Rh(III) di-Me phosphonate complexes. The Ir derivative catalyses the stereoselective reduction of 4-t-Bu C_6H_7O. Rhodium (I) promoted oxygenation of terminal olefins to methylketones have also been observed.[35]

The interaction of anionic halo and nitro complexes of Rh and Ir with HNO_3 has been followed. Coordination of nitrate is preceded by formation of an intermediate aqua complex. Isomerization of $Ir-NO_2$ to Ir-O-NO was noted.[36] Trans-$[M(OClO_3)(CO)(Ph_3P)_2]$ (M=Ir, Rh) were prepared from the treatment of the chloride complexes with $AgClO_4$. The compounds undergo solvolytic displacement in polar solvents.[37] An electrochemical study of the interaction of ClO_4^- with $Rh(en)_3^{3+}$ (en=ethylenediamine) and other metal complexes, was shown to be inversely proportional to the size of the ion pair.[38] A method for the preparation of Ru formate from the reaction of RuO_4, $[Ru(H_2O)Cl_5]^{2-}$ or $[Ru_2OCl_{10}]$ with boiling formic acid has been reported,[39] as has the convenient synthesis of [RhL(CO)Z] from the reaction of $R(CO)Z_2Cl$ with ML_2 (L=3-hydroxycoumarin; $Z=PPh_3$, CO; M=Cu,Ni,Pd) (85% yield).[40] Sequential ligand replacement reactions, involving halides, oxalate and ethylenediamine, have been used as an easy route to the preparation of bis(ethylenediamine) halogen Rhodium (III) complexes.[41] Chiral rhodium β-diketonates were prepared by heating $RhCl_3{\cdot}nH_2O$ with (+)-3-acetyl camphor. The isomers could be separated from each other on silica gel.[42] The four diastereoisomers of the ruthenium(III) analogue retain configuration upon electrolytic oxidation and reduction.[43] Chiral rhodium complexes were used in the asymmetric hydrosilyation of ketones (30-46% optical purity).[44] The tridentate ligand, diacetylmonoxime thiosemicarbazone, has been used to prepare a series of rhodium(III) complexes.[45]

Sulfur Donors: Various geometrical isomers of iridium(III) dimethyl sulfoxide complexes have been prepared and their structures assigned, including the catalytically active dichlorotris (dimethyl sulfoxide) hydridoiridium(I).[46-7] Varying the ratio of polar solvents changed isomeric ratios. The crystal structures of $[Rh(DMSO)_3Cl_3]$ and $Na[Rh(DMSO)_2Cl_4]$ have been reported. The former structure possesses one o-bonded sulfoxide group.[48-9] The rate constants for the cleavage of the metal-C bond in $\eta C_5H_5Ru(CO)_2R$ ($R=Me,Ch_2Ph$) were investigated in neat SO_2,at 18-65°,by IR and nmr spectroscopy. In general, methyl complexes undergo scission more rapidly than benzyl derivatives, as noted in other electrophilic scission reactions.[50] A novel photochemical rearrangement of $[\eta\text{-}C_5H_5Ru(CO)SC_6F_5]_2$ to a three bridged dinuclear complex, containing 2 S and 1 CO bridges was reported[51] and the preparation and properties of monothio and monoselenocarbamate complexes of Rh(I) studied. The IR data revealed monodentate coordination through the S or Se atom.[52] Dithiocarbamate (dtc) complexes of rhodium and ruthenium have been synthesized and their solution and solid state stereochemistry observed.[53-5] The geometry of the RuS_6 core was typical of other $M(dtc)_3$ complexes in that it was intermediate between trigonal-antiprismatic and trigonal-prismatic. One ligand with potential bidentate S-N coordination, pyridine-2-thiol, has been demonstrated to form adducts, $[Rh_2OAc_4L_2]$ (OAc=acetate) where the Rh-S and N-H stretching frequencies confirmed S coordination only.[56] A number of bidentate $(S\text{-}S)^-$, (S-N) and (S-O) ligand-metal reactions have been studied. The syntheses and crystal structure of Rh(I) dithiolene complexes have been reported.[57-8] The complexes react with MeI or $PhCh_2Br$ to yield neutral species in which akylation occurs only at one of the S atoms, as shown by the X-ray data.

In solution, the adducts are fluxional. Complexes of Rh, Ir and Ru with the (S-S)$^-$ ion [(S-S)$^-$=S_2CNMe_2, S_2PMe_2, S_2PAr_2, S_2COEt] have been thoroughly investigated.[59-62] Their preparation was generally by ligand (halide) displacement, the use of polar solvents at room temperatures favoring initial monodentate coordination. The geometry of the isomer formed depended on the reaction conditions. [Ru(S_2PEt_2)$_2$(PMe_2Ph)$_2$] was shown to have a distorted octahedral geometry.[62] Bidentate ligands containing S and N or S and O have been used to prepare Rh(III) and Ru(III) complexes.[63-4] The stability constants have been measured for the thioglycolic, thiolactic and thiomalic derivatives of Ru(III).[64] Two macrocycles 1,4,8,11-tetrathia-cyclo-tetradecane and 13,14-benzo-1,4,8,11-tetra-thia cyclopentadecane have been used to synthesize Rh(III) complexes. The tetradentate ligand appeared to coordinate in a folded configuration, and the ligand field strength of the thioether donor was comparable to that of a secondary amine.[65] Rh(R_2S)$_3$ Cl_3 (R=Me,Et) reacted with 2,2'-bipyridine (L) in nonpolar solvents to form a mixture of RhL(Me_2S)Cl_3 isomers at elevated temperatures. Excess L displaced the thioether.[66]

Reaction of 3:1 molar CSe_2 and 1:1 molar areneselenols with trans-IrCl(CO)(PPh_3)$_2$ yielded the oxidative addition products IrCl(CO)CSe_2(PPh_3)$_2$ and IrHCl(CO)(Se aryl) (PPh_3)$_2$. The configuration of the CSe_2 adduct is similar to that of CS_2.[67]

GROUP V DONORS:

Nitrogen Donors: A number of complex reactions and spectral properties of dinitrogen have been reported. Phosphine stabilized complexes of [$RuH_2(N_2)(PPh_3)_3$] can be prepared by passing a stream of N_2, through a solution of [RuHCl(PPh_3)$_3$].[68] A similar procedure has been used for the preparation of trans-[RhClA{P(C_6H_{11})$_3$}$_2$] (A=O_2,N_2,C_2H_4,CO,H_2). The spontaneous formation of the nitrogen complex has been ascribed to steric requirements and electronic properties of the phosphine ligand.[69] An alternate route utilizes ligand replacement, in this case ethylenediamine (en) is displaced by alkaline NO to form [Ru(en)$_2N_2H_2O$]$^{2+}$.[70] The relative affinities of N_2 and pentene for Ru have been measured,[71] and spectral assignments for Ru-N_2 and Os-N_2 made through the use of ^{15}N.[72-3]

Mixed ligand nitridoosmium(VI) and nitridoosmium(VIII) complexes were prepared by the reaction of K[$OsNX_4$(H_2O)] with L in HX or solvent (X=Br,Cl; L=bipyridyl, 1,10-phenanthroline, 8-hydroxyquinoline, biguanide).[74] The synthesis of sulfamatepentaamineiridium(III) or ruthenium(III) salts was accomplished by using sulfite ion as the trapping agent for [M(NH_3)$_5$ (NH)]$^{3+}$, (M=Ru,Ir) generated in aqueous solutions.[75] A number of new osmium and ruthenium complexes containing coordinated [(R_3P=N)$^-$], (R=Ph,Et,Me) have been generated using a method established last year. The oxidation by chlorine of the osmium complex, [Os(NPR_3)Cl_3(PR_3)$_2$ to the nitrido species, [$OsNCl_3$ (PR_3)$_2$], in methanol was observed.[76]

The nitrosyl complexes at Ir, Os, Rh, Ru have all been synthesized in high yield preparations. The use of alcoholic solvents insured the rapid precipitation of the products.[77] Cationic rhodium nitrosyls, $[Rh(NO)L_4][PF_6]_2$ were prepared *via* ligand substitution with $NOPF_6$, and reacted with uninegative and neutral ligands to form a variety of nitrosyl complexes. Minor changes in the other ligands influenced the interchange between linear and bent NO.[78] Coordinated nitrosyl in $[Ru(bipy)_2(NO)X]^{2+}$ can function as an electrophile in aromatic substitution reactions,[79] and certain dinitrosyl complexes of Ir and Rh are unexpectedly reactive, catalysing the conversion of NO and CO into N_2O and CO_2.[80] The relative strengths of the bonds *cis*- and *trans*- to the nitrosyl in $[RuCl_5NO]$ have been calculated[81] and variety of spectroscopic measurements on Ru, Rh, and Os nitrosyl complexes made.[82-5] Three crystal structures have been reported: $[Ir_4(NO)(PPh_3)_3]ClO_4$;[86] $[Ru(NO)_2(PPh_3)_2]$ 1/2 C_6H_5,[87] and $[Ru(NH_3)_5NO]\ Cl_3 \cdot H_2O$,[88] the dinitrosyl possessed an N-Ru-N angle of 139.2°.

The reactions of 15 different amines (am) with the polymeric $[RuCl(diene)]_n$ produced a new stereoisomeric series of $[RuCl_2(am)_2(diene)]$ and $[RuClH(am)_2diene]$ complexes at generally low temperatures (0-80°), in both polar and nonpolar solvents.[89] The amines used varied from primary aliphatics to tertiary (pyridine). The convenient synthesis of $[Os(NH_3)_6]I_3$ was accomplished by reducing $(NH_4)_2OsCl_6$ with Zn dust in NH_3, followed by precipitating the product with KI. Metathetical preparations of $[Os(NH_3)_6]X$ (X=Cl, Br, $HClO_4$) used the iodide as starting material.[90] Linear tetramines, (L), $NH_2(CH_2)_n\ NH(CH_2)_m$-$NH(CH_2)_n\ NH_2$ (n=2,3; m=3,2,4) form *cis* and *trans*- $[RhLCl_2]Cl$ complexes, when n=2, m=3. In the remaining cases, only the *trans*-isomers were isolated.[91] The fundamental metal ligand vibrational modes for ruthenium, rhodium, osmium and iridium acidopentaammines were distinguished, and assigned by deuteration or halogen substitution experiments.[92]

The crystal structure of optically active, (+)-tris L(-)-*trans*-1,2,-diaminocyclohexane Rh(III) has been determined. The Rh atom is surrounded nearly octahedrally by the 6 N atoms, the central C-C bond in the puckered C_6 ring inclined at an angle of 65.9° to the threefold axis.[93]

An unusually stable cation, $[Ru(NH_2N{:}CMe_2)(P\{OMe\}_3)_4][BPh_4]_2$, has had its stability ascribed to hydrogen bonding between the *trans*-hydrazones and the phosphite oxygens. The inference was drawn from the short N-O distances (2.82Å).[94]

The preparation, configurations, and reactivities of a series of amino acid complexes with Rh(III) have been presented. The amino acids used were alanine, valine, lucine, proline, and threonine.[95]

The syntheses of nitrogen and carbon bound imidazole complexes of ruthenium ammines has been reported. For pentaammine ruthenium (II and III) guanine complexes, coordination is believed to occur through N_7;[96] for the comparable imidazole series, good yields of complexes having the ligand bound to Ru through N_3 were achieved through the treatment of $[Ru(NH_3)_5H_2O]^{2+}$ with the respective imidazole under Ar, followed by air oxidation. When C_4 and C_5 were substituted, e.g., benzimidazole, or 4,5-dimethyl imidazole, coordination at C_2 took place. The electronic and ^1H nmr spectral features of these complexes are described.[97] New rhodium complexes of dipyrromethenes were produced by refuxing $[Rh(CO)_2Cl]_2$ with the ligand in benzene.[98] A number of complexes of 1,10-phenanthroline and 2,2'-bipyridyl, including their derivatives with Ru(II), Ir(I), Ir(III) and Rh(III) have been reported.[99-102] Optically active tris (o-phenanthroline) Ru(II) complexes were prepared, and the oxidative addition reactions of $[Ir(diene)L]^+$ (diene=cyclooctadiene, 2,5-norbornadiene; L=o-phenanthroline, 2,2'-bipyridyl) followed. The structure of $[IrL_3]^{3+}$(L=bipyridyl) was established through ^{13}C and ^1H nmr techniques, as well as by elemental analyses. The successful synthesis was carried out by boiling K_3IrCl_6 with $K_2S_2O_8$ and $KHSO_4$ in water, followed by fusion at ca 250°. 2,2'-bipyridine was added and the mixture fused once more in a CO_2 atmosphere. Removal of the K_2SO_4 was accomplished by neutralizing with $KHCO_3$ and extracting the complex into methanol. The bidentate, ortho-substituted pyridines, 2-amino methyl pyridine, 2-pyridinalimine, 2-pyridine carboxaldehyde (L), have been used to synthesize a series of $[Ru(NH_3)_4\ L]^{2+}$ complexes from the reaction of the free ligand with either $Ru(NH_3)_5H_2O$ or cis-$[Ru(NH_3)_4\ H_2O]$ in aqueous solutions. Air oxidation of the 2-amino methyl pyridine complex yielded the 2-pyridinalimine homologue.[103] $[Ir(\eta_3\text{-}2\text{-}MeC_3H_4)(CO)(PPh_3)_2]$ reacted with trifluoroacetonitrile in benzene to form a 6-membered chelate structure. Similar treatment of the 1-methylallyl and π-allyl analogues produced, respectively structures I and II.[104]

I. II.

The use of acetonitrile resulted in an electrophilic attack on the π-allyl in [M(diene)(π-allyl)$_2$] (M=Ru,Os; diene=norbornadiene, 1,5-cyclooctadiene) to yield cis-[M(diene)(π-allyl)(Me CN)$_2$]$^+$. The labile CH_3CN could be replaced by neutral bidentate ligands; its stereospecific exchange was followed.[105]

The oxidative addition of RX to the macrocyclic rhodium(I) complex $[Rh\ C_{11}H_{18}N_4O_2]BF_2$, (R=Me,Et,Bu, entyl; X=I,Br,OTS) proceeded through a cationic Rh(III) complex in the presence of LiCl, to yield only a trans-chloro rhodiumalkyl. Under similar conditions, the trans-bromo alkyl did not exchange with LiCl.[106] Similar reactions were observed for some salicylaldimine complexes at Rh.[107]

An nmr study of 2-imino-4-pentanone chelates of bis (ethylene) rhodium(I) has demonstrated greater lability of the ethylene trans to the N than of that trans to the oxygen. The effect of N substitution on the exchange rate was followed.[108-9]

Linkage isomerism of the isocyanate group was observed in Rh(I) complexes through manipulation of the solvent used. Treatment of $Rh(PPh_3)_3Cl$ with Ph_4As/CH_3CN yielded $[Rh(PPh_3)_3NCO]$. Addition of EtOH to the reaction mixture reversed the bonding.[110] Thiocyanato derivatives of ruthenium, $[Et_4N]_2[Ru_2(CO)_4(NCS)_6]$, $[R_4N][Ru(CO)_2(NCS)_4]$ and trans-$[RuNO(NH_3)_4NCS)_2$, however, are bonded through nitrogen when prepared under acidic conditions (HCl/KSCN).[111-12]

Phosphorus, Arsenic and Antimony Donors: The activation of small molecules by phosphine, arsine or stilbene stabilized complexes of osmium, ruthenium, rhodium and iridium have been investigated by a variety of techniques.[68] Proton decoupled Fourier transform ^{31}P nmr of $RuCl_2(PR_3)_4$ indicated that the complex was essentially completely dissociated to pentacoordinate $RuCl_2(PR_3)_3$ at 30° and underwent further dissociation to $[RuCl_2(PR_3)_2]$ (R=Ph) at -90°, explaining the interchangeability of $RuCl_2L_3$ and $RuCl_2L_4$ in synthetic procedures.[113] Similarly, a chloro bridged intermediate was shown to be involved in the formation of the final products, $MXAL_n$, MAL'L where: $A=C_2H_4$,CO; $L=PPh_3$, $AsPh_3$, $Sb(o\text{-}MeC_6H_4)_3$, etc.; L'=acac, COD; M=Rh,Ir; n=2 or 3, depending on the solvent, and the size of the aryl substituent on Sb. Pentacoordination was favored in the solid state.[114-5] The reaction of $SbPh_3$ with $RhCl_3 \cdot 3H_2O$ in EtOH gave mer-$[Rh(SbPh_3)_3Cl_3]$ only. Mer-$[Rh(SbPh_3)_3Cl_3X]$ could be obtained by refluxing the above with NaX (X=Br,I,NCS).[117] Rh and Ir halides also react with tertiary arsines, (L), to form hexacoordinate $MX_2(CO)AL_2$, $MX_3(CO)L_2$, $MX(CO)A_2L_2$, MX_3L_3 and $MX(CO)L_2$ (A=H,O). The iridium hydride, $HRh(diop)_2$, which acts as an efficient asymmetric hydrogen catalyst was prepared by stirring $RhCl_3 \cdot 3H_2O$ with 2,3-O-isopropylidene-2, 3-dihydroxy-1,4-bis(diphenyl phosphino)butane, (diop), in EtOH, followed by the rapid addition of H_2CO/KOH.[119]

Ligand exchange reactions performed in polar solvents provided a convenient route to the formation of a series of cationic phosphite, phosphonite and phosphinite, (L), complexes of Ru, Os, Rh, and Ir complexes. The same general procedure was used for all complexes, which may be isolated as the Ph_4B^+ salts. The new complexes were prepared either by direct reaction of the appropriate ligand with metal halides or metal carbonyl/olefin halides.[120-2] The stereochemically temperature dependent rigid/nonrigid ML_5^+ (M=Rh,Ir; L=phosphite) were prepared by olefin displacement with excess L in the starting dimer $[M(OL)_2Cl]_2$ ($OL=C_2H_4$ or C_8H_{14}; M=Rh or Ir respectively) and studied by variable temperature nmr.[123-4] The reaction of NaBH and excess $P(OEt)_3$ with $[RuCl_2\{P(OEt)_3\}_4]$ in THF provided an easy route to the O-sensitive dihydride.[125]

A variety of Ru, Ir and Rh complexes with phosphine or arsine containing bidentate ligands have been isolated and in one case the kinetics of the reaction followed. Ligands used were: ethylene-1,2-bis(diphenyl arsine),(L_1;[126-8] o-styryldiphenyl phosphine/arsine (L_2);[129] o-diphenyl phosphino N,N-dimethylaniline (L_3);[130] and, o-dimethylarsino-N, N-dimethyl aniline.[131] The yields were high for all preparations. The complexes, $RuX_2(L_x)_2$ (X=Cl,Br) were oxidatively and thermally stable, whereas the Rh(I) complex $[Rh(L_3)_2]X$ (X=Cl,PF_6, SbF_6, BPh_3,ClO_4) oxidatively added Cl_2 or MeI, to form the trans-product. The anticipated low valent Ir(I) complex was not formed under the reaction conditions employed. Instead, $[IrHCl(L_3)_2]PF_6$ (90% yield) was isolated. Bubbling CO through a solution of $2-MeOC_6H_4P(CMe_3)_2$(L) in boiling i-PrOH containing $IrCl_6^{3-}$ produced an intermediate 4-coordinate carbonyl complex which upon further refluxing in nonpolar solvents gave ∿50% (III), as well as the pentagonal hydride.[132] The structure was confirmed by X-ray crystallography.

III.

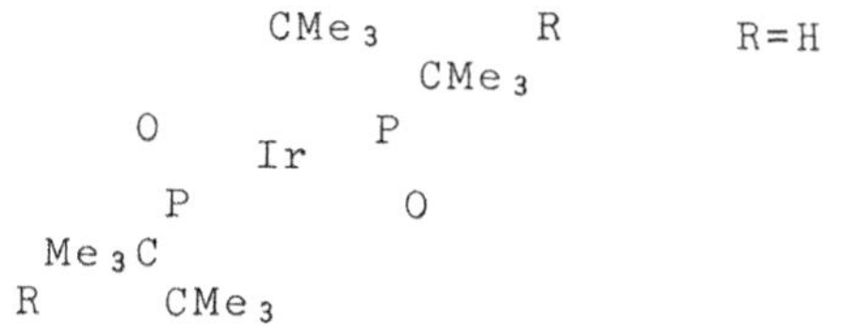

The preparation of iridium complexes with multidentate arsenic or phosphorus containing ligands has been a subject of continuing interest. The tridentate ligands, bis[3-(dimethylarsino) propyl] phenyl arsine/phosphine, (L_2L'), reacted with trans-$[Ir(CO)(EPh_3)_2X]$ (E=P,As; X=F,Cl,Br,I) to form both $[Ir(CO)L_2L']X$ and $[Ir(CO)(L_2L')X]$[133] while $(o-Ph_2LC_6H_4)_2LPh$ (L=As,P) was used to synthesize isomeric octahedral $[IrX_3L_2L']$ complexes.[134] The tetradentate ligands, 1,3-propanebix(3-dimethylarsino-propylphenylarsine (L_3L')[135] and $(o-Ph_2LC_6H_4)_3L'Ph$(L,L'= As,P)[136] tended to form $[Ir(CO)L_3L']Cl$ or $[IrXL_cL']$ products. An unusual product, $[Ir(Ph_3P)L_3L'][BPh_4]$ was isolated from a reaction mixture with the latter ligand. The yields varied from 20-85%. Nmr and other spectroscopic techniques were used to characterize the structures.

GROUP IV DONORS:

Carbon Dioxide and Carbonyl Complexes: Several new rhodium complexes with CO_2 have been synthesized by the reaction of

CO_2 with $[RhPPh_3(CO)_2]_2 \cdot C_6H_6$ in the presence of excess PPh_3 and nonpolar solvents.[137-8] The magnetic properties of $Rh_2(PPh_3)_3(CO)_n(CO_2)_m$ (n=2, m=1,2) were reported[138] and the insertion of CO_2 into a rhodium-carbon bond established by X-ray crystallography.[139] Carbon dioxide was also used for the quantitative preparation of the carbonyl complex $[Rh(PPh_3)_2(CO)Cl]$ in the presence of $HSi(OEt)_3$.[140]

The reaction kinetics for the metal-metal bonded cyclo-tris (tetracarbonyl ruthenium) (3Ru-Ru) with some phosphorus- and arsenic-donors are pseudo-first-order in decalin.[141]

The elusive binuclear carbonyl complexes, $M_2(CO)_8$, have been synthesized by 15°K matrix cocondensation of M with pure CO, under conditions of high metal concentration (M=Rh,Ir). They were also formed by matrix dimerization of of $2M(CO)_4$ at 45-50°K. The reaction products were verified by IR.[142] Black, diamagnetic $RH_8(CO)_{19}C$ crystals were obtained from the oxidation of $K_2Rh_6(CO)_{15}C$ by ferric ammonium alum, under CO. $[H_3O]$ $[Rh_{15}(CO)_{28}C_2]^{15}$ was also recovered when the oxidation was carried out in N_2.[143]

Five coordinate carbonyl complexes of ruthenium(II) and osmium (II) can be prepared by treating the halides with P $(C_6H_{11})_6$,(L), in i-PrOH. The products, $MHX(CO)L_2$ and $MX_2(CO)L_2$ activate small molecules such as SO_2, CS_2 and CO to form new complexes, e.g. $MHX(CO)L_2SO_2 \cdot nC_6H_6$, $M(CO)XL_2(CS_2H)$, (M=Ru,Os;X=Cl,Br).[144-5] A detailed investigation of the ligand exchange reactions of the salt, $[Ph_3(CH_2PH)P][RuCl_3(CO)(OL)]$(OL=bicycloheptadiene) with group V B and VI B donors led to three classes of new isomeric products. When the ligand, (L), was Me_2S, Me_2SO or CH_2:CHC⋮N, displacement of the olefin occurred and the isomeric $[RuCl_3(CO)-L_2]$ (V), complexes were isolated (75-80% yield). However, when L= $AsPh_2$,$SbPh_3$, or C_6H_5N, halide displacement was more prevalent, resulting in 60% yields of $[RuCl_2(CO)(OL)L]$ and 20% V. Excess $SbPh_3$, PPh_3, Me_2S displaced the olefin as well. The bidentate ligands, 2,2'bipyridyl or 1,10-phenanothroline reacted similarly to form polymeric $[RuCl_2(CO)L]_n$, while Me_2PhP, or $MePh_2P$ displaced the carbonyl in preference to the olefin.[146]

Rh(I) complexes $[RhLL(CO)_2]^+$ and $[RhLL(CO)X]$ can be prepared by ligand exchange of CO in $[RhLL(CO)_3]$, using LiX, (LL=bipyr., o-phenan.).[147] The complexes undergo oxidative additions with alkyl halides to form the normal 6-coordinate species, as do the trans-$[MCl(CO)L_2]$, (L=tertiary; M=Rh, Ir), when they undergo rapid addition of HCl, Cl_2, MeX, phosphine CCl_4 or PhCOCl, to give the respective adducts. Allyl chloride added reversibly to the rhodium complex in benzene, but in polar solvents rapidly isomerized to the acyl product. The tertiary phosphines used contained at least one o-methoxyphenyl group, whose oxygen directly interacted with the metal, lowering the energy of activation for the reactions.[148] The alkoxycarbonyl complex of Ir(I) could be prepared by reacting CO, MeOH and NEt_3 with cis-[IrX_2L_2] to give the trans-product.[150]

A dicarbonyl derivative, cis-[$RhCl(CO)_2PO(C_6H_{11})_5$], which was stable to loss of CO and subsequent dimerization, was formed from the reaction of the dimer with a 1:2 ratio of $(C_6H_5)_3PO$.[151]

A few new 6-coordinate [$Rh(CO)XICl(EPh_3)_2$], (VI), complexes were prepared from the addition of ICl to a benzene solution of trans-[$Rh(CO)XL_2$], (X=Cl, Br, I; E=P, As, Sb).[152] When X=I or Cl, ICl could be replaced by Cl_2, or I_2, to form complexes isomeric with VI.[152]

Monomeric thiocarbonyl ruthenium complexes, M'[$RuCl_3(CS)(PPh_3)_2$], are formed from the bridge splitting reaction of [$RuCl_2(CS)(PPh_3)_2$]$_2$ with MCl/HCl (M'=Ph_4As,Et_4N,R_4P). The structural analysis of a previously unreported dimer, CS $(PPh_3)_2RuCl_3Ru(PPh_3)_2Cl$, resulting from the reaction of CS with $RuCl_2(PPh_3)_2$ revealed a distorted octahedral coordination surrounding each Ru, and, 3 chlorine bridges.[153-4] The analogous 6-coordinate Rh(III) and Ir(III) complexes [$MCl_3(CS)L_2$] were prepared by adding $CSCl_2$ to solutions of [$IrCl(N_2)L_2$], [$Ir(CO)L_2MeCN$] or [$RhClL_3$](L=PPh_3), respectively.[155]

Olefin Complexes: O-styryldiphenylphosphine (SP) has been used in the synthesis of a variety of stable products. In the presence of $Ru_3(CO)_{12}$ and refluxing n-octane, SP reacted to form 5-coordinate bidentate olefinic species of the type [$Ru(CO)_{4-n}(SP)_n$] (n=1,2). Increasing the temperature of the reaction coupled the ligand producing a tetradentate butadienyl complex, and several isomeric tridentate monoolefinic Ru(II) complexes which were the first stable Ru(II) monoolefins isolated.[156] Mixed, ethylenetetrafluoroethylene I (I) complexes can be obtained in good yields from the reaction of [$IrCl(C_2H_4)_2$]$_2$ with C_2F_4. A number of neutral ligands can be added to these complexes with or without alkene substitution.[157] Mixed carbonyl cyclooctene Rh(I) complexes have been reported from the reaction, [$Rh(CO)_2Cl$]$_2$ + [$Rh(C_8H_{14})_2Cl$]$_2$.[158] Rh(I) acrolein (OL) complexes were synthesized via olefin exchange in the presence of Group Vb bidentate ligands, [$Rh(COCl)(C_2H_4)(OL)L$] and [$Rh(OL)_2L'$] were the products.[159] The reaction of polymeric [$RuCl_2$(1,5-COD]$_n$ with amines, acetonitrile or Group VIB bidentate ligands resulted in a variety of new monomeric, neutral, or cationic, olefinic complexes.[160-2] Rh(I) reacts readily with carvone to form diastereoisomeric planar compounds containing two diolefins.[163] Rhodium complexation has also been used as a mechanism for resolving optical isomers of substituted 1,5-hexadiene.[164]

With the exception 2,3-dimethyl-1,3-butadiene, butadienic ligands (L) were isomerized and rearranged in the presence of $Ru_3(CO)_{12}$ to form $HRu_3(CO)_9(LH)$.[165]

Dimeric arene complexes were prepared by the dehydrogenation of cyclohexadienes with $RuCl_3$ in EtOH. The dimers underwent bridge splitting reactions with a variety of ligands to form [RuX_2(arene)L].[166] The complexes were active as catalysts in the hydrogenation of several olefins.[167]

The cyclooctatetraenes, RC_8H_7, underwent transannular cyclization on reaction with $Ru_3(CO)_{12}$ in refluxing heptane to form pentalene complexes, $Ru_3(CO)_8(RC_8H_5)$.[168] Ring contraction was also reported for trans-cyclododeca-1,5,9-triene, 1,3,5- and 1,3,6-cyclooctatrienes in the presence of $[Ru(MMe_3)(CO)_4](M{=}Ge,Si)$. The major reaction product for the latter two trienes was $[Ru(CO)_3(C_8H_{10})]$.[169-170] A number of bicyclic triene and tetraene complexes of Rh(I) have been reported.[171] Protonation of $[(C_8H_8)M(C_5H_5)]$ (M=Rh(I),Ir(I)) led to a stable $[(C_8H_9)M(C_5H_5)]^+$ cation which was coordinated to the metal via a π-allyl- and a metal-olefin bond.[172]

Alkynes: Heating $[Os(CO)_{12-n}(M'MePh)_n]$,(n=1,2;M'=P,As) gave C_6H_4-bridged fluxional complexes of the type $[HOs_3(C_6H_4)(M'Me_2)(CO)_{9-n}L_n]$.[173]

Reactions of activated disubstituted acetylenes with $IrH(CO)(PPh_3)_3$ produced some novel metallocyclopentadienes containing two different acetylenic units incorporated into the metallocyclic ring. The complexes were themselves good catalysts for the cyclotrimerization of the acetylenes.[174] Cyclodi- or trimerization of disubstituted acetylenes was also accomplished with $(\eta\text{-}C_5H_5)Rh(CO)_2$. The major products of the reaction were $(\eta\text{-}C_5H_5)Rh(C_6[CF_3]_6)$ and $(\eta\text{-}C_5H_5)Rh(C_5[CH_3]_4O)$.[175] Macrocyclic alkadiynes, e.g., 1,7-cyclododecadiyne, react with $C_5H_5Rh(CO)_2$ in boiling C_8H_{16} to give pale yellow, volatile, tricyclic cyclobutadiene derivatives.[176]

π-Allyl and π-Cyclopentadiene Complexes: Treatment of $\pi\text{-}C_6H_6MCl_2$ with $RCH{:}CRCH_2HgCl$ in aqueous MeOH provided 50-70% yields of π-allyl-π-benzene metal chlorides (M=Os,Ru; R=H,Me).[177] The use of the counter ion, BF_4^-, in the preparation of the cationic bis(π-allyl)Rh(III) and Ir(III) complexes enhanced their solubility to the point that could be studied spectroscopically. The vacant coordination site in $[Rh(\pi\text{-allyl})(PPh_3)_2]$ complexes can be occupied by small neutral ligands, e.g., CO, F_3P, Me_2NF_2.[180] A 60% yield of VII was obtained by refluxing o-styryldiphenylphosphine with $RhCl_2$.

VII.

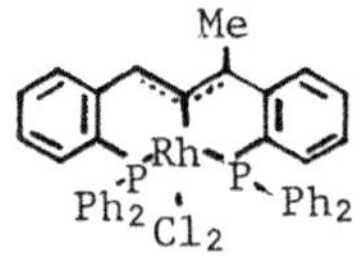

The analogous reaction with $IrCl_4$ produced only the dimerized monoolefinic $(IrCl_3SPSP)$ complex.[181]

The reaction of 1,3- or 1,5-COD with $Ru_3(CO)_{12}$, followed by treatment of the complex with Ph_3CBF_4, resulted in hydride abstraction and formation of $[(C_8H_{11})Ru(CO)_3]^+$. A similar reaction occurred with bicyclo[3.2.1]octa-2, 6-diene as the starting alkene, however in this case, coordination to the metal was through a π-allyl- and metal-carbon σ-bond.[182]

The zwitterionic $(\pi\text{-}C_5H_5)Ru(BPh_4)$ has been prepared by refluxing $(\pi\text{-}C_5H_5)Ru(PPh_3)_2Cl$ with $NaBPH_4$ in MeOH. Ru is sandwiched between the cyclopentadienyl ring and one of the phenyl rings of BPh_4^-.[183] The 1-metallocenyl-2-methyl propylcations, (M=Os,Ru), were found to be chiral using nuclei diastereotopism,[184] and $(\pi\text{-}C_7H_7)(\pi\text{-}C_7H_9)Ru_3(CO)_6$ exhibited fluxional behavior down to -100°.[185] Reaction of $RuH(PPh_3)_2(\pi\text{-}C_5H_5)$ and $C_2(CF_3)_2$ in

Et_2O afforded the cis-vinyl- and butadienyl Ru complexes; both characterized by IR, nmr and mass spectrometry.[186] Diazocyclopentadienes insert into the rhodium halogen bond of halogen bridged dimers to give new halo-substituted pentahapto-cyclopentadienyl complexes.[187] $(\eta\text{-}C_5H_5)Rh(PPh_3)_2$ undergoes both oxidative addition and alkene substitution, with a variety of small molecules, e.g., alkyl halides, olefins, acetylenes, CS_2.[188]

Iso-Nitrile and Carbenoid Complexes: The reactivity and syntheses of some tetrakis(aryl- and alkyl isocyanide) Rh(I) and Ir(I) complexes have been discussed. Only the Rh(I) analogue will add $C_2(CN)_4$ and only $C_2(CN)_4$ was found to form adducts. Both $[Ir(COD)Cl_2]$[189] and $[Ir(CO)_3Cl]_n$[190] were used as starting materials in the reaction. Stable 5-coordinate $[Ir(CNR)(LL)_2]^+$ cations (R=Me,p-MeC_6H_4; LL=$(CH_nPPh_2)_2$ (n=1,2); have been reported in which the 5 donor atoms will not undergo ligand exchange reactions. Oxidative additions, $[ML_5]^+ + XY \rightarrow [M(Y)L_5]^{2+} + X^-$, did occur when Y=H,Cl,I or HgCl.[191]

The complex, $[IrCl(CHF_2)(OCOCF_2Cl)(CO)L_2]$, which was isolated as the initial product of the reaction between $IrCl(CO)L_2$ and $CF_2ClCOONa$, was thermally decomposed to $[IrCl_2(CHF_2)(CO)L_2]$. Attempts to trap CF_2, generated in the decomposition process were unsuccessful.[192] New Rh(III) carbene syntheses have been described using imidoyl- and chloromethyleneammonium chlorides as the synthetic reagents in nonpolar solvents. The products were obtained in 42-99% yields.[193-4]

Several new cyclic carbene complexes have been synthesized, including: $[Ru(C\{OCH_2CH_2CH_2\})(\pi\text{-}C_5H_5)(CO)_2]^+$;[195] $[Ir\{C[N(CH_3)CH(Me)CH_2S]\}(CO)(PPh_3)_2]^+$;[196] and trans-$[Rh(N{:}C(CF_3)_2)\{C[N(R)CH_2]_2\}(PPh_3)_2]$.[197] Chelated dicarbene complexes of Rh(III) were the product of the reaction between primary amines and the secondary carbene $[RHMe(I)\{C(NHPr)(NHCMe_3)\}(Me_3CNC)_3]^+$.[198] A 5-membered metallocyclic Ru dicarbene was formed through the addition of the bifunctional nucleophiles hydrazine or hydroxylamine to a pair of cis-isocyanide ligands.[199]

Metalation Reactions and Metal-Carbon Bonds: Metalation of fluorinated azobenzenes by $RuMe(PPh_3)_2(\pi\text{-}C_5H_5)$ at 100° in nonpolar solvents proceeded smoothly to yield diphenyl 2-(η-cyclopentadienylphenyl) phosphine nonafluoro(phenylazo)phenyl-2C,N'-ruthenium.[200] 2-(phenylazo)phenyl complexes of Ir(III) were obtained by oxidative addition of the azobenzene to $Ir(CO)(PPh_3)_2$.[201] Aniline undergoes o-metalation in the presence of $Os(CO)_{12}$ and refluxing excess aniline to form $H_2Os_3(HNC_6H_4)(CO)_8$.[202] Benzo[h]quinoline,(L), reacted with Ir(III) to form $[IrCl_2L_2]$, in which coordination was through the N and C_{10} positions;[203] and, $RhCl_3$ metalated the methyl group in 8-methylquinoline.[204]

Hydrosilyation of $RuH_2(PPh_3)_4$ with RR'R"SiH gave $RuH_3(SiRR'R")(PPh_3)_3$ where R,R'R" were specific combinations of akyls, phenyl and H.[205] Novel methyliridium III compounds were synthesized by the addition of $MeOSO_2F$ or $MeOSO_2CF_3$ to $IrCl(CO)(PPh_3)_2$. The products were highly labile; however, substitution of various anions for OSO_2F^- or $OSO_2CF_3^-$ provided a convenient route to a series of Ir(III) complexes which were inaccessible via direct oxidative addition.[206] Grignard reactions of $CpM(Ph_3P)ICF_3$ with MeMgI, or PhMgI, provide a route to the formation of $CpMRI(PPh_3)$ (Cp=π-C_5H_5;M=Rh,Ir).[207]

Literature cited:

1. R. F. Heck, Organotransition Metal Chemistry, Acad.Press, New York and London, 1974.
2. M. Herberhold, METAL PI-COMPLEXES, 2, 508 (1974).
3. M. J. Cleare, PLATINUM MET. REV., 18, 122 (1974).
4. E. Miki and T. Ishimori, KAGAKU NO RYOIKI, 28, 503 (1974).
5. D. Banerjea, J. INDIAN CHEM. SOC., 51, 90 (1974).
6. M. Wrighton, CHEM. REV., 74, 401 (1974).
7. G. Csontos, B. Heil, and L. Marko, ANN. N.Y. ACAD. SCI., 239, 47 (1974).
8. P. Uguagliati,G. Deganello and U. Belluco, INORG. CHIM. ACTA, 9, 203 (1974).
9. R. Cramer, INORG. SYNTH., 15, 14 (1974).
10. A. Wojcicki, ADV. IN ORGANOMETAL. CHEM., 12, 32 (1974).
11. F. Bonati,G. Minghetti, INORG. CHIM. ACTA, 9, 95 (1974).
12. I. Bernal, Ed., ANN. N.Y. ACAD. SCI., 239, part 1 (1974).
13. G. Dolcetti and N. W. Hoffman, INORG. CHIM. ACTA, 9, 269 (1974).
14. K. B. Yatsimirskii, J. INDIAN CHEM. SOC., 51, 32 (1974).
15. N. Bartlett and A. Tessaud, C. R. ACAD. SCI., SER. C, 278, 1501 (1974).
16. W. Preetz, A. Scheffler and H. Homborg, Z. ANORG. ALLG. CHEM., 406, 92 (1974).
17. W. Preetz, and H. Homborg, Z. ANORG. ALLG. CHEM., 407, 1 (1974).
18. J. E. Fergusson and P. F. Heveldt, AUST. J. CHEM., 27, 661 (1974).
19. E. N. Yurchenko, L. M. Plyasova and N. Katal, IZV. SIB. OTD. AKAD. NAUK SSSR, SER. KHIM. NAUK, 5, 137 (1974).
20. L. Vaska, and J. Peone, Jr., INORG. SYNTH., 15, 64 (1974).
21. P. W. Armit, and T. A. Stephenson, J. ORGANOMETAL. CHEM., 73, C33 (1974).
22. A. Maisonnat, P. Kalck and R. Poilblanc, J. ORGANOMETAL. CHEM., 73, C36 (1974).
23. A. Maisonnat, P. Kalck and R. Poilblanc, INORG. CHEM., 13, 2996 (1974).
24. G. Palyi, A. Vizi-Orosz, L. Marko, F. Marcati and G. Bor, J. ORGANOMETAL. CHEM., 66, 295 (1974).
25. P. Garrou, and G. E. Hartwell, J. CHEM. SOC., CHEM. COMMUN., 10, 381 (1974).
26. A. W. Sleight, MAT. RES. BULL., 9, 1177 (1974).
27. F. Abraham, G. Nowogrocki and D. Thomas, C. R. ACAD. SCI., SER.C., 278, 421 (1974).
28. J. Darriet, ACTA CRYSTALLOGR., SECT. B., 30, 1459 (1974).
29. A. B. Nikol'skii and Y. I. D'yachenko, ZH. NEORG. KHIM., 19, 1889 (1974).
30. L. O. Atovmyan and O. A. D'yachenko, ZH. STRUKT. KHIM., 15, 831 (1974).
31. A. B. Nikol'skii, Y. I. D'yachenko and L. A. Myund, ZH. NEORG. KHIM., 19, 2506 (1974).

32. S. I. Ginzburg, T. A. Fomina and O. N. Evstaf'eva, ZH. NEORG. KHIM., 19, 1358 (1974).
33. A. Spencer and G. Wilkinson, J. CHEM. SOC., DALTON TRANS., 786 (1974).
34. M. A. Bennett and T. R. B. Mitchell, J. ORGANOMETAL. CHEM., 70, C30 (1974).
35. C. W. Dudley, G. Read and P. J. C. Walker, J. CHEM. SOC., DALTON TRANS., 1926 (1974).
36. L. K. Shubochkin, E. F. Shubochkina, O. V. Popov, L. D. Sorokina and M. A. Golubnichaya, IZV. SIB. OTD. AKAD. NAUK SSSR, SER. KHIM. NAUK, 2, 24 (1974).
37. J. Peone, B. R. Flynn and L. Vaska, INORG. SYNTH., 15, 68 (1974).
38. A. K. Pyartman, N. P. Kolobov, L. E. Merkul'eva and V. E. Mironov, ZH. NEORG. KHIM., 19, 1691 (1974).
39. V. I. Berezhnoi, and A. N. Nesmeyanov, VESTN. MOSK. UNIV. KHIM., 15, 363 (1974).
40. L. M. Ryzhenko, M. L. Khidekel and A. D. Shebaldoya, IZV. AKAD. NAUK SSSR, SER. KHIM., 6, 1418 (1974).
41. A. W. Addison, R. D. Gillard, P. S. Sheridan, and L. R. H. Tipping, J. CHEM. SOC., DALTON TRANS., 709 (1974).
42. G. W. Everett and A. Johnson, INORG. CHEM., 13, 489 (1974).
43. G. W. Everett, and R. R. Horn, J. AM. CHEM. SOC., 96, 2087 (1974).
44. R. J. P. Corriu, and J. J. E. Moreau, J. ORGANOMETAL. CHEM., 64, C51, (1974).
45. A. V. Ablov, N. I. Belichuk and O. A. Bologa, ZH. NEORG. KHIM., 19, 2774 (1974).
46. Y. M. Y. Haddad, H. B. Henbest and J. Trocha-Grimshaw, J. CHEM. SOC., PERKIN TRANS., 592 (1974).
47. Y. M. Y. Haddad, H. B. Henbest, J. Husbands, T. R. B. Mitchell, and J. Trocha-Grimshaw, J. CHEM. SOC., PERKIN TRANS., 596 (1974).
48. V. I. Sokol, N. D. Rubtsova, A. Y. Gribenyuk and M. A. Porai-Koshits, ZH. STRUKT, KHIM., 15, 716 (1974).
49. V. I. Sokol, N. D. Rubtsova, A. Y. Gribenyuk, ZH. STRUKT. KHIM., 15, 318 (1974).
50. S. E. Jacobson and A. Wojcicki, J. ORGANOMETAL CHEM., 72, 113 (1974).
51. R. K. Graham and A. Pryde, J. ORGANOMETAL. CHEM., 74, 105 (1974).
52. F. W. Pijpers, A. H. Dix and J. G. M. Van der Linden, INORG. CHIM. ACTA., 11, 41 (1974).
53. L. R. Gahan, and M. J. O'Connor, J. CHEM. SOC., CHEM. COMMUN., 68 (1974).
54. B. L. S. Edgar, UNIV. MICROFILMS, ANN ARBOR, MICH. ORDER NO. 74-17,297; D. J. Duffy, IBID, ORDER NO. 74-17, 242.
55. L. H. Pignolet, INORG. CHEM., 13, 2051 (1974).
56. I. P. Evans and G. Wilkinson, J. CHEM. SOC., DALTON, 946 (1974).

57. D. G. VanDerveer and R. Eisenberg, J. AMER. CHEM. SOC., 96, 4994 (1974).
58. I. Bernal, A. Clearfield and J. S. Ricci, Jr., J. CRYST. MOL. STRUCT., 4, 43 (1974).
59. D. J. Cole-Hamilton and T. A. Stephenson, J. CHEM. SOC., DALTON, 1818 (1974).
60. R. N. Mukherjee and M. D. Zingde, INDIAN J. CHEM., 12, 848 (1974).
61. R. Kirmse, W. Dietzsch and E. Hoyer, INORG. NUCL. CHEM. LETT., 10, 819 (1974).
62. J. D. Owen and D. J. Cole-Hamilton, J. CHEM. SOC., DALTON, 1867 (1974).
63. P. K. Maudal, INDIAN J. CHEM., 12, 845 (1974).
64. C. Bremard, REV. CHIM. MINER., 11, 354 (1974).
65. K. Travis and H. Daryle, INORG. CHEM., 13, 2591 (1974).
66. A. P. Kochetkova, L. B. Sveshnikova, and V. M. Stepanovich, ZH. NEORG. KHIM., 19, 1420 (1974).
67. I. Kawakami, Y. Ozaki, T. Tanaka, J. ORGANOMETAL. CHEM., 69, 151 (1974).
68. W. H. Knoth, INORG. SYNTH., 15, 31 (1974).
69. H. L. M. Van Gaal, F. G. Moers and J. J. Steggerda, J. ORGANOMETAL. CHEM., 65, C43, (1974).
70. S. Pell and J. N. Armor, J. CHEM. SOC., CHEM. COMMUN., 259, (1974).
71. F. Pennella, J. ORGANOMETAL. CHEM., 65, C17, (1974).
72. S. Pell, R. H. Mann, H. Taube and J. N. Armor, INORG. CHEM., 13, 479 (1974).
73. M. W. Bee, S. F. A. Kettle and D. B. Powell, SPECTROCHIM. ACTA PART A, 30A, 585 (1974).
74. N. Ta, P. Pramanik and D. Sen, J. INDIAN CHEM. SOC., 51, 374 (1974).
75. T. R. Weaver and F. Basolo, INORG. CHEM., 13, 1535 (1974).
76. D. Pawson and W. P. Griffith, INORG. NUCL. CHEM. LETTERS, 10, 253 (1974).
77. N. Ahmad, J. J. Levison, S. D. Robinson and M. F. Uttley, INORG. SYNTH., 15, 45 (1974).
78. N. G. Connelly, M. Green and T. A. Kuc, J. CHEM. SOC., CHEM. COMMUN., 542 (1974).
79. W. L. Bowden, W. L. Little and T. J. Meyer, J. AM. CHEM. SOC., 96, 5605 (1974).
80. B. L. Haymore, J. A. Ibers, J. AMER. CHEM. SOC., 96, 3325 (1974).
81. A. B. Nikol'skii, V. I. Baranovskii, N. V. Ivanova and O. V. Sizova, ZH. NEORG. KHIM., 19, 263 (1974).
82. A. B. Nikol'skii, N. V. Ivanova, I. V. Vasilevskii and S. M. Nikiforov, ZH. NEORG. KHIM., 19, 2510, (1974)(Russ).
83. K. G. Caulton, INORG. CHEM., 13, 1774 (1974).
84. K. A. Bolshakov, N. M. Sinitsyn, V. F. Travkin and A. A. Svetlov, DOKL. AKAD, NAUK SSSR, 216, 554 (1974).
85. N. M. Sinitsyn, T. M. Gorlova, V. F. Travkin and E. G. Sin'ko, ZH. NEORG. KHIM., 19, 2515 (1974).

86. G. R. Clark, J. M. Waters and K. R. Whittle, INORG. CHEM., 13, 1628 (1974).
87. A. P. Gaughan, Jr., B. J. Corden, R. Eisenberg and J. A. Ibers, INORG. CHEM., 13, 786 (1974).
88. F. Bottomley, J. CHEM. SOC. DALTON, 1600 (1974).
89. C. Potvin and G. Pannetier, BULL. SOC. CHIM. FRANCE, 783 (1974).
90. F. Bottomley, S. Tong, INORG. CHEM., 13, 243 (1974).
91. R. W. Halliday, R. H. Court, CAN. J. CHEM., 52, 3469 (1974).
92. M. W. Bee, S. F. A. Kettle, D. B. Powell, SPECTROCHIM. ACTA, PART A, 30, 139 (1974).
93. R. Kuroda, Y.Sasaki and Y. Saito, ACTA CRYSTALOGR., SECT. B, B30, 2053 (1974).
94. M. J. Nolte and E. Singleton, J. CHEM. SOC., DALTON, 2406 (1974).
95. R. D. Gillard, S. H. Laurie, D. C. Price, D. A. Phipps, and C. F. Weick, J. CHEM. SOC., DALTON TRANS., 1385 (1974).
96. M. J. Clarke and H. Taube, J. AM. CHEM. SOC., 96, 5413 (1974).
97. R. J. Sundberg, R. F. Bryan, I. F. Taylor, Jr., and H. Taube, J. AM. CHEM. SOC., 96, 381 (1974).
98. M. W. Roomi, TETRAHEDRON LETT., 1131 (1974).
99. R. D. Gillard, and R. E. E. Hill, J. CHEM. SOC., DALTON TRANS., 1217 (1974).
100. G. Mestroni, A. Camus and G. Zassinovich, J. ORGANOMETAL. CHEM., 73, 119 (1974).
101. C. M. Flynn, Jr., and J. N. Demas, J. AM. CHEM. SOC., 96, 1959 (1974).
102. S. P. Ghosh, P. Bhattacharjee and L. K. Mishra, J. INDIAN CHEM. SOC., 51, 308 (1974).
103. V. E. Alvarez, R. J. Allen, T. Matsubara and P. C. Ford, J. AM. CHEM. SOC., 96, 7686 (1974).
104. M. Green, S. H. Taylor, J. J. Daly, and F. Sanz, J. CHEM. SOC., CHEM. COMMUN., 361 (1974).
105. R. R. Schrock, B. F. G. Hohnson and J. Lewis, J. CHEM. SOC., DALTON TRANS., 951 (1974).
106. J. P. Collman, M. R. MacLaury, J. AM. CHEM. SOC., 96, 3019 (1974).
107. C. A. Rogers and B. O. West, J. ORGANOMETAL. CHEM., 70, 445 (1974).
108. K. Bouchal, J. Kriz and F. Hrabak, COLLECT. CZECH. CHEM. COMMUN., 39, 439 (1974).
109. J. Kriz and K. Bouchal, J. ORGANOMETAL. CHEM., 64, 255 (1974).
110. S. J. Anderson, A. H. Norbury, and J. Songstad, J. CHEM. SOC., CHEM. COMMUN., 37 (1974).
111. A. R. M. Abbadi, M. A. Shehadeh and J. V. Kingston, J. INORG. NUCL. CHEM., 36, 1173 (1974).
112. N. M. Sinitsyn, V. V. Borisov and L. A. Pshenichnikova,

ZH. NEORG. KHIM., 19, 3050 (1974).
113. K. G. Caulton, J. AM. CHEM. SOC., 96, 3005 (1974).
114. C. A. Tolman, P. Z. Meakin, D. I. Lindner, and J. P. Jesson, J. AMER. CHEM. SOC., 96, 2762 (1974).
115. P. E. Garrou, and G. E. Hartwell, J. CHEM. SOC., CHEM. COMMUN., 381 (1974).
116. P. E. Garrou and G. E. Hartwell, J. ORGANOMETAL. CHEM., 69, 445 (1974).
117. W. E. Hill, and C. A. McAuliffe, INORG. CHEM., 13, 1524 (1974).
118. G. K. N. Reddy and N. M. N. Gowda, INDIAN J. CHEM., 12, 185 (1974).
119. W. R. Cullen, A. Fenster and B. R. James, INORG. NUCL. CHEM. LETTERS, 10, 167 (1974).
120. D. A. Couch and S. D. Robinson, INORG. CHIM. ACTA, 9, 39 (1974).
121. D. A. Couch and S. D. Robinson, INORG. CHEM., 13, 456 (1974).
122. D. A. Couch, S. D. Robinson and J. N. Wingfield, J. CHEM. SOC., DALTON, 1309 (1974).
123. P. Meakin and J. P. Jesson, J. AMER. CHEM. SOC., 96, 5751 (1974).
124. J. P. Jesson and P. Meakin, J. AMER. CHEM. SOC., 96, 5760 (1974).
125. W. G. Peet and D. H. Gerlach, INORG. SYNTH., 15, 38 (1974).
126. R. K. Poddar, U. Agarwala and P. T. Manoharan, J. INORG. NUCL CHEM., 36, 2275 (1974).
127. R. K. Poddar and U. Agarwala, J. INORG. NUCL. CHEM., 36, 575 (1974).
128. S. S. Sandhu and A. K. Mehta, INDIAN J. CHEM., 12, 834, (1974).
129. M. A. Bennett, R. N. Johnson and I. B. Tomkins, INORG. CHEM., 13, 346 (1974).
130. T. B. Rauchfuss and D. M. Roundhill, J. AMER. CHEM. SOC., 96, 3098 (1974).
131. A. Peloso and L. Volponi, J. CHEM. SOC. DALTON, 278 (1974).
132. R. Mason, K. M. Thomas, H. D. Empsall, S. R. Fletcher, P. N. Heys, E. M. Hyde, C. E. Jones and B. L. Shaw, J. CHEM. SOC., CHEM. COMMUN., 612 (1974).
133. C. A. McAuliffe and R. Pollock, J. ORGANOMETAL. CHEM., 74, 463 (1974).
134. D. G. E. Kerfoot, R. J. Mawby, A. Sgamellotti and L. M. Venanzi, INORG. CHIM. ACTA, 8, 195 (1974).
135. S. T. Chow and C. A. McAuliffe, J. ORGANOMET. CHEM., 77, 401 (1974).
136. R. J. Nynott, E. F. Trogu and L. M. Venanzi, INORG. CHIM. ACTA, 8, 201 (1974).

137. K. S. Kolomnikov, T. S. Belopotapova, T. V. Lysyak and M. E. Vol'pin, J. ORGANOMETAL. CHEM., 67, C25 (1974).
138. V. T. Kalinnikov, G. M. Larin, V. M. Novotortsev, T. S. Belopotapova, T. V. Lysyak, and I. S. Kolomnikov, DOKL. AKAD NAUK SSSR, 217, 1345 (1974).
139. A. I. Gusev and U. T. Struchkov, ZH. STRUKT. KHIM., 15, 282 (1974) (Russ).
140. P. Svoboda, T. S. Belopotapova and J. Hetflejs, J. ORGANOMETAL. CHEM., 65, C37 (1974).
141. A. Poe and M. V. Twigg, J. CHEM. SOC. DALTON, 1860 (1974).
142. L. A. Hanlan and G. A. Ozin, J. AMER. CHEM. SOC., 96, 6324 (1974).
143. V. G. Albano, P. Chini, S. Martinengo, M. Sansoni, and D. Strumolo. J. CHEM. SOC., CHEM. COMMUN., 299 (1974).
144. F. G. Moers, R. W. Hoedt, M. Ten and J. P. Langhout, J. ORGANOMETAL. CHEM., 65, 93 (1974).
145. F. G. Moers, R. W. M. Ten Hoedt and J. P. Langhout, J. INORG. NUCL. CHEM., 36, 2279 (1974).
146. L. Ruiz-Ramirez and T. A. Stephenson, J. CHEM. SOC. DALTON, 1640 (1974).
147. G. Mestroni, A. Camus and G. Zassinovich, J. ORGANOMETAL. CHEM., 65, 119 (1974).
148. E. M. Miller and B. L. Shaw, J. CHEM. SOC. DALTON, 480 (1974).
149. H. D. Empsall, E. M. Hyde, C. E. Jones and B. L. Shaw, J. CHEM. SOC. DALTON, 1980 (1974).
150. H. C. Clark, and K. Von Werner, SYNTH. REACT. INORG. MET.-ORG. CHEM., 4, 355 (1974).
151. G. Bandoli, D. A. Clemente, G. Deganello, G. Carturan, P. Uguagliati and U. Belluco, J. ORGANOMETAL. CHEM., 71, 125 (1974).
152. K. Goswami and M. M. Singh, INDIAN J. CHEM., 12, 858 (1974).
153. T. A. Stephenson, E. S. Switkes and (in part) P. W. Armit, J. CHEM. SOC. DALTON, 1134 (1974).
154. T. A. Stephenson, E. S. Switkes and (in part) P. W. Armit, J. CHEM. SOC. DALTON, 1139 (1974).
155. M. Kubota and C. J. Curtis, INORG. CHEM., 13, 2277 (1974).
156. M. A. Bennett, R. N. Johnson, and I. B. Tomkins. J. AMER. CHEM. SOC., 96, 61 (1974).
157. H. L. M. Van Gaal and A. Van der Ent, INORG. CHIM. ACTA 7, 653 (1973).
158. Yu. S. Varshavskii, T. G. Cherkasova, N. A. Buzina, V. A. Kormer, J. ORGANOMET. CHEM. 77, 107 (1974).
159. M. Hidai, K. Ishimi, M. Iwase, Y. Uchida, NIPPON KAGAKU KAISHI , 732 (1974).
160. C. Potvin, G. Pannetier, BULL. SOC. CHIM. FR., 783 (1974).
161. P. Powell, J. ORGANOMETAL. CHEM. 65, 89 (1974).
162. T. V. Ashworth, E. Singleton, J. ORGANOMET. CHEM, 77, C31, (1974).

163. V. Schurig, J. ORGANOMETAL CHEM. 74, 457 (1974).
164. V. Aris, J. M. Brown, B. T. Golding, J. CHEM. SOC. PERKINS II, 700 (1974).
165. O. Gambino, M. Valle, S. Aime, G. A. Vaglio, INORG. CHIM. ACTA, 8, 71 (1974).
166. M. A. Bennett, A. K. Smith, J. CHEM. SOC., DALTON TRANS., 233 (1974).
167. R. Iwata, I. Ogata, TETRAHEDRON, 29, 2753 (1973).
168. J. A. K. Howard, S. A. R. Knox, V. Riera, F.G.A.Stone, P. Woodward, J. CHEM. SOC. CHEM. COMMUN., 452, (1974).
169. S. A. R. Knox, R. P. Phillips, F. G. A. Stone, J.C.S. DALTON, 658 (1974).
170. A. C. Szary, S. A. R. Knox, F. G. A. Stone, J.C.S. DALTON, 662 (1974)
171. R. Grigg, J. L. Jackson, TETRAHEDRON 29, 3903 (1973).
172. J. Evans, B. F. G. Johnson, J. Lewis, D. J. Yarrow, J.C.S. DALTON, 2375 (1974).
173. A. J. Deeming, R. E. Kimber and M. Underhill, J. CHEM. SOC., DALTON TRANS., 2589 (1973)
174. W. H. Baddley and G. B. Tupper, J. ORGANOMETAL. CHEM, 67, C16 (1974).
175. R. S. Dickson and H. P. Kirsch, AUST. J. CHEM., 27, 61 (1974).
176. R. B. King and M. N. Ackermann, J. ORGANOMETAL. CHEM., 67, 431 (1974).
177. A. Z. Rubezhov, A. S. Ivanov, and S. P. Gubin, IZB. AKAD. NAUK SSSR, SER. KHIM., 1902 (1974); ibid, 931 (1974).
178. J. W. Faller, B. V. Johnson, T. P. Dryja, J. ORGANOMETAL. CHEM. 65, 395 (1974).
179. M. Green, G. Parker, J. CHEM. SOC., DALTON TRANS. 333 (1974).
180. D. A. Clement, J. F. Nixon, J. S. Poland, J. ORGANOMETAL CHEM. 76, 117 (1974).
181. M. A. Bennett, R. N. Johnson, G. B. Robertson, I. B. Tomkins, and P. O. Whimp, J. ORGANOMET. CHEM., 77,C43 (1974).
182. A. J. Deeming, S. S. Ullah, A. J. P. Domingos, B. F. G. Johnson, J. Lewis, J. CHEM. SOC. DALTON, 2093 (1974); b) B. F. G. Johnson, D. J. Yarrow, ibid. 1054 (1974); c) A. J. P. Domingos, B. F. G. Johnson, J. Lewis, ibid 145 (1974).
183. G. J. Kruger, A. L. du Preez, R. J. Haines, J.C.S. DALTON, 1302 (1974).
184. V. I. Sokolov, P. V. Petrovskii, A. A. Koridze and O. A. Reutov, J. ORGANOMETALLIC CHEM. 76, C15 (1974).
185. T. H. Whitesides, R. A. Budnik, J. CHEM. SOC. CHEM. COMMUN. 302 (1974).
186. T. Blackmore, M. L. Bruce and F. G. A. Stone, J.C.S. DALTON, 106 (1974).
187. V. M. Day, B. R. Stults, K. J. Reimer, A. Shaver, J. AMER. CHEM. SOC. 96, 1227 (1974).

188. Y. Wakatsuki and H. Yamazaki, J. ORGANOMETAL. CHEM., 64, 393 (1974).
189. T. Tanaka, K. Kawakami, T. Kaneshima, and M. Haga, ANNALS NEW YORK ACADEMY OF SCIENCES 23955 (1974).
190. F. Faraone, R. Pietropaolo, and E. Rotondo, J. CHEM. SOC. DALTON 2262 (1974).
191. W. M. Bedford and G. Rouschias, J.C.S. DALTON, 2532 (1974).
192. A. J. Schultz, J. V. McArdle, G. P. Khare, R. Eisenberg, J. ORGANOMETAL. CHEM. 72, 415 (1974).
193. M. F. Lappert and A. J. Oliver, J.C.S. DALTON, 65 (1974).
194. B. Cetinkaya, M. F. Lappert, J.C.S. DALTON, 1591 (1974).
195. C. H. Game, M. Green, J. R. Moss, J.C.S. DALTON 351 (1974).
196. P. S. Fraser, W. R. Roper, F. G. A. Stone, J. CHEM. SOC. DALTON 760 (1974); ibid 102 (1974).
197. M. J. Doyle, M. F. Lappert,G. M. McLaughlin, J. McMeeking, J. CHEM. SOC. DALTON TRANS. 1494 (1974).
198. P. R. Branson, R. A. Cable, M. Green, M. K. Lloyd, J. CHEM. SOC. CHEM. COMMUN, 364 (1974).
199. D. J. Doonan, A. L. Balch, INORG. CHEM. 13, 921 (1974).
200. M. I. Bruce, R. C. F. Gardner, B. L. Goodall, F. G. A. Stone, J. Doedens, and J. A. Moreland, J. CHEM. SOC., CHEM. COMMUN., 185 (1974).
201. M. I. Bruce, B. L. Goodall, L. Brian, F. G. A. Stone, and B. J. Stone, and B. J. Thomson, AUST. J. CHEM., 27, 2135 (1974).
202. C. C. Yin and A. J. Deeming, J CHEM. SOC., DALTON TRANS., 1013 (1974).
203. M. Nonoyama, BULL. CHEM. SOC. JAP., 47, 767 (1974).
204. M. Nonoyama, J. ORGANOMETAL. CHEM., 74, 115 (1974).
205. H. Kono, and Y. Nagai, CHEM. LETT. 931 (1974).
206. D. Strope and D. F. Shriver, INORG. CHEM. 13 2652 (1974).
207. S. A. Gardner and M. D. Rausch, INORG. CHEM. 13, 997 (1974).

The Platinum Metals

PART II: Palladium and Platinum

P. D. Kaplan

College of Medicine
University of Cincinnati
Cincinnati, Ohio

INTRODUCTION

Several monographs on organotransition metal and platinum chemistry along with a number of reviews have appeared.[1-4] The subject of reaction mechanisms has been discussed[1] and the topic of monolefin pi-complexes updated.[4] Ligand exchange reactions involving mono-and bidentate ligands containing N, O, and S donars have been summarized[5,6] as have insertion reactions involving a variety of platinum complexes.[7,8] Isocyanide complexes of platinum and palladium were covered in a general review of the subject reflecting the current interest in oxidative addition, insertion, and carbene syntheses.[9]

Methods of producing a variety of platinum and palladium catalysts have been reviewed.[10]

<u>GROUP VII DONORS</u>: Two new halide complexes have been synthesized. Platinum dichloride (Pt_6Cl_{12}) was reacted with bromine to form a new mixed halide, α-$Pt_6Cl_{12}Br_2$, which on heating in inert solvents gave β-$Pt_6Cl_2Br_2$.[11] A mixture of PdF_2 and MF (M = K, Rb, Tl) was heated at 400-500°C under argon to produce $MPdF_3$(and K_2PdF_6).[12]

The crystal and molecular structure of the novel polynuclear complexes $[Et_4N]Hg_2PtCl_8$,$[Et_4N]Hg_3MCl_{10}$ (M=Pt,Pd) have been determined. The centrosymmetric trinuclear anion is linked by Cl bridges involving terminal HgCl bonds, the latter Hg_3MCl_{10} anion contains additional HgClM bridging Cl atoms.

Several interesting counter ions have been used in the synthesis and subsequent reactions of $[R_4N]_2[MX_4]^{2-}$ complexes, (X = halide; R = octyl, dodecyl, cetyl). Spectral data indicated intramolecular H bonding between N-H...Cl in these salts[14] as well as in the Pt (IV) salts, $H_2[PtCl_6]$ L_2 (where L = pyridine or quinoline).[15] The synthesis and cleavage of the dimer salts $(R_3C_3)_2^+[M_2Cl_6]^{2-}$, (R = Me_2N, Pr)(M = Pt, Pd), has been reported.[16] Cleavage of the dimer by the olefin, C_8H_{14}, proceeded smoothly to produce $2[(OL)MCl_3]^-$ when the counter ion was N-methyl substituted cyclopropenium. When the counter ion had R = propyl, the reaction proceeded in two steps with an intermediate, partially cleaved dimer observable by its visible spectrum.[17]

The Russians have continued their interest in the syntheses of specific geometrical isomers of the types $ML^1L\ X^1\ X]$ or $[ML^1L\ (X^1)_2(X)_2]$ (where L^1,L = NH_3, NO_2, ethylenediamine, C_2H_4, DMSO; and X^1,X = Cl, Br, I) with the synthesis of nitrochlorobromo chlorodiammine platinum from the oxidative chlorination of cis-$Pt(NH_3)_3\ NO_2Br$.[18] Correlations were made between ligand lability and cis and trans effects in several related reactions.[19-21]

Thermal isomerization of the complex $[PtL_2][PtCl_4]$ at 150-160° produced the complex $[PtLCl_2]$. The ligands used all exert a strong trans effect, e.g., L = $RSCH_2CH_2SR'$, R = Et,Pr,iso-Pr.[22-3]

GROUP VI DONORS:

Oxygen Donors: The synthesis and crystal structures of several Pd and Pt oxides have been reported. Na_2O_2 or Na_2O were reacted with PdO under argon at 900° to produce $Na_2Pd_3O_4$[24], while K_2PdO_3 was prepared by the oxidation of K_2PdO_2 at 450°C in oxygen.[25] Both crystal structures belong to the orthorhombic space group. The reaction of Pd or PdO with Bi_2O_3 at ca 700° in air yielded tetragonal Bi_2PdO_4.[26] Single crystals of the pyrochlore, $M_2Pt_2O_7$, were prepared by crystallization in flux under O pressure (M = Y,Pb,Nd,Gd,Ho).[27] The magnetic moments increased from 3.35 to 9.32 $\mu\beta$ in this series.

A tetragonal palladium hydroxide, $Na_2[Pd(OH)_4]$, identified by IR, thermal analysis and x-ray diffraction was produced by treating $PdCl_2$ with aqueous NaOH.[28] Complexes of the type, $[PtH_2L(OH)_2]$, and $[PtH_2L(OH)X]$ were prepared by treating aqueous $[PtH_4LCl_2.5H_2O]$ with 30% H_2O_2 (H_4L = ethylenediaminetetraacetic acid; X = Cl, Br).[29] A palladium hydroxide is being used as a catalyst for the efficient hydrogenation of unsaturated aromatic hydrocarbons and heterocycles[30] and $PdO_2.4CaO$, synthesized from CaO and hydrated PdO_2 at high temperatures, was used as a catalyst for the oxidation of NH_3 to NO (50% yield).[31]

A new procedure for the conversion of ketoximes to the corresponding ketone employs a phosphine stabilized di-oxygen complex of palladium.[32]

Independent of the oxidation state of the starting material, a mixed valence, orthophosphate complex of platinum (PtL_2),

(L = orthophosphate) was generated by reacting H_3PO_4 with $K_2[Pt(X)_4]$ (X = NO_2,Cl) and $HClO_4$ at 150-300°.[33] Carboxylic acid anhydrides undergo oxidative additions to Pt(0) complexes to form trans[$PtL_2[C(O)R][O_2CR]$] and cis-[$PtL_2[C(O)CF_2CF_2CO_2]$], (R = Me,CF_3,C_2F_5).[34]

The syntheses of several bidentate, o-bonded, complexes of Pd or Pt have been reported.[35-39] In one instance, a variety of palladium β-diketonate complexes were used as starting materials for the efficient synthesis of tetrakis-(triarylphosphine) palladium.[35] The crystal structure of acetylacetonatoacetylacetonyltriphenylphosphinepalladium(II) benzene solvate confirmed that bonding was through oxygen and a γ-carbon.[36] A number of the geometrical isomers that can arise out of this type of linkage isomerism have been isolated from the reaction of $[Pt(CH_3)_2Br_2]_n$ with thallium acetate in benzene.[38] A dimeric intermediate, $[Pt(CH_3)_2AcacBr]_2$ which appeared to have two isomeric forms was treated with neutral ligands (L) to produce the final isomeric products $[Pt(CH_3)_2AcacLBr]$. The use of $PtCl_4$ or PtX_6 salts as the starting material for the preparation of a series of Pt(IV)acetylacetonates yielded two unusual oxidation/reduction products, an acetylacetonate condensate(I), acting as a counterion to $[PtCl_6]^{2-}$,

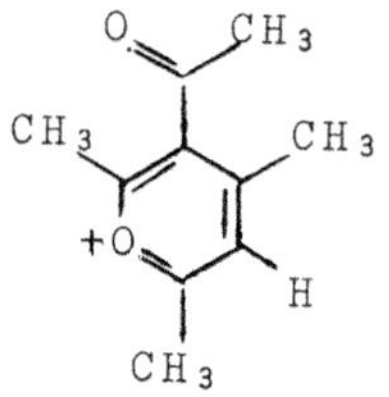

I.

and $[KPt(Acac)_2X]$ (X=Cl,BrI).

The bi-dentate ligand isonitrosoacetonimine and its derivatives have been used for the preparation of a series of palladium complexes having O- and N-coordinated isonitroso groups.[40] By allowing isonitrosoacetylacetone, the appropriate RNH_2 and $PdCl_2$ to react in $CHCl_3$ and then treating the mixture with base, one obtained the desired PdLL′ product.[41]

The crystal structures of two very similar molecules, chloro-1-(2-thiazolylazo)-2-naptholatopalladium(II) and 1-(2-thiazolylazo)-2-naphtholatopalladium(II), established that the terdentate ligand, 1-(2-thiazolylazo)-2-naphtholato, is coordinated to Pd through the phenolic O atom, azo N and thiazole N atom. The fourth site is occupied by Cl.[42-3] Several bidentate mercapto acids, e.g., α-and β- mercaptopropionic acid (L) have been used to synthesize ML_2 complexes, where M = Pd or Pt, that were probably s-bridged polymers.[44-5]

Several amino acid complexes, including a cyclic tetrakis derivative, have been synthesized in keeping with the present interest in antitumor agents.[46-7]

Sulfur Donors: The preparation and crystal structure of PdPS, which is isostructural with PdPSe, has been reported.[48]

A 90% yield of platinum and palladium sulfide complexes of the type [M(SEt_2)$_2$] has been reported for the reaction of sodium ethylmercaptide with potassium tetrachloroplatinate or palladate.[49] The reaction of Pt dimethyl sulfide complexes with hydroxylamine and two oximes (L) has been followed and the acid dissociation constants measured for *cis*- and *trans*- [Pt(MeS) LCl_2].[50]

An nmr study on selenium or tellurium analogues of [M L_2 X_2] where M = Pt, Pt; X = Cl,Br,I; and L = Et_2Se or Et_2Te, demonstrated unambiguous inversion of configuration about the Se or Te bonds.[51] Such inversion of configuration about sulfur in similar platinum complexes had been previously followed by nmr techniques.

The manipulation of the stoichiometries involved in the oxidative addition reaction of CSe_2 or arenselenols to PtL_4 (L = PPh_3, PPh_2Me) changes the nature of the adduct.[52] A 3:1 ratio of CSe_2:PtL_4 leads to Pt(CSe_2)L_2 with a similar configuration to that observed for the corresponding CS_2 complex. Using the same 3:1 ratio of areneselenols to PtL_4 affords the corresponding bis(aryl selenide) complexes, whereas changing the molar ratios to 1:1 yields the insertion products, *cis*- and *trans*- [PtH(SeR)L_2](R = Ph, tolyl). The Pt hydrides are susceptible to dissociative exchange of the phosphine ligands in solution and therefore exist as a mixture of *cis*- and *trans*- isomers. An attempted oxidative addition of the type $XY + ML_4$ [M(X)(Y)L_2] between SCl_2 and [Pt(CO)$_2$$PPh_3$)$_2$], was not successful, instead, SCl_2 acted as a chlorinating agent and the resultant products were [$PtCl_2$(CO)PPh_3] and a mixture of *cis*- and *trans*- [$PtCl_2$(PPh_3)$_2$].[53]

The preparation of [M_2(SCF_3)$_2$(PR_3)$_4$]X_2 complexes (M = Pd or Pt; R = Et or Ph; X = BF_4 or ClO_4) in high yields has been described.[54] ^{19}F nmr was used to assign structures and demonstrate the existence of *syn*- and *anti*- isomers. The SCF_3 bridges are weak and readily undergo cleavage reactions with L = PR_3, pyridine, p-toluidine, Cl^-) to give the monomeric *cis*- [Pt(SCF_3)L(PR_3)$_2$] cation or neutral complex. The choice of solvent was critical in some cases. Changing from acetone to CH_2Cl_2 or nitromethane in the reaction of [Pt_2Cl_2(PPh_3)$_4$]-(ClO_4)$_2$ with $AgSCF_3$ produced [Pt(SCF_3)(ClO_4)(PPh_3)$_2$] rather than the expected product.

The strong bridging abilities of alkyl and aryl sulfides is apparent in the preparation of a series of platinum(II) anions $[Pt_2X_6(SMe_2)]^{2-}$ (X=Cl,Br) containing SMe_2 as the sole bridging group. The anions, $[Pt_2X_5(SMe_2)]^-$ contain both a halo and sulfide bridging group. The sulfur-bridged $[Pt_2Cl_5SMePh]^-$ was prepared but $[Pt_2X_4(SPh_2)_2]$ complexes are halide bridged.[55]

The syntheses and subsequent kinetic study of a series of mono- and bidentate sulfoxide complexes of Pt(II) has been reported.[56-7] Complexes of Pd or Pt with either $CH_2(SH)CH(SH)CH_2SO_3H$, $CH_2(SH)CH(SH)CH_2(X)CH_2SO_3H$ or ethyl(carboxymethylene)-ethylene disulfide (X=O,S or SO_2) have also been prepared and their acid dissociation constants measured.[58-9] Both neutral and cationic species were formed depending on the ligand and the reaction conditions. The reaction of one equivalent of $NaMCl_4$ with two equivalents of aqueous $RC_6H_4SeO_2Na$ yielded $[M(RC_6H_4SeO_2)_2(H_2O)]$ (M=Pt,Pd;R=CH_3,H,Cl,Br,NO_2). As with the sulfur analogues, selenium is bound to the metal.[60]

X-ray photoelectron spectroscopy of the new paramagnetic dithiene complexes, $[Et_4N][M(PhC(S^-):C(S)Ph)_2]$ (M = Pt,Pd), places the extra electronic charge principally on S of the ligand. The charge on the particular metal remains essentially constant throughout a series of neutral, anionic and dianionic complexes that were prepared for this study.[61]

A number of new complexes of the type trans-$[M(SS)L_2]$, and $[M(SS)_2]$ (SS=$SeSCNR_2$,S_2CNR_2,S_2COR,S_2PR_2,$\overline{S_2CO}$) have been described.[62-70] The (N,N-dimethyldichalcogenocarbamato)bis(tert-phosphine)palladium(II) and platinum(II)dimethyltin(IV) trichlorides were characterized by IR and nmr. There is very little difference in the trans effect exerted by either Se or S. A platinum dithiocarbonimidato complex was obtained via abstraction of sulfur from phenyl or methyl isothiocyanate in the reaction with $[Pt(allene)L_2]$ (L = PPh_3).[64] Preliminary x-ray structural analysis of $Ph_4As[Pt(S_2COEt)_3]$ shows that Pt(II) bonds to one bidentate and 2 monodentate xanthate ions.[66] Polymeric Pd complexes of a dithiophosphinic acid were synthesized and used for the separation of Lewis bases by chemisorption chromatography.[68] Isocyanate replacement by SO_2 or CS_2 was observed in the reaction,

$$[(R_3P)_2M(NCO)_2] + 2\ SO_2 \text{ or } CS_2 + 4\ MeOH \xrightarrow{MeOH} (R_3P)_2M(SO_2OMe)_2, \text{ or } (R_3P)_2M(SC{:}S\ OMe)_2, + 2H_2NCO_2Me$$

(R = Ph,Et). The latter alkoxydithiocarbonato intermediate lost one thiocarbonato ligand to become

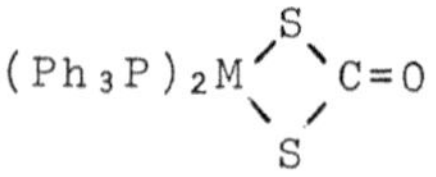

(M = Pd,Pt).

Monothio derivatives of fluorinated β-diketones and a mixed

thioseleno derivative, 1,1,1-trifluoro-4-(2-thienyl)-4-seleno-3-butene-2-one, have been used to prepare stable Pd(II) and Pt(II) complexes.[69-70] Palladium reacts with dithiomalonamide and N,N'-diphenyldithiomalonamide somewhat analogously; however, these ligands also form $M(HL)_2X_2$ and $M(HL)X_2$ complexes (X = Cl, Br, I) in which S, N bidentate coordination is suggested from the ir stretching frequency for Pd-N and Pd-S.[71]

The x-ray study of an interesting thiocyanato-bridged species $Pd_2(SCN)_2[Ph_2PO)_2H]_2$ has revealed the existence of a hydrogen bis (diphenylphosphinate) anion(II).[72]

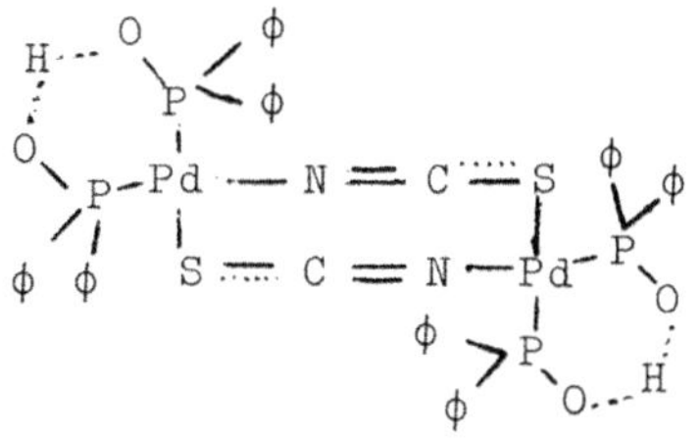

II.

A symmetrical hydrogen bond has been observed in several other phosphinate species and may be of general significance in complexes of the type $[MX(RO)_2PO\}\{(RO)_2POH\}]$.

Several other crystal structure studies of Pt or Pd thiocyanate complexes have corroborative evidence that coordination of the thiocyanate is controlled completely by steric effects as long as the Π-bonding capabilities are kept constant. Thus, M-N=C=S bonding was attributed to steric hindrance of S-bonding.[73-5] Counter ion interactions in solution as well as in the solid state have reversed the bonding order for Pd complexes; however, for the corresponding Pt complexes, the S-bonded species were more stable in both solution and solid state.[76] Thirty five Pt and Pd thiocyanate complexes have been prepared using a bidentate di-arsine or a bidentate arsine/phosphine ligand. The five coordinate $[PdL_2(NCS)](CNS)$ (L=O-phenylenebis-(diphenylarsine), (o-diphenylarsinophenyl)diphenylphosphine were isolated from the reaction of $LiClO_4$ with $PdL_2(CNS)_2$ in CH_2Cl_2.[77]

Pentacoordinate complexes of Pd(II) with substituted thioureas and heterocyclic amines have been reported from the reaction of naphthyl thiourea (Nta) with $PdCl_2$ in MeOH to give $[Pd(Ntu)_4Cl]$. Ligand substitution replaces the chloride with a variety of amines.[78] Aqueous $K_2[PdCl_4]$ was treated with $P(OEt)_3$ followed by thiourea (L) to give $[Pd[P(OEt)_3]L_3]Cl_2$.[79] N,N'-ethylene thiourea (L) reacted with K_2PtCl_4 to produce a halogen bridged dimer $[PtCl_2L]_2$ and the bidentate, N-allylthiourea (L') formed the complex $[PtL'Cl_2]$.[80]

The crystal structures of three palladium complexed macrocycles were reported.[81-3] $[Pd(C_{10}H_{22}N_2OS_2]$, exhibited a pyramidal distortion from the square plane due to a strong O-Pd interaction.

GROUP V DONORS:

Nitrogen Donors: An attempt to prepare dinitrogen complexes of zerovalent platinum and palladium containing triethyl phosphine, $N_2M(PEt_3)_3$, was unsuccessful, although the nickel homologue was isolated.[84]

Addition of equimolar quantities of $NaNO_2$ under N_2 to a weakly acidic solution of K_2PtCl_4, followed by the addition of excess $(Et_4N)Cl$ yields a dark green air sensitive novel dimer whose crystal structure has been determined[85] (Figure III).

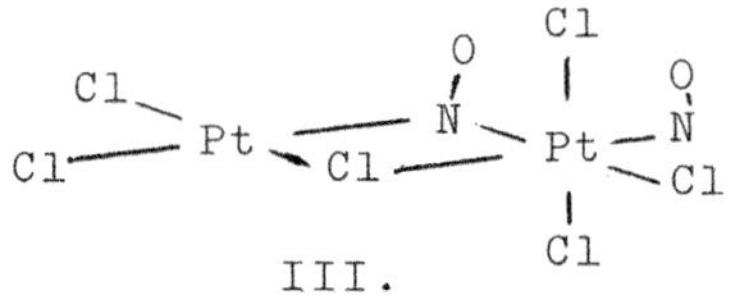

III.

A 5-coordinate nitrosyl species, $[C_9H_7][PtNOCl_4]$ has also been reported[86] as well as a trinuclear Pd complex with bridging NO's.[87]

A useful summary of the preparative procedures for complexes of platinum(II) with nitro groups and sulfur containing ligands has appeared[88], and linkage isomerizim of the type, $M-NO_2 \longrightarrow M-O-NO$, (M = Pd,Pt), has been discussed in terms of the oxidation reduction, process occurring at the ligand.[89]

A chiral square planar palladium complex of (S,S)-2,7-diisopropyl-3,6-diazaoctanedioic acid has been made. One of the two carboxylic groups remains uncoordinated to the metal.[90]

A novel arylazoplatium dimer(V), which is readily hydrogenated has been prepared by the displacement of diphenylacetylene (OL) with the aryldiazonium salt in $[Pt\ L_2(OL)]$ (L = P Ph_3). When the reaction was performed under H_2 followed by the addition of dry HCl, an arylhydrazonium salt could be isolated. Structure IV has been postulated on the basis of the ir and nmr assignments.[91]

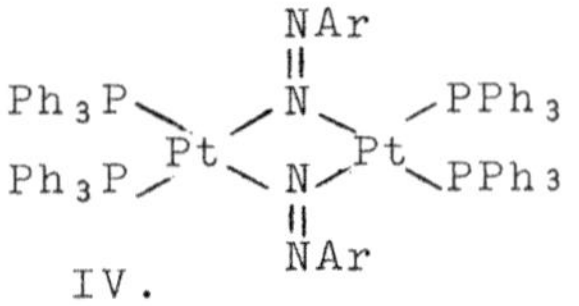

IV.

A series of arylazodiazonium, cationic, monomers, trans-$[Pt(N_2Ar)(PEt_3)_2]^+$ were prepared by treating the neutral complex, trans-$[PtCl(N_2Ar)(PEt_3)_2]$ with $NaBPh_4$ or $NaClO_4$ in polar solvents followed by adding a neutral ligand (where L = pyridine, PEt_3, NH_3, ethylisonitrile). Protonation of the complexes yielded the corresponding aryldiimide, and reduction of the protonated species in the presence of catalyst gave the arylhydrazine complex.[92] An attempt to make trans-$[Pt(PEt_3)_2(N_2C_6H_4F)(CO)]$ was unsuccessful. The ^{19}F nmr and ir of the m- and p-fluorobenzenediazo derivatives, trans-$[Pt(PEt_3)_2(N_2C_6H_4F)Cl]$ yield evidence for decreased M → N_2Ar Π-bonding in going from m- to p-fluorosubstitution. Back bonding has not been considered to be large in these doubly bent structures.[93] The structure of the protonated (aryldiazene) trans-$[Pt(PEt_3)_2(HNNC_6H_4F)Cl]^+$ has been established by X-ray analysis as having the bent configuration with a Pt-N bond length of 1.973Å.[94]

Mercurial derivatives of 2-(Arylazo)aryl compounds were used as arylating agents for MCl_2 to produce trans-$M(R_3P)_2(C_6H_4N_2Ph)$ (M = Pt,Pd).[95] A somewhat similar cyclopalladated hydrazone has been used to synthesize new bimetallic compounds in reactions with $Co(CO)_4^-$ or $[Mo(CO)_3$-n-$C_5H_5]$.[96]

An investigation into the kinetics of the reactivity of amines with dichloro[biacetylbis(N-methyl, N-phenyl)osazone]platinum(II) indicated that the reactivity of the osazone complex could be related to the labilizing effect of the interaction between one ortho-hydrogen of one phenyl ring of the ligand and the (leaving) chloride in the transition state. Under the appropriate conditions, orthometallation of the complex occurred.[97]

Complexes from the reaction of 4-arylhydrazono derivatives of 3,5-dioxopyrozolidines(H_2L) with K_2PdCl_4 in several polar solvents have been reported.[98] The ligand could be made bidentate by substituting a hydroxyl group for one of the orthoprotons on the phenyl ring and using a basic solvent.

1,3-diaryltriazines, (ArN:NN(H)Ar) react with $M(PPh_3)_4$(M = Pt or Pd) in benzene to form cis- and trans-bis(1,3-di-p-tolyltriazenido)bistriphenylphosphine) metal within a few minutes. The cis-isomer of both metals was less soluble. Cis → trans isomerization occurred in solution (60-75%).[99] A similar SP^2 hybridized system containing sulfur, RN:S:NR, was used to synthesize trans-$[PtCl_2(RN:S:NR)L]$, (L = PR_3, AsR_3SbR_3,Sr_2,SeR_2,TeR_2) (80% yield). The dimer $[PtCl_2L]_2$ was used as the starting material in this case. Coordination isomerization from N to N, as well as rotation around the N-R bond could be observed.[100]

N-boryl and N-silylketimines react with $Pd(NCPh)_2Cl_2$ to form air sensitive, $Pd(Ph_2C:NR)Cl_2$ where R = Et_3B, Me_3Si, in 67 to 73% yields. The "B nmr and ir data would seem to indicate Pd-

(CN) coordination.[101]

Platinum has been shown to abstract $(CN)^-$ from BH_3CN^- to form $[PtH(CN)L_2]$ (L = tertiary phosphines) in an inert atmosphere,[102] and nitrile complexes of Pd or Pd have been reported to undergo several interesting reactions.[103-4] The complexes of MeCN, EtCN and PhCN with $PdCl_2$ hydrolyse with difficulty to carbamides[105] and the reaction of $NaCH(CN)_2$ in MeOH with $[M(PPh_3)_2MeI]$ or $[M(PPh_3)_2(CH_2COR)Cl]$ yielded complexes;

V. 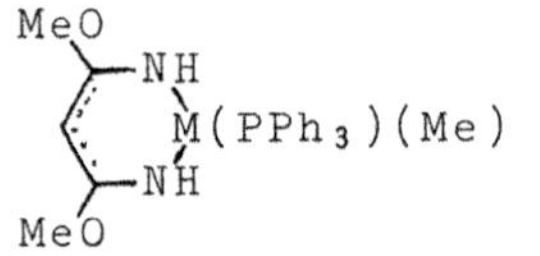 VI. 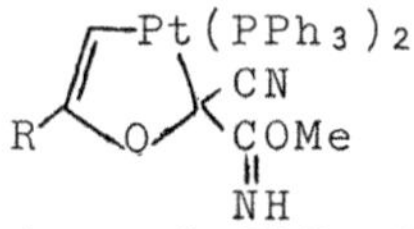

Changing the solvent in the reaction of $ClCH_2CN$ with $Pt(PPh_3)_4$ changed the ratios of the product, *trans*- and *cis*- $[Pt(PPh_3)_2(CH_2CN)Cl]$, the use of acetone favoring the *cis*-product.[104-6] Both the nmr chemical shift and the $J^{195}Pt-{}^1H$ were indicative of the isomeric structure.

The reaction of $Pd_2[PhCH:CH)CO]_3$ with bipyridine (BiPy) and PhCONCO in benzene afforded Pd(BiPy) (PhCONCO) and, the analogous thio-containing metallocyclic structure was the product of the reaction with PhCSNCO.[107] A similar novel structure has been proposed as the possible product of the reaction of $Pt(PPh_3)_3CO$ and toluene-*p*-sulphonyl azide in dry benezene; however, the formation of the urea derivative was the more typical response. The urea derivative could also be obtained from $M(PPh_3)_4$ (M=Pt, Pd) and the corresponding isocyanate.[108]

A variety of bidentate substituted pyridine ring complexes have been synthesized. 2-(alkenyl)pyridines can be bidentate or monodentate depending on whether acetone or benzene is used as the solvent, and whether bulky groups are present at the olefin.[109] 2-(2'pyridyl)benzimidazole was used to synthesize 47 new complexes including those containing Pd and Pt.[110] Palladium complexes of 2-[β-(diphenylphosphino/arsino) ethyl pyridine contain either monodentate or bidentate coordination through the P/As atom.[111] The possibility of a fluxional 5-coordinate, 1,10-phenanthroline complex, $[Pt(PEt_2)_2phenCl]$ in solution has been suggested from nmr and conductance measurements; however, in the solid state phenanthroline is monodentate.[112] A bidentate 2, 6-diaminopyridine complex of palladium has been synthesized with antiviral, antimitogenic properties.[113]

The terdentate N-(2-pyridylmethyl)-2-(dimethylamino)acetamide coordination to Pt and Pd results in crowding of the *cis* halogen and 6-hydrogen on the pyridine ring.[114]

CD and mass spectrometry have been used to investigate the interaction of *cis*- and *trans*- $Pt(ammine)_2Cl_2$ antineoplastic complexes with several nucleosidic analogues.[115-16] The synthesis and crystal structures of bis(6-mercapto-9-benzylpurine) palladium(II) has been reported. Both purine ligands are coordinated to Pd through the S and N_7 atoms.[117]

Benzoxazole, benzothiazole, benzotriazoles and 5-methyl or 5-trifluoromethyl tetrazole complexes of palladium and platinum have been synthesized and studied this year.[118-19] Linkage and cis-trans isomerism were the subject of investigation for 11 complexes of the type $Pd(R_3P)_2(tet)_2$(tet = tetrazole). It was possible to observe N_1N_1, N_2N_2 and N_1N_2 coordination spectrophotometrically via ^1H or ^{19}F nmr. The equilibrium thermodynamic data collected indicated cis-trans isomerization was a function of solvent dipole moment, temperature and the phosphine used. Tetrazole exchange was not rate controlling.[120]

The discovery of the antitumor activity of cis-$[PtCl_2(am)_2]$, where am = NH_3, stimulated the synthesis and testing of a wide variety of related complexes and has been the subject of two recent reviews.[121-22] Less attention has been paid to the correlation of antitumor activity with the basicity, stereochemistry or lability of the coordinated amines,[123-24] although considerable numbers of studies are being performed on ligand exchange reactions involving amines as well as phosphines. The kinetics for a variety of substitution reactions have been recorded.[123-25,128] It is clear from nmr evidence that ligand exchange, in the presence of adventitious ligand, is a common phenomenon.[130] Some evidence for a solvated intermediate has appeared,[126-27] and an efficient method for the synthesis of $[PtCl_2(am)]_2$ and cis-$[PtCl_2am_2]$ (am = isoquinoline, piperidine) by (UV) irradiation was the byproduct of one study.[130]

The syntheses of some mixed, triamine complexes of platinum(II) has been achieved through the treatment of aqueous trans-$[Ptam_2(NO_2)_2]$ with the appropriate base, followed by precipitating the cation with $K_2Pt(NO_2)_4$.[131] Improved yields in the synthesis of Pt(II) cis and trans- $[Ptam_2x_2]$ (X = Cl,No) from the reduction of K_2PtCl_6 with $N_2H_4 \cdot H_2SO_4$ are obtained when the intermediates in the reaction, K_2PtCl_4 and $Ptam_4Cl_2$ are not isolated.[132] The photochemical reduction of $[Ptam_2NO_2Br_3]$, its aquation product and the hydroxide took place readily in aqueous solutions. In all cases $[Ptam_2NO_2Br]$ was the end product.[133] Mixed halide Pt(IV) complexes were synthesized from a solution of hot, aqueous $Ptam_2NO_2X$ (X = Br or Cl) through which Cl_2 or excess liquid Br_2 was passed.[134]

An interesting chiral Pt(II) complex, cis-dichloro[(R)-α-methyl benzylamine] (S)-1,2,2,trimethylpropyl(R)-vinylether] platinum(II) in which both the amine and the olefin have optically active centers was synthesized by olefin replacement of cis-$[Pt(am)(C_2H_4)Cl_2]$.[135]

Histidine complexes of platinum are usually unidentate but can become bidentate by treatment of the complex with KOH.[136] A variety of bidentate, ethylenediamine or propylenediamine, (Pn), Pt(IV) complexes have been synthesized and the kinetics of _cis-trans-_ photoisomerization in $[PtPn_2Cl_2]Cl_2$ studied.[137-141] CD spectra of the optically active amminedichloronitro(ethylenediamine) platinum chloride complexes show that inversion at the amino group occurs during deprotonation of the complex.[139] The crystal structure analyses of three Pd(II) complexes with _cis_-3,5-diaminopiperidine demonstrate a square planar coordination around the metal with the piperidine rings _syn_ or _anti_ to each other, depending on whether the structure is deprotonated or protonated at the ring nitrogens.[142]

Phosphorus, Arsenic and Antimony Donors: Nine new ternary arsenides of the type MM^IAs were prepared by solid state diffusion from the powdered elements. PdTiAs and PdCrAs had hexagonal Fe_2P type structures.[143]

Two coordinate, $Pd[PRR'_2]_2$ and $Pd[PR_3]_2$(R = phenyl; R' = t-butyl or cyclohexyl) were prepared from PRR'_2 and $Pd(\eta\text{-}C_5H_5)(\eta\text{-}C_3H_5)$ in hexane at room temperature for several hours. The respective crystals separated out upon cooling. A three-centered non-bonded interaction between the orthohydrogens on the phenyl rings and Pd were observed at -71°C in ^1H nmr spectra and inferred by the short Pt-H bond distances. Under mercury lamp irradiation at 0°C, $Pd[P(t\text{-}Bu)_3]$ absorbs H_2 to become a polymeric hybrid, while $Pd[PPh(t\text{-}Bu)_2]_2$ is unreactive to H_2 but absorbs O_2, even in the solid state, to form green crystalline stable $PdO_2[PPh(t\text{-}Bu)_2]_2$.[144-45]

Stereospecific syntheses of _cis-_ and _trans-_ $[PtClLPEt_3]^+$ and and cis-$[PtClX(PEt_3)_2]$ (L = Py, PPh_3, $PClPh_2$, $P(OPh)_3$; X = SCN, NO_2, NO_3, N_3, NCO, Br) were developed through the use of bridge cleavage reactions on the respective dimers.[146] Bridge cleavage reactions in square planar Pt(II) complexes occur with retention of configuration.

New arsine and stilbene complexes of the type, $[M[As(L'_2)]_2X_2]$ and $[Pd(SbL_3)\ X\ ]$, where M = Pt, Pd; L = aryl L' = alkyl and X = halide or NO_2, have been described.[147-48] The syntheses were performed in polar solvents at low temperatures, and the stilbene dimer $Pd_2(NO_2)_4(SbL_3)_2$ had bridging O-N-O groups. Some mononuclear bidentate stilbene/arsine/phosphine complexes have been synthesized. O-phenylenebis(dimethylstilbene) formed the expected MX_2L (M = Pd, Pt) complexes.[149] The polycyclic 2, 11-bis(diphenylphosphinomethyl) benzophenanthrene reacted with metal halides or isocyanates to form the _trans_-Pt(II) or Pd(II) complex.[150] Stable O-and C-metallated derivatives of 2-alkoxy phenylphosphines and platinum have been prepared.[151] Metallation was accomplished by heating the _trans_-complex $[PtCl_2L_2]$ in solvent, or directly under high vacuum (68-88% yield).

Metal complexes of tris(aziridino)phosphine and bis (aziridino) morpholinophosphine have been prepared from the reaction of metal benzonitrile complexes (M = Pt,Pd) with the ligands as part of an effort to synthesize anticancer metal analogues of the akylating agents bearing aziridino rings.[152]

Pyramidal five-coordinate platinum and palladiam compounds of phospholes and phosphindolines have been compared using X-ray crystallography.[153-54] The ML_3X_2 structures exhibit a tetragonal distortion of the basal bonds due to bond length and minimum non-bonded distance requirements forcing adjacent Pd-P bonds to be inclined at angles >90°.

Some interesting cationic, phosphinite, phosphonite and phosphite complexes of platinum and palladium have been conveniently prepared by ligand displacement reactions in polar solvents. The same general procedure was used for the preparation of all $[ML_5]^{2+}$. The dicatonionic products were characterized by elemental analysis ^1H nmr and conductivity measurements.[155] A series of $[ML_5]^{2+}$ (L = $P(OR)_3$, R = Me, Et; M = Pd, Pt) were prepared by a different route, the phosphite was added to a suspension of MCl_2 in ROH under N_2 and stirred until all reactants went into solution, the counterion was added.[156] All complexes of this type are stereochemically nonrigid at ambient temperatures and stereochemically rigid on the nmr time scale at -100-150°. Rearrangements are shown to be intra molecular by retention of the $^{195}Pt-^{31}P$ and $^{31}P-^{31}P$ coupling at elevated temperatures; however, intermolecular exchange from reaction, $ML_5^{2+} \rightleftharpoons ML_4^{2+}$ + L, can occur when excess L is present.[156-57]

GROUP IV DONORS:

Carbonyl Complexes: A new series of inorganic, diamagnetic oligomers, $[Pt_3(CO)_3(\mu_2\text{-}CO)_3]_n^{2-}$ n = 2-5, have been prepared via two routes: a) reduction of $Pt(CO)_2Cl_2$ with alkali metals in the presence of CO, and b) reduction of Na_2PtCl_6 with CO and methanolic NaOH. Structural investigation of the series for n = 2,3,5 indicates that all are based upon the polymerization of a common triplatinum hexacarbonyl component containing direct Pt-Pt bonding in a trigonal-prismatic array.[158]

Vibrational spectral analysis of the dimer $[Pd_2X_4(CO)_2]$ indicates that the bridging carbonyl is bent about the CO bridge in solution. The oligomer, $(PdXCO)_n$, was shown by ir spectra to have alternative CO and X bridges between Pd atoms.[159]

The reaction of a dihalogenide bis(tertiary phosphine) palladium or platinum complex with CO, methanol and triethylamine gave methoxycarbonyls in good yields when the phosphine ligands had at least one phenyl group.[160] An unusual trans-relationship of the alkoxycarbonyl ligands was found in trans-bis(ethoxycarbonyl) bis(triphenylphosphine) platinum.[161]

Mono- and Di-Olefin Complexes: The organometallic [2.2.1] propellane, bis(triphenylphosphine)platinum(II) $\Delta^{1,4}$-bicyclo [2.2.0]hexane, has been prepared by olefin exchange under N_2 at -80° in toluene.[162] Upon warming to -20° and adding ether, the air stable crystals of the complex precipitated out. The structure (VII) was confirmed by X-ray crystallography (C_1-C_4= 1.55Å).

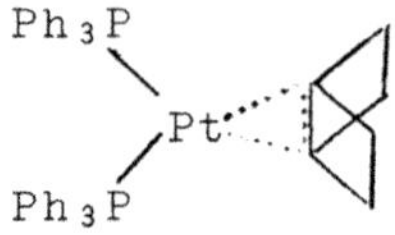

VII.

The platinum(II)vinyl alcohol complex,[$Pt(C_2H_4O)$(acac)] (acac=acetylacetone) has been found to be less stable than its β-oxoethyl derivative, which could be formed directly by treating a THF solution of [$Pt(C_2H_4$(acac)Cl] with aqueous KOH and excess acetaldehyde at 0°. Protonation of the σ-complex with HCl yielded the Π-ethenol. The analogous reactions took place when the ethylene complex was treated with acetone and aqueous alkali; however, attempts to obtain stable Π-enol complexes employing a ligand system other than acetylacetonate failed.[163] It is of interest that a similar treatment of olefinic complexes e.g., cis-[$PPhEt_2)PtCl_2$(olefin)], with methoxide in CH_2Cl_2 resulted primarily in [$(PPhEt_2)PtCl_2$] along with varying amounts (max. 30%) of the methylvinylether.[164] A study of the oxidative hydrolysis of ethylene in the presence of [Pd(ethylenediamine) $(H_2O)_2]^{2+}$ or [Pd(diethylenetriamine) $H_2O]^{2+}$ indicated that wherever possible, prior coordination of the hydroxide ion cis to the olefin was necessary. Direct attack on the olefin by the hydroxide was slow in comparison.[165] By contrast the nucleophilic attack on olefins by amines in palladium (and platinum) complexes is direct and proceeds trans. The configuration of the σ-Pd complexes were detected by $LiAlD_4$ reduction.[166] Reversible amination has been reported for 1,5-cyclooctadiene and 1,5-hexadiene palladium(II) chloride.[167]

The crystal structures of the cyclopentene and cycloheptene platinum complexes indicate virtually no change in structure except for a slight elongation of the olefinic bond.[168] Instability of cyclic olefinic complexes of platinum has been noted in another study measuring the free energies of activation for rotation of substituted olefins in cis-[$PtCl_2$(L)(olefin)] and [PtCl(acac)(olefin)], where L is a tertiary phosphine or arsine.[169] The coordinated olefinic C-C separation in bis(triphenylphosphine)allene palladium was 1.401Å. The other bond remained unchanged (1.304A) in length.[170]

The bromo derivatives of the uncharged diolefinic complexes M(diolefin)Br, as well as the unstable dicationic analogues $[M(diolefin)L_n](PF_6)_2$,(M=Pd,Pt, diolefin = 1,5-hexadiene, cycloocta-1,5-diene and 2,5-norbornadiene; L=Ch_3CN, dipyridyl, $(Ph_2P)_2C_2H_4$; n=2,1) were prepared, the latter cations at -15° in $CHCl_2$.[171-2] They are all susceptible to nucleophilic attack, reacting with methoxide ion to form the 2-alkoxy-5-enyl derivatives, or, when the weakly coordinated acetonitrile was present cis to norbornadiene, [2+2]-and [2+4]-cycloadditions of propylene on norbornadiene occurred without dimerization of norbornadiene itself. One thermally forbidden [2+2]-addition product, 3-methyl tricyclo[4.2.1.$0^{2,5}$]nonene-7 formed 55% of the organic extract.

Several trans (R) palladation reactions with diolefins have been reported to occur readily when R=Cl, aryl, OMe[173-6], the most interesting of which involved the exocyclic double bond in 7-methylene bicyclo[2.2.1]hept-2-ene and the complex formed from the addition to endo-5-vinyl-2-norbornene in nonpolar solvents.

The preparation of dibenzylacetone (DBA) derivatives of Pd(o) Pt(o) have been summarized recently and reported to undergo a variety of interesting reactions.[177] The value of the dissociation energy of the platinum carbon Π-bond in $Pt(DBA)_2$, (273kJ/mol), is indicative of a relatively strong bond. Decomposition of the palladium complex to unidentified products, prevented comparison of the two.[178] Several groups have completed their crystal structure studies of bis(dibenzylideneacetone)dipalladium solvates, which show an s-cis, trans conformation for two of the ligands and an s-cis, cis form for the third.[179-80] Ligand exchange, oxidative addition and complex formation reactions of M(DBA)n n=2,3 with activated olefins, acetylenes and quinones have been presented and several novel products isolated in varying yields. Palladiacyclopentadiene complexes were formed in high yields from $Pd(DBA)_2$, or $Pd_2(DBA)_3$, and RC_2R (R = COOMe, Ph) in polar solvents. The complexes underwent further reactions to form substituted butadienes, furans or benzenes depending on the reactants. The platinum analogue $(Pt(C_4R_4)I_n$ was not successfully characterized, however, its adduct with phosphine, $[Pt(C_4R_4)]$ $[(PPh_3)_2]$, was.[181-6]

Π-allyl complexes: The preparation of Π-allyl complexes from Me_2C:$CHCH_2Me$, MeCH:CHCH:CHMe and $NaPdCl_4$ or $Pd(NCPh)_2Cl_2$ has been described[187] and a general high yield synthesis of Π-allyl palladium chloride dimers from cycloalkanes or alkylidene cycloalkanes presented.[188] Selective alkylatian of the methyl groups of _cis_- and _trans_- $Me_2C:CHCH_2CH_2CMe:CHCH_2CH_2COMe$ without protection of the carbonyl group was carried out via the Π-allyl palladium complexes which were separated by chromatography.[189] Allyloxycarbonyl complexes of Pt(II) undergo decarboxylation in refluxing benzene to give the corresponding Π-allyl complex,[190] and allyl palladium carbonyl halides were prepared by treating the allyl palladium halide with CO under an inert gas atmosphere.[191] Vinyl silanes and vinyl germanes dimerized under Ar to form the corresponding Π-allyl palladium chloride complex in 26% yield.[192]

Acetylene Complexes: Several reports have appeared on polypyrazolylborate ($HB(pz)_3$) stabilizes 5-coordinate complexes of platinum(II) containing acetylenes,[193] olefins, allene and carbonyl.[194] All were synthesized by adding the unsaturated ligand to what is believed to be the polymeric $[Pt(CH_3)[HB(pz)_3]]$ in CH_2Cl_2. Crystallographic results on Pt(II) and Pt(o) acetylenic complexes confirm the trigonal bipyramidal structure indicated by nmr and corroborate the presence of a strong platinum-acetylene bond.[195-6] Interesting Π-phosphinacetylene dimers of Pd(o) or Pt(o) (VIII) were prepared in high yields by

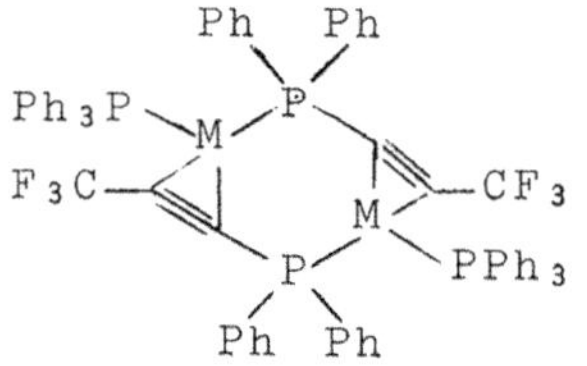

VIII.

treating the cis-$[MCl_2(Ph_2PC:CCF_3)_2]$ derivatives with $NaBH_4$ in dry THF.[197] X-ray crystallography confirmed the structure. Hydrolysis of the chloro analogue, dichloro-bis-(trifluoromethyldiphenyl)phosphinoacetylene) palladium(II), produced an unusual chloro bridged diphenylphosphinate palladium dimer containing a symmetrically displaced H^+ between each two O atoms.

Isocyanide and Carbenoid Complexes: The reactions of platinum and palladium with isocyanides has been recently reviewed.[198] Several reports this year have been concerned with the synthesis of bidentate carbenoid ring systems, while a number of monodentate cyclic and acyclic complexes have been made.[201-5] Thus, new bidentate (and one tridentate) chelating ligands were prepared from the respective addition of amidines, amines, and biacetylhydrazone, to platinum (II) and palladium (II) _iso_cyanides in aqueous solutions.[199-200] In general, these new ligands, which contain both a trigonal carbon donor and an imine nitrogen donor, and their complexes are very stable to acidic, and oxidizing conditions. Isocyanide (fluorocarbon) platinum and palladium complexes served as the precursors to carbene complexes of the type: (IX) $[R'_2C{:}CR''_2\ M\ C(NR,R_2)_2\ (CNR_3)]$; (X) $[CR'_2{\cdot}O{\cdot}CR'_2OM\{C(NR,R_2)_2\}(CNR_3)]$ and (XI), $[\{(R'_4C)C_4\ Pd - \{C(NR,R_2)_2\}(CNR_3)]$ where R' and R" were electron withdrawing and R_1R_2electron donating. The crystal structure of X, (M=Pd), places the plane of the carbene ligand perpendicular to the coordination plane of Pd, with a Pd-C bond distance of 2.07 Å, indicating virtually no back donation from Pd to the carbene.[207] By comparison, the crystal structures of *cis*- and *trans*-isomers of $[PtCl_2\{C(PhNCh_2)_2\}(PEt_3)]$, indicate that the plane of the carbene varies from 90° for the *cis*-isomer to 71° for the *trans*-isomer.[208] The insertion of isocyanide into an olefinic complex, $[Pt(C_{10}H_{12}OCH_3)\ (Ph_3P)Cl]$, was followed and found to occur in a stepwise fashion.[206] Palladium(II) isocyanides, *cis*-$[Pd(RNC)L\ Cl_2]$ (R=aryl, L=phosphine, nitrile) react with bifunctional amines, $R'NH_2$, to give neutral carbenes, *cis*-$[Pd\{C(NHR)\ NHR'\}LCl_2]$, in which R' is not bonded to the central metal. Coordination of R' occurs only by treatment with $NaClO_4$ or fluoroborate. Bis(carbene) derivatives were also reported that underwent chelation or deprotonation with perchlorate or KOH, respectively.[203]

The ready formation in high yields of mono-, di-, or tri-carbene complexes of Pt(II) and Pd(II) with the electron rich 1,1',3,3'-tetramethyl-2,2'-bi-imidazolidinylidene (L) has been studied and several new complexes, *cis*- or *trans*-$[MR(PPh_3)_{3-n}L_n]$ (M=Pd or Pt; R=Me) synthesized. Cationic cyclic carbene complexes of the type 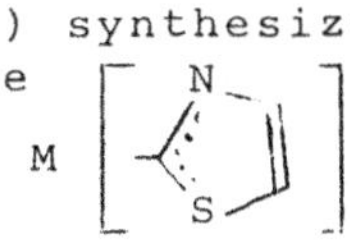 and 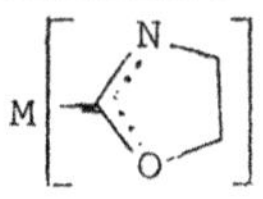 (M=Pd or Pt) were prepared via oxidative addition of 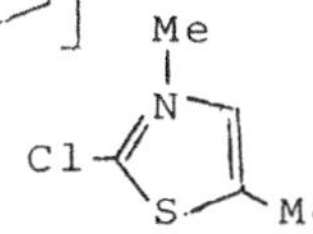 to the zero-valent ML_4(L=Phosphine) complexes,[210] or by reaching $HO(CH_2)_nNC$ (n=2 and 3) with the metal halide.[204]

Ylides: Several ylide complexes of palladium and platinum are reported. Π-allylphosphonium ylide complexes of Pd(II) were readily prepared by the reaction of the corresponding allylphosphonium hexahalodipalladate with sodium acetate,[211] and keto-stabilized sulfonium ylides displace styrene and benzonitrile from their $PdCl_2$ adducts to produce a stable $(ylide)_2PdCl_2$ complex.[212] Both Pt(II) and Pd(II) halide complexes react with methyl phenyl sulfonium p-chlorophenacylide in CH_2Cl_2 to form a stable sulfur ylide, $[MX_2(R_2S)\ (CH_3(C_6H_5)SCHC(O)C_6H_4Cl)]$ (M=Pd, Pt; X=Cl,Br,I) (25% yield for Pt, 60% yield for Pd). Coordination to the methine carbon was established by IR and nmr. The $PtCl_2$-and $PtBr_2$ complexes have a *cis*-configuration in solution while the PtI complex is *trans*. The corresponding Pd complexes were too insoluble for nmr studies.[213]

<u>Alkyl, Aryl and Related Complexes</u>: Palladium atoms oxidatively insert aryl, alkyl, and acyl carbon-halogen bonds at -196° to form a new series of reactive intermediates ArPdX, RPdX, and RCOPdX. Efficiency of the insertion is in the order C-I > C-Br > C-Cl, and the stability of the intermediate varies greatly with the organic group attached. These intermediates can be readily trapped with tertiary phosphines to form the $[(R_3P)_2Pd(R)X]$ (R'=Ar, R or RCO) adducts.[214-5] A number of preparations employing oxidative additions of RX to Pt(o) and Pt(II) complexes have been described,[216-7] and new 5-coordinate platinum(II) complexes $[Pt\ X_2L_3]$ are reported from the reactions of various sec. alkyl and benzyl halides with complexes of the type PtL_n (n = 3 or 4; L = a phosphine).[218] The mechanisms for these additions have been a current topic of investigation with evidence being presented for: one electron processes observed in the reaction of $M(PEt_3)_3$ (M=Pt,Pd) and isopropyl iodide <u>via</u> nmr CIDNP effects;[219] nonradical (S_n2) processes in the configuration inversion of optically active tertiary halides in their reaction with $(Ph_3P)_4Pd$;[220-1] radical chain processes occuring in the reaction of alkyl halides with $M(PEt_3)_3$(M=Pt,Pd) for less reactive halides e.g., 1-bromoethane.[222]

Dimethyl- and diphenyl magnesium are as reactive as organo lithium compounds towards $PtCl_2L_2$ complexes (L=mono- or bidentate tertiary phosphines) and yield Pt R_2L_2 rapidly at 20° in THF.[223] Organomercurials, RHgX (R=F_3C, p-tolyl, $(P\text{-}MeOC_6H_5)_2$-C:CH, MeO_2CCH_2; X = Br, Cl) provide a useful route to the σ-bonded organo platinum(II) derivatives $[MRX(PPh_3)_2]$.[224-5] One step reduction of $PtCl_2$ to platinum(o) cluster complexes $[Pt_4(CO)_5L_4$-[4] followed by oxidative addition with MeI provides a new route to platinum(II), [Pt(CO)L(I)Me] complexes where L is a tertiary phosphine. These react with uncharged ligands, L' to form the carbonyl insertion or substitution products.[226] The kinetic data for the reaction, Pt(CO)LXR]+L' ⟶ [Pt LL'X(COR)], suggest a two step mechanism with the rate determining step being the formation of the acyl intermediate $[Pt(PPh_3)I(COMe)]$.[227]

A number halovinyl complexes of platinum and palladium have been reported from the reaction of halogenated olefins with the zerovalent metal. Reactivity could be correlated with ease of C-halogen bond breakage, the bromo derivatives being more reactive than the corresponding chloro compounds.[228-9] There was evidence for stereospectific addition since retention of configuration was observed.[230]

The reaction of fluorohalobenzenes with zerovalent $M(PEt_3)_4$ also yield the oxidative addition products, <u>trans</u>-$MX(C_6H_4F)$-$(PEt_3)_2$.[231] Cationic fluorophenyl complexes of palladium and platinum can be easily prepared by treating the perchlorate complex with a neutral ligand.[232-3] Hexafluoro-2-butyne and dimethylacetylene dicarboxylate insert into the M-R bonds in

$M(Me)_2Cl_2L_2$ (M=Pd,Pt; L_2 = bipyridyl, tertiary phosphines) to form the vinyl derivatives[234-5]. Hexafluoro butadiene complexes of platinum react with SnX_2 (X=Cl,Br) in benzene to yield the trans-Pt(X)(CF:CFCF:CF_2)L_2 product[236]. Reaction of [Pt(trans-stilbene)$(PEt_3)_2$] with $C_6(CF_3)_6$ affords the complex [Pt{$C_6(CF_3)_6$}$(PEt_3)_2$], shown by variable temperature ^{19}F nmr to be a fluxional molecule. Displacement of stilbene with excess hexafluoro-2-butyne yields a substituted hexakis(trifluoromethyl)platinacyclohepta-cis, trans, cis - triene[237]. Treating terminal acetylenes with zerovalent Pd$(PPh_3)_n$ (n=3,4) gave air stable colorless trans -$(R_3P_2Pd(C:CR)_2$[238]. Ligand substitution of 1-(chloromethyl) naphthalene for stilbene in Pt(OL)$(PPh_3)_2$ resulted in the chloroaryl adduct, which when treated with MeLi followed by refluxing in toluene yielded the metallated dihapto [(8-methylene)-1-naphthyl]bis(triphenylphosphine) platinum(II)][239]. Internal metallation reactions involving [MX_2LEMe_2(1-naphthyl)$_2$] were studied. When M=Pt; X = Cl, I, there was no tendency to undergo metallation; however, when X=Me, E=As,P, the napthyl group was metallated in the 8-position. The Pd complexes remained unchanged or decomposed under the same reaction conditions[240]. Condensation of hydrazine Pd complexes with carbonyl compounds R'R"CO (R'=H,Me; R"=Ph) yielded cyclopalladated dimers which could by converted to the monomeric species with R_4NX[241-2]. Metallation via the resonance stabilized ylide, 1-benzoylimino pyridinum betaine, occurs with Pd or Pt salts[243].

The metallocyclobutane complexes of platinum and palladium have been prepared from the reaction of 1,1,2,2-tetracyanocyclopropane (and its 3-alkyl derivatives) with ML_n (n = 3,4; L = Tertiary phosphines/arsine). The structure determination showed that Pt had inserted into the C_1C_2 bond of the ring. Ligand exchange reactions were followed for the neutral L's[244-5]. A similar reaction has been described between tetracyanooxirane and PtL_4 to form metallooxacyclobutane[246].

Heterometallic Complexes: Synthesis of a stable metallabicycle (XII) has been achieved from thereaction of di(2-pyridyl)acetylene with bis or tetrakis triphenyl phosphine platinum(o) followed by treating the complex with anhydrous $CoCl_2$[247].

Bis(benzonitrile) palladium chloride reacts with ferrocene and vinyl ferrocene in nonpolar solvents at room temperature to form $[C_5H_5FeC_5H_5]^+PdCl_3^-$ and $[C_5H_5FeC_5H_4CH:CH_2PdCl_2]_2$.[248] Heteronuclear triangular cluster compounds containing two cobalt atoms bonded cis to the platinum were formed from the reaction of 2 tetracarbonyl cobaltate ions with $PtCl_2(XPh_2)_2(CH_2)_2$ (X=As,P) in purified THF.[249] The reaction between equimolar quantities of $PtCl_2$ $\{(Ph_2P)_2CH_2\}$and $Hg(SiMe_3)_2$ in warm benzene produced [Pt$(SiMe_3)_2$-$\{(Ph_2P)CH_2\}$($HgSiMe_3$)]. A hexagonal structure was assigned on the bases of ^1H nmr spectroscopy.[250] The crystal structures of $FePt_2(CO)_5$ $[P(OPh)_3]_3$ and $Fe_2Pt(CO)_9(PPh_3)$ indicate a closo-arrangement.[251-2]

Literature Cited

1. R. F. Heck, Organotransition Metal Chemistry, Acad. Press N.Y., N.Y. (1974).
2. U. Belluco, Organometallic and Coordination Chem. of Platinum, Acad. Press, N.Y., N.Y. (1974).
3. J. Chatt, ADV. IN ORGANOMETAL. CHEM. 12, 2 (1974).
4. M. Herberhold, Metal Pi-Complexes, Vol. 2: Complexes with Monoolefinic Ligands, Pt. 2: Specific Aspects, Elsevier Press, New York, N.Y., 508 (1974).
5. P. Uguagliati, G. Carturan, U. Belluco, CHIM. IND. (Milan) 56, 287 (1974).
6. D. Banerjea, J. INDIAN CHEM. SOC. 51, 90 (1974).
7. M. Graziani, M. Lenarda, U. Belluco, CHIM. IND. (Milan) 56, 294 (1974).
8. A. Wojcicki, ADV. IN ORGANOMETAL. CHEM. 12, 32 (1974).
9. F. Bonati and G. Minghetti, INORG. CHIM. ACTA 9, 95 (1974).
10. M. J. Cleare, PLATINUM MET. REV. 18, 122 (1974).
11. V. Bonora, M. Jawork, M. F. Pilbrow, J. CHEM. SOC., CHEM. COMMUN. 15, 616 (1974).
12. E. Alter and R. Hoppe, Z. ANORG. ALLG. CHEM. 408, 115 (1974).
13. R. M. Barr, M. Goldstein, D. N. G. Hairs, M. McPartin, A. J. Markwell, J. CHEM. SOC. CHEM. COMMUN. 221 (1974).
14. K. A. Bolshakov, N. M. Sinitsyn, V. F. Borbat, L. I. Selina, M. V. Rubtsov, ZH. NEORG. KHIM. 19, 122 (1974).
15. D. Craciunescu, C. Ghirvu, A. Doadrio, Y. Shohet, REV. CHIM. MINER. 11, 85 (1974).
16. D. C. Harris and H. B. Gray, INORG. CHEM. 13, 2250(1974).
17. J. C. Thibeault, R. F. Ziolo, H. B. Gray, CRYST. STRUCT. COMMUN. 3, 473 (1974).
18. A. G. Samatov, N. N. Zheligovskaya, I. N. Odin, V. I. Spitsyn, IZV. AKAD. NAUK SSSR, SER. KHIM. 9, 2142(1974).
19. N. N. Zheligovskaya, S. V. Loginov, L. V. Popov, V. I. Spitsyn, IZV. AKAD. NAUK SSSR, SER. KHIM. 6, 1411(1974).
20. M. I. Gelfman, ZH. NEORG. KHIM. 19, 2826 (1974).
21. A. I. Stetsenko, B. S. Lipner, ZH. OBSHCH. KHIM. 44, 2289 (1974).
22. Y. N. Kukushkin, G. S. Krylova, S. I. Bakhireva, ZH. NEORG. KHIM. 19, 1694 (1974).
23. Y. N. Kukushkin, N. P. Kiseleva, S. I. Bakhireva, M. A. Kirillova, ZH. OBSHCH. KHIM. 44, 1741 (1974).
24. N. Wilhelm, R. Hoppe, Z. ANORG. ALLG. CHEM. 409, 60 (1974).
25. H. Sabrousky, W. Bronger, D. Shmitz, Z. NATURFORSCH TEIL B. 24, 10 (1974).
26. V. B. Lazarev, I. S. Shaplygin, ZH. NEORG. KHIM. 19, 2388 (1974).
27. J. Ostorero, H. Makram, J. Loriers, C. R. ACAD. SCI., SER. C 278, 737 (1974).

28. B. N. Ivanov-Emin, L. D. Borzoza, D. Sudzhben, N. N. Ivanova, A. I. Ezhov, ZH. NEORG. KHIM. 19, 1880 (1974).
29. I. F. Markova, N. N. Zheligovskaya, V. I. Spitsyn, IZV. AKAD. NAUK SSSR, SER. KHIM. 2, 271 (1974).
30. L. L. Benezra, GER. OFFEN. 2407096, 29 August 1974.
31. A. Kolaezkowski, J. Andrzej, J. Szustakowski, CHEM. STOSOW, 18, 3 (1974).
32. K. Maeda, I. Moritani, T. Hosokawa, S. Murahashi, TETRAHEDRON LETT. 10 797 (1974).
33. S. I. Ginzburg, M. I. Yuz'ko, Y. Y. Kharitonov, ZH. NEORG. KHIM. 19, 464 (1974).
34. D. M. Blake, S. Shields, L. Wyman, INORG. CHEM. 13, 1595 (1974).
35. S. Migamoto, P. A. JAPAN, 73, 38688.
36. M. Horike, Y. Kai, N. Yasuoka , N. Kasai, J. ORGANOMETAL. CHEM. 72, 441 (1974).
37. B. E. Reichert and B. O. West, J. ORGANOMETAL. CHEM. 71, 291 (1974).
38. J. R. Hall and G. A. Swile, J. ORGANOMETAL. CHEM. 67, 455 (1974).
39. C. Oldham and A. P. Ketteringham, INORG. CHIM. ACTA 9, 127 (1974).
40. B. C. Sharma, K. S. Bose, C. C. Patel, SYN. REACTIV. INORG. METAL-ORG. CHEM. 4, 61 (1974).
41. M. Kurahashi, BULL. CHEM. SOC. JPN. 47, 2045 (1974).
42. M. Kurahashi, CHEM. LETT. 1, 63 (1974).
43. S. K. Srivastava, K. B. Pandeya, H. L. Nigam, INDIAN J. CHEM. 12, 530 (1974).
44. M. Chandrasekharan, M. R. Udupa, G. Aravamudan, J. INORG. NUCL. CHEM. 36, 1153 (1974).
45. M. Chandrasekharan, M. R. Udupa, G. Aravamudan, J. INORG. NUCL. CHEM. 36, 1417 (1974).
46. F. Hogue, H. Frye, INORG. NUCL. CHEM. LETT. 10, 505 (1974).
47. L. M. Volshtein and O. P. Slyudkin, ZH. NEORG. KHIM. 19 131 (1974).
48. W. Jeitschko, ACTA CRYSTALLOGR., SECT. B, B30, 2565, (1974).
49. Y. N. Kukuskin, S, Simanova, V. K. Krylov, L. G. Melgunova, ZH. NEORG. KHIM. 18, 2569 (1973).
50. Y. N. Kukushkin, A. I. Stetsenko, V. G. Duibanova, S. G. Strelin, ZH. NEORG. KHIM. 19, 2803 (1974).
51. R. J. Cross, T. H. Green, R. Keat, J. CHEM. SOC., CHEM. COMMUN. 207 (1974).
52. K. Kawakami, Y. Ozaki, T. Tanaka, J. ORGANOMETAL. CHEM. 69, 151 (1974).
53. W. H. Baddley and D. S. Hamilton, INORG. NUCL. CHEM. LETT. 10, 143 (1974).
54. K. R. Dixon, K. C. Moss, M. A. R. Smith, J. CHEM. SOC., DALTON TRANS. 971 (1974).

55. P. L. Goggin, R. J. Goodfellow, F. J. S. Reed, J. CHEM. SOC., DALTON TRANS., 576 (1974).
56. Y. N. Kukushkin, E. A. Andronov, G. P. Gur'Yanova, V. P. Yustratov, ZH. NEORG. KHIM. 19, 146 (1974).
57. Y. N. Kukushkin, G. S. Yatskevich, K. M. Trusova, ZH. NEORG. KHIM. 19, 128 (1974).
58. A. T. Pilipenko, O. P. Ryabushdo, L. A. Krivokhizhina, N. L. E chenko, UKR. KHIM. ZH. (Russ. Ed.) 40, 73 (1974).
59. Y. N. Kukushkin, E. A. Andronov, V. P. Yustratov, M. A. Kirillova, ZH. OBSHCH. KHIM. 44, 1905 (1974).
60. C. Preti, G. Tosi, D. DeFilippo, G. Verani, INORG. NUCL. CHEM. LETT. 20, 541 (1974).
61. S. O. Grim, L. Matienzo, W. E. Swartz, Jr., INORG. CHEM. 13, 447 (1974).
62. N. Sonoda, S. Araki, T. Onishi, T. Tanaka, J. INORG. NUCL. CHEM. 36, 1985 (1974).
63. C. G. Sceney, R. J. Magee, INORG. NUCL. CHEM. LETT. 10, 323 (1974).
64. F. L. Bowden, R. Giles, R. N. Haszeldine, J. CHEM. SOC., CHEM. COMMUN. 15, 578 (1974).
65. M. C. Cornock, D. F. Steele, T. A. Stephenson, INORG. NUCL. CHEM. LETT. 10, 785 (1974).
66. C. G. Sceney and R. J. Magee, CHEM. SCR. 6, 47 (1974).
67. R. N. Mukherjee and M. D. Zingde, INDIAN J. CHEM. 12, 848 (1974).
68. W. Kuchen, J. Delventhal, H. Keek, CHEM. BER. 107, 2938 (1974).
69. M. Das and S. E. Livingstone, AUST. J. CHEM. 27, 2109 (1974).
70. T. Honjo, CHEM. LETT. 481 (1974).
71. G. C. Pellacani, CAN. J. CHEM. 52, 3454 (1974).
72. D. V. Naik, G. J. Palenik, S. Jacobson, A. J. Carty, J. AMER. CHEM. SOC. 96, 2286 (1974).
73. Y. S. Wong, S. Jacobson, P. C. Chieh, A. J. Carty, INORG. CHEM. 13, 284 (1974).
74. S. Jacobson, Y. S. Wong, P. C. Chieh, A. J. Carty, J. CHEM. SOC., CHEM. COMMUN. 13, 520 (1974).
75. G. J. Palenik, W. L. Steffen, M. Mathew, M. Li, D. W. Meek, INORG. NUCL. CHEM. LETT. 10, 125 (1974).
76. J. L. Burmeister, R. L. Hassel, K. A. Johnson, J. C. Lim, INORG. CHIM. ACTA 9, 23 (1974).
77. W. Levason and C. A. McAuliffe, J. CHEM. SOC., DALTON TRANS., 2238 (1974).
78. M. M. Khan, J. INORG. NUCL. CHEM. 36, 299 (1974).
79. A. D. Trottskaya, G. A. Levshina, I. N. Andreev, K. D. Levshin, E. V. Chistyakov, ZH. OBSHCH. KHIM. 44, 1835 (1974).
80. Y. U. Kukushkin, V. V. Sibirskaya, T. K. Mikhal'Chenko, L. V. Konovalov, U. V. Fadeev, O. I. Arkhangel'Skaya, ZH. OBSHCH. KHIM. 44, 1912 (1974).

81. R. Louis, D. Pelissard, R. Weiss, ACTA CRYSTALLOGR., SECT. B B30, 1889 (1974).
82. B. Metz, D. Moras, R. Weiss, J. INORG. NUCL. CHEM. 36, 785 (1974).
83. R. Louis, J. C. Thierry, R. Weiss, ACTA CRYSTALLOGR., Sect. B, 30, 753 (1974).
84. C. A. Tolman, D. H. Gerlach, J. P. Jesson, R. A. Schunn, J. ORGANOMETAL. CHEM. 65, C23 (1974).
85. J. M. Epstein, A. H. White, S. B. Wild, A. C. Willis, J. CHEM. SOC., DALTON, 4, 436 (1974).
86. T. S. Khodashova, V. S. Sergienko, A. I. Stetsenko, M. A. Porai-Koshits, L. A. Butman, ZN. STRUKT. KHIM. 15, 471 (1974).
87. M. Calligaris, T. Yoshida, S. Otsuka, INORG. CHIM. ACTA 11, L15 (1974).
88. G. V. Derbisher, O. N. Evstaf'eva, A. V. Babaeva, ZH. NEORG. KHIM. 19, 1433 (1974).
89. L. K. Shubochkin, E. F. Shubochkina, O. V. Popov, M. A. Golubnichaya, L. O. Sorokina, IZV. SIB. OTD. AKAD. NAUK SSSR, SER. KHIM. NAUK 2, 24 (1974).
90. R. J. Magee, W. Mazurek, M. J. O'Connor, A. T. Phillip, AUST. J. CHEM. 27, 1885 (1974).
91. L. Toniolo, J. A. McGinnety, T. Boschi, G. Deganello, INORG. CHIM, ACTA 11, 143 (1974).
92. A. W. B. Garner and M. J. Mays, J. ORGANOMETAL. CHEM. 67, 153 (1974).
93. W. E. Carroll, M. E. Deane, F. J. Lalor, J. CHEM. SOC., DALTON TRANS. 16, 1837 (1974).
94. S. D. Ittel and J. A. Ibers, J. AMER. CHEM. SOC. 96, 4804 (1974).
95. R. J. Cross and N. H. Tennent, J. ORGANOMETAL. CHEM. 72, 21 (1974).
96. P. Braunstein, J. Dehand, M. Pfeffer, J. ORGANOMETAL. CHEM. 76, C35 (1974).
97. G. Annibale, L. Caltalini, L. Maresca, G. Michelm, G. Natile, INORG. CHIM. ACTA 10, 211 (1974).
98. S. A. D'Yachenko, B. L. Moldaver, V. G. Yakutovich, I. A. Nagibina, ZH. OBSHCH. KHIM. 44, 2280 (1974).
99. K. R. Laing, S. D. Robinson, M. F. Uttley, J. CHEM. SOC. DALTON TRANS. 1205 (1974).
100. J. Kuyper and K. Vrieze, J. ORGANOMETAL. CHEM. 74, 289 (1974).
101. G. Schmid and L. Weber, CHEM. BER. 107, 547 (1974).
102. D. G. Holah, A. N. Hughes, B. C. Hui, K. Wright, CAN. J. CHEM. 52, 2990 (1974).
103. H. Yamamoto and K. Suzuki, J. ORGANOMETAL. CHEM. 71, C38 (1974).
104. K. Suzuki, H. Yamamoto, S. Kanie, J. ORGANOMETAL. CHEM. 73, 131 (1974).
105. S. Paraskewas, SYNTHESIS 8, 574 (1974).

106. R. Ros, J. Renaud, R. Roulet, J. ORGANOMET. CHEM. 77, C4 (1974).
107. S. Hasegawa, K. Itoh, Y. Ishii, INORG. CHEM. 13, 2675 (1974).
108. W. Beck, W. Rieber, S, Cenini, F. Porta, G. LaMonica, J. CHEM. SOC., DALTON TRANS. 3, 298 (1974).
109. B. T. Heaton and D. J. A. McCaffrey, J. ORGANOMETAL. CHEM. 70, 455 (1974).
110. S. P. Ghosh, P. Bhattacharjee, L. K. Mishra, J. INDIAN CHEM. SOC. 51, 308 (1974).
111. E. Uhlig and S. Keiser, Z. ANDRG. ALLG. CHEM. 406, 1 (1974).
112. G. W. Bushnell, K. Dixon, M. A. Khan, CAN. J. CHEM. 52, Pt. 1, 1367 (1974).
113. H. A. Tayim, A. H. Malakian, A. B. Bikhazi, J. PHARM. SCI. 63, 1469 (1974).
114. M. Nagoya, K. Sato, K. Yamasaki, NIPPON KAGAKU KAISHI, 696 (1974).
115. I. A. Roos, A. J. Thomson, S. Mansy, J. AM. CHEM. SOC. 96, 6484 (1974).
116. I. A. Roos, A. J. Thomson, J. Eagles, CHEM.-BIOL. INTERACTIONS 8, 421 (1974).
117. H. I. Heitner and S. J. Lippard, INORG. CHEM. 13, 815 (1974).
118. J. E. House, Jr. and P. S. Lau, J. INORG. NUCL. CHEM. 36, 223 (1974).
119. B. Singh, R. Singh, K. P. Thakur, M. Chandra, INDIAN J. CHEM. 12, 631 (1974).
120. D. A. Redfield, J. H. Nelson, R. A. Henry, D. W. Moore, H. B. Jonassen, J. AM. CHEM. SOC. 96, 6298 (1974).
121. K. P. Beamont and C. A. McAuliffe, INORG. CHIM. ACTA 8, 105 (1974).
122. L. N. Essen, INORG. SYNTH. 15, 93 (1974).
123. P. D. Braddock, R. Romeo, M. L. Tobe, INORG. CHEM. 13, 1170 (1974).
124. R. Romeo and M. L. Tobe, INORG. CHEM. 13, 1991 (1974).
125. A. V. Belyaev, S. Y. Dvurechenskaya, B. I. Peshchevitskii, ISV. SIB. OTD. AKAD. NAUK SSSR, SER. KHIM. NAUK. 5, 54 (1974).
126. G. Faraone, V. Ricevuto, R. Romeo, M. Trozzi, J. CHEM. SOC.,DALTON TRANS. 1377 (1974).
127. V. Ricevuto, R. Tomeo, M. Trozzi, J. CHEM. SOC., DALTON TRANS. 927 (1974).
128. R. D. Gillard, J. CHEM. SOC., DALTON TRANS. 2320 (1974).
129. R. J. Cross and N. H. Tennent, J. CHEM. SOC., 1444 (1974).
130. F. Pesa and M. Orchin, J. ORGANOMET. CHEM. 78, C26 (1974).
131. Y. N. Kukushkin, I. V. Pakhomova, E. N. Kalyukova, ZH. OBSHCH. KHIM. 44, 16 (1974).

132. U. N. Kukushkin, V. F. Budanova, V. F. Chvala, ZH. NEORG. KHIM. 19, 135 (1974).
133. N. N. Zheligovskaya, A. G. Samatov, I. F. Markova, R. G. Kileev, VESTN. MOSK. UNIV., KHIM. 15, 174 (1974).
134. A. G. Samatov, N. N. Zheligovskaya, V. I. Spitsyn, IZV. AKAD. NAUK SSSR, SER. KHIM., 274 (1974); ibid 1467 (1974).
135. F. Sartori, L. Leoni, R. Lazzaroni, P. Salvadori, J. CHEM. SOC., CHEM. COMMUN. 322 (1974).
136. L. M. Volshtein, L. D. Dikanskaya, ZH. NEORG. KHIM. 19, 150 (1974).
137. N. N. Zheligovskaya, N. Kamalov, V. I. Spitsyn, VESTN. MOSK. UNIV., KHIM. 15, 102 (1974).
138. H. D. Shelton, R. Desiderato, A. A. Syamal, CRYST. STRUCT. COMMUN. 3, 43 (1974).
139. O. N. Andrianova, I. F. Golovaneva, IZV. AKAD. NAUK SSSR, SER. KHIM. 533 (1974).
140. G. A. Shagisultanova, A. V. Loginov, S. V. Krupitskii, KHIM. V.Y.S. ENERG. 8, 467 (1974).
141. D. J. A. De Waal, W. Robb, INT. J. CHEM. KINET. 6, 309 (1974); ibid 323 (1974).
142. H. Manohar, D. Schwarzenbach, HELV. CHIM. ACTA 57, 519 (1974)
143. B. Deyris, J. Roy-Montreuil, A. Rouault, A. Krumbuegel-Nylund, J. P. Senateur, R. Fruchart, A. Michel, C. R. ACAD. SCI., SER. C 278, 237 (1974).
144. M. Matsumoto, H. Yoshioka, K. Nakatsu, T. Yoshida, S. Otsuka, J. AMER. CHEM. SOC. 96, 3322 (1974).
145. A. Immirzi, A. Musco, J. CHEM. SOC., CHEM. COMMUN. 400 (1974).
146. K. R. Dixon, K. C. Moss, A. R. M. Smith, CAN. J. CHEM. 52, 692 (1974).
147. N. M. N. Gowda, G. K. N. Reddy, AUST. J. CHEM. 27, 2557 (1974).
148. D. Negoiu, L. Paruta, REV. ROUM. CHIM. 19, 57 (1974).
149. E. Shewchuk, S. B. Wild, J. ORGANOMETAL. CHEM. 71, C1 (1974).
150. N. J. Destefano, D. K. Johnson, L. M. Venanzi, ANGEW. CHEM. 86, 133 (1974).
151. C. E. Jones, B. L. Shaw, B. L. Turtle, J. CHEM. SOC., DALTON TRANS. 992 (1974).
152. R. B. King, O. Von Stetten, INORG. CHEM. 13, 2449 (1974).
153. K. M. Chui, H. M. Powell, J. CHEM. SOC., DALTON TRANS. 1879 (1974).
154. K. M. Chui, M. H. Powell, J. CHEM. SOC., DALTON TRANS. 2117 (1974).
155. D. A. Couch, S. D. Robinson, INORG. CHEM. 13, 456 (1974).

156. P. Meakin, J. P. Jesson, J. AM. CHEM. SOC. 96, 5751 (1974).
157. J. P. Jesson, P. Meakin, J. AM. CHEM. SOC. 96, 5760 (1974).
158. J. C. Calabrese, L. R. Dahl, P. Chini, G. Longoni, S. Martinengo, J. AMER. CHEM. SOC. 96 2614 (1974).
159. P. L. Goggin, J. Mink, J. CHEM. SOC., DALTON TRANS. 534 (1974).
160. H. C. Clark, K. Von Werner, SYNTH. REACT. INORG. MET.-ORG. CHEM. 4, 355 (1974).
161. P. L. Bellon, M. Manassero, F. Porta, M. Sansoni, J. ORGANOMET. CHEM. 80, 139 (1974).
162. M. E. Jason, J. A. McGinnety, K. B. Wiberg, J. AM. CHEM. SOC. 96, 6531 (1974).
163. J. Hillis, J. Francis, M. Ori, M. Tsutsui, J. AMER. CHEM. SOC. 96, 4800 (1974).
164. F. R. Hartley, G. W. Searle, J. ORGANOMETAL. CHEM. 69, 315 (1974).
165. C. Burgess, F. R. Hartley, G. W. Searle, J. ORGANOMET. CHEM. 76, 247 (1974).
166. B. Akermark, J. E. Backvall, K. Siiralahansen, K. Sjoberg, K. Zetterberg, TETRAHEDRON LETT. 1363 (1974).
167. C. Agami, J. Levisalles, F. Rose-Munch, J. ORGANOMETAL. CHEM. 65, 401 (1974).
168. J. Bordner, D. W. Wertz, INORG. CHEM. 13, 1639 (1974).
169. J. Ashley-Smith, Z. Souek, B. F. G. Johnson, J. Lewis, J. CHEM. SOC., DALTON TRANS. 128 (1974).
170. K. Okamoto, Y. Kai, N. Yasuoka, N. Kasai, J. ORGANOMETAL. CHEM. 65, 427 (1974).
171. I. A. Zakharova, L. A. Leites, V. T. Aleksanyan, J. ORGANOMETAL. CHEM. 72, 283 (1974).
172. R. Roulet, R. Vouillamoz, HELVETICA CHIMICA ACTA 57, 2139 (1974).
173. G. Wiger, G. Albelo, M. F. Rettig, J. CHEM. SOC., DALTON TRANS. 2242 (1974).
174. W. T. Wipke, G. L. Goeke, J. AM. CHEM. SOC. 96, 9 (1974).
175. A. Kasahara, T. Izumi, K. Endo, T. Takeda, M. Ookita, BULL. CHEM. SOC. JPN. 47, 1967 (1974).
176. G. Parker, A. Salzer, H. Werner, J. ORGANOMETAL. CHEM. 67, 131 (1974).
177. Y. Ishii, ANN. N. Y. ACAD. SCI. 239, 115 (1974).
178. S. J. Ashcroft, A. Maddock, G. Beech, J. CHEM. SOC., DALTON TRANS. 462 (1974).
179. T. Ukai, H. Kawazura, Y. Ishii, J. J. Bonnet, J. A. Ibers, J. ORGANOMETAL. CHEM. 65, 253 (1974).
180. C. G. Pierpont, M. C. Mazza, INORG. CHEM. 13, 1891 (1974).

181. T. Ito, S. Hasegawa, Y. Takahashi, Y. Takahashi, Y. Ishii, J. ORGANOMETAL. CHEM. 73, 401 (1974).

182. Y. Ishii, S. Hasegawa, S. Kimura, K. Itoh, J. ORGANO-METAL. CHEM. 73, 411 (1974).

183. W. J. Cherwinski, B. F. G. Johnson, J. Lewis, J. CHEM. SOC., DALTON TRANS. 1405 (1974).

184. K. Mosley, P. M. Maitlis, J. CHEM. SOC., DALTON TRANS. 169 (1974).

185. J. Lukas, INORG. SYNTH. 15, 75 (1974).

186. B. M. Trost, P. E. Strege, TETRAHEDRON LETT. 2603 (1974).

187. B. M. Trost, T. J. Dietsche, T. J. Fullerton, J. ORG. CHEM. 39, 737 (1974).

188. H. Kurosawa, R. Okawara, J. ORGANOMETAL. CHEM. 71, C35 (1974).

189. V. A. Golodov, E. O. Fisher, U.S.S.R. 415272, 15 Feb. (1974).

190. K. Yamamoto, K. Shinohara, T. Ohuchi, M. Kumada, TETRAHEDRON LETT. 1153 (1974).

191. H. C. Clark, L. E. Manzer, INORG. CHEM. 13, 1291 (1974).

192. B. W. Davies, N. C. Payne, INORG. CHEM. 13, 1843 (1974).

193. H. C. Clark, L. E. Manzer, INORG. CHEM. 13, 1996 (1974).

194. B. W. Davies, N. C. Payne, INORG. CHEM. 13, 1848 (1974).

195. S. Jacobson, A. J. Carty, M. Mathew, G. J. Palenik, J. AM. CHEM. SOC. 96, 4330 (1974).

196. D. V. Naik, G. J. Palenik, S. Jacobson, A. J. Carty, J. AMER. CHEM. SOC. 96, 2286 (1974).

197. P. Kong, D. Roundhill, J. CHEM. SOC., DALTON TRANS. 187 (1974).

198. F. Bonati, G. Minghetti, INORG. CHIM. ACTA 9, 95 (1974).

199. A. L. Balch, J. E. Parks, J. AMER. CHEM. SOC. 96, 4114, (1974).

200. B. Crociani, R. L. Richards, J. CHEM. SOC., DALTON TRANS. 693 (1974).

201. C. H. Davies, C. H. Game, M. Green, F. G. A. Stone, J. CHEM. SOC., DALTON TRANS. 357 (1974).

202. L. E. Manzer, J. CHEM. SOC., DALTON TRANS. 1535 (1974).

203. R. Zanella, T. Boschi, B. Crociani, U. Belluco, J. ORGANOMETAL. CHEM. 71, 135 (1974).

204. K. Bartel, W. P. Fehlhammer, ANGEW. CHEM. 86, 588 (1974).

205. M. H. Chisholm, D. A. Couch, J. CHEM. SOC., CHEM. COMMUN. 42 (1974).

206. R. Zanella, G. Carturan, M. Graziani, U. Belluco, J. ORGANOMETAL. CHEM. 65, 417 (1974).

207. A. Modinos, P. Woodward, J. CHEM. SOC., DALTON TRANS. 2065 (1974).

208. L. Muir, K. Muir, J. CHEM. SOC., DALTON TRANS. 2427 (1974).

209. B. Cetinkaya, P. Dixneuf, M. F. Lappert, J. CHEM. SOC., DALTON TRANS. 1827 (1974).
210. P. J. Fraser, W. R. Roper, F. G. Stone, J. CHEM. SOC., DALTON TRANS. 102 (1974).
211. K. Itoh, H. Nishiyama, T. Ohnishi, Y. Ishii, J. ORGANOMETAL. CHEM. 76, 401 (1974).
212. P. Bravo, G. Fronza, C. Ticozzi, G. Gaudiano, J. ORGANOMETAL. CHEM. 74, 143 (1974).
213. H. Koezuka, G. Matsubayashi, T. Tanaka, INORG. CHEM. 13, 443 (1974).
214. K. J. Klabunde, J. Y. F. Low, J. AM. CHEM. SOC. 96, 7674 (1974).
215. K. J. Kalbunde, J. Y. F. Low, H. F. Efner, J. AMER. CHEM. SOC. 96, 1984 (1974).
216. J. R. Hall, G. A. Swile, J. ORGANOMET. CHEM. 76, 257 (1974).
217. T. G. Appleton, H. C. Clark, L. E. Manzer, J. ORGANO-METAL. CHEM. 65, 275 (1974).
218. R. G. Pearson, W. Louw, J. Rajaram, INORG. CHIM. ACTA 9, 251 (1974).
219. A. V. Kramer, J. A. Osborn, J. AM. CHEM. SOC. 96, 7832 (1974).
220. P. K. Wong, K. S. Y. Lau, J. K. Stille, J. AM. CHEM. SOC. 96, 5956 (1974).
221. K. S. Y. Lau, R. W. Fries, J. K. Stille, J. AMER. CHEM. SOC. 96, 4983 (1974).
222. A. V. Kramer, J. A. Osborn, J. AM. CHEM. SOC. 96, 7832 (1974).
223. F. Glockling, T. McBride, R. J. I. Pollock, INORG. CHIM. ACTA 8, 77 (1974).
224. V. S. Petrosyan, A. B. Permin, S. G. Sakharov, O. A. Reutov, J. ORGANOMETAL. CHEM. 65, C7 (1974).
225. V. I. Sokolov, V. V. Bashilov, L. M. Anishchenko, O. A. Reutov, J. ORGANOMETAL. CHEM. 71, C41 (1974).
226. C. J. Wilson, M. Green, R. J. Mawby, J. CHEM. SOC., DALTON TRANS. 421 (1974).
227. C. J. Wilson, M. Green, R. J. Mawby, J. CHEM. SOC., DALTON TRANS. 1293 (1974).
228. B. F. G. Johnson, J. Lewis, J. D. Jones, K. A. Taylor, J. CHEM. SOC., DALTON TRANS. 34 (1974).
229. G. Oehme, H. Pracejus, Z. CHEM. 14, 24 (1974).
230. J. Rajaram, R. G. Pearson, J. A. Ibers, J. AMER. CHEM. SOC. 96, 2103 (1974).
231. G. W. Parshall, J. AMER. CHEM. SOC. 96, 2360 (1974).
232. R. Uson, P. Royo, J. Gimeno, J. ORGANOMETAL. CHEM. 72, 229 (1974).
233. R. Uson, P. Royo, J. Fornies, SYN. REACTIV. INORG. METAL-ORG. CHEM. 4, 157 (1974).
234. N. Chaudhury, M. G. Kekre, R. J. Puddephatt, J. ORGANO-METAL. CHEM. 73, C17 (1974).

235. T. G. Appleton, M. H. Chisholm, H. C. Clark, K. Yasufuku, J. AM. CHEM. SOC. 96, 6600 (1974).
236. P. K. Maples, M. Green, F. G. A. Stone, J. L. Spencer, J. CHEM. SOC., DALTON TRANS. 1194 (1974).
237. J. Browning, M. Green, J. L. Spencer, F. G. A. Stone, J. CHEM. SOC., DALTON TRANS. 97 (1974).
238. J. H. Nelson, A. W. Verstuyft, J. D. Kelly, H. B. Jonassen, INORG. CHEM. 13, 27 (1974).
239. J. A. Duff, B. L. Shaw, B. L. Turtle, J. ORGANOMETAL. CHEM. 66, C18 (1974).
240. J. M. Duff, B. E. Mann, B. L. Shaw, B. Turtle, J. CHEM. SOC., DALTON TRANS. 139 (1974).
241. P. Braunstein, J. Dehand, M. Pfeffer, INORG. NUCL. CHEM. LETT. 10, 521 (1974).
242. P. Braunstein, J. Dehand, M. Pfeffer, INORG. NUCL. CHEM. LETT. 20, 581 (1974).
243. S. A. Dias, A. W. Downs, W. R. McWhinnie, INORG. NUCL. CHEM. LETT. 10, 233 (1974).
244. M. Lenarda, R. Ros, M. Graziani, V. Belluco, J. ORGANOMETAL. CHEM. 65, 407 (1974).
245. D. J. Yarrow, J. A. Ibers, M. Lenarda, M. Graziani, J. ORGANOMETAL. CHEM. 70, 133 (1974).
246. R. Schlodder, J. A. Ibers, M. Lenarda, M. Graziani, J. AM. CHEM. SOC. 96, 6893 (1974).
247. G. R. Newkome, G. L. McClure, J. AMER. CHEM. SOC. 96, 617 (1974).
248. M. S. Nametkin, S. P. Gubin, I. K. Dobrov, V. D. Tgurin, A. V. Nefed'ev, R. A. Stukan, IZV. AKAD. NAUK SSSR, SER. KHIM. 1439 (1974).
249. J. Dehand, J. F. Nennig, INORG. NUCL. CHEM. LETT. 10, 875 (1974).
250. A. Glockling, R. J. I. Pollock, J. CHEM. SOC., DALTON TRANS. 2259 (1974).
251. V. G. Albano, G. Ciani, J. ORGANOMETAL. CHEM. 66, 311 (1974).
252. R. Mason, J. A. Zubieta, J. ORGANOMETAL. CHEM. 66, 279 (1974).

COPPER

John R. Wasson

Department of Chemistry
University of Kentucky
Lexington, Kentucky 40506

GENERAL

Interactions of histidine and other imidazole derivatives with transition metal ions in chemical and biological systems have been reviewed (1) as have the oxidation states and reversible redox reactions of metalloporphyrins (2). Structural aspects and biological function of erythrocuprein (3), magnetic resonance studies of copper metalloenzymes (4), models for copper-protein interaction (5) and the synthesis and ion binding of synthetic multidentate macrocyclic compounds have been reviewed (6). The formation constants of Cu(II) complexes have been considered mechanistically (7). Factors influencing formation of mixed complexes of Cu(II) have been discussed and investigated experimentally (8).

Copper metalloboranes and metal-boron bonding (9) and pressure effects on the properties of a number of copper compounds (10) have been reviewed. The Cu(I)-S stretching frequency in $[Cu((CH_3)_3PS)Cl]_3$ at 314 cm^{-1} has been assigned (11) on the basis of Cu isotope labeling. An infrared study of isotopically enriched bistropolonatocopper(II) has led to the assignment of bands at 309 and 332 cm^{-1} to Cu-O stretching modes (12). Mass spectra of several Cu(II) Schiff base complexes have been described (13). The optical and magnetic properties of the CuO_6 chromophore in lithium phosphate glass have been reported (14) as have the NQR spectra of Cu(I) N,N-diethyl-, -dipropyl- and -dibutyl-dithiocarbamates (15). The Cu(II) catalyzed oxidation of hydroxylamine by peroxydisulfate has been studied (16). The oxidation states of copper involved in such reactions remain unresolved. Homogeneous catalysis by copper compounds and other transition metal complexes has been reviewed (17). Linear free energy relationships have been shown (18) to exist for the free energy of complex formation of d^{10} ions, e.g., Cu(I), with the free energy changes expressed as standard reduction potentials. An interesting correlation of the electronic absorption spectra of Cu(II) complexes and the "rule of average environment" has been presented (19).

An X-ray diffraction investigation of a saturated solution of $CuCl_2$ in 12.2N hydrochloric acid has been interpreted (20) in terms of $Cu_4Cl_{10}(H_2O)_8^{2-}$ as being the average solute species. The structure of K_2CuF_4 has been re-determined (21) as has that of Ba_2CuF_6 (22) and both found to be tetragonally elongated in accord with other experimental data

(23). The "10/3 Effect" in the ESR spectra of a number of Cu(II)-chloro complexes has been explored (24). The X-band ESR spectrum of $^{63}Cu^{2+}$ doped zinc Tutton's salt enriched to 16 at.% in ^{17}O has been examined (25) and a dynamic, rather than a static Jahn-Teller effect proposed for the system. The determination of structural properties of dimeric transition metal ion complexes from EPR spectra has been reviewed (26). Copper(I) complexes (27), the organic chemistry of copper (28) and oxidative coupling via organocopper compounds (29) have been reviewed.

COPPER(I) COMPOUNDS

Reaction of equimolar quantities of cyanoacetic acid and Cu(I) t-butoxide in tetrahydrofuran followed by evaporation under reduced pressure gives (30) white Cu(I) cyanoacetate. Upon heating the latter in dimethylformamide cyanomethylcopper(I) is obtained. The reaction of trifluoroacetic anhydride with Cu_2O in dry benzene at 60° for 1 hour under nitrogen in an H-tube gives crystals of tetrakis(copper(I)trifluoroacetate)dibenzene. At room temperature indene (C_9H_8) smoothly replaces benzene to to form the tetrameric compound $Cu(TFA)\cdot C_9H_8$ (31). Cu(I) acetate was prepared (32) by reducing Cu(II) acetate with copper foil in an acetonitrile-acetic acid-acetic anhydride mixture under nitrogen or in pyridine under nitrogen (33).

The reaction of copper metal with $AsCl_3$ has been shown (34) to yield CuCl, As and Cu_3As as the only products. Copper(I) perrhenate can be prepared by heating elemental copper with $Cu(ReO_4)_2$ in vacuum at 700°K or from $Cu(ReO_4)_2$ in the Knudsen cell of a mass spectrometer (35). Reaction of lithium dimethylcuprate(I) at low temperature with lithium aluminum hydride in diethyl ether yields (36) highly pyrophoric lithium dihydridocuprate(I), $LiCuH_2$. Reactions of olefins (37), alkyl halides (38) and arylthallium(III) (39) compounds with Cu(I) and -(II) thiocyanates have been described. Addition of CuI to an ethereal solution of isopropylthioallyllithium at -78° under nitrogen leads to the formation of a white insoluble polymeric species with allyl-Cu bonding. A similar complex results with t-butylthioallyllithium (40). Ethylene forms a stable 1:1 complex with Cu(I) in Y zeolite (41). Complexation with homogeneously generated CuCl as well as displacement of 1,5-cyclooctadiene from its CuCl complex lead to the preparation of $(C_6H_5)_2Si(CH{=}CH_2)_2\cdot 2CuCl$ and $(C_6H_5)Si(CH{=}CH_2)_3\cdot 2CuCl$ (42). O-ethylboranocarbonatobis(triphenylphosphine)copper(I) results from the reaction of $\{(C_6H_5)_3P\}_3CuCl$ with potassium O-ethylborancarbonate in chloroform (43). Yellow brown $t\text{-}C_4H_9OCu\cdot C{\equiv}O$ results upon passing carbon monoxide through a benzene solution of t-butoxycopper(I). Similar treatment of t-butoxycopper(I) in benzene with t-butylisocyanide yields $t\text{-}C_4H_9OCu\cdot CNC_4H_9$ (44). Treatment of alkylmercuric halides in succession with 1 equivalent of iodo(tri-n-butylphosphine)copper(I) and 3 equivalents of t-butyllithium at -78° in tetrahydrofuran results in formation of a reactive ate complex which incorporates the alkyl group originally bonded to mercury. The reactions of this alkyl group resemble those of a lithium dialkylcuprate (45). The extraction of CuCl from aqueous solution with tertiary phosphites(S) leads to formation of complexes of the types $S\cdot CuCl$ and $2S\cdot CuCl$. Tertiary phosphites complexed with Cu(I) are resistant to hydrolysis in aqueous solution (46). White $Cu_2B_{10}H_{10}$ results upon interaction of copper(II) acetate with $K_2B_{10}H_{10}$ in aqueous trifluoroacetic acid. The perdeuterated compound can also be obtained. The reaction of $Cu_2B_{10}H_{10}$ with triphenylphosphine in ethanol affords $[\{(C_6H_5)_3P\}_2Cu]_2B_{10}H_{10}$ (47). Colorless tetrafluoroboratotris(triphenylphosphine)copper(I) is prepared (48) by treating a suspension of $Cu(P\{C_6H_5\}_3)_3BH_4$ in ethanol with ethanolic tetrafluoroboric acid. Reaction of cis-$[ReCl(CO)_3(PPh_3)_2]$ and the arylacetylide $CuC_2C_6F_5$ yields (49) $[CuCl(PPh_3)]_4$, $[Re(C_2C_6H_5)(CO)_3(PPh_3)_2]$ and $[ReCu(C_2C_6F_5)_2(CO)_3(PPh_3)_2]$, the latter of which has been structurally characterized. A NH_4Cl-CuCl mixture can be used as a catalyst for the synthesis of ketazine directly from benzophenone, ammonia and oxygen (50). A mixture of $ZnCl_2$ and CuCl can also be employed for the synthesis of ketazine (51). Polymeric 2-(dimethylamino)phenylcopper, $(ArCu)_n$ (Ar = o-$(CH_3)_2NC_6H_4$), reacts with CuX (X = Cl, Br) to yield $Ar_4Cu_6X_2$ cluster compounds. $Ar_4Cu_6I_2$ can be obtained by treating the corresponding chloro compound with LiI in benzene. Lithium arylacetylides react with $Ar_4Cu_6X_2$ compounds to yield the $Ar_4Cu_6R_2$ (R = C_6H_5-C≡C, etc.) mixed organocopper cluster compounds (52). The reaction of p-nitrobenzenediazonium tetrafluoroborate with tetrakisacetonitrilecopper(I) perchlorate has been examined (53). White tetrakis(methyl diphenylphosphinite)copper(I) tetrafluoroborate has been prepared (54) by refluxing methyl diphenylphosphinite and CuCl in methanol and then adding sodium tetraphenylborate. The corresponding ethyl diphenylphosphinite compound was similarly prepared. Bistrimethylsilylamidocopper, prepared in situ from hexamethyldisilazane, n-butyllithium and CuI, couples with aryl iodides to give moderate yields of silyl-protected amines (55). $[(C_6H_5)_3P]_3Cu_2Cl_2$ is one of the products of the hydrochloric acid decomposition of $[(C_6H_5)_3P]_2Cu(BH_4)$ (56).

The compounds Cu_3NbS_3Se, $Cu_3NbS_2Se_2$, Cu_3NbSSe_3, Cu_3TaS_3Se, Cu_3TaSSe_3 and $CuTaS_2Se_2$ have been prepared (57) from stoichiometric mixtures of the elements heated to 550°C. Thiosulfate reduction of Cu(II) in the presence of potassium selenocyanate yields CuSeCN. The reaction of potassium selenocyanate with $Cu(NO_3)_2 \cdot 3H_2O$ can be employed (58) to prepare $Cu(SeCN)_2$ and $KCu(SeCN)_3$. A variety of Cu(I) trithiocarbonates, e.g., $KCuCS_3$ and Cu_2CS_3, have been prepared (59) by a number of procedures, e.g., $CsCuCS_3$ from the reaction of $CuCl_2$ with aqueous CsOH and carbon disulfide. Copper(I) halide complexes of stoichiometry CuL_nX have been prepared (60) where L is a diaryltelluride (n = 3, L = Ph_2Te, X = Cl; n = 2, L = Ph_2Te, X = Br; n = 2, L = (p-tolyl)$_2$Te, X = Cl, Br, I; n = 2, L = (p-$EtOC_6H_4$)$_2$Te, X = Cl, Br; n = 1, L = Ph_2Te or (p-$EtOC_6H_4$)$_2$Te, X = I). Stoichiometric mixtures of reactants in ethanolic or hydrohalic acid solutions were employed to obtain the compounds. Yellow cyclo-di-μ-{bis-[2-(N,N-dimethylamino)ethyl]disulfide}dicopper(I) tetrafluoroborate was synthesized (61) by direct reaction between Cu(I) and the disulfide ligand and also by disulfide ligand reduction of Cu(II) tetrafluoroborate. {Bis[2-(2-pyridyl)ethyl]disulfide}copper(I) perchlorate is prepared (62) by reaction of copper(II) perchlorate with bis[2-(2-pyridyl)ethyl]disulfide in methanol. Thiosalicylamide (TSA) interacts with Cu(I) and Cu(II) to form $Cu_2(TSA)_4Cl_2$ which can be employed for analytical determination of the metal (63). Interaction of copper(I) chloride with a hot ethanol solution of 1-phenyl-3-(2-pyridyl)-2-thiourea(L) (1:4 ligand-metal ratio) yields (64) CuL_2Cl. The Cu(II)-Cu(I)-(S)-2'-aminoethylcysteine system has been investigated but solids were not characterized (65). Colorless μ_4-ethylenethiourea-cyclo-tetrakis-μ-(ethylenethiourea)tetrakis[ethylenethioureacopper(I)] nitrate hexahydrate, $[Cu_4(ettu)_9(NO_3)_4]\cdot 6H_2O$, results upon dropwise addition of aqueous $Cu(NO_3)_2$ to hot aqueous ettu solution (66). Tetrakis(tetraphenylphosphonium)hexakis(1,2-dithiosquarato)octacuprate(I) can be obtained by adding the tetraphenylphosphonium salt of the ligand to tetrakisacetonitrilecopper(I) perchlorate in acetonitrile (67). Tris(s-dimethylthiourea)copper(I) tetrafluoroborate were readily obtained by slow evaporation of a solution of $Cu(BF_4)_2\cdot 6H_2O$ and s-dimethylthiourea. A slightly more complicated procedure was necessary for the preparation of tris(thiourea)copper(I) tetrafluoroborate (68). Triphenylarsine sulfide, ϕ_3AsS, trimethylphosphine sulfide, Me_3PS, and tris-(dimethylamino)phosphine sulfide, $(Me_2N)_3PS$, react with Cu(II) tetrafluoroborate and $[Cu(NCCH_3)_4]ClO_4$ to yield three-coordinate Cu(I) complexes, e.g., $[Cu(Me_3PS)_3]BF_4$. Related complexes can be similarly obtained in ethanol solutions. Complexes of the type $Cu(Me_3PS)X$ (X = Cl, Br, NCS) result when the parent tetrafluoroborate complex is treated with the appropriate lithium salt in ethanol (69).

COPPER(II) COMPOUNDS

Ba_2CuTeO_6 can be obtained by heating copper(II) and barium nitrates with tellurium dioxide (70). The compounds MV_2O_6 ($M = Cu^{2+}$, Ni^{2+}, etc.) are prepared (71) by high temperature-high pressure reactions of appropriate mixtures of the metal oxides. $CuAl_2O_3$ can be obtained (72) from CuO and Al_2O_3 by solid state reaction at 1000°C. Yellow green Cu(II) antimonate, $CuSb_2O_6$, is obtained (73) by heating CuO and Sb_2O_3 at 750° for 24 hours. Mixed crystals of the types $Cs_2Ba(Cu^{II}, Co^{II})(NO_2)_6$ and $Cs_2Ba(Ni^{II}, Cu^{II})(NO_2)_6$ have been prepared (74) by modification of previously described (75) methods and the semiconductor properties investigated. The zeolite, CuNaY, is obtained by treating NaY-zeolite with copper(II) nitrate solution (76). Olive green Cu(II) chelates can be prepared (77) by interaction of the metal acetate with a variety of 2-hydroxy-4-alkoxybenzophenones in 50% ethanol.

Bright green crystals of triaquanitratocaffeinecopper(II) nitrate have been prepared (78) by slow evaporation of a solution of Cu(II) nitrate and caffeine. Eleven 2-hydroxy-4 or 5-substituted acetophenonato Cu(II) complexes have been obtained (79) by interaction of aqueous Cu(II) chloride with a methanolic solution of potassium hydroxide and the appropriate ligand. The amine adducts of nitroketone ligand chelates: $[Cu(nta)_2(2\text{-Mepy})]$, $[Cu(nca)_2(2\text{-Mepy})]$, $[Cu(nap)_2(2,6\text{-Me}_2py)]$, $[Cu(nta)_2(2,6\text{-Me}_2py)]$, $[Cu(nca)_2(2,6\text{-Me}_2py)]$, $[Cu(nca)_2(quin)]$, $[Cu(na)_2(quin)]$, $[Cu(nta)_2(quin)]$, $[Cu(nca)_2(quin)]$ and $[Cu(nta)_2(4\text{-Mepy})]$ where na = nitroacetone, nap = 2-nitroacetophenone, nta = 3,3-dimethyl-1-nitrobutan-2-one, nca = 3-nitrocamphor, quin = quinuclidine, 2-Mepy = 2-methylpyridine, 4-Mepy = 4-methylpyridine and $2,6\text{-Me}_2py$ = 2,6-dimethylpyridine, can be prepared (80) by interaction of the CuO_4 chelates with the bases in acetone or hexane solutions. The bis-pyridine adduct of the

Cu(II) chelate of acetoacetylferrocene results upon evaporation of a pyridine solution of the metal chelate (81). Blue or pale green Cu(II) bis-complexes with para- and meta-substituted benzeneselenic acids precipitate immediately upon addition of the sodium salt of the acid to $CuCl_2 \cdot 2H_2O$ dissolved in the minimum quantity of water (82). Stability constants for Cu(II) complexes with oxamide-N,N'-diacetic acid were determined but solids were not characterized (83). Interaction of β-diketones with copper(II) acetate yields the appropriate β-diketonate chelate (84). The distributions of complexes in Cu(II)-methoxyacetate, -ethoxydiacetate, and -thiodiacetate systems have been re-assessed (85). Nitrilotriacetate (NTA) anion forms a variety of Cu(II) chelates, $NaCuNTA \cdot H_2O$, $H_4Cu(NTA)_2 \cdot 2H_2O$, $HCuNTA \cdot 2H_2O$ and $Cu_2Cu(NTA)_2 \cdot 2H_2O$, depending on the details of the method of preparation (86). Stability constants for Cu(II) chelates with eight N-arylhydroxamic acids in 50 vol. % dioxane-water have been reported (87) but solids were not isolated. The greenish brown Cu(II) chelate of 4-hydroxy-3-oximinomethylazobenzene by refluxing a copper(II) salt with a water-ethanol solution of the ligand (88). The extraction of aqueous Cu(II) by di(2-ethylhexyl)phosphoric acid in n-dodecane has been investigated (89) as has the stability constant of the Cu(II) chelate of 2,4-dihydroxyacetophenone oxime in various solvents (90) but solid complexes were not isolated. Extraction of Cu(II) from aqueous solution into organic phases containing long-chain carboxylic acids has also been described (91). Dihydrated complexes of 2-quinoxalinecarboxylic acid and its 1- and 4-N-oxides, $Cu(2\text{-}QxCOO)Cl \cdot 2H_2O$, $Cu(2\text{-}QxCOO\ 1\text{-}oxide)_2 \cdot 2H_2O$ and $Cu(2\text{-}QxCOO\ 4\text{-}oxide)_2 \cdot 2H_2O$, were prepared (92) by interaction of the metal chloride with the appropriate ligand in ethanol. The stability constants for Cu(II)-citrate complexes in solution have been measured (93) but solid species were not characterized. Cu(II) ion binding by polyacrylic acid has been examined potentiometrically (94) but no solids were isolated. Anhydrous $Cu(SO_3CH_3)_2$ has been prepared (95) by the reaction of methanesulfonic acid with the anhydrous chloride salt at 160-170°C under suction. Green-to-blue six-coordinate complexes $Cu(acac)_2(en)$ and $Cu(etac)_2$-(diamine) (acac = acetylacetonate; etac = ethyl acetoacetate; diamine = ethylenediamine (en), trimethylendiamine, 2,2'-bipyridine or 1,10-phenthroline) have been isolated (96) by the reactions of the respective parent β-diketonate with the corresponding diamine in dry dichloromethane. The reaction of $Cu(acac)_2$ with en in 1:2 mole ratio in dichloromethane saturated with water yields purple $Cu(acac)_2(en)_2 \cdot 2H_2O$ (96). Anhydrous bis(diethoxyphosphato)copper(II) has been prepared (97,98) by the reaction of triethylphosphate with the anhydrous metal chloride at 170-180°C. Copper(II) complexes of the type $CuLX_2$ (X = NO_3 or NO_2 and L = 1,2-dimorpholinoethane and 1,2-dipiperidinoethane) can be prepared (99) by adding the ligand in ethanol to a solution of the hydrated metal nitrate in ethanol. The nitrite complexes are obtained by adding sodium nitrite to the metal nitrate solution prior to addition of the amine. The reactions of 2-, 3- and 4-picoline N-oxides with copper(II) nitrate yield a variety of products depending on details of reaction conditions (100). Light green $[Cu(phenNO)_2](ClO_4)_2$ results upon reaction of 1,10-phenanthroline N-oxide(phenNO) in ethanol-triethyl orthoformate solution with the hydrated metal salt (101). Complexes of the type $CuL_2(NO_3)_2$ (L = 1,10-phenanthroline N-oxide and 2,2'-bipyridine N-oxide) are prepared

(102) by interaction of the hydrated metal nitrate in triethyl orthoformate-ethanol (2:3 v/v) mixture with the appropriate ligand.

A variety of Cu(II) chelates of phenyl-2-pyridyl ketone (Py-CO-C_6H_5) e.g. [Cu(Py-CO-C_6H_5)$_2$]X_2 (X = Cl, NO_3), [Cu(Py-CO-C_6H_5)(Py-C(OH)(C_6H_5)O]-NO_3·$2H_2O$ can be obtained (103) by adjusting the pH of an ethanolic solution of the ligand and the appropriate Cu(II) salt. The ligand does not form a Schiff base with ethylenediamine. However, addition of ethylenediamine to alcoholic solutions of Cu(II) salts and Py-CO-C_6H_5 yields a number of mixed complexes with exchange-coupled Cu(II) ions. Di-2-pyridylketoxime (dpkx) reacts with $CuCl_2$ in acetone to yield Cu(dpkx)Cl_2. Dissolving the latter in hot nitromethane produces Cu(dpkx)Cl_2·CH_3NO_2 (104). The stabilities of complexes of various linear phosphate ions with the $Cu(NH_3)_3^{2+}$ cation have been investigated (105) but solids were not isolated. Blue amorphous Cu[(R,R)-ebc]·H_2O and Cu[(S,S)-ebhc] ([R,R)-ebc] = 2,9-diamino-4,7-dithiadecanedioic acid and [(S,S)-ebhc] = 2,11-diamino-5,8-dithiadodecanedioic acid) have been isolated (106) subsequent to interaction of aqueous Cu(II) chloride solutions with methanolic solutions of the potassium salts of the ligands. Blue Cu[(S,S)-ebv]·$0.5H_2O$ (S,S-ebv = (S,S)-2,7-diisopropyl-3,6-diazooctanedioate) results upon addition of a methanolic solution of the acid to a methanolic solution of copper(II) chloride dihydrate (107). Bis(iminodiacetamide)copper(II) perchlorate results upon interaction of the ligand and metal perchlorate in water (108). A variety of Cu(II) complexes with potentially terdentate N-substituted picolinamides has been described (109). Stability constant data for Cu(II) chelates with diglycine-N,N-diacetic acid, triglycine-N,N-diacetic acid and tetraglycine-N,N-diacetic acid have been reported (110) but solids were not isolated. Ethyl picolinate (EP) forms complexes of the type Cu(EP)X_2 (X = Cl, ClO_4) by reaction of the appropriate metal salt with excess ligand (111). Blue crystals of Cu(II) pyridine 2,6-dicarboxylate, Cu(dipic)·$2H_2O$ separate upon cooling a mixture of copper acetate monohydrate and pyridine 2,6-dicarboxylic acid in boiling water (112). Various adducts of Cu(dipic) of the type Cu(dipic)·L and Cu(dipic)L·$2H_2O$ result upon interaction of aqueous or nonaqueous ligands, e.g., pyridine, with Cu(dipic)·$2H_2O$. The interaction of copper(II) acetate and 2,2',2"-terpyridyl in water followed by addition of sodium perchlorate yields Cu(terpy)(acetate)ClO_4·H_2O (112). Equilibrium constants have been determined (113) for parent and mixed complexes of Cu(II) with aspartic and glutamic acids and glycine but mixed complexes were not isolated. A related type of study including various diamine ligands has been described (114). Bis(isocyanurato)diamminecopper(II), Cu($C_3N_3O_3H_2$)$_2$-(NH_3)$_2$, results upon interaction of copper(II) chloride with boiling aqueous cyanuric acid followed by addition of aqueous ammonia (115). Stability constants of mixed complex species with aspartic and glutamic acids and glycine have been reported (116) but solid complexes were not isolated. Similar studies of Cu(II)-pyrimidine (uracil, thymine and cytosine) interactions have been described (117). The reactions of aroylhydrazines with Cu(II) salts have been described (118). With nitrate and sulfate salts only bis complexes were obtained. With the chloride both mono and bis complexes were obtained at low temperature while at high temperature oxidation-reduction reactions occurred with formation of complexes containing mixed oxidation states of copper. $Cu(N_2H_5)_2(SO_4)_2$ is prepared (119) by the interaction of aqueous hydrazinium sulfate and copper sulfate. Formation constants have been determined for the Cu(II) complexes formed with a styrene-vinylpicolinic acid copolymer (120). Copper(II) complexes with 6-methylpyridine-2-carboxylic, pyridine-2-carboxylic and pyridine-3-carboxylic acids have been characterized (121). Equilibrium constants have been reported for Cu(II) coordination of L,L-("pure") and D,L("mixed") diastereomers of the dipeptides Ala-Ala, Ala-Phe, Leu-Leu and Leu-Tyr but solids were not isolated (122). Blue-violet $CuL_2 \cdot 2H_2O$ (L = deprotonated 1-carbamyl-3,4,5-trimethylpyrazole) forms upon addition of aqueous ammonia to a solution of a copper(II) salt and 1-carbamyl-3,4,5-trimethylpyrazole (123). Mixed Cu(II) chelates with N,N,N',N'-tetramethylethylenediamine (TMEN) and various β-diketonate ions of the general type [Cu(TMEN)(β-dik)]$X \cdot nH_2O$ (X = ClO_4^- or NO_3^-; n = 1/2 or 0) can be prepared (124) by interaction of the β-diketonate with dimeric $[Cu_2(OH)_2(TMEN)_2]X_2$. Cu(2,2'-bipyridyl)$^{2+}$ charge transfer complexes with the purine moiety of adenosine, inosine, AMP^{2-}, IMP^{2-}, ATP^{4-} and ITP^{4-} have been characterized (125) but solids were not isolated. Pyrazine-3-carboxamide-2-carboxylic acid forms a bis-complex tetrahydrate with Cu(II). It is formed by refluxing aqueous Cu(II) chloride and pyridine-2,3-dicarboxamide (126).

N-substituted salicylaldiminates with tertiary phosphine or arsine groups in the side chain form 1:2 complexes with Cu(II) in which no Cu-P or Cu-As bonding occurs (127). When N-methylmercapto groups are attached to the side chain a similar lack of interaction of Cu with the thiomethyl groups occurs (128). A number of Schiff base complexes of Cu(II) result upon addition of Cu(II) acetate and potassium ethoxide to an ethanol solution of pyridoxal hydrochloride and the appropriate amine (129). The seven co-ordinate diaquo(2,6-diacetylpyridinebis(semicarbazone)copper(II), CuDAPSC, cation is found in Cu(DAPSC)$(NO_3)_2 \cdot 3H_2O$ which is prepared by heating a slurry of DAPSC with $Cu(NO_3)_2 \cdot 3H_2O$ in ethanol (130). The dark brown Schiff base (formed by refluxing aniline, o-toluidine, m-toluidine and p-toluidine and 2-hydroxy-5-methylacetophenone in ethanol) complexes of copper(II) with a number of bidentate ligands have been prepared (131) by refluxing the Schiff base with copper(II) acetate in ethanol for 2-3 hours. One-to-one Cu(II) Schiff base complexes derived from (a) 2,5-dihydroxypropiophenone and ethylenediamine (en), (b) 2,5-dihydroxybenzophenone and en as well as 1,2-propanediamine, and (c) 2,5-dihydroxyacetophenone and 1,2-propanediamine, prepared in the conventional manner, have been reported (132). Sky blue $CuSB \cdot H_2O$ and greenish blue Cu(SB'H)$\cdot 2H_2O$ result upon refluxing copper(II) acetate with ethanolic solutions (133) of ethyl 3-[(o-arsonophenyl)imino]butyrate(SBH_2) and N-(1-methyl-3-oxobutylidene)-o-arsanilic($SB'H_3$). Salicylaldehyde salicylhydrazone (SSH) complexes Cu(SSH-H)$_2$, Cu(SSH-2H) and Cu^I(SSH-H) can be prepared (134) by refluxing the metal chloride with an ethanolic solution of the ligand. The condensation of 2-acetylpyridine with 3,3'-iminobispropylamine yields a Schiff base ligand, L, which forms a square-pyramidal Cu(II) complex, $CuL(ClO_4)_2$, upon interaction with an ethanolic solution of the metal perchlorate (135). The metal hydroxide which precipitates in the procedure can be dissolved with a few drops of acetic acid. A variety of salicylaldehyde Schiff base complexes of Cu(II) have been prepared (136) by refluxing the appropriate substituted salicylaldehyde chelates with amines or diamines in ethanol. Green Cu(II) salicyl-N-o-aminobenzenesulfonylglycine is obtained by condensation of bis(salicylaldehydato)copper(II) with o-aminobenzenesulfonylglycine in methanol. The Schiff base complex can also be obtained by interaction of CuO and the Schiff base itself (137). Macrocyclic Schiff base chelate compounds of Cu(II) can be prepared (138) by refluxing 2,6-diacetylpyridine (or 2-acetyl-6-formylpyridine) with 3,3'-diamino-di-n-propylamine with the metal salt in 50% v/v aqueous ethanol.

Stability constants for Cu(II) interactions with asparagine and glutamine (139) and cyclopentylamine and cyclohexylamine (140) have been reported but solids were not isolated. The distribution of complexes in Cu(II)-pyrazole systems has been re-assessed (141). The macrocyclic ligand 5,7,12,14-tetraethyl-7,14-dimethyl-1,4,8,11-tetra-azacyclotetradeca-4,11-diene(L) can be isolated as a perchlorate salt which reacts with copper(II) acetate to yield $[CuL](ClO_4)_2$ (142). The perchlorate salt of copper(II) tetra(4-N-methylpyridyl)porphine can be obtained (143) by refluxing Cu(II) perchlorate with an aqueous solution of the ligand. Blue-green bis(N,N-dimethyl-1,3-propanediamine)copper(II) bromide results upon interacting the diamine and metal bromide in methanol (144). Deep violet $[CuL]ZnCl_4 \cdot H_2O$ (L = 1,4,8,11-tetraazaundecane) results upon addition of $ZnCl_2$ to a solution of $CuCl_2 \cdot 2H_2O$ and L in methanol (145). Blue-violet diperchloratotetrakis(imidazole)copper(II) can be obtained from an aqueous solution containing imidazole, sodium and copper(II) perchlorates (146). Copper(II) phthalocyanine can be obtained by heating the metal with phthalodinitrile for several hours at 260°C (147). The electrochemical reduction of Cu(II) tetrabenzo[b,f,j,n][1,5,9,13]tetraazacyclohexadecine complexes has been investigated and the synthesis of formally univalent complexes described (148). Stability constants have been determined (149) for Cu(II) chelates of racemic <u>cis</u>- and diequatorial <u>trans</u>-2,3-diamino-<u>trans</u>-decalins, <u>cis</u>- and <u>trans</u>-2,3-diaminotetralins and 3,3-dimethyl-1,2-diaminobutane but solids were not isolated. Deep blue bis(cyanotrihydroborato)-1,1,4,7,7-pentamethyldiethylenetriaminecopper(II) results (150) from the reaction of sodium cyanotrihydroborate in ethanol with a mixture of Cu(II) bromide and 1,1,4,7,7-pentomethyldiethylenetriamine in aqueous ethanol. Bispyridinebis(5,5'-diethylbarbiturato)copper(II) can be obtained (151) by mixing dilute aqueous solutions of $CuCl_2 \cdot 2H_2O$, pyridine and sodium barbital in the molar ratio 1:2:2. The mauve solution of pH 7.2 deposits the product as hexagonal plates. Deep blue bis(1,3-diaminopropane)isothiocyanatocopper(II) perchlorate is obtained (152) by adding $AgClO_4$ to a hot aqueous solution of the stoichiometric amount of the diamine-thiocyanate complex. Purple crystals of bis(5,5'-diethylbarbiturato)bis(β-picoline)copper(II) dihydrate are deposited from aqueous solutions of $CuCl_2 \cdot 2H_2O$, sodium barbital and β-picoline in the molar ratio 1:2:2 (153). Bis(5-nitrouracil)copper(II) dihydrate and bis(5-iodouracil) copper(II) dihydrate were prepared (154) by careful addition of an aqueous solution of the uracil, sodium hydroxide and copper(II) acetate. N-(2-aminoethyl)salicylamide (Hens) and N-(2-(2-aminoethyl)aminoethyl)salicylamide (Hdns) form Cu(ens)$NH_3 \cdot 2H_2O$, Cu(ens)$\cdot H_2O$, and Cu(dns)H_2O by interaction of the appropriate ligand with the metal halide in water or ammoniacal solution (155). A number of Cu(II) complexes with pentadentate linear acid amides have been obtained by mixing the metal salt, ligand and appropriate amount of potassium hydroxide in a water-ethanol mixture (156).

Bis(N-isopropyl-2-methylpropane-1,2-diamine)copper(II) nitrate dihydrate results upon dropwise addition of ethanolic $Cu(NO_3)_2 \cdot 3H_2O$ to ethanolic amine (157). A variety of five-co-ordinate (diamine)(triamine)-Cu(II) complexes, prepared by interaction of the amines with $Cu(NO_3)_2 \cdot 3H_2O$ in methanol, have been described (158). The dihydrate of the N-benzyl-DL-valine Cu(II) complex crystallizes from an aqueous methanol solution at pH 8.5 as lilac plates (159). The copper(II) derivative of the macrocyclic Schiff base derived from heptane-2,4,6-trione and 1,2-diaminoethane has been described (160). Solution equilibria involving Cu(II) and asymmetric diaminoamides have been reported (161) but no solids were isolated. Bis(1,3-propanediamine)copper(II) m-chlorobenzoate was prepared (162) by heating CuO, 1,3-propanediamine and m-chlorobenzoic acid together in an ethanol-water mixture. The reaction of copper acetate with 2-aminopyridine and phthalonitrile in refluxing methanol affords the 1,3-bis(2-pyridylimino)isoindoline complex, CuL($C_2H_3O_2$) (163). Thiocyanato, cyanato and selenocyanato complexes of Cu(II) with propylenediamine can be obtained by controlling solution concentrations of copper(II) nitrate, the potassium pseudohalide salt and propylenediamine (164). Association constants for complexes of Cu(II) and other ions with pentaammine-(pyrazine)ruthenium(II) have been determined but solids were not isolated (165). The thermodynamics of chelate formation of the Cu(II) ion with 1,48-triazaoctane was reported (166) but solids were not isolated. Complexes of the type $Cu(NCO)_2L_2$ (L = aniline and substituted-anilines) can be obtained (167) by adding a methanolic solution of the ligand to an aqueous solution containing $Cu(NO_3)_2$ and KNCO. Complexes with heterocyclic ligands can be similarly prepared. It has been found possible to dope zinc, cadmium and mercury pentacyanonitrosylferrates with Cu(II) ions in low concentrations (168). A variety of Cu(II) complexes containing one or two 2-(2'-pyridyl)benzimidazole (PBH) molecules has been reported (169). The complexes containing one molecule of ligand are obtained when a methanolic solution of Cu(II) halide and ligand solution in acetone is mixed and refluxed for a few minutes. The diisothiocyanato complex Cu(PBH)$(NCS)_2$ is obtained as a deep green precipitate when an aqueous-ethanol solution of the corresponding dichloro complex is refluxed with an excess of KNCS on a steam bath. A variety of Cu(II) complexes with 2-aminobenzimidazole and 2-aminomethylbenzimidazole have been described (170). Both ligands can be deprotonated and can form inner complexes. Reaction conditions determine the products obtained. The formation of Cu(II) species with N,N'-bis(2-picolyl)ethylenediamine and exchange with the corresponding Ni(II) chelate have been investigated (171). Bis-2-aminoethyl ether (oden) forms Cu(oden)$^{2+}$, Cu(oden)$_2^{2+}$ and Cu(OH)(oden)$^+$ in solution (172) but solids were not characterized. Formation constants for the Cu(II)-adenine system were reported (173). Copper(I) and -(II) complexes with thiomorpholin-3-one can be prepared by reaction of the appropriate copper salt in a little excess of the molten ligand (174).

Copper(II) chelates of the type $Cu(R,R'C{=}N{-}NCSSCH_3)_2$ (where R = R' = CH_3; R = R' = C_5H_{10}; R = CH_3, R' = C_6H_5) are obtained by mixing ethanolic solutions of copper acetate and the hydrazine S-methyl dithiocarboxylate Schiff base ligands (175). The reaction of tetramethyldiphosphine-disulfide, $Me_2P_2S_2$ with $CuCl_2{\cdot}2H_2O$ in ethanol at room temperature affords a small amount of brown $(Me_2P_2S_2)CuCl_2$ as well as white $[(Me_2P_2S_2)CuCl]_2$ (176). The preparation and properties of the selenoxine chelate of Cu(II) and related compounds have been reported (177). The Cu(II)-thiourea (tu) system in 0.1M nitric acid has been studied by ESR and visible spectroscopy and the complex $Cu(tu)_3NO_3{\cdot}H_2O$ isolated (178). Dark violet $Cu(L\text{-}H)_2$ (L = 2-methyl-3-hydroxy-4-thiopyrone) can be prepared (179) by interaction of a benzene solution of the ligand with the metal halide in methanol. Bis(2,2'-bipyridyl)thioureacopper(II) perchlorate can be obtained as deep green crystals from an aqueous solution containing bis(bipyridyl)-copper(II) perchlorate and thiourea (180). Copper(II) thiomorpholine and N-methylpiperazine dithiocarbamates are readily obtained (181) by interaction of the sodium salts of the ligands with aqueous copper(II) chloride. Yellowish green diaquo(2-mercaptopyrimidin-4-olato(2-))-copper(II) is prepared (182) by adding an aqueous solution ($pH \sim 7$) of 2-thiouracil to an aqueous solution of copper(II) nitrate. Polymeric 1:1 Cu(II) complexes with dithiooxamide, N,N'-dimethyl- and N,N'-dicyclohexyldithiooxamide can be prepared (183) by dropwise addition of an ethanol solution of the ligand to a warm ethanol solution of Cu(II) acetate. Controlled potential electrolysis or iodine can be employed to oxidize the maleonitriledithiolene chelate of Cu(II) in the presence of perylene to form the $(perylene)_2[CuS_4C_4(CN)_4]$ molecular complex (184). Bis-diethanoldithiocarbamatocopper(II) results upon reacting aqueous copper chloride with ammonium diethanoldithiocarbamate (185). Stable copper(II) dicyclohexyldithiophosphinate is prepared (186) by addition of a stoichiometric amount of ammonium dicyclohexyldithiophosphinate to methanolic copper(II) chloride. Brown bis[N-benzyl-N-(2-oxo-3,4,5,6-tetrahydroxyhexyl)dithiocarbamato]copper(II) monohydrate results from the reaction of aqueous $Cu(NO_3)_2$ with 3-benzyl-5-hydroxy-5-(1,2,3,4-tetrahydroxy-n-butyl)thiazolidine-2-thione (187). Yellow bis(thiazolidine-2-thionato)copper(II) is prepared by the reaction of thiazolidine-2-thione in ethanol-dimethyl sulfoxide with ethanolic $Cu(NO_3)_2{\cdot}3H_2O$ (187). Water insoluble $(Bz\phi_3P)_2Cu(DED)$ ($Bz\phi_3P$ = benzyltriphenylphosphonium; DED = 1,1-dicarboethoxy-2,2-ethylenedithiolate) is precipitated from brown-red dilute aqueous solutions of $K_2Cu(DED)_2$ upon addition of $Bz\phi_3PCl$. In the presence of a 10-15% excess of Cu(II) ions or H_2O_2 aqueous solutions of $K_2Cu(DED)_2$ afforded a water insoluble green complex $KCu(DED)_2$ (188). The complexes of Cu(II) with 1-(pyridyl-2-methylene)-4,4-dimethylthiosemicarbazide, 1-(pyridyl-2-methylene)-2-methylthiosemicarbazide and 1-(6-methyl-2-pyridylmethylene)-2-methylthiosemicarbazide are of the form $CuLX_2$ (X = Cl, Br). The N-substituted thiosemicarbazide Schiff base ligands and copper halides are stoichiometrically combined in ethanol (189). The monothio-β-diketonate $Cu(\phi CSCHCOCF_3)_2$ was prepared by interaction of the β-diketone with copper(II) acetate; thermodynamic data for formation of pyridine and 2,2'-bipyridine were also reported (190). The Schiff base of

S-methylthiosemicarbazide with pyridine-2-carbaldehyde (pmts) forms (191) stable Cu(II) complexes of the compositions: Cu(pmts)Cl_2, Cu(pmts)Br_2 and Cu(pmts)$(NO_3)_2 \cdot 0.5H_2O$ by interaction of the ligand with ethanolic solutions of the appropriate metal salts. Copper(II) complexes of formula $CuLX_2$ (X = Br, I, L = 2,2',2"-triaminotriethylamine (tren), triethylenetetramine (trien), 3,3'-diaminodipropylamine (dpt)) have been prepared as have Cu(tren)$Cl_2 \cdot H_2O$, Cu(trien)$(ClO_4)_2$ and Cu(trien)$(ClO_4)_2 \cdot H_2O$ (192).

Dark brown $CsCuNiF_6$ and light grey $CsCuTiF_6$ have been prepared (193) by fluorination of simple starting materials. Fluorination of a 1:1 mixture of CuCl and $RhCl_3$ in a Monel autoclave at 500°C yields brown $CuRhF_6$ (194). Emerald green bis(diethylammonium tetrachlorocuprate(II) was prepared (195) by the interaction of the amine hydrochloride and copper chloride dihydrate in absolute ethanol. Dark crystalline bis(2,2'-bipyridyl)copper(II) bis[dichlorocuprate(I)] can be obtained upon addition of KCl and CuCl to an aqueous solution of bis(2,2'-bipyridyl)copper(II) chloride (196). Crystalline anilinium tetrachlorocuprate(II) can be obtained (197) by evaporating a concentrated hydrochloric acid solution of $CuCl_2 \cdot 2H_2O$ and aniline hydrochloride (1:6 molar ratio). Trichloro{[2-(diphenylphosphenyl)ethyl]dimethylammonium}copper(II) can be isolated (198) as golden orange crystals from an ethanol solution containing anhydrous $CuCl_2$, [2-(diphenylphosphinyl)ethyl]dimethylamine and concentrated hydrochloric acid. Green bis(isopropylammonium)tetrachlorocuprate(II), $[(CH_3)_2CHNH_3]CuCl_4$, is prepared (199) by evaporation of a methanol solution of $CuCl_2$ and $(CH_3)CHNH_2 \cdot HCl$ in 1:2 molar ratio. Blue-green 1-(2-aminoethyl)-piperazinecopper(II) chloride results upon refluxing an ethanolic solution of the ligand and copper(II) chloride (200). Iminodiacetonitrile (idn) and iminodipropionitrile (ipn) form (201) Cu(II) complexes of the type CuL_2X_2 (X = Cl, Br, L = idn or ipn) which are prepared by interaction of anhydrous metal halides with the ligands in ethanol. 6-Nitro-2,3-di-(2-pyridyl)quinoxaline (dqnq) reacts with equimolar quantities of Cu(II) halides in refluxing ethanol to yield Cu(dqnq)X_2 (S = Cl, Br) (202). A green 1:1 cinchonine-$CuCl_2$ complex precipiates upon addition of tetrahydrofuran solutions of copper chloride and cinchonine (203). CuL_2X_2 (X = Cl, Br, NCS) and CuL_4X_2 (X = Cl, Br, NO_3, ClO_4, NO_3, $B\phi_4$) complexes where L = N-ethyl- and N-propyl-imidazole have been obtained (204) by adding the appropriate ligand and metal salt in hot methanol. Metathetical reactions were employed to obtain tetrafluoroborate and NCS^- salts. The complexes were recrystallized from a methanol-isopropanol mixture. Greenish-yellow Cu(isoxazole)Cl_2 results (205) upon mixing the metal salt and ligand in stoichiometric ratio in acetone, methanol or acetone-methanol mixtures. Yellow crystals of dichloro(N,N,N',N'-tetramethylpyridine-2,6-dicarboxamide)copper(II) were obtained (206) upon addition of acetonitrile solutions of anhydrous $CuCl_2$ and the dicarboxamide ligand.

DIMERIC AND POLYMERIC COMPOUNDS

Pyridine and quinoline adducts of copper(II) acetate can be obtained (207) by adding the appropriate base to a solution of anhydrous copper(II) acetate in acetonitrile. Saturated heterocyclic amine adducts, e.g., morpholine, of copper(II) acetate and propionate were obtained (208) by dissolving the hydrated carboxylate in the appropriate amine and cooling the solution to 0°C. Piperazine forms 1:1 polymeric complexes while the other amines form 1:2 complexes. Dimeric Cu(II) N-p-tolylsalicylidene-amine propionate is readily prepared by the interaction of the Schiff Base complex and propionate salt in hot ethanol (209). Acetato- and aryl-carboxylato-copper(II) adducts with antipyrine have been prepared (210) by interaction of the components in hot n-butanol. The Cu(II) compounds of composition CuX_2A (X = $CH_3CH_2CO_2^-$, $ClCH_2CH_2CO_2^-$, $CH_3CHClCO_2^-$ or $CH_3CCl_2CO_2^-$ and A = 1-phenyl-2,3-dimethyl-5-pyrazolone(antipyrine)) were prepared (211) by treating antipyrine (in excess) with Cu(II) propionate, α-chloropropionate, β-chloropropionate or α,α'-dichloropropionate and a small amount of the corresponding free acid in hot methanol solution. Dark blue (β-alanyl-L-histidinato)copper(II) dihydrate was prepared (212) by interacting carnosine with freshly prepared $Cu(OH)_2$, filtering off the excess hydroxide and adjusting the final pH to 10. A variety of complex Cu(II) ammine oxalates have been prepared and characterized (213). The materials obtained depend on reaction conditions. The reaction of copper(II) acetate monohydrate with 2,2-dimethoxypropane in boiling methanol followed by addition of a 4-substituted pyridine N-oxide yields (214) the corresponding dimeric bis-adduct of copper(II) acetate. Copper(II) acetate and chloroacetates can be prepared (215) by treating a slight excess of basic copper carbonate with aqueous solutions of the acids. The light green dimeric quinoline adduct of Cu(II) trifluoro-acetate is obtained (216) by heating the monomeric violet bis-quinoline adduct of Cu(II) trifluoroacetate. Green tetranuclear $[Cu_2OH(O_2CCF_3)_3(quinoline)_2]_2$ is prepared by dissolving anhydrous $Cu(O_2CCF_3)_2$ in 95% ethanol and adding quinoline (217).

Binuclear $Cu_2(fsac\text{-}R)_2$ complexes, where $H_2fsac\text{-}R$ denotes Schiff bases prepared from 3-formylsalicylic acid and alkylamines, can be obtained (218) by interaction of the Schiff bases with copper(II) acetate in water. Green Cu(II) 3-formylsalicylate is prepared by heating an aqueous mixture of 3-formylsalicylic acid, sodium carbonate and Cu(II) acetate (219). Copper(II) complexes of N-(2-pyridylmethyleneamino)-2-pyridinecarboxamide (Hpap), $Cu_2(OH)(pap)SO_4 \cdot 4H_2O$ and $Cu_2Cl_3(pap)$, are prepared (220) by interaction of the ligand with aqueous or ethanolic solutions of the appropriate metal salts. Black tetranuclear di-μ-chloro-bis(chloro[N,N'-ethylenebis(salicylideneiminato)copper(II)]copper(II) can be prepared (221) by slowly diffusing $CuCl_2$ in aqueous ethanol and N,N'-ethylenebis-(salicylaldiminato)copper(II) in chloroform into one another. Structures of related complexes have also been described (222). Violet crystals of dimeric [3-(salicylideneamino)-1-propanolato]copper(II) can be obtained upon cooling a refluxed solution of the material in N,N-dimethylformamide (223). Dimeric Cu(II) complexes of N-(hydroxyalkyl)pyrrole-2-carbox-aldimines (alkyl - ethyl, propyl, isopropyl) have been prepared (224) by refluxing the Schiff base with copper(II) acetate in anhydrous methanol. Blue trinuclear complexes of the type $[Cu_3[R_2N(CH_2)_2NH(CH_2)_3O]_2(OH)_2]$-$(ClO_4)_2$ (R = CH_3, C_2H_5) were prepared (225) by adding a methanolic solution of the N,N-disubstituted diaminoalcohol and sodium hydroxide to a hot ethanol solution of hydrated Cu(II) perchlorate. Corresponding green binuclear complexes of the type $[Cu_2(R_2N(CH_2)_2NH(CH_2)_3O)_2](ClO_4)_2$ result upon changing the Cu(II) perchlorate and sodium hydroxide concentrations. Thiocyanato-bridged binuclear Cu(II) complexes with Schiff base ligands derived from 2,6-diformyl-4-methylphenol and N,N-dimethylethylenediamine, N,N-diethylethylenediamine and 2-pyridylmethylamine have been described (226) as have related compounds with related Schiff bases prepared from 4-substituted-o-aminophenol (227). Bis[bromo-, chloro-, and nitrato-(N-n-butyl-5-chloro-α-phenyl-2-hydroxybenzylidene)aminato-μ-O-copper(II)] were prepared (228) by the reaction of copper(II) nitrate or halide in absolute ethanol with bis[N-butyl(5-chloro-α-phenyl-2-hydroxybenzylidene)-aminato]copper(II) which in turn was prepared by the reaction of copper(II) acetate with the Schiff base ligand. Reddish purple prisms of the binuclear hetero-metal complex CuNi(fsacen)·$3H_2O$ (where H_4fsacen = N,N-ethylenebis(3-carboxysalicylaldimine)) was prepared by reaction of copper(II) acetate with the ligand followed by the reaction of the Cu chelate, Cu(H_2fsacen) with lithium hydroxide and aqueous nickel(II) chloride (229). Nine alkoxo-bridged Cu(II) complexes with N,N-dialkyl-diaminealcohols $[Cu(R_2N(CH_2)_nNH(CH_2)_mO]X$ ($R = CH_3$ or C_2H_5, n = 2 or 3, m = 2 or 3, X = ClO_4^- or $B(C_6H_5)_4^-$) and with N,N-dialkylaminoalcohols, $[Cu((CH_3)_2N(CH_2)_3O)X]$ (X = Cl, Br) have been prepared and characterized (230). N,N'-1,3-propylene-bis(salicylideneiminato)copper(II) reacts (231) with some uranyl salts to form bi- and trinuclear complexes. A variety of complex Cu(II) carbonates, e.g., $K[Cu_2(CO_3)_2OH]\cdot 2H_2O$, $Na_3[Cu_2(CO_3)_2OH]\cdot 4H_2O$ and $Na_2[Cu(CO_3)_2]\cdot 3H_2O$ can be obtained (232) by reactions of copper acetate or nitrate with alkali carbonate and bicarbonate salts. The products obtained depend on reaction conditions. Reaction of $Cu(hfac)_2\cdot H_2O$ [hfac = $(CF_3CO)_2CH$] with equimolar amounts of pyrazine (pyz) in CCl_4 gives $[Cu(hfac)_2(pyz)]_3$(I). Recrystallization from hot CCl_4 gives I together with $[Cu(hfac)_2]_2$pyz. A 1:2 mixture of reactants gives $Cu(hfac)_2(pyz)_2$. I is a linear coordination polymer (233). Black crystalline dinuclear bis[heptanetrionato(2-)]bispyridinedicopper(II), $Cu_2(hpt)_2py_2$, results upon crystallizing $Cu_2(hpt)_2$ from a hot pyridine solution (234). Compounds of the type $[CuA_2\cdot H_2O]$ and $[CuA_2\cdot 4H_2O]_2$, where A = N-acetylglycine (aceturic acid), can be prepared (235) by interaction of aqueous solutions of the sodium salt of the ligand with Cu(II) perchlorate. Amine adducts of the type CuA_2B (B = piperazine, 2,2'-bipyridine and 1,10-phenanthroline) and CuA_2B_2 (B = N-methylpiperazine, piperidine, morpholine, pyridine and ethylenediamine) were obtained by reaction of the amines with $[CuA_2\cdot 4H_2O]_2$ (235).

Dimeric blue $[Cu_2(tren)_2(CN)_2](B\phi_4)_2$ (tren = 2,2',2"-triaminotri-ethylamine, $B\phi_4$ = tetraphenylborate) can be prepared (236) by addition of NaCN and $NaB\phi_4$ to an aqueous solution of $CuSO_4$ and tren. Crystals can be obtained from an acetonitrile solution of the product. Dimeric species with the formulation $[Cu_2(tren)_2X_2](B\{C_6H_5\}_4)_2$ (X = OCN^-, SCN^- and $SeCN^-$; tren = 2,2',2"-triaminotriethylamine) can be prepared by interaction of the ligand and appropriate salts in aqueous solution. Crystals of the compounds may be obtained by evaporating acetonitrile solutions of the complexes (237). Reactions between $(\pi\text{-}C_5H_5)Fe(CO)_2X$ (X = Cl, Br) and CuC_2R (R = $C_6H_5^-$, $p\text{-}CH_3C_6H_4^-$, $p\text{-}FC_6H_4^-$ or $C_6F_5^-$) afford the dimeric complexes

$\{(\pi-C_5H_5)Fe(CO)_2(C_2R)CuX\}_2$. The air stable orange crystalline products are moderately soluble in benzene, ether or acetone (238). Condensation of biacetylmonoxime with ethylamine in the presence of Cu(II) perchlorate yields (239) $[Cu_3L_3O]ClO_4$ where L = $CH_3C(NO^-)CH_3C{=}NC_2H_5$. Similar reaction with N,N-diethylethylenediamine instead of ethylamine yields $Cu_2L'_2(ClO_4)_2 \cdot H_2O$ where L' = $CH_3C(NO^-)CH_3C{=}NCH_2CH_2N(C_2H_5)_2$ (239). Brownish green crystals of tetrachloro-μ-[bis(dimethylglyoximato)copper(II)-dicopper(II), $[Cu(DMG)_2][CuCl_2]_2$ is obtained upon adding ethanolic $CuCl_2 \cdot 2H_2O$ to $Cu(DMG)_2$ in ethanol. $[Cu(DMG)_2][Cu(OH_2)_2]_2Br_4$ and $[Cu(DMG)_2]_2Cu(NO_3)_2$ are similarly prepared (240). Bright green dichloro-monoacetylacetonedioximecopper(II) results upon addition of ethanolic $CuCl_2$ to the dioxime in ethanol - it further reacts upon addition of Cu(II) acetate to yield dichloro-bisaqua-μ-[bis-acetyl-acetonedioximato-copper(II)]dicopper(II) (240). Light green $[Cu(LNO)_2SO_4]_2$ (LNO = 2,6-lutidine N-oxide) is prepared (241) by reacting the ligand with a methanol solution of the metal sulfate. The Cu(II) complex of a polyoxime, $\{[CH_2]_{1.6}[C(=NOH)C(=NOH)]_{0.8}[C=O]_{0.2}\}$, is prepared by adding excess cupric acetate to the polymer in 1:1 dioxan-isopropanol under N_2. The green insoluble complex so obtained catalyzes a number of reactions (242). Binuclear Cu(II) complexes in which N,N'-bis-(2-pyridylmethyl)oxamidato-copper(II) and N,N'-bis-(2-pyridylethyl)oxamidatocopper(II) function as bidentate ligands have been described (243). A number of complexes of the type Cu(acac)X where X = OCH_3, OC_2H_5, $OCH_2C_6H_5$, N_3^- and pyrazolate anion and $[Cu_2(acac)_2(OH)(pyrazolate)]$ have been prepared and characterized (244). 3,3'-(4,4'-Biphenylenebisazo)bis(2,4-pentanedione) reacts with copper acetate to give a 3:1 (metal-ligand) complex in which copper acetate is bound to the chelated ligand through the azo groups (245). Polymeric complexes of the type $Cu(pipdtc)_2(CuBr)_n$ (n = 4 and 6) and $Cu(pipdtc)_2(CuCl)_4$ have been prepared (246) by the reaction of bis-(piperidyldithiocarbamato)copper(II), $Cu(pipdtc)_2$ and Cu(II) halide in a chloroform-ethanol mixture. A variety of complexes with phenylethylamine (PEA) and mescaline (M) - $Cu_4OCl_6(PEA)_4$, $Cu_4OCl_6(M)_4$, $Cu(PEA)_2Cl_2$, $Cu(M)_2Cl_2 \cdot H_2O$, $Cu(PEA)_3Br_2$, $Cu(PEA)_4Cl_2$, $Cu(PEA)_4Br_2$, $Cu(M)_4Br_2$, $(PEAH)_2(CuCl_4)$, and $(MH)CuCl_4$ - have been prepared (247). The products obtained depend on the details of the reaction conditions and the original paper should be consulted. Golden-yellow trans-tetra-μ-[chlorodichloro-diaquocopper(II)]copper(II)bis(tetramethylene sulfone), $Cu_3Cl_6(H_2O)_2 \cdot 2TMSO_2$, can be prepared (248) by dissolving anhydrous $CuCl_2$ in ethanol and and adding tetramethylene sulfone to the solution. Red-brown plates of tetraphenylphosphonium trichlorocuprate(II), which contains non-planar dimeric $Cu_2Cl_6^{2-}$ units, precipitate immediately upon addition of equimolar portions of tetraphenylphosphonium chloride and $CuCl_2$ in absolute ethanol (249).

REFERENCES

1. R. J. Sundberg and R. B. Martin, CHEM. REVS., 74, 471 (1974).
2. J.-H. Fuhrhop, STRUCTURE AND BONDING, 18, 1 (1974).
3. U. Weser, STRUCTURE AND BONDING, 17, 2 (1973).
4. P. J. Quilley and G. A. Webb, COORD. CHEM. REVS., 12, 407 (1974).
5. R. Osterberg, COORD. CHEM. REVS., 12, 309 (1974).

6. J. J. Christensen, D. J. Eatough and R. M. Izatt, CHEM. REVS., 74, 351 (1974).
7. M. Tanaka, J. INORG. NUCL. CHEM., 36, 151 (1974).
8. R. B. Martin and R. Prados, J. INORG. NUCL. CHEM., 36, 1665 (1974).
9. N. N. Greenwood and I. M. Ward, CHEM. SOC. REVS., 3, 231 (1974).
10. E. Sinn, COORD. CHEM. REVS., 12, 185 (1974).
11. J. Kincaid, K. Nakamoto, J. A. Tiethof and D. W. Meek, SPECTROCHIM. ACTA, 30A, 2091 (1974).
12. B. Hutchinson, D. Eversdyk and S. Olbricht, SPECTROCHIM. ACTA, 30A, 1605 (1974).
13. L. T. Taylor and J. G. Dillard, INORG. CHEM., 13, 2620 (1974).
14. M. Berretz and S. L. Holt, J. INORG. NUCL. CHEM., 36, 49 (1974).
15. T. J. Bastow and H. J. Whitfield, J. INORG. NUCL. CHEM., 36, 97 (1974).
16. R. Swaroop and Y. K. Gupta, J. INORG. NUCL. CHEM., 36, 169 (1974).
17. G. Dolcetti and N. W. Hoffman, INORG. CHIM. ACTA, 9, 269 (1974).
18. R. D. Hancock, N. P. Finkelstein and A. Evers, J. INORG. NUCL. CHEM., 36, 2539 (1974).
19. E. J. Billo, INORG. NUCL. CHEM. LETT., 10, 613 (1974).
20. D. L. Wertz and J. L. Tyvoll, J. INORG. NUCL. CHEM., 36, 3713 (1974).
21. R. Haegele and D. Babel, Z. ANORG. ALLGEM. CHEM., 409, 11 (1974).
22. H. G. von Schnering, Z. ANORG. ALLGEM. CHEM., 400, 201 (1973).
23. C. Friebel and D. Reinen, Z. ANORG. ALLGEM. CHEM., 407, 193 (1974).
24. R. Scaringe and G. Kokoszka, J. CHEM. PHYS., 60, 40 (1974); L. Mauriello, R. Scaringe and G. Kokoszka, J. INORG. NUCL. CHEM., 36, 1565 (1974).
25. D. Getz and B. L. Silver, J. CHEM. PHYS., 61, 630, 638 (1974).
26. T. D. Smith and J. R. Pilbrow, COORD. CHEM. REVS., 13, 173 (1974).
27. F. H. Jardine, ADV. INORG. CHEM. RADIOCHEM., 17 (1974).
28. A. E. Jukes, ADV. ORGANOMETALLIC CHEM., 12 (1974).
29. T. Kauffmann, ANGEW. CHEM. INTERNAT. EDN., 13, 291 (1974).
30. T. Tsuda, T. Nakatsuka, T. Hirayama and T. Saegusa, J. C. S. CHEM. COMM., 557 (1974).
31. P. F. Rodesiler and E. L. Amma, J. C. S. CHEM. COMM., 599 (1974).
32. D. A. Edwards and R. Richards, INORG. NUCL. CHEM. LETT., 10, 945 (1974).
33. R. D. Mounts, T. Ogura and Q. Fernando, INORG. CHEM., 13, 802 (1974).
34. M. Cartwright and A. A. Woolf, J. INORG. NUCL. CHEM., 36, 3858 (1974).
35. K. Skudlarski and W. Lukas, ROCZNIKI CHEM., 48, 745 (1974).
36. E. C. Ashby, T. F. Korenowski and R. D. Schwartz, J. C. S. CHEM. COMM., 157 (1974).
37. A. Onoe, S. Uemura and M. Okano, BULL. CHEM. SOC. JAPAN, 47, 2818 (1974).
38. N. Watanabe, M. Okano and S. Uemura, ibid., 47, 2745 (1974).
39. S. Uemura, S. Uchida, M. Okano and K. Schikawa, ibid., 47, 3254 (1974).
40. K. Oshima, H. Yamamoto and H. Nozaki, J. AMER. CHEM. SOC., 95, 7926 (1973).

41. Y.-Y. Huang and D. E. Mainwaring, J. C. S. CHEM. COMM., 584 (1974).
42. J. W. Fitch and C. A. Kettner, SYN. REACT. INORG. METAL-ORG. CHEM., 4, 1 (1974).
43. J. C. Bommer and K. W. Morse, J. AMER. CHEM. SOC., 96, 6222 (1974).
44. T. Tsuda, H. Habu, S. Horiguchi and T. Saegusa, J. AMER. CHEM. SOC., 96, 5930 (1974).
45. D. E. Bergbreiter and G. M. Whitesides, J. AMER. CHEM. SOC., 96, 4937 (1974).
46. T. V. Healy, C. J. Barton and I. Windeatt, J. INORG. NUCL. CHEM., 36, 2575 (1974); T. V. Healy and C. J. Barton, ibid., 36, 2579 (1974).
47. T. E. Paxson, M. F. Hawthorne, L. D. Brown and W. N. Lipscomb, INORG. CHEM., 13, 2772 (1974).
48. A. P. Gaughan, Jr., Z. Dori and J. A. Ibers, INORG. CHEM., 13, 1657 (1974).
49. O. M. Abu Salah, M. I. Bruce and A. D. Redhouse, J. C. S. CHEM. COMM., 855 (1974).
50. H. Hayashi, K. Kawasaki and T. Murata, CHEM. LETT., 1079 (1974).
51. H. Hayashi, K. Kawasaki and T. Murata, CHEM. LETT., 89 (1974).
52. G. Van Koten and J. G. Noltes, J. C. S. CHEM. COMM., 575 (1974).
53. T. Cohen, R. J. Lewarchik and J. Z. Tarino, J. AMER. CHEM. SOC., 96, 7753 (1974).
54. D. A. Couch and S. D. Robinson, INORG. CHEM., 13, 456 (1974).
55. F. D. King and D. R. M. Walton, J. C. S. CHEM. COMM., 256 (1974).
56. D. F. Lewis, S. J. Lippard and P. S. Welcker, J. AMER. CHEM. SOC., 96, 3805 (1974).
57. A. Muller and W. Sievert, Z. ANORG. ALLGEM. CHEM., 406, 80 (1974).
58. E. Soderback, ACTA CHEM. SCAND., A28, 116 (1974).
59. G. Gattow and C. W. Voigt, Z. ANORG. ALLGEM. CHEM., 409, 185 (1974).
60. W. R. McWhinnie and V. Rattanaphani, INORG. CHIM. ACTA, 9, 153 (1974).
61. T. Ottersen, L. G. Warner and K. Seff, INORG. CHEM., 13, 1904 (1974).
62. L. G. Warner, T. Ottersen and K. Seff, INORG. CHEM., 13, 2819 (1974).
63. S. C. Shome, M. P. Ray and M. Mazumdar, J. INDIAN CHEM. SOC., 51, 256 (1974).
64. P. K. Mandal, INDIAN J. CHEM., 12, 845 (1974).
65. C. V. Philip and D. W. Brooks, INORG. NUCL. CHEM. LETT., 10, 739 (1974).
66. A. L. Crumbliss, L. J. Gestant, R. C. Rickard and A. T. McPhail, J. C. S. CHEM. COMM., 545 (1974).
67. F. J. Hollander and D. Coucouvanis, J. AMER. CHEM. SOC., 96, 5646 (1974).
68. I. F. Taylor, Jr., M. S. Weininger and E. L. Amma, INORG. CHEM., 13, 2835 (1974).
69. J. A. Tiethof, A. T. Hetey and D. W. Meek, INORG. CHEM., 13, 2505 (1974).
70. P. Kohl and D. Reinen, Z. ANORG. ALLGEM. CHEM., 409, 257 (1974).
71. M. Gondrand, A. Collomb, J. C. Joubert and R. D. Shannon, J. SOLID STATE CHEM., 11, 1 (1974).

72. H. Paulsson and E. Rosen, Z. ANORG. ALLGEM. CHEM., 401, 172 (1973).
73. J. B. Wooten, G. G. Long and L. H. Bowen, J. INORG. NUCL. CHEM., 36, 2177 (1974).
74. J. C. Barnes and J. M. Thomas, INORG. CHIM. ACTA, 9, 171 (1974).
75. J. C. Barnes and R. D. Peacock, J. CHEM. SOC. (A), 558 (1971).
76. K.-H. Steinberg, H. Bremer and F. Hofmann, Z. ANORG. ALLG. CHEM., 407, 162 (1974).
77. E. Uhlig and U. Dinjus, Z. ANORG. ALLGEM. CHEM., 397, 297 (1973).
78. M. B. Cingi, R. Borromei and L. Oleari, INORG. CHIM. ACTA, 11, 167 (1974).
79. M. Haruta, Z. Yoshida and H. Ogoshi, BULL. CHEM. SOC. JAPAN, 47, 3029 (1974).
80. D. Attanasio, I. Collamati and C. Ercolani, J. C. S. DALTON TRANS., 2442 (1974); 1319 (1974).
81. H. Imai and T. Ota, BULL. CHEM. SOC. JAPAN, 47, 2497 (1974).
82. C. Preti, G. Tosi, D. De Filippo and G. Verani, INORG. NUCL. CHEM. LETT., 10, 541 (1974).
83. F. Sprta and M. Bartasek, COLL. CZECH. CHEM. COMM., 39, 2023 (1974).
84. H. Kolind-Andersen, S.-O. Lawesson and B. Folkesson, REC. TRAV. CHIM., 93, 123 (1974).
85. M. H. Mihailov, V. Ts. Mihailova and V. A. Khalkin, J. INORG. NUCL. CHEM., 36, 127 (1974).
86. F. J. M. Rajabalee, J. INORG. NUCL. CHEM., 36, 557 (1974).
87. Y. K. Agrawal and S. G. Tandon, J. INORG. NUCL. CHEM., 36, 869 (1974).
88. P. N. Mohandas, O. P. Sunar and C. P. Trivedi, J. INORG. NUCL. CHEM., 36, 937 (1974).
89. R. Grimm and Z. Kolarik, J. INORG. NUCL. CHEM., 36, 189 (1974).
90. V. Seshagiri and S. B. Rao, J. INORG. NUCL. CHEM., 36, 353 (1974).
91. M. J. Jaycock, A. D. Jones and C. Robinson, J. INORG. NUCL. CHEM., 36, 887 (1974).
92. Y. Kidani, K. Ohira and H. Koike, BULL. CHEM. SOC. JAPAN, 47, 2040 (1974).
93. T. B. Field, J. L. McCourt and W. A. E. McBryde, CAN. J. CHEM., 52, 3119 (1974).
94. J. A. Marinsky, N. Imai and M. C. Lim, ISRAEL J. CHEM., 11, 601 (1973).
95. R. C. Paul, V. P. Kapila, N. Palta and S. K. Sharma, INDIAN J. CHEM., 12, 825 (1974).
96. T. Kurauchi, M. Matsui, Y. Nakamura, S. Ooi, S. Kawaguchi and H. Kuroya, BULL. CHEM. SOC. JAPAN, 47, 3049 (1974).
97. R. C. Paul, V. P. Kapila, R. S. Battu and S. K. Sharma, INDIAN J. CHEM., 12, 827 (1974).
98. C. M. Mikulski, L. L. Pytlewski and N. M. Karayannis, Z. ANORG. ALLGEM. CHEM., 403, 200 (1974).
99. A. L. Lott, INORG. CHEM., 13, 667 (1974).
100. N. M. Karayannis, C. M. Mikulski, L. L. Pytlewski and M. M. Labes, INORG. CHEM., 13, 1146 (1974).
101. A. N. Speca, L. L. Pytlewski and N. M. Karayannis, J. INORG. NUCL. CHEM., 36, 1227 (1974).

102. A. N. Speca, L. L. Pytlewski, N. M. Karayannis and C. Owens, J. INORG. NUCL. CHEM., 36, 3751 (1974).
103. B. Kirson, ISRAEL J. CHEM., 12, 711 (1974).
104. M. W. Blackmore, C. G. Sceney, D. R. O'Neill and R. J. Magee, J. INORG. NUCL. CHEM., 36, 1170 (1974).
105. H. Waki, K. Yoshimura and S. Ohashi, J. INORG. NUCL. CHEM., 36, 1337 (1974).
106. R. J. Magee, W. Mazurek, M. J. O'Conner and A. T. Phillip, AUST. J. CHEM., 27, 1629 (1974).
107. R. J. Magee, W. Mazurek, M. J. O'Connor and A. T. Phillip, AUST. J. CHEM., 27, 1885 (1974).
108. M. Sekizaki, BULL. CHEM. SOC. JAPAN, 47, 1447 (1974); M. Sekizaki, Y. Yamada and H. Takahashi, NIPPON KAGAKU KAISHI, 1057 (1974).
109. M. Nonoyama and K. Yamasaki, INORG. CHIM. ACTA, 7, 677 (1973).
110. R. J. Motekaitis and A. E. Martell, INORG. CHEM., 13, 550 (1974).
111. R. C. Paul, R. S. Chopra, R. K. Bhambri and G. Singh, J. INORG. NUCL. CHEM., 36, 3703 (1974).
112. D. P. Murtha and R. A. Walton, INORG. CHIM ACTA, 8, 279 (1974).
113. I. Nagypal, A. Gergely and E. Farkas, J. INORG. NUCL. CHEM., 36, 699 (1974).
114. H. Sigel, R. Caraco and B. Prijs, INORG. CHEM., 13, 462 (1974).
115. P. G. Slade, M. Raupach and E. W. Radoslovich, ACTA CRYST., B29, 279 (1973).
116. I. Nagypal, A. Gergely and E. Farkas, J. INORG. NUCL. CHEM., 36, 699 (1974).
117. M. M. Taqui Khan and C. R. Krishnamoorthy, J. INORG. NUCL. CHEM., 36, 711 (1974).
118. M. F. Iskander, S. E. Zayan, M. A. Khalifa and L. El-Sayed, J. INORG. NUCL. CHEM., 36, 551 (1974).
119. H. T. Witteveen and J. Reedijk, J. SOLID STATE CHEM., 10, 151 (1974).
120. R. Paton, E. C. Watton and L. R. Williams, AUST. J. CHEM., 27, 1185 (1974).
121. V. M. Ellis, R. S. Vagy and E. C. Watton, AUST. J. CHEM., 27, 1191 (1974).
122. R. Nakon and R. J. Angelici, J. AMER. CHEM. SOC., 96, 4178 (1974).
123. S. N. Poddar and S. R. Chowdhury, INDIAN J. CHEM., 12, 1107 (1974).
124. Y. Fukuda, A. Shimura, M. Mukaida, E. Fujita and K. Sone, J. INORG. NUCL. CHEM., 36, 1265 (1974).
125. C. F. Naumann and H. Sigel, J. AMER. CHEM. SOC., 96, 2750 (1974).
126. A. Tenhunen, FINN. CHEM. LETT., 55 (1974).
127. E. Uhlig and B. Machelett, Z. ANORG. ALLGEM. CHEM., 409, 320 (1974).
128. E. Uhlig and B. Machelett, Z. ANORG. ALLGEM. CHEM., 408, 83 (1974).
129. S. Yamada, Y. Kuge and T. Yamayoshi, INORG. CHIM. ACTA, 8, 29 (1974).
130. D. Webster and G. J. Palenik, J. AMER. CHEM. SOC., 96, 7565 (1974).
131. M. N. Patel, C. B. Patel and R. P. Patel, J. INORG. NUCL. CHEM., 36, 3868 (1974).
132. L. D. Dave and V. B. Mohankumar, INDIAN J. CHEM., 12, 984 (1974).
133. S. S. Sandhu and G. K. Sandhu, INDIAN J. CHEM., 12, 830 (1974).

134. K. K. Narang and A. Aggarwal, INORG. CHIM. ACTA, 9, 137 (1974).
135. R. H. Prince and D. A. Stotter, INORG. CHIM. ACTA, 10, 89 (1974).
136. M. Chikura, H. Yokoi and T. Isobe, BULL. CHEM. SOC. JAPAN, 47, 2208 (1974).
137. N. N. Ghosh and M. Dasgupta, J. INDIAN CHEM. SOC., 51, 206 (1974).
138. R. H. Prince, D. A. Stotter and P. R. Woolley, INORG. CHIM ACTA, 9, 51 (1974).
139. A. C. Baxter and D. R. Williams, J. C. S. DALTON TRANS., 1117 (1974).
140. G. K. R. Makar and D. R. Williams, J. C. S. DALTON TRANS., 1121 (1974).
141. M. H. Mihailov, V. Ts. Michailova and V. A. Khalkin, J. INORG. NUCL. CHEM., 36, 141 (1974).
142. N. F. Curtis, J. C. S. DALTON TRANS., 347 (1974).
143. R. F. Pasternack, E. G. Spiro and M. Teach, J. INORG. NUCL. CHEM., 36, 599 (1974).
144. R. Nasanen, M. Nasakkala and A. Lehtinen, FINN. CHEM. LETT., 139 (1974).
145. N. F. Curtis and N. B. Milestone, AUST. J. CHEM., 27, 1167 (1974).
146. G. Ivarsson, ACTA CHEM. SCAND., 27, 3523 (1973).
147. J. F. Boas, P. E. Fielding and A. G. MacKay, AUST. J. CHEM., 27, 7 (1974).
148. N. Takvoryan, K. Farmery, V. Katovic, F. V. Lovecchio, E. S. Gore, L. G. Anderson and D. H. Busch, J. AMER. CHEM. SOC., 96, 731 (1974).
149. T. Yano, H. Kobayashi and K. Ueno, BULL. CHEM. SOC. JAPAN, 47, 3033 (1974).
150. B. G. Segal and S. J. Lippard, INORG. CHEM., 13, 822 (1974).
151. M. R. Caira, G. V. Fazakerley, P. W. Linder and L. R. Nassimbeni, ACTA CRYST., B29, 2898 (1973).
152. M. Cannas, G. Carta and G. Marongiu, J. C. S. DALTON TRANS., 550 (1974).
153. G. V. Fazakerley, P. W. Linder and L. R. Nassimbeni and A. L. Rodgers, INORG. CHIM. ACTA, 9, 193 (1974).
154. J. F. Villa, H. C. Nelson, J. Arnold and D. T. Barnett, SYN. REACT. INORG. METAL-ORG. CHEM., 4, 113 (1974).
155. M. Nonoyama, SYN. REACT. INORG. METAL-ORG. CHEM., 4, 365 (1974).
156. M. Nonoyama, INORG. CHIM. ACTA, 10, 59 (1974).
157. R. Nasanen, M. Nasakkala and H. Jokela, FINN. CHEM. LETT., 93 (1974).
158. H. Ojima and K. Nonoyama, Z. ANORG. ALLGEM. CHEM., 401, 207 (1974).
159. G. G. Aleksandrov, Yu. T. Struchkov, A. A. Kurganov, S. V. Rogozhin and V. A. Davankov, INORG. NUCL. CHEM. LETT., 10, 959 (1974).
160. D. E. Fenton and S. E. Gayda, J. C. S. CHEM. COMM., 960 (1974).
161. M. M. Muir and P. R. Rechani, INORG. CHIM. ACTA, 11, 137 (1974).
162. R. Uggla, O. Orama, M. Sundberg, E. Tirronen and M. Klinga, FINN. CHEM. LETT., 185 (1974).
163. W. O. Siegl, INORG. NUCL. CHEM. LETT., 10, 825 (1974).
164. R. N. Dash and D. V. R. Rao, J. INDIAN CHEM. SOC., 51, 211 (1974).
165. M. S. Pereira and J. M. Molin, INORG. CHEM., 13, 386 (1974).
166. R. Barbucci, L. Fabbrizzi and P. Paoletti, J. C. S. DALTON TRANS., 2403 (1974).

167. M. Quastlerová-Hvastijova, J. Kohout and J. Gazo, Z. ANORG. ALLGEM. CHEM., 396, 341 (1973); MONATSH., 104, 779 (1973).
168. M. C. R. Symons, D. X. West and J. G. Wilkinson, INORG. NUCL. CHEM. LETT., 10, 995 (1974).
169. S. P. Ghosh, B. Bhattacharjee and K. K. Mishra, J. INDIAN CHEM. SOC., 51, 308 (1974).
170. M. J. M. Campbell, R. Grzeskowiak and S. G. Juneja, J. INORG. NUCL. CHEM., 36, 2485 (1974).
171. R. K. Steinhaus and C. L. Barsuhn, INORG. CHEM., 13, 2922 (1974).
172. R. Barbucci and A. Vacca, J. C. S. DALTON TRANS., 2363 (1974).
173. G. K. R. Makar and D. R. Williams, J. INORG. NUCL. CHEM., 36, 1675 (1974).
174. D. DeFilippo, F. Devillanova and G. Verani, J. INORG. NUCL. CHEM., 36, 1017 (1974).
175. L. El-sayed, M. F. Iskander and A. El-Toukhy, J. INORG. NUCL. CHEM., 36, 1739 (1974).
176. F. A. Cotton, B. A. Frenz, D. L. Hunter and Z. C. Mester, INORG. CHIM. ACTA, 11, 111, 119 (1974).
177. Y. Mido, I. Fujiwara and E. Sekido, J. INORG. NUCL. CHEM., 36, 537 (1974) and references therein.
178. H. M. Ratajczak and L. Pajdowski, J. INORG. NUCL. CHEM., 36, 459 (1974).
179. E. Uhlemann, H. Motzny and G. Wilke, Z. ANORG. ALLGEM. CHEM., 401 255 (1973).
180. K. T. McGregor and W. E. Hatfield, J. C. S. DALTON TRANS., 2448 (1974).
181. G. Marcotrigiano, G. C. Pellacani and C. Preti, J. INORG. NUCL. CHEM., 36, 3709 (1974).
182. I. P. Khullar and U. Agarwala, AUST. J. CHEM., 27, 1877 (1974).
183. L. Menabue, G. C. Pellacani and G. Peyronel, INORG. NUCL. CHEM. LETT., 10, 187 (1974).
184. L. Alcacer and A. H. Maki, J. PHYS. CHEM., 78, 215 (1974).
185. B. Annuar, J. O. Hill and R. J. Magee, J. INORG. NUCL. CHEM., 36, 1253 (1974).
186. H. J. Stoklosa, G. L. Seebach and J. R. Wasson, J. PHYS. CHEM., 78, 962 (1974).
187. W. D. Basson and A. L. du Preez, J. C. S. DALTON TRANS., 1708 (1974).
188. F. J. Hollander, M. L. Caffery and D. Coucouvanis, J. AMER. CHEM. SOC., 96, 4682 (1974).
189. M. A. Malik and D. J. Phillips, J. INORG. NUCL. CHEM., 36, 2229 (1974).
190. D. R. Dakternieks and D. P. Graddon, AUST. J. CHEM., 27, 1351 (1974).
191. M. A. Malik and D. J. Phillips, AUST. J. CHEM., 27, 1133 (1974).
192. M. Massacesi and G. Ponticelli, GAZZ. CHIM. ITAL., 104, 789 (1974).
193. R. Jesse and R. Hoppe, Z. ANORG. ALLGEM. CHEM., 403, 143 (1974).
194. V. Wilhelm and R. Hoppe, Z. ANORG. ALLGEM. CHEM., 407, 13 (1974).
195. D. R. Hill and D. W. Smith, J. INORG. NUCL. CHEM., 36, 466 (1974).
196. J. Kaiser, G. Brauer, F. A. Schröder, I. F. Taylor and S. E. Rasmussen, J. C. S. DALTON TRANS., 1490 (1974).

197. K. P. Larsen, ACTA CHEM. SCAND., A28, 194 (1974).
198. M. G. Newton, H. D. Caughman and R. C. Taylor, J. C. S. DALTON TRANS., 1031 (1974).
199. D. N. Anderson and R. D. Willett, INORG. CHIM. ACTA, 8, 167 (1974).
200. R. L. Dotson, INORG. NUCL. CHEM. LETT., 10, 879 (1974).
201. G. W. Watt and P. H. Javora, J. INORG. NUCL. CHEM., 36, 1745 (1974).
202. D. F. Colton and W. J. Geary, J. INORG. NUCL. CHEM., 36, 1499 (1974).
203. M. M. Khan, S. M. F. Rahman and N. Ahmad, INDIAN J. CHEM., 12, 111 (1974).
204. M. Massacesi and G. Ponticelli, J. INORG. NUCL. CHEM., 36, 2209 (1974).
205. A. Cristini, G. Ponticelli and C. Preti, J. INORG. NUCL. CHEM., 36, 2473 (1974).
206. J. G. H. du Preez, B. van Brecht, J. F. de Wet and J. Koortz, INORG. NUCL. CHEM. LETT., 10, 935 (1974).
207. F. E. Mabbs, J. K. Porter and W. R. Smail, J. INORG. NUCL. CHEM., 36, 819 (1974).
208. G. Marcotrigiano, G. C. Pellacani and C. Preti, Z. ANORG. ALLGEM. CHEM., 408, 313 (1974).
209. T. Tokii, S. Emori and Y. Muto, BULL. CHEM. SOC. JAPAN, 47, 2887 (1974).
210. J. Kratsmar-Smogrovic, J. Kohout, K. Staiger and S. Surka, Z. ANORG. ALLGEM. CHEM., 409, 273 (1974).
211. M. Melnik, FINN. CHEM. LETT., 189 (1974).
212. J. F. Villa, H. C. Nelson and R. T. Curran, Jr., INORG. CHEM., 13, 1255 (1974).
213. M. Melnik, H. Langfelderova, J. Garaj and J. Gazo, INORG. CHEM. ACTA, 7, 669 (1973).
214. D. Hibdon and J. H. Nelson, INORG. CHIM. ACTA, 7, 629 (1973).
215. J. A. Faniran and K. S. Patel, J. INORG. NUCL. CHEM., 36, 2261 (1974).
216. J. A. Moreland and R. J. Doedens, J. C. S. CHEM. COMM., 28 (1974).
217. R. G. Little, J. A. Moreland, D. B. W. Yawney and R. J. Doedens, J. AMER. CHEM. SOC., 96, 3834 (1974).
218. M. Tanaka, H. Okawa, T. Tamura and S. Kida, BULL. CHEM. SOC. JAPAN, 47, 1669 (1974).
219. M. Tanaka, H. Okawa, I. Hanaoka and S. Kida, CHEM. LETT., 71 (1974).
220. M. Nonoyama, INORG. CHIM. ACTA, 10, 133 (1974).
221. C. A. Bear, J. M. Waters and T. N. Waters, J. C. S. DALTON TRANS., 1059 (1974).
222. J. M. Epstein, B. N. Figgis, A. H. White and A. C. Willis, J. C. S. DALTON TRANS., 1954 (1974).
223. P. J. Nassiff, E. R. Boyko and L. D. Thompson, BULL. CHEM. SOC. JAPAN, 47, 2321 (1974).
224. C. R. Panley and L. J. Theriot, INORG. CHEM., 13, 2033 (1974).
225. Y. Nishida and S. Kida, CHEM. LETT., 339 (1974).
226. T. Ichinose, Y. Nishida, H. Okawa and S. Kida, BULL. CHEM. SOC. JAPAN, 47, 3045 (1974).

227. H. Okawa, I. Ando and S. Kida, BULL. CHEM. SOC. JAPAN, 47, 3041 (1974).
228. P. Gluvchinsky, G. M. Mockler, P. C. Healy and E. Sinn, J. C. S. DALTON TRANS., 1156 (1974).
229. H. Okawa, M. Tanaka and S. Kida, CHEM. LETT., 987 (1974).
230. Y. Nishida, F. Numata and S. Kida, INORG. CHIM. ACTA, 11, 189 (1974).
231. G. Condorelli, I. Fragala and S. Giuffrida, CHEM. LETT., 683 (1974).
232. A. K. Sengupta and A. K. Nandi, J. INORG. NUCL. CHEM., 36, 2479 (1974).
233. R. C. E. Belford, D. E. Fenton and M. R. Truter, J. C. S. DALTON TRANS., 17 (1974).
234. A. B. Blake and L. R. Fraser, J. C. S. DALTON TRANS., 2554 (1974).
235. G. Marcotrigiano and G. C. Pellacani, CAN. J. CHEM., 52, 3607 (1974).
236. D. M. Duggan and D. N. Hendrickson, INORG. CHEM., 13, 1911 (1974).
237. D. M. Duggan and D. N. Hendrickson, INORG. CHEM., 13, 2929 (1974).
238. O. M. Abu Salah and M. I. Bruce, J. C. S. DALTON TRANS., 2302 (1974).
239. J. G. Mohanty, S. Baral, R. P. Singh and A. Chakravorty, INORG. NUCL. CHEM. LETT., 10, 655 (1974).
240. C. B. Singh and B. Sahoo, J. INORG. NUCL. CHEM., 36, 1259 (1974).
241. N. M. Karayannis, C. M. Mikulski, L. L. Pytlewski and M. M. Labes, INORG. CHIM. ACTA, 10, 97 (1974).
242. S. J. Kim and T. Takizawa, J. C. S. CHEM. COMM., 356 (1974).
243. H. Ojima and K. Nonoyama, Z. ANORG. ALLGEM. CHEM., 401, 195 (1973).
244. C. G. Barraclough, R. W. Brookes and R. L. Martin, AUST. J. CHEM., 27, 1843 (1974).
245. A. M. Talati and R. N. Kapadia, INDIAN J. CHEM., 12, 767 (1974).
246. R. M. Golding, A. D. Rae, B. J. Ralph and L. Sulligoi, INORG. CHEM., 13, 2499 (1974).
247. G. W. Watt and M. T. Durney, BIOINORG. CHEM., 3, 315 (1974).
248. D. D. Swank and R. D. Willett, INORG. CHIM. ACTA, 8, 143 (1974).
249. M. Textor, E. Dubler and H. R. Oswald, INORG. CHEM., 13, 1361 (1974).

SILVER AND GOLD

B. Jack McCormick

Department of Chemistry
West Virginia University
Morgantown, West Virginia 26505

INTRODUCTION

The chemistry of compounds containing gold-carbon bonds continues to be developed at a rapid pace and it seems clear that additional research in this area will be fruitful and exciting. General articles of interest include a survey of the organogold chemistry being pursued in the USSR (1) and a discussion of the potential activity of compounds with gold-antimony bonds in homogeneous catalysis (2). An article (3) dealing with the electrorefining of silver has appeared which may be of some interest to electrochemists and others.

COMPOUNDS CONTAINING (Ag,Au)-C BONDS

Alkyl, Aryl, and Related Groups: An important area of research that has developed in the past few years deals with synthesis and reactions of alkyl and aryl compounds and it is apparent that some general synthetic techniques have been developed. Thus, it is possible to prepare a wide variety of gold(I) and silver(I) complexes of the type RML (where R is an alkyl or aryl group, M is gold or silver, and L is a tertiary phosphine) by treating XML, where X is a halide, with appropriate organolithium or Grignard reagents. Specific examples of these syntheses in gold chemistry and provided by Kochi and co-workers (4,5) and others (6). Some related silver compounds can be synthesized in much the same way, but they are unstable and decompose in the range -50° to 0° by a mechanism that is thought to involve concerted carbon-silver bond cleavage and carbon-carbon bond formation (7). Kochi's group has also prepared a series of Au(III) complexes of the type $R(CH_3)_2AuP(C_6H_5)_3$ ($R=CH_3$, C_2H_5, n-C_3H_7, t-C_3H_7, i-C_4H_9, t-C_4H_9) in which the stereochemistry is well defined (4,5); the rich reaction chemistry of the alkyl gold(I,III) complexes has been extensively investigated. It should also be noted that a new method for the preparation of $[(CH_3)_2AuI]_2$ has been developed (4) that gives better yields (70%) than does the previously reported procedure (8).

In a study of the reactions of $(CH_2{=}CH)AuP(C_6H_5)_3$ with a variety of electrophilic reagents, it was found that the Au-C bond is easily cleaved to give a series of not unusual triphenylphosphine gold(I) derivatives (9).

Gold(I) ketenide, Au_2C_2O, has been obtained from keten and chloro(1,4-oxathian)gold(I) in acetonitrile containing amines under argon at -10° (10). Base adducts also were prepared.

Several gold compounds of the type $\{R[Au(P(C_6H_5)_3)_2]\}BF_4$, where $R=C_6H_5$, $p\text{-}CH_3C_6H_4$, $CH_2=CH$, $C_5H_5FeC_5H_4$, $C_5H_5FeC_5H_3Cl$, $C_5H_5FeC_5H_3OCH_3$, or $C_5H_5FeC_5H_3CH_2N(CH_3)_2$, have been obtained from $RAuP(C_6H_5)_3$ and HBF_4 or from the treatment of $CH_3AuP(C_6H_5)_3$ with appropriate organomercury compounds and HBF_4 (11). These compounds are thought to contain the

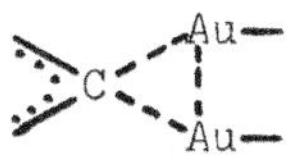

moiety and are related to previously reported compounds derived from ferrocene (12). Some additional work involves manganese in compounds that are related to the ferrocene series (13). A few compounds of the type $(RC_5H_4)AuPR'_3$, where R=H or CH_3 and $R'=C_6H_5$ or C_2H_5, have been prepared from halogoldphosphines and either the corresponding organolithium or organothallium reagent (14). These compounds are thought to contain monohapto RC_5H_4 groups and their syntheses are closely related to the general method described at the beginning of this section.

Details of the reactions of $R_4N[AuX_4]$ with various phenylhydrazines have now been published (15). The arylgold(III) compounds so formed, $R_4N[R'C_6H_4AuX_3]$, are neither air nor moisture sensitive and are unusual in that they do not contain stabilizing ligands such as phosphines. The synthesis of 2-[(dimethylamino)methyl]phenylgold also has been reported (6). This compound, which can be obtained in very good yield, is insoluble in hydrocarbon and ether solvents; consequently, a polymeric structure was suggested. Several papers dealing with compounds having the pentafluorophenyl group bonded to gold were published during the year. The key synthetic reaction (16,17) is the formation of perchloratobis(pentafluorophenyl)triphenylphosphinegold(III) (or the corresponding arsine complex) according to eq. (1).

$$[ClAu(C_6F_5)_2L] + AgClO_4 \longrightarrow [ClO_4Au(C_6F_5)_2L] + AgCl \qquad (1)$$

Although the perchlorato complex is not a stable solid, it is a very useful synthetic intermediate because of the ease with which the perchlorate moiety can be displaced by a variety of anionic and neutral monodentate ligands. Of the several reagents tested, only mono-olefins and very weak σ-donors failed to react. Other reactions investigated include those with diphosphines and diarsines, bidentate N- and O-donors, and cyclopolyolefins. In the latter two cases reduction to Au(I) takes place. Sulfate can be substituted for perchlorate without a change in most of the reactions.

The reactions of several Au(I) compounds with bis(pentafluorophenyl)-thallium(III) halides have been studied and a variety of products are formed, depending on the groups bonded to gold in the starting materials (18). The original paper should be checked for details. Preparative procedures for dihaloperhaloaryl gold(III) complexes have been published (19). In general, the preparations involve oxidation of perhaloaryl-Au(I) compounds with halogens or $TlCl_3$ (19). If the more strongly oxidizing halogens are used, the Au(I)-C bond is cleaved and the desired Au(III) compounds are not obtained. Finally, it should be mentioned that $(C_6H_5)_2Zn$ reacts with $AuCl_3$ or Au(CO)Cl to give the thermally stable compound $(C_6H_5)_3AuZn$, which is thought to be a dimer containing bridging phenyl groups (20).

Isocyanide, Carbene, and Related Complexes: Several new papers dealing with isocyanide complexes of gold and their conversion to carbene complexes have been published. Parks and Balch (21) have reported the details of their work dealing with the reaction of isocyanides, amines, and tetrachloroaurate, which produces carbenes of the form $Au\{C(NR_2)-NR'H\}_2^+$. It now seems well established that the treatment of appropriate gold(I) complexes with isocyanides and base is a general route to gold carbene complexes. The synthetic techniques leading to the related trimeric(alkoxy)(alkylimino)methylgold(I) compounds, $[(RO)(R'N=)CAu]_3$, are described by Minghetti and Bonati (22) and others (21). Other work with isocyanides involves the interesting reaction of $AuCl_4^-$ in H_2O or CH_3OH with 2-and 3-hydroxyalkylisocyanides (23). Although the experimental details are not given, it is reported that the isocyanides rearrange instantaneously to give Au(III)-oxazolidinylidine complexes.

These Au(III) complexes are slowly reduced in excess isocyanide to Au(I)-oxazolidinylidines of type I or, ultimately, II.

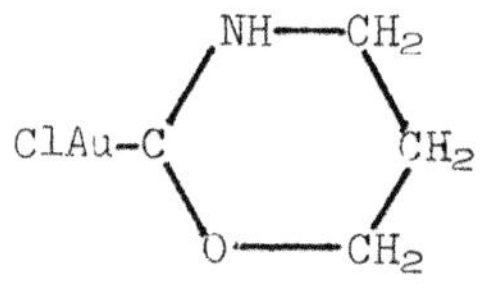
I

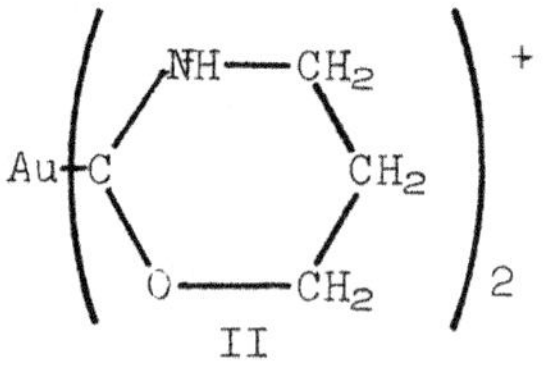
II

Details of the reaction (24) of the "electron rich" olefin 1,1′,3,3′-tetramethyl-2,2′-bi-imidazolidinylidene with a variety of metal complexes, including those of gold, to form carbene complexes have appeared (25). Finally, Minghetti and Bonati (26) have reported that trans-$[(p\text{-}MeC_6H_4\text{-}NH)_2C]_2AuI_2ClO_4\cdot(C_2H_5)_2O$ is an unusually reactive carbene complex and that it may be a synthetically useful precursor of other carbene derivatives.

Another paper (27) dealing with the previously reported (28) stabilization of Ag-C bonds in compounds containing phosphonium groups (derived from ylids) was published during the year.

Phosphines and Related Ligands: Several compounds of potential synthetic interest having the general formula ClAuL, where L is a phosphine or arsine, were prepared quite straightforwardly from $ClAuS(CH_3)_2$

and L in acetone (4). The mixed halide complexes $ClBr_2AuL$ ($L=P(C_6H_5)_3$ or $P(CH_3)_3$) were reported in the same paper. A brief investigation (29) of the reaction of $(C_6H_5)_3PAuCl$ with excess phosphine has been carried out. It was found that $[(C_6H_5)_3P]_2AuCl\cdot{}^1/_2C_6H_6$ can be isolated as a crystalline solid and a X-ray crystallographic examination was carried out. The remaining new phosphine and arsine work reported (30) deals with the reactions of $AgNO_3$ with several diphosphines and diarsines in ethanol. Compounds of the type $[(AgNO_3)_2L]_2$, in which the nitrate ligands appear to be monodentate, are formed. It is suggested that the diphosphine or diarsine ligands bridge between two metal atoms to form the dimer.

Several papers dealing with the coordination of phosphite, phosphonite, and phosphinite ligands have appeared. The caged phosphite 4-ethyl-2,6,7-trioxa-1-phosphabicyclo[2.2.2]octane can be used to prepare cationic complexes of the forms $[AuL(CH_3CN)]ClO_4$, $[AuL_2]ClO_4$, and $[AuL_4]ClO_4$ (31). Unstable complexes analogous to the latter compound are formed when chloro(cyclooctene)gold is treated with trimethyl or triethylphosphite in benzene (32); thus, the strong ligational properties of the caged phosphites may be utilized to confer stability in the gold-phosphite series. The caged phosphite mentioned above has also been used to synthesize the neutral complexes XAuL, where X=Br or Cl (31). Some additional studies on previously known (32) trialkylphosphitesilver complexes were also reported during the year (33). Finally, it should be mentioned that tri-n-butylphosphite has been used as an extracting agent (water to kerosene) for both silver and gold (34).

Robinson and co-workers have described the results of studies of the ligational properties of dialkylphenylphosphonites and alkyldiphenylphosphinites with respect to gold and silver (35,36). Neutral complexes of the types ClAgL, where L is a phosphonite or phosphinite, were prepared from AgCl and L in ethanol. In one case a dimer, $\{AgCl[P(OC_2H_5)(C_6H_5)_2]\}_2$, formed. Only one neutral gold complex, $AuCl[P(OC_2H_5)(C_6H_5)_2]$, which is very soluble, was prepared and the yield was low (35). Several cationic complexes of the type $[ML_4]B(C_6H_5)_4$, where M is Ag or Au, also were reported (36).

Gold(I) forms 1:1 complexes with 2,4,6-triphenylphosphabenzene in which the two-coordinate gold is bonded to the heteroatom of the phosphabenzene ring (37).

SIMPLE SALTS AND OTHER COMPLEXES

Simple Salts: A new method for the preparation of anhydrous gold(III) chloride, Au_2Cl_6, in good yield from tetrachloroauric acid and thionyl chloride has been described (38). A useful preparation that gives ClAu(CO) quantitatively from Au_2Cl_6 in a room temperature, atmospheric pressure synthesis is described in the same paper, along with several other syntheses that employ Au_2Cl_6 as a starting material. A new synthesis of silver hyponitrite that gives relatively low yields, but which is said to be shorter and more convenient than other procedures, has been published (39). Other salts that may be of interest are Rb_2AgF_4 and Cs_2AgF_4 (40),

Ag_2NiMF_7 (M=Cr, Fe, Al, In) (41), $Ag_6(SO_4)(SiO_4)$ (42), and the silver(I) salts of N,N-bis(fluorosulfuryl)ureide (43) and 3-thioindole (44).

Nitrogen Donors: Anhydrous gold(III) chloride has been utilized in the synthesis of several pyridine (38) and nitrile (38,45) complexes of the type $AuCl_3L$, where L is the nitrogen donor. It was concluded (45) that aromatic nitriles do not form ring-metallated products as was originally reported (46). The useful starting material $ClAuS(CH_3)_2$ has been employed once again in the synthesis of N-pyrazolylgold(I) derivatives (47), which are thought to be cyclic oligomers involving nine-membered rings, analogous to be silver compounds reported in 1973. A reinvestigation (48) of the interaction of Ag(I) with pyridine carboxylic acids and 8-hydroxyquinoline has led to the conclusion that previous formulations of complexes derived from these ligands are incorrect. These systems appear to be rather complex and the original articles should be consulted for details. Cationic complexes of quinoline and isoquinoline with Ag(I) have been prepared in which the counter ion is a chromium complex anion (49). Both Ag(I) and Au(III) are reported (50) to form complexes with 3-hydroxy-1-mercaptopyrimidine. The Ag(I) complex is thought to involve O,N-coordination, whereas the Au(III) complex is considered to be N,S-bonded. Other silver(I) complexes that may be of some interest include those derived from $AgBF_4$ and oxaziridines (51), $AgNO_3$ and pyrazine-N-oxides (52), and racemic p,p′-diamino-2,3-diphenylbutane (53,54). Lastly, some gold(III) complexes derived from dithiooxamide ligands have been synthesized and studied with the conclusion that at least some of the compounds involve N,S-donor sites (55).

Chalcogen Donors: Dimethylsulfoxide solvates of $AgNO_3$ and $AgClO_4$ have been briefly investigated and it should be noted that the perchlorate is explosive (56). Sulfur donors are involved in a number of complexes including $ClAu[S{=}P(C_6H_5)_3]$ (again prepared from $ClAuS(CH_3)_2$) (4) and a very novel compound in which silver(I) is bonded to the sulfur atom of coordinated thiocarbonyl (57). Monothio-and monoselenocarbamate salts react readily with $ClAuP(C_6H_5)_3$ to give complexes in which the carbamate is bonded to Au through sulfur or selenium and nmr studies on the complexes so formed have provided a measure of the barrier to rotation about the C∺N bond (58). Other gold-chalcogen bonds are found in several new complexes of N,N-dimethylthiourea (59) and substituted dithiooxamides (55).

MISCELLANEOUS COMPOUNDS

A silver complex of O-ethylborancarbonate, $(C_6H_5)_3PAgH_3BCOOC_2H_5$, has been prepared along with an analogous copper compound (60). These compounds, which involve bonding to the metal atom through hydrogen atoms, are the first reported metal complexes of this ligand. In a solution study of ketone complexes of silver fluoroborate an adduct of p-methylacetophenone was isolated as a solid (61). Other adducts reported include those of silver(I) trifluoromethanesulfonate with various aromatic molecules (62) and silver perchlorate with naphthalene (63) and anthracene (64). In the latter two cases crystal structures are reported. The cyclooctatetraen (COT) complexes (COT)AuCl and (COT) Au_2Cl_4 have been reported and characterized (65).

REFERENCES

1. A. N. Nesmeyanov, E. G. Perevalova, K. I. Grandberg, and D. A. Lemenovskii, BULL. ACAD. SCI. (USSR), DIV. CHEM. SCI., 23, 1068 (1974).
2. M. C. Muller, GOLD BULLETIN, 7, 39 (1974).
3. A. J. Parker, J. W. Diggle, and J. Avraamides, AUST. J. CHEM., 27, 721 (1974).
4. A. Tamaki and J. K. Kochi, J. ORGANOMETAL. CHEM., 64, 411 (1974).
5. A. Tamaki, S. A. Magennis, and J. K. Kochi, J. AMER., CHEM. SOC. 96, 6140 (1974).
6. G. Van Koten and J. G. Noltes, J. ORGANOMETAL. CHEM., 80, C56 (1974).
7. G. M. Whitesides, D. E. Bergbreiter, and P. E. Kendall, J. AMER. CHEM. SOC., 96, 2806 (1974).
8. F. H. Brain and C. S. Gibson, J. CHEM. SOC., 762 (1939).
9. K. I. Grandberg, E. I. Smyslova, and A. N. Kosina, BULL. ACAD. SCI. (USSR), DIV. CHEM. SCI., 22, 2721 (1974).
10. E. T. Blues, D. Bryce-Smith, I. W. Lawston, and G. D. Wall, J. CHEM. SOC., CHEM. COMMUN., 513 (1974).
11. A. N. Nesmeyanov, E. G. Perevalova, K. I. Grandberg, D. A. Lemenovskii, T. V. Baukova, and O. B. Afanassova, J. ORGANOMETAL. CHEM., 65, 131 (1974).
12. V. G. Andrianov, Y. T. Struchkov, and E. R. Rossinskaja, J. CHEM. SOC., CHEM. COMMUN., 338 (1973).
13. A. N. Nesmeyanov, E. G. Perevalova, T. V. Baukova, and K. I. Grandberg, BULL. ACAD. SCI. (USSR), DIV. CHEM. SCI., 23, 830 (1974).
14. G. Ortaggi, J. ORGANOMETAL. CHEM., 80, 275 (1974).
15. P. Braunstein and R. J. H. Clark, INORG. CHEM., 13, 2224 (1974).
16. R. Uson, A. Laguna, and J. L. Sanjoaquin, J. ORGANOMETAL. CHEM., 80, 147 (1974).
17. R. Uson, P. Royo, and A. Laguna, SYNTHESIS IN INORGANIC AND METAL-ORGANIC CHEM., 3, 237 (1973).
18. R. Uson, P. Royo, and A. Laguna, J. ORGANOMETAL. CHEM., 69, 361 (1974).
19. R. Uson, A. Laguna, and J. Pardo, SYNTHESIS IN INORGANIC AND METAL-ORGANIC CHEM., 4, 499 (1974).
20. P. W. J. DeGraaf, J. Boersma, and C. J. M. Van Der Kerk, J. ORGANOMETAL. CHEM., 78, C19 (1974).
21. J. E. Parks and A. L. Balch, J. ORGANOMETAL. CHEM., 71, 453 (1974).
22. G. Minghetti and F. Bonati, INORG. CHEM., 13, 1600 (1974).
23. K. Bartel and W. P. Fehlhammer, ANGEW. CHEM. INT. ED., 13, 599 (1974).
24. B. Cetinkaya, P. Dixneuf and M. F. Lappert, J. CHEM. SOC., CHEM. COMMUN., 206 (1973).
25. B. Cetinkaya, P. Dixneuf, and M. F. Lappert, J. CHEM. SOC., DALTON TRANS., 1827 (1974).
26. G. Minghetti and F. Bonati, J. ORGANOMETAL. CHEM., 73, C43 (1974).

27. H. Schmidbaur, J. Adlkofer, and M. Heimann, CHEM. BER., 107, 3697 (1974).
28. H. Schmidbaur, J. Adlkofer, and W. Buchner, ANGEW. CHEM. INT. ED., 12, 415 (1973).
29. N. C. Baenziger, K. M. Dittemore, and J. R. Doyle, INORG. CHEM., 13, 805 (1974).
30. W. Levason and C. A. McAuliffe, INORG. CHIM. ACTA, 8, 25 (1974).
31. G. P. Fenske and W. R. Mason, INORG. CHEM., 13, 1783 (1974).
32. K. J. Coskran, R. D. Bertrand, and J. G. Verkade, J. AMER. CHEM. SOC., 89, 4535 (1967).
33. D. A. Couch and S. D. Robinson, INORG. CHEM. ACTA, 9, 39 (1974).
34. T. V. Healy, C. J. Barton, and I. Windeatt, J. INORG. NUCL. CHEM., 36, 2575 (1974).
35. D. A. Couch, S. D. Robinson, and J. N. Wingfield, J. CHEM. SOC., DALTON TRANS., 1309 (1974).
36. D. A. Couch and S. D. Robinson, INORG. CHEM., 13, 456 (1974).
37. K. C. Dash, J. Eberlein, and H. Schmidbaur, SYNTHESIS IN INORGANIC AND METAL-ORGANIC CHEM., 3, 375 (1973).
38. D. Belli Dell'Amico and F. Calderazzo, GAZZ. CHIM. ITAL., 103, 1099 (1973).
39. G. D. Mendenhall, J. AMER. CHEM. SOC., 96, 5000 (1974).
40. R. H. Odenthal, D. Paus, and R. Hoppe, Z. ANORG. ALLEG. CHEM., 407, 144 (1974).
41. J. M. Dance, J. Grannec, C. Jocoboni, and A. Tressaud, COMPT. REND., SER. C., 279, 601 (1974).
42. H. L. Keller and H. Muller-Buschbaum, Z. ANORG. ALLEG. CHEM., 408, 205 (1974).
43. R. E. Noftle and J. Crews, INORG. CHEM., 13, 3031 (1974).
44. T. D. Lee, M. V. Pickering, and G. D. Daves, J. ORG. CHEM., 39, 1106 (1974).
45. F. Calderazzo and D. Belli Dell'Amico, J. ORGANOMETAL. CHEM., 76, C59 (1974).
46. M. S. Kharasch and T. M. Beck, J. AMER. CHEM. SOC., 56, 2057 (1939).
47. F. Bonati, G. Minghetti, and G. Banditelli, J. CHEM. SOC., CHEM. COMMUN., 88 (1974).
48. A. D. Baker, M. Brisk, and A. Storch, J. INORG. NUCL. CHEM., 36, 1755 (1974).
49. P. K. Mathur, J. INORG. NUCL. CHEM., 36, 943 (1974).
50. I. P. Khullar and U. Agarwalla, INDIAN J. CHEM., 12, 1096 (1974).
51. D. R. Christ, G. J. Jordan, and J. A. Hashmall, J. AMER. CHEM. SOC., 96, 4927 (1974).
52. P. J. Huffman and J. E. House, J. INORG. NUCL. CHEM., 36, 2618 (1974).
53. N. P. Marullo, J. F. Allen, G. T. Cockran, and R. A. Lloyd, INORG. CHEM., 13, 115 (1974).
54. G. T. Cochran, J. F. Allen and N. P. Marullo, INORG. CHEM. ACTA., 1, 109 (1967).
55. A. C. Fabretti, G. C. Pellacini, G. Peyronel, and B. Scapinelli, J. INORG. NUCL. CHEM., 36, 1067 (1974).
56. S. Ahrland and N.-O. Bjork, ACTA CHEM. SCAND., A28, 823 (1974).

57. R. D. Dombek and R. J. Angelici, J. AMER. CHEM. SOC., 96, 7568 (1974).
58. F. W. Pijpers, A. H. Dix and J. G. M. Van der Linden, INORG. CHEM. ACTA., 11, 41 (1974).
59. G. Marcotrigiano, R. Battistuzzi, and P. Morini, INORG. NUCL. CHEM. LETTERS, 10, 641 (1974).
60. J. C. Bommer and K. W. Morse, J. AMER. CHEM. SOC., 96, 6222 (1974).
61. D. R. Crist, Z.-H. Hsieh, G. J. Gordon, F. P. Schinco, and C. A. Maciorowski, J. AMER. CHEM. SOC., 96, 4932 (1974).
62. M. B. Dines, J. ORGANOMETAL. CHEM., 67, C55 (1974).
63. E. A. H. Griffith and E. L. Amma, J. AMER. CHEM. SOC., 96, 743 (1974).
64. E. A. H. Griffith and E. L. Amma, J. AMER. CHEM. SOC., 96, 5407 (1974).
65. P. Tauchner and R. Huttel, CHEM. BER., 107, 3761 (1974).

ZINC, CADMIUM AND MERCURY

J. B. SMART

NUCLEAR DIAGNOSTICS, INC.
TROY, MICHIGAN 48084

GENERAL INORGANIC COMPOUNDS

The utilization of reactions at elevated temperatures to provide new compounds and as a means of synthesizing compounds of high purity was the subject of several papers. The direct synthesis of CdI_2 was accomplished by placing iodine and cadmium metal at opposite ends of an evacuated ($1x10^{-5}$ torr) ampule (1). The iodine is heated to 70-75°C and the cadmium to 400°C while the ampule is continuously rotated. The reaction requires about four to six hours. Heating cadmium and platinum in an oxygen atmosphere (one atmosphere pressure) at a temperature of 500-550°C for one week a platinum bronze of the composition $CdPt_3O_4$ is obtained (2). If the reaction temperature is held at 550-590°C $Cd_xPt_3O_4$ (x less than one) is obtained. Above 590°C the product is $Cd_xPt_3O_6$ while no reaction occurs below 500°C.

Zinc cyanamide of very high purity was prepared by treating ZnO with three equivalents of urea at 135°C (under vacuum) followed by calcining at 600°C (3). High purity zinc chloride is obtained through passing HCl over molten zinc and distillation of the product at 650°C (4). The major contaminate was oxygen (100 ppm). Selenides, sulfides and tellurides of zinc, cadmium and mercury (except mercuric sulfide) of semi-conductor purity can be obtained by heating a mixture of stoichiometric quantities of the elements in an evacuated quartz tube directly with an oxy-gas torch (5). The authors caution to heat only part of the ampule to avoid the possibility of an explosion. Somewhat more drastic conditions were necessary to synthesize cadmium, zinc and mercury hexafluororhenium IV. To accomplish this it was necessary to maintain a mixture of $RhCl_3$ and the appropriate metal carbonate (or oxide) at 500°C under 400 atmospheres of fluorine gas for five days (6). Zinc fluorosilicate hexahydrate can be obtained by evaporation of a solution of zinc metal dissolved in $HSiF_6$ (7). The corresponding fluoroplumbate hexahydrate derivatives of zinc and cadmium precipitate when the metal fluoride is added to a solution of lead tetraacetate dissolved in 40% HF (8).

Under high vacuum conditions, fluorosulfuryl isocynate (FSO_2NCO) in the presence of acetonitrile reacts with HgO to eliminate CO_2 and form mercury(II)fluorosulfurylamide and mercury(II)N,N'bis(fluorosulfuryl)ureide (9). When the ratio of FSO_2NCO to HgO is two:one, no CO_2 gas is evolved and a compound with alternating Hg-N bonds is obtained (I):

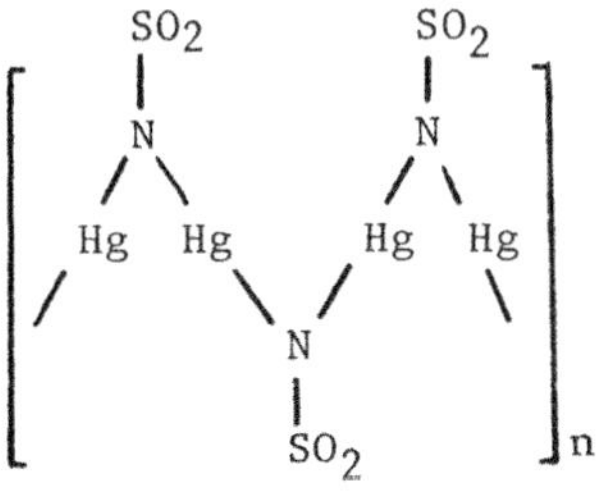

I

Dialkyl benzylphosphonates readily react with $Hg(OAc)_2$ to yield a Hg-P bond (10):

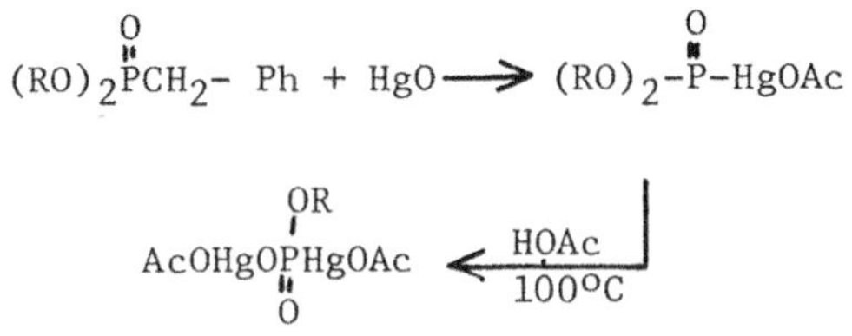

Ashby has reported an easy preparation of zinc hydride (11). Treatment of $ZnBr_2$ or Zn I_2 with lithium or sodium hydride in THF yields a white precipitate of ZnH_2 which is stable for several days. When KH is used the resulting precipitate is a mixture of potassium-zinc halogen-hydride complexes.

COMPLEX AND COORDINATION COMPOUNDS

While large numbers of complexes of zinc, cadmium and mercury continue to be isolated, few new synthetic processes have appeared. The concept of "template synthesis", where a metal ion apparently assists the combination of several smaller organic molecules into one larger species coordinated to the metal ion has received some attention. Often these new ligands are stable only when coordinated to the metal ion. The synthesis of chelate complexes of 1,3-bis(arylimino)isoindoles(II) described by Siegle (12) is a typical example. The ligand (II) does not form in the absence of the metal ion. Both Cd^{++} and Zn^{++} have been utilized in the rearrangement of 2,6-bis (2-methyl-2-benzothianzolinyl)pyridine (III) to form the shiff base complexes (13). Again, the complex (III) is stable only in the presence of the cation. In an analogous fashion, Silver has demonstrated that malononitrile reacts with ethanol in the presence of zinc or cadmium halides to from the corresponding chelates of 2,4-diamino-5-cyano-6-ethoxylpyridine (14).

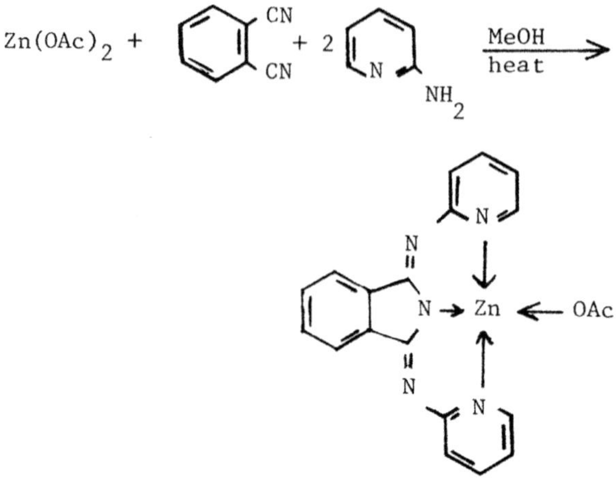

II

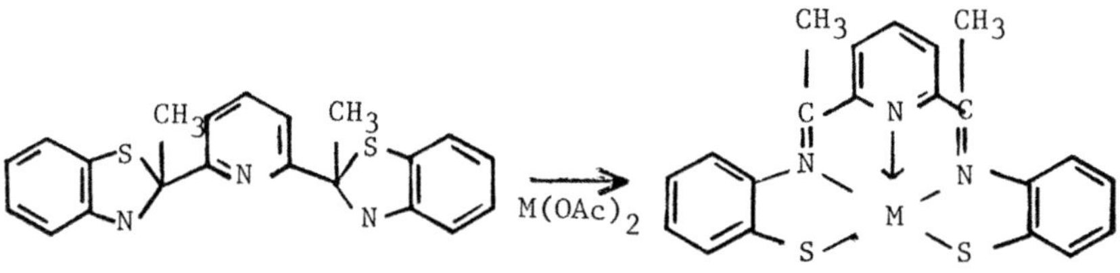

III

The synthesis of some mixed complexes has been reported. The optically active (phenylalaninato) (salicylaldehydrato)-zinc (II) complex (IV) can be obtained by heating an HCl/ethanol solution of salicylaldehyde with d- or l-phenylalanine in the presence of zinc acetate (15). Refluxing bis-(diethyldithiocarnato) zinc (II) with pyridine or pyridine derivatives, in $HCCl_3$, results in the coordination of pyridine (or derivative) to the form the corresponding mixed complex about the zinc ion (16).

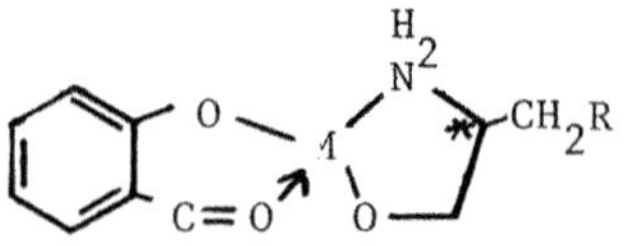

IV

Silver, (17,14) reports an efficient method for preparing mixed chloro-bromo complexes of zinc, cadmium and mercury. Mixing a phosponium bromide, such as 1-(3,3-dibenzoyl)propyltriphenyl-phosphonium bromide, $(DBP)^+Br^-$, with cadmium chloride yields $(DBP)_2(CdCl_2Br_2)$.

Alkylphosphines readily form adducts with mercuric halides (18). Three products (V through VII) can be isolated from the reaction between 1,1-bis (diphenylphosphino)ferrocene and HgX_2 in methanol (19).

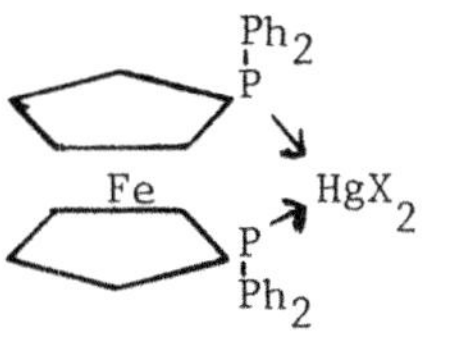

V

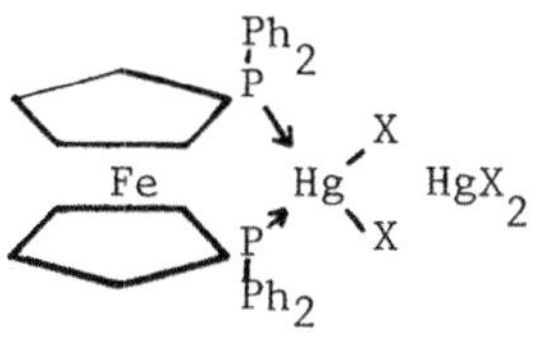

VI

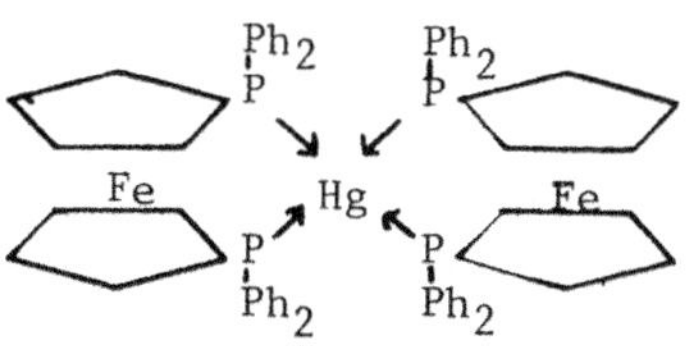

VII

Reactions of zinc or cadmium chloride with the sodium salt of p-phenylene-bis(p-methoxyphenyldithiophosphinic acid) yields poly [p-phenylene-bis(p-methoxyphenyldithiophosphidato)] metal complex (VIII) (20).

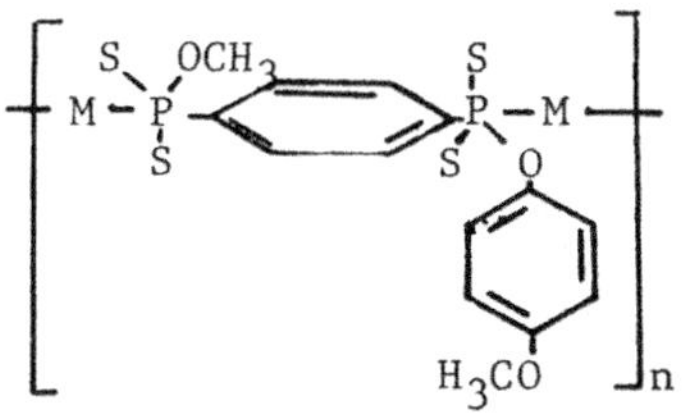

VIII

Corbin (21) reports that the coordinated group of the Shiff's base-zinc complex derived from various 2-substituted benzothiozolines are readily reduced with sodium borohydrides to the corresponding amines:

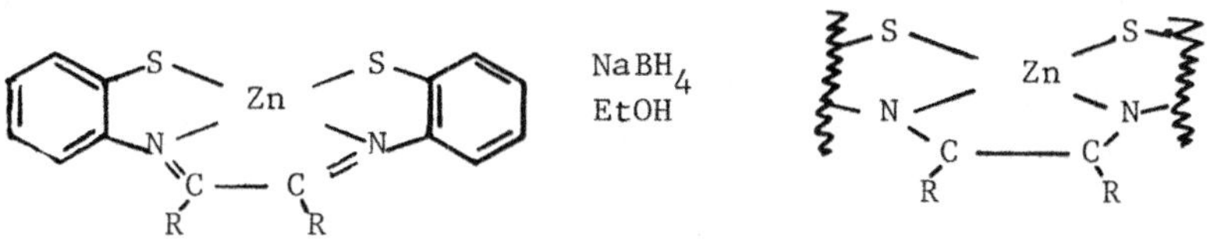

Aromatic amines such as p-fluoroanaline form adducts with $Hg_2(NO_3)_2$ to give N-Hg(I) coordination (IX) (22):

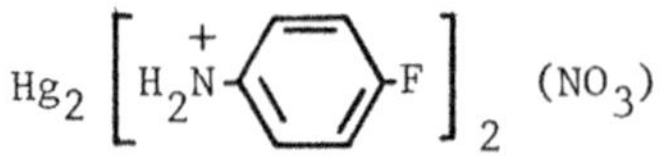

IX

Other pyridines such as 2-ethylpyridine or 4-benzylpyridine form complexes with $Hg_2(ClO_4)_2$ in anhydrous methanol at 5°C (23). The general formula for these complexes is $(Hg_2L_4)(ClO_4)_2$. However, less basic pyridines (eg., 3-methyl pyridine) do not form complexes.

Mikheeva et al. (24) and Mal'tseve et al. (25) report that Li, Na and Rb borohydrides form borohydride zincates, which are coordinated to ethers (reaction solvent). Attempts to form complexes of Cd $(BF_4)_2$ with the ligand 3,5-dimethylpyrazole resulted in a compound with fluorine bridges (26):

$$CdCO_3 + HBF_4(35\%) + DMP \xrightarrow{\text{methanol}}$$

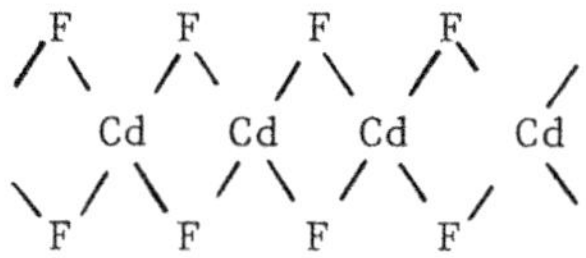

Fluorine also provides stability to an unusual mercury containing cation (27):

$$HgO + 2NiF_2 + HgF_2 \xrightarrow[400-500^{\circ}]{} [Hg_2O]^{++}[NiF_6]^{-2}.$$

Replacing HgO in this reaction with HgS yields a corresponding complex of the $[Hg_2S]^{++}$ cation.

ORGANOMETALLIC COMPOUNDS

Organo-cadmium and zinc: Wardell (28) has reviewed the literature on the synthesis of Group II organometallic compounds for the year 1972. The utilization of organozinc compounds in organic synthesis has also been reviewed (29), however, this work is mainly concerned with the synthetic utility of organozinc compounds and not with their isolation.

Denton (30) has described the synthesis of methyl-cadmium and zinc derivatives of hexaborane (10).

$$B_6H_{10} + (CH_3)_2M \xrightarrow{\text{THF}} (CH_3{+}M{-}B_6H_9)\cdot 2THF + CH_4$$

When M = Hg no reaction occurs. The addition of dicrotylzinc to a terminal double bond has been found to involve addition of the total zinc moiety across the double bond (31):

The cadmium salt of tetrafluoroterephthalic acid releases CO_2 when heated to form a linear polymeric derivative (32):

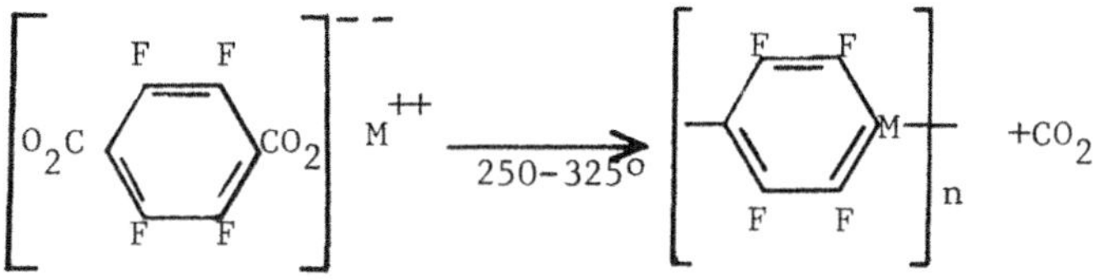

The mercury salt behaves in a similar fashion, but the zinc salt does not decarboxolate. Powdered zinc or cadmium metals when mixed in THF for several days with $Co_2(CO)_8$, $[C_5H_5\text{-}Fe(CO)_2]_2$ or $[C_5H_5Mo(CO)_3]_2$ form the respective products $[Co(CO)_4]_2M$, $[C_5H_5Fe(CO)_2]_2M$ and $[C_5H_5Mo(CO)_3]_2M$ (33).

<u>Organomercury:</u> Seyfreth has reviewed the literature on the synthesis of organomercury compounds for 1972 (34) and 1973 (35).

Clathration of dimethyl mercury offers a unique method of handling this toxic substance (36). Thus recrystallization of 4-p-hydroxyphenyl-2, 2,4-trimethyl thiochromin from neat dimethyl mercury yields such a clathrate. The ratio of the host to guest is 6:1 and the mercury compound is not released from the crystal lattice when pumped under high-vacuum. Crushing or dissolving the crystals in an appropriate solvent does release the dimethylmercury species.

The synthesis of fluorinated organo-mercurials is still receiving a good deal of attention. Thus Flegler (37) describes a synthesis of CF_3-HgI (X) through heating F_3CI and Hg to 325-340°C (in a "bomb"). The product, X, in turn can be converted to several derivatives of the general formula F_3CHgY, where Y = NCO^-, N_3^-, SCF^-, $1/2(C_2O_4)$, through the reaction of X with the appropriate AgY salt. Two other ether type compounds were also obtained:

$$X + Ag_2O + Ag_2CO_3 \xrightarrow{TlOH} F_3CHg\text{-}O\text{-}HgCF_3$$

$$X + Tl_2CS_3 \longrightarrow F_3CHg\text{-}S\text{-}HgCF_3$$

In a somewhat analogous fashion, a mixed species can be obtained (38):

$$F_3CHg\text{-}O_2C_2F_3 + Me_3Sn\,CF_3 \xrightarrow{\text{benzene}} F_3CHgMe$$

Under basic conditions (sodium t-butoxide), the proton of pentaflurobenzene is readily mercurated by the tetrabromomercurate (II) ion, phenylmercuric chloride and p-tolylmercuric chloride forming the corresponding pentafluoromercury compounds (39). Polyfluoro- and polychloro-substituted arylmercury compounds are readily formed on the thermolysis (120-240°C under vacuum) of the appropriate mercuric polyhalogen arenesulphonates (40). Both the mono- and diarylmercury compounds are obtained.

Larock has published several articles concerning his investigations of the synthesis of organomercurials through the use of boranes. Thus geminal diboranes, obtained from hydroboration of terminal acetylenes can be used to form a geminal dimercurial (41).

$$R\text{-}CH_2CH[B(OMe)_2]_2 + HgCl_2 \xrightarrow{\text{NaOH}} RCH_2CH(HgCl)_2$$

It is interesting to note that this reaction does not proceed if alkyl groups are attached to the boron atom (in place of the alkoxide groups). The presence of an alkoxide group also appears to be important in the formation of secondary mercury compounds from boranes (42):

$$(R_2CH)_3B + Hg(OAc)_2 \xrightarrow[0^o;\ THF]{^-OR} R_2CHHgOR$$

If $Hg_2(OR)_2$ is used in the reaction with boranes it is possible to transfer two alkyl groups (43):

$$R_3B + 2Hg_2(OR')_2 \xrightarrow{^-OR'} 2RHgOR'$$

Akagi (44) has reinvestigated the formation of methylmercuric acetate through the ultra-violet catalized reaction of mercuric acetate in acetic acid and has found that a stoichiometric amount of free sulfur inhances the yield by 260 fold. Diethyldiphenyltin reacts with mercuric acetate in refluxing chloroform to give an almost quantitative yield of diphenylmercury in one to two hours. The reaction proceeds as well when substituted phenyl groups are used (45). Anhydrous mercuric acetate when heated to 150°C for 20 hours with acetonitrile forms $(AcOHg)_3CCN$ (46). Treatment of this with silver acetate yields $AcOHgCH_2CN$.

Heal (47) has described the synthesis of a heptasulfurimide derivative of phenylmercury which is stable for several hours at room temperature and is a good intermediate for transfer of the heptasulfurimide group:

$$C_6H_5HgOAc \quad + \quad S_7NH \xrightarrow{MeOH} S_7NHgC_6H_5$$

The first synthesis of a mercury phosphineimide (XI) has been reported by Lorberth (48):

$$MeHgN(SiMe_3)_2 \quad + \quad HNP\ (NMe_2)_3 \xrightarrow[60-90^{o}]{} MeHgNP(NMe_2)_3$$

XI

Although aminomercuration reactions are not new they are slow and often yield oils. By performing the reactions under pressure it has been demonstrated (49) that the reactions occur rapidly forming stable crystalline substances. For example, the reaction

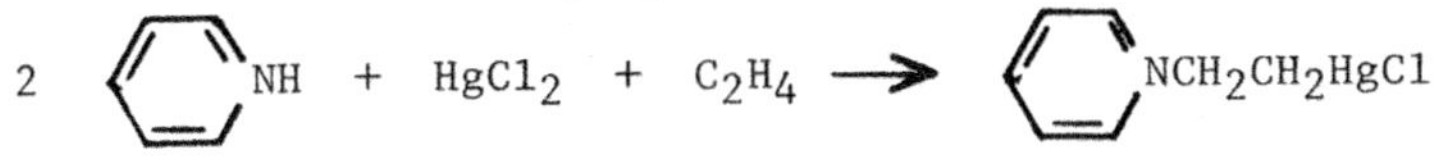

requires 16 hours under normal conditions but only one hour at 600psi of ethylene. At atmospheric pressure the reaction between diethylamine, mercuric chloride and ethylene yields only an unstable yellow liquid. At 600psi of ethylene the same reaction produces a tan, stable crystalline product, N-(2-chloromercuriethyl)diethyl amine.

Mercuration of 1,5-hexadiene with mercuric acetate yields the cyclic ether derivative (XII) when carried out in water (50), but in methanol a straight chain product is obtained.

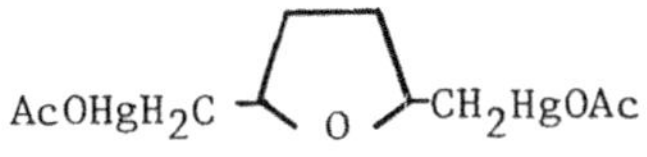

XII

Mercuration of 1,2-diphenylcyclopropane with mercuric acetate results in opening of the ring and 1-3 addition by the mercuric acetate group (51). Seyfreth has utilized the well known mercuration of an active hydrogen to incorporate mercury-carbon bonds into phosphonate esters (52):

$$X_2CH_2 \quad + \quad Hg(OAc)_2 \xrightarrow[\Delta]{} X_2C(HgOAc)_2$$

X = a phosphonate ester.

The formation of phosphonium ylide salts can be achieved by two routes (53):

$$MeHgCl \ + \ Me_3P{=}CH_2 \xrightarrow[-75^\circ]{THF} [Me_3PCH_2HgMe]^+ \ Cl^-$$

R.T.

$$HgCl_2 \ + \ 2Me_3P{=}CH_2 \xrightarrow{THF} [Me_3PCH_2HgCH_2PMe_3]^+ \cdot 2Cl^-$$

Using a similar reaction scheme silylmethylphosphonium ylide salts result when $Me_3P{=}CHSiMe_3$ is used in the reaction. Methyl dithiofluorophosphonic acid reacts slowly in $HCCl_3$ with Me_2Hg to from crystals of MeHgSPSFMe (54).

Direct mercuration of ferrocene with mercuric acetate in refluxing methanol/ether gives, in addition to the monochloromercury species, 1,2- and 1,1'-bis(chloromercuri)ferrocene. The latter species can be separated from the others through extraction with butanol (the 1,1'- derivative being insoluble) (55). Mercuration of iodoferrocene, followed by symmetrization through elution of the reaction mixture from an alumina column with a skellysolve-B/benzene mixture results in bis(2-iodoferrocenyl)mercury (56). Tricarbonylbenzene chromium (II) is readily mercurated by mercuric acetate to yield tricarbonyl [(chloromercurio)benzene] chromium (II) (57).

Dean (58) has described a general synthesis of arene complexes of Hg(I). Treatment of mercurous hexafluoroarsinate with an arene (benzene, naphthalene, etc.) in SO_2 immediately precipitates the corresponding complex. Sterically hindered silicon phenols readily react with phenyl mercuric hydroxide to yield polymeric species on hydrolysis (59):

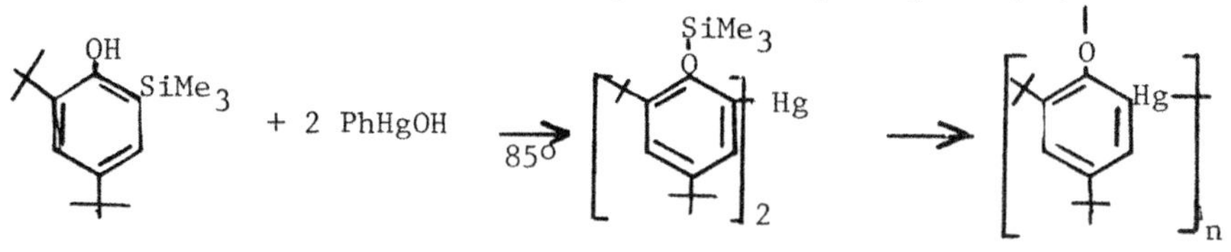

Fusion of mercury trifluoroacetate with sodium benzoate, nitrobenzene, benzamide or phenylmercuric chloride rapidly forms the per-mercuriated arene (as $-HgOCOCF_3$) (60). The original functional group (eg., $-NO_2$) is left intact.

A simple method of reducing RHgX compounds has been described by Aranda (61).

$$PhNHCH_2CH_2HgBr \xrightarrow[2)\,MeOH]{1)\,Li/THF} PhNHCH_2CH_3$$

COMPOUNDS WITH METAL-METAL BONDS

Group IV metals: Reduction of dialkyl derivatives with alkyl stannanes and germanes have produced $[(Me_3SiCH_2)_3M]_2$ Cd (M= Sn or Ge) (62), $[(Me_3CCH_2)_3Sn]_2Cd$ (63), several $(R_3Sn)_2Hg$ derivatives (64) and the first synthesis of an optically active germyl-mercury derivative, bis(methyl-1-naphthylphenylgermyl)mercury (65). Reduction of diethylzinc with triphenylgermane in bis-2-methoxyethyl ether yields either $EtZnGePh_3$ or $(Ph_3Ge)_2Zn$ depending on the stoichiometry of the starting materials (66).

A compound with a tin-mercury-germanium grouping was obtained in the following manner (67):

$$(C_6F_5)_3SnBr + Et_3GeHgGe(C_6F_5)_3 \xrightarrow[uv.]{} (C_6F_5)_3SnHgGe(C_6F_5)_3$$

If a 1:1 stoichiometry is exceeded than the bis[tri(perfluoro)-phenyltin] mercury derivative is obtained. This latter product is also obtained from the reaction of tri(perfluoro)phenyltin bromide with bis(triethylgermyl)-mercury (68).

Triethylgermylmercury derivatives of 1,2-dicarbaclosododecaborane(12) and 1,7-dicarbacloso-dodecaborane(12) result from the exchange reactions between bis(triethylgermyl)mercury and the appropriate chloromercury carborane (69):

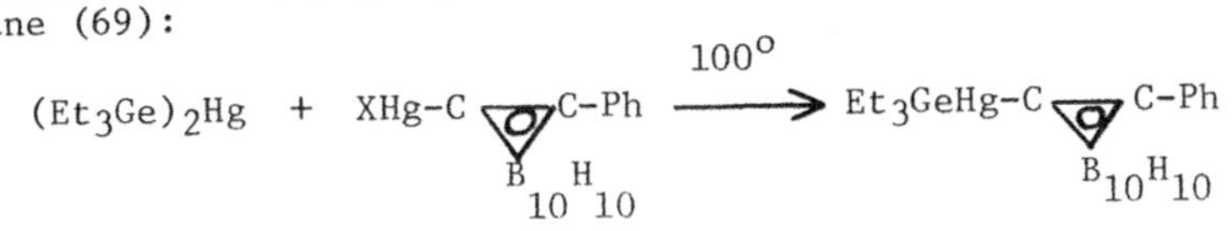

Other groups: Arsenic-mercury bonds can be obtained by reduction of mercury derivatives with arsine (70) in isopropyl alcohol:

$$MeHgX + 3MeHgO\text{-}t\text{-}C_4H_9 + AsH_3(\text{in } N_2) \longrightarrow [As(HgMe)_4]^+X^-$$

If the arsine is not diluted with nitrogen then tris(methylmercury)arsinide is obtained. Reduction of alkyl mercuric derivatives with selenium or tellurium hydrides yields the predicted compounds (71):

$$MeHgNO_3 + SeH_2(N_2) \xrightarrow{ether} [Se(HgMe)_3]^+NO_3^-$$

$$MeHgBr + TeH_2(N_2) \xrightarrow{MeOH} Te(HgMe)_2$$

Haupt reports that bis(pentacarbonylmanganese)mercury can be obtained by heating $Mn_2(CO)_{10}$ and metallic mercury in a "bomb" to 170°C for seven hours (xylene solvent) (72). Finally, an interesting insertion of iron into a mercury silicon bond has been reported (73):

$$Hg(SiMe_3)_2 + Fe(CO)_5 \xrightarrow[Vac.\ h\nu]{hexane} \left[(CO)_4\text{-}\underset{\substack{|\\ Me_3Si}}{Fe}\text{—}\right]_2 Hg$$

REFERENCES

1. D. P. Belotskii, Ya. A. Lyuter and T. N. Sushkevich, UKR. KHIM. ZH., 40, 989 (1974).
2. D. Cahen, J. A. Ibers and J. B. Wagner, Jr., INORG. CHEM., 13, 1377 (1974).
3. N. A. Sokolov, A. M. Pavlov, V. S. Golov and Yu. I. Dergunov, TR. KHIM. KHIM. TEKHNOL., 24 (1973).
4. L. E. Topv and S. J. Yosim, SYN. INORG. METAL. ORG. CHEM., 3, 47 (1973).
5. P. Polivka, and V. Kablu, CHEM. LISTY., 68, 75 (1974).
6. V. Wilhelm and R. Hoppe, Z. ANORG. ALLG. CHEM., 407, 13 (1974).
7. S. Ray, A. Zalkin and D. H. Templeton, ACTA CRYSTALLOGR. SECT. B, 29, 2741 (1973).
8. R. L. Davidovich, T. F. Levchishina and T. A. Kaidalova, ZH. NEORG. KHIM, 18, 625 (1973).
9. R. E. Nuftle and J. Crews, INORG. CHEM., 13, 3031 (1974).
10. W. I. Awad, M. El-Deek and E. El-Sawi, TETRAHEDRON LETT., 47, 4663 (1973).
11. J. J. Watkins and E. C. Ashby, INORG. CHEM., 13, 2350 (1974).
12. W. V. Siegl, INORG. NUCL. CHEM. LETT., 10, 825 (1974).
13. L. F. Lindoy and D. H. Busch, INORG. CHEM., 13, 2494 (1974).
14. J. L. Silver, DISS. ABSTR. INTR. B, 33, 5201 (1973).
15. E. G. Rukhadze, G. P. Talyzenkova and Yu. A. Malentin, ZH. OBSHCH. KHIM., 43, 1087 (1973).
16. D. L. N. Sarma and D. V. Ramana Rao, CURR. SCI., 43, 576 (1974).
17. J. L. Barmeister, J. L. Silver, E. T. Weleski, Jr., E. E. Schqeizer and C. M. Kopay, SYN. INORG. METAL-ORG. CHEM., 3, 339 (1973).
18. S. O. Grim, P. J. Lui and R. L. Keiter, INORG. CHEM., 13, 342 (1974).
19. K. R. Mann, W. H. Morrison, Jr. and D. N. Hendrickson, INORG. CHEM., 13, 1180 (1974).
20. W. Kuchen, J. Delventhal and H. Keck, CHEM. BER., 107, 2938 (1974).
21. J. L. Corbin and D. E. Work, CAN. J. CHEM., 52, 1054 (1974).
22. K. Brodersen and N. Hacke, CHEM. BER., 107, 3260 (1974).
23. D. L. Kepert and D. Taylor, AUST. J. CHEM., 27, 1199 (1974).
24. V. I. Mikheeva, N. S. Kedrova and N. N. Mal'tseva, IZV. AKAD. NAUK. SSSR SER. KHIM., 23, 512 (1974).
25. N. N. Mal'tseva, N. S. Kedrova and V. I. Milcheeva, ZH. NEORG. KHIM., 18, 1989 (1973).
26. M. A. Guichelaar, J. A. M. van Hest and J. Reedijk, INORG. NUCL. CHEM. LETT., 10, 999 (1974).
27. F. Champlon, D. Bernard, J. Pannetier and J. Lucas, C. R. ACAD. SCI. SER. C, 278, 1185 (1974).
28. J. L. Wardell, ORGANOMETAL. CHEM., 2, 16 (1973).
29. J. Furukawa and N. Kawabata, ADVAN. ORGANOMETAL CHEM., 12, 83 (1974).
30. S. E. Hayes, DISS. ABSTR. INT. D, 34, 119 (1973).
31. H. Lehmkuhl and H. Nehl, J. ORGANOMETAL. CHEM., 60, 1 (1973).
32. P. Sartori and H. J. Frohn, CHEM. BER., 107, 1195 (1974).
33. S. E. Hayes, DISS. ABSTR. INT. D, 34, 119 (1973).
34. D. Seyfreth, J. ORGANOMETAL. CHEM., 62, 33 (1973).
35. D. Seyfreth, J. ORGANOMETAL. CHEM., 75, 13 (1974).

36. R. J. Cross, J. J. McKendrick and D. D. MacNicol, NATURE (LON.), 245, 146 (1973).
37. V. K. Flegler, A. Hass and H. Massman, CHEM. ZTG., 97, 661 (1973).
38. V. S. Petrosyan, A. B. Permin, S. G. Sacharov and O. A. Reutov, J. ORGANOMETAL. CHEM., 65, C7 (1974).
39. H. B. Albrecht, G. B. Deacon and M. J. Tailby, J. ORGANOMETAL. CHEM., 69, 313 (1974).
40. P. G. Cookson and G. B. Deacon, AUST. J. CHEM., 26, 1893 (1973).
41. R. C. Larock, J. ORGANOMETAL. CHEM., 61, 27 (1973).
42. R. C. Larock, J. ORGANOMETAL. CHEM., 67, 353 (1974).
43. R. C. Larock, J. ORGANOMETAL. CHEM., 72, 35 (1974).
44. H. Akagi and E. Takabatake, CHEM. LETT., 761 (1974).
45. O. P. Syutkina, E. M. Panov, Y. I. Lodochnikova and K. A. Kocheshkov, DOLK. AKAD. NAUK. KHIM. SSSR. 216, 333 (1974).
46. J. W. Grimm, K. C. Roeber, G. Oehme, J. Alm, H. Mennega, and H. Pracejns, J. PRAKT. CHEM., 316, 557 (1974).
47. H. G. Heal and R. J. Ramsay, J. INORG. NUCL. CHEM., 36, 950 (1974).
48. J. Lorberth, J. ORGANOMETAL. CHEM., 71, 159 (1974).
49. R. F. DeBrule and G. G. Hess, SYNTHESIS, 197 (1974).
50. A. V. Gomez, M. J. Barluenga and M. Yus Astiz, REV. ACAD. CIENC. EXACTAS, FIS.-QUIM. NATUR. ZARAGOZA, 28, 225 (1973), CHEM. ABSTR. 80, 83164g (1974).
51. Yu. S. Shabarov and S. N. Burenko, ZH, OBSHCH. KHIM., 43, 2330 (1973).
52. D. Seyfreth and R. S. Marmor, J. ORAGNOMETAL. CHEM., 59, 231 (1973).
53. H. Schmidtbaur and K. H. Rathlein, CHEM. BER., 107, 104 (1974).
54. H. W. Roesky, M. Diefl and A. H. Norbury, Z. NATURFURSCH TEIL B, 28, 707 (1973).
55. M. D. Rausch, L. P. Klemann, A. Seigel, R. F. Kovar and T. H. Gund, SYN. INORG. METAL-ORG. CHEM., 3, 193 (1973).
56. P. V. Roling and M. D. Rausch, J. ORG. CHEM., 39, 1420 (1974).
57. G. K. I. Magomedov, V. G. Syrkin and A. S. Frenkel, ZH. OBSHCH, KHIM., 42, 2450 (1972).
58. P. A. Dean, D. G. Ibbott and J. B. Stothers, J. C. S. CHEM. COMM. 626 (1973).
59. G. A. Razuvaev, N. S. Vasleiskaya and N. N. Vavilina, ZH. OBSHCH, KHIM., 44, 135 (1974).
60. G. B. Deacon and G. F. Furouhurson, J. ORGANOMETAL. CHEM., 67, C1 (1974).
61. V. Aranda, J. Barluenga, A. Ara and G. Asensio, SYNTHESIS, 135 (1974).
62. G. S. Kalinina, O. A. Kraglaya, B. J. Petrov, E. A. Shchupak and N. S. Vyazankin, ZH. OBSHCH. KHIM., 43, 2224 (1973).
63. B. V. Fedot'ev, O. A. Kruglaya and N. S. Vyazankin, IZV. AKAD. NAUK, SSSR SER. KHIM., 23, 713 (1974).
64. U. Blaukat and W. P. Neumann, J. ORGANOMETAL. CHEM., 63, 27 (1973).
65. V. M. Vodolazskaya, B. V. Fedot'ev, Yu. I. Baakov, O. A. Kruglaya and N. S. Vyazankin, J. ORGANOMETAL. CHEM., 63, C5 (1973).
66. V. T. Bychkov, N. S. Vyazankin and G. A. Razuvaev, ZH. OBSHCH, KHIM., 43, 793 (1973).
67. M. N. Bochkarev, L. P. Maiorova and N. S. Vyazankin, ZH. OBSHCH. KHIM., 42, 2348 (1972).
68. M. N. Bochkarev, S. P. Korneva, L. P. Maiorova, V. A. Kuznetsov and

68. M. N. Bochkarev, S. P. Korneva, L. P. Maiorova, V. A. Kuznetsov and N. S. Vyazankin, ZH. OBSHCH, KHIM., 44, 308 (1974).
69. O. A. Kruglaya, L. I. Zakharkin, V. I. Petrov, G. S. Kalinina and N. S. Vyazankin, SYN. INORG. METAL-ORG. CHEM., 3, 63 (1973).
70. D. Breitinger and G. P. Arnold, INORG. NUCL. CHEM. LETT., 10, 517 (1974).
71. D. Breitinger and W. Morell, INORG. NUCL. CHEM., LETT., 10, 409 (1974).
72. H. J. Haupt and F. Newmann, J. ORGANOMETAL. CHEM., 74, 185 (1974).
73. W. Jetz and W. A. G. Graham, J. ORGANOMETAL. CHEM., 69, 383 (1974).

PHOTOCHEMICAL SYNTHESES OF INORGANIC COMPOUNDS

John R. Wasson

Department of Chemistry
University of Kentucky
Lexington, Kentucky 40506

INTRODUCTION

Since 1948, a time when Noyes and Boekelheide (1) presented an assessment of the state of the art, organic photochemistry has achieved something akin to a level of true sophistication. Organic photochemical syntheses have become routine even if the products are not always those anticipated. In putting together this review of synthetic inorganic photochemistry the author has repeatedly recalled the remarks of Noyes and Boekelheide (1) concerning the primitive state of their subject. Indeed, despite certain developments the present view of synthetic inorganic photochemistry is one of a vast rather unexplored area; if for no other reason, there are more elements of interest than those of concern to organic chemists.

Although a bountiful literature on inorganic photochemistry exists, the number of studies exclusively concerned with syntheses is rather small. Few attempts have been made to develop photochemical syntheses of various classes of materials. Perforce the present effort is made to provide a guide to the literature and equipment, review photochemical reactions of compounds of various elements and suggest profitable directions of inquiry. Space limitations preclude exposition of basic photochemistry and an exhaustive survey of the literature. Since the spectroscopy and photochemistry of inorganic compounds are necessarily considered together, a number of articles more relevant to electronic spectroscopy have also been cited. However, it is hoped that this article encourages additional studies of the photochemistry of inorganic compounds, particularly with regard to syntheses. Synthetic photochemical studies can be expected to yield new and unusual inorganic compounds, provide improved syntheses of some known compounds and pave the way for basic investigations of photophysical and photochemical processes.

THE LITERATURE

Besides texts well-known to organic chemists (2,3) a number of elementary texts (4,5) can be recommended. Several monographs (6-12), although not necessarily of immediate interest to inorganic chemists, are worthy of perusal. The text by Balzani and Carassiti (13) is of particular interest. A forthcoming book (14) should also be of appeal to inorganic chemists. The review literature (Table 1) is very valuable. For experimental details the thesis literature is generally superior to journal articles, reviews and monographs. The novice will quickly note that in spite of the abundance of literature, there is a need for an elementary practical text on inorganic photochemistry as well as assorted well-described photochemical syntheses of inorganic materials. Presently, the novice must start simply, experience several frustrations and slowly grope his way to higher levels of efficiency and sophistication. Nevertheless, many useful and important photochemical syntheses of inorganic compounds can be developed with rather simple and inexpensive equipment and a relative disregard for quantum yields and details of photochemical processes.

Table 1. Guide to the Photochemical Literature

Series

Specialist Periodical Reports: Photochemistry, The Chemical Society, London, Vol. 5, 1974.

Advances in Photochemistry, J. Wiley, N.Y., Vol. 8, 1971.

Annual Survey of Photochemistry, J. Wiley, N.Y., Vol. 3, 1971.
E. C. Lim, ed., "Excited States", Academic Press, N.Y., Vol. 1, 1974.
J. B. Birks, ed., "Organic Molecular Photophysics", J. Wiley, N.Y., Vol. 1, 1973.

Journals

Molecular Photochemistry, Vol. 6, 1974.

Photochemistry and Photobiology, Vol. 21, 1975.

Journal of Photochemistry, Vol. 3, 1975.

Journal of Luminescence, Vol. 9, 1975.

Reviews	Reference
Photochemistry in interstellar space	15
Relationship between mass spectrometric, thermolytic and photolytic reactivity	16
Luminescence of molecular systems	17
Principles governing the luminescence of organic molecules	18
Structures of electronically excited molecules	19, 20
Charge-transfer complexes in the excited state	21
Excimers	22
Chemically produced excited states	23
Quantum processes in photosynthesis	24
Photophysical primary steps in solutions of aromatic compounds	25
Radiationless transitions in gases and liquids	26
The triplet state and molecular electronic processes in organic molecules	27

Photochemistry of coordinated azides	74
Charge transfer photochemistry of coordination complexes in solution	75
Solid-state photochemical reactions of coordination compounds	76
Photochemistry of hexacoordinate complexes of the heavier transition metals	77
Excited states of metal complexes and their reactions	78

The list of books and reviews in this article is not exhaustive. Additional bibliographies can be found in Schönberg's text (55) and in PHOTOCHEM. PHOTOBIOLOGY 20, 309 (1974).

EXPERIMENTAL TECHNIQUES

Equipment for photochemical syntheses has been described in several texts (1-5) and reviews (30, 40-45). Light sources, filters and filter solutions, reaction vessels and quantum yield measurements have all been discussed in better detail elsewhere. Table 2 lists a number of sources for descriptions of photochemical procedures and techniques.

Table 2. Photochemical Apparatus and Procedures

Quantum Yield Measurements	Reference
Apparatus for quantitative and preparative photolysis: The Wisconsin Black Box	79
"Merry-Go-Round" quantum yield apparatus	80
Complications in measuring quantum yields using cylindrical sample cells	81
The potassium ferrioxalate actinometer	82
The malachite green cyanide actinometer	83
The uranyl oxalate actinometer	84
The Reinecke salt actinometer	85
The o-Nitrobenzaldehyde actinometer	86
Review of analytical methods and techniques in actinometry	87, 93
Quantum yields for solid-state photochemical reactions	88
Determination of absolute emission quantum yields of powdered samples	89
Procedure for determination of quantum yields	94
A thermal actinometer for flash photolysis	95
Apparatus	
Electrodelessly discharged photochemical lamps	90
The bromine (1633Å) UV lamp	91
A simple filter holder for a commercial immersion well photolysis apparatus	92
A simple apparatus for quantitative photochemical studies	96
An inexpensive pyrex photochemical reactor	97
Construction of irradiation bins	98
A series of sharp-cut and band-pass glass filters for the range 252-800 nm	99

An apparatus and method for preparative scale mercury-sensitized vapor phase photochemistry	100
Light sources and light filters	55
Preparative photochemical instrumentation	1-5,40,55, 101-103
Vacuum ultraviolet techniques in photochemistry	104
Vacuum techniques in photochemistry	105
Photochemistry of solid layers	106
Electron microscopy and the photochemical reactions of solids	107
Diffuse reflectance spectroscopy and photochemistry of coordination compounds	108

In synthetic work polychromatic radiation of high intensity is often employed. This is generally provided by the rather common 450-W high pressure Hg lamp used with immersion well assemblies. In some of our own work the apparatus illustrated in Figure 1 has been employed. It consists of a low pressure Hg immersion lamp with an output of 2.5-W at 2537Å. About 90% of the light emitted is at 2537Å and the long-lived lamps rarely heat solutions beyond 40°C. The lamps (of varying lengths) are available from ACE GLASS and other commercial sources. The reaction vessel consists of a manifold which holds the immersion lamp, a reflux condenser, a thermometer and a connection to a gaseous nitrogen line. The solution to be photolyzed is placed in bottom sample chamber with a spin bar. The system is connected and flushed with nitrogen (to prevent ozone formation) before turning on the lamp. The system is continually flushed with nitrogen and stirred magnetically during the photolysis period. The Pyrex reaction vessels can be conveniently assembled by a local glassblower or obtained commercially (Mateyka Scientific Glassblowing, 3345 Post Road, #117, Lexington, Kentucky 40503). In addition to conventional immersion wells and apparatus of the type just described numerous special design and commercial photochemical reactors can be obtained. Rayonet photochemical reactors (The Southern New England Ultraviolet Co., 954 Newfield St., Middletown, Conn. 06457) of the "merry-go-round" type are commonly employed for organic reactions and have equal utility for inorganic syntheses. More extensive descriptions of commercial photochemical apparatus are available (109,110).

Although preparative photochemical apparatus of the types mentioned above will be useful for synthetic work for some time to come, new equipment will necessarily need be developed as laser light sources become increasingly available, tunable, and economical. In principle, the laser offers the possibility of preparing a chosen reactant with a given vibrational level of a particular electronic excited state. The study of laser-induced photochemical reactions is still in its infancy. The possibility of selective photochemical reactions in gas mixtures by means of excitation of predissociation states of the specific molecules by laser radiation has been considered theoretically (111). Laser-induced chemical reactions have recently been the subject of a stimulating review (112). The tremendous synthetic possibilities offered by laser light sources will require a much more sophisticated appreciation of the spectroscopy of

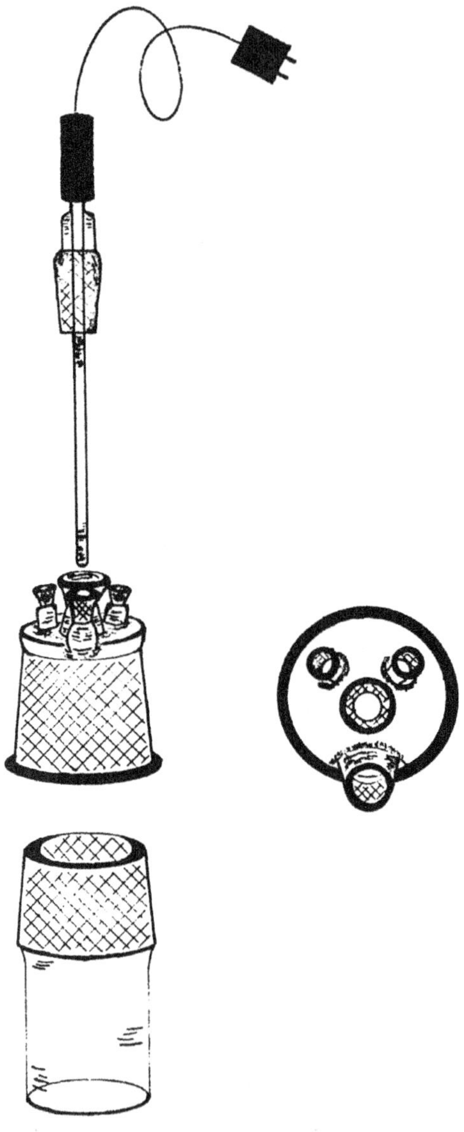

Figure 1

molecules by the preparative chemist who wishes to fully exploit the capabilities of the available technology.

PHOTOCHEMICAL REACTIONS AND SYNTHESES

In the following the author has provided a non-exhaustive potpourri of photochemical reactions and syntheses of inorganic compounds. Many additional reactions are listed in the reviews and texts cited earlier. The number of photochemical reactions of inorganic compounds known greatly exceeds the number of well-described photochemical syntheses. The spectroscopic states associated with the various reactions are generally ignored since they are discussed in the references cited or have not been elucidated in any detail. Thorough spectroscopic (absorption and luminescence spectra) characterization is of importance in understanding photochemical reactions. Nevertheless, products can be obtained in the absence of such information. It is regrettable that extensive spectroscopic <u>and</u> photochemical studies are available for so few systems.

Within recent years much work has been expended on the development of models for predicting the photoreactions and relative quantum yields of transition metal and organometallic complexes (113-118). Additional effort of this nature is needed for the understanding of the photochemical behavior of non-transition metal species.

LITHIUM

Ethyl lithium undergoes photolysis (2537Å) to yield ethane, ethylene, lithium metal and lithium hydride (119). The photolysis of o-anisyllithium in diethyl ether with a high pressure Hg arc yields, after hydrolysis, 2-methoxybiphenyl (120). Other products can also be obtained. Methyllithium adds photochemically to anthracene to yield substantial amounts of the 2- and 9-methylated products (121). UV spectra of alkyl lithium compounds and molecular orbital calculations for tetrameric and hexameric methyllithium (122) should be of assistance in elucidating the photochemistry of these systems.

GROUP IIIA ELEMENTS

Photolysis of benzene solutions of BBr_3 and BI_3 with a Hg vapor lamp results in the formation of phenylboron dibromide and phenylboron diiodide, respectively. No photochemical reaction is observed with BCl_3 (123). Radiation with a CO_2 laser (1.5 W) initiates a chain reaction with diborane (200 torr) which results in the formation of icosaborane(16), $B_{20}H_{16}$, as the main product. Visible luminescence accompanies the reaction (124). At pressures below 50 torr B_2H_6 and a laser power of 8 W no chain reaction can be started. The reaction is initiated by controlled vibration excitation of a BH_2 wagging mode. Photolysis of diborane with 1849Å ultraviolet radiation leads to the formation of H_2 and B_4H_{10}. This and related photochemical reactions of diborane have been reviewed (125). MO calculations for B_4Cl_4, $B_6H_6^{2-}$, $B_9H_9^{2-}$, $B_{10}H_{10}^{2-}$, $B_{12}H_{12}^{2-}$ and $B_{12}Cl_{12}^{2-}$ are of potential photochemical interest (126). When aqueous solutions of $CsB_{10}H_9IC_6H_5$ are irradiated with a medium-pressure Hg lamp with a Corex (290 nm cutoff) filter $B_{10}H_9I^{2-}$, O_2, C_6H_6 and a trace amount of biphenyl are obtained. When $CsB_{10}H_9IC_6H_5$ is irradiated in a water-propanol mixture, the reaction proceeds according to: $B_{10}H_9IC_6H_5 + (CH_3)_2CHOH \xrightarrow{h\nu} B_{10}H_9I^{2-} + H^+ + (CH_3)_2CO + C_6H_6$ (127). The metallocarborane μ-[(π-C_5H_5)Fe(CO)$_2$]$C_2B_4H_7$ upon UV irradiation produces the sandwich compounds (π-C_5H_5)Fe^{II}H(π-2,3-$C_2B_4H_6$) and (π-C_5H_5)Fe^{III}(π-2,3-$C_2B_4H_6$) (128). Gaseous $H_2B_2O_3$ photochemically decomposes to yield B_2O_3 and H_2 and reacts with B_2H_6 to yield boroxine, $H_3B_3O_3$ (129).

Photolysis of borazine with 1849Å radiation produces H_2, borazanaphthalene, diborazinyl and a non-volatile polymer. A mechanism has been proposed involving internal conversion from an excited singlet state of borazine to a vibrationally excited ground state. Borazyne, $B_3N_3H_4$ is a probable photochemical intermediate. In equimolar mixtures of borazine and borazine-d_6 rapid photochemical hydrogen-deuterium exchange is observed (130). B-Monochloroborazine may be obtained by the photolysis (at 1849Å) of low pressure mixtures of borazine with HCl, CH_3Cl, $CHCl_3$, CCl_4 or HSO_3Cl (131). B-Monodeuterated borazine can be prepared by the 1849Å photolysis of borazine in the presence of D_2 (132). Photolysis of hexafluoroacetone-borazine mixtures with radiation above 2200Å yields B-(2H-hexafluoro-2-propoxy)borazine and B-(perfluoro-tert-butoxy)borazine (133). These products can also be prepared by the 1849Å photolysis of borazine with the appropriate alcohol precursor. The main products of the photochemical (∿1850Å) reactions of borazine-oxygen and borazine-water mixtures are $B_3N_3H_5OH$ (B-monohydroxyborazine) and $(B_3N_3H_5)_2O$ (diborazinyl ether). $B_3N_3H_5NH_2$ (B-monoaminoborazine) results from the photochemical reaction of borazine with ammonia (134). The reactions:

$$B_3N_3H_6 + NH_3 \longrightarrow B_3N_3H_5NH_2 + H_2$$

$$B_3N_3H_6 + CH_3Br \longrightarrow B_3N_3H_5Br + CH_4$$

occur upon irradiation with 1849Å light. Small amounts of borazanaphthalene, $B_5N_5H_8$, can also be obtained and minor amounts of other products can be found depending on the details of the reaction conditions (135).

Molecular orbital calculations relevant to photochemical reactivity have been described for bis(amino)diborane(4) and (amino)vinylborane (136). Irradiation of a 0.10 M methylcyclohexane solution of trans-(benzylamino)benzylphenylborane either at 2537Å in a Rayonet reactor or with a Corex-D filtered Hg arc for 10 min at -22°C produces an aminoborane

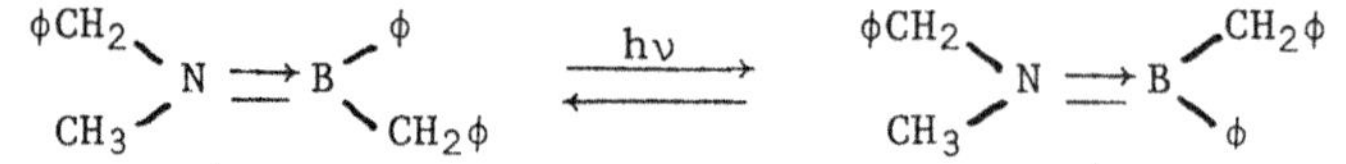

trans cis

isomer mixture containing about 15% of the cis-isomer (137). The quantum yield for the trans-cis photoisomerization at 2537Å is 0.36±0.07. Boron-alkyl photocleavage can occur in carbon tetrachloride solutions of (N-methylanilino)phenylalkylboranes (138). Tetrakis(dimethylamino)diborane-(4) reacts photochemically (300 nm) in carbon tetrabromide to yield chlorotris(dimethylamino)diborane(4) in 74% yield as well as bis(dimethyl-amino)methane, tris(dimethylamino)borane and chloroform. In chloroform irradiation of tetrakis(dimethylamino)diborane(4) gives chlorotris(di-methylamino)diborane(4) as well as smaller amounts of tris(dimethylamino)-borane, bis(dimethylamino)borane and methylene chloride. In methylene chloride, tris(dimethylamino)borane is the major product with some chlorotris(dimethylamino)diborane(4), bis(dimethylamino)borane and methyl chloride being formed (139). Photolysis of sodium tetraphenylborate at 2537Å yields biphenyl, assorted dienes as well as terphenyls while tri-phenylboron, irradiated in cyclohexane under nitrogen at 2537Å, yields no hydrocarbon products. Workup of irradiated triphenylboron yields only phenol and phenylboronic acid (140). Irradiation at 254 nm of sodium tetraphenylborate in anhydrous tetrahydrofuran or 1,2-dimethoxyethane yields biphenyl, a solution reagent having the properties of sodium di-phenylborate and a mixture of sodium arylborohydrides (141). Irradiation of neutral or slightly basic solutions of 3- and 4-pyridineboronic acids results in the formation of borate ions and pyridine (142). The photo-induced deboronations of a number of benzeneboronate anions has also been investigated (142).

The Hg-sensitized photolysis of triethylaluminum in aliphatic hydro-carbon solvents proceeds according to (143):

$$2(C_2H_5)_3Al \xrightarrow[2537Å]{Hg} C_2H_6 + [(C_2H_5)_2Al]_2CHCH_3$$

whereas the corresponding photolysis in benzene yields the metal and ethane and ethylene. Upon irradiation of various ethers in the presence of trialkylaluminum compounds cleavage of the ether can occur and the hydroxy group of benzilic alcohols can be replaced by alkyl groups (144). The reaction:

$$Tl^{+} + 2H^{+} + O_2 \xrightarrow{h\nu} Tl^{3+} + H_2O_2$$

has been examined (145) and is suggestive of synthetic applications.

GROUP IVA ELEMENTS

The photochemistry of organic compounds adsorbed on silica surfaces has received some attention (146) and similar work for inorganic species appears warranted. The photochemical syntheses of silicon compounds have been the subject of a number of papers. Rhenium carbonyl reacts with trichlorosilane and RCl_2SiH ($R = CH_3$, C_6H_5) in cyclo- or n-hexane or benzene under Hg lamp irradiation to form the silyl hydride complexes $Cl_3SiHRe_2(CO)_9$ and $RCl_2SiHRe_2(CO)_9$. Irradiation of $Re_2(CO)_{10}$ with $(C_6H_5)_3SiH$ affords $HRe_3(CO)_{14}$ (147). Photochemical reaction of an excess of trichlorosilane with 2-chloro-1,1-difluoroethylene gives trichloro-(2, 2-difluoroethyl)silane as the major product. Photochemical reaction of trichlorosilane with 2-bromo-1,1-difluoroethylene gives the reduced olefin 1,1-difluoroethylene and 2,2-difluoroethyl- and 2-bromo-2,2-difluoroethyl-trichlorosilanes (148). Photolysis of 2-phenylheptamethyltrisilane in the presence of 2,3-dimethylbutadiene at 253.7 nm yields a number of products, e.g.,

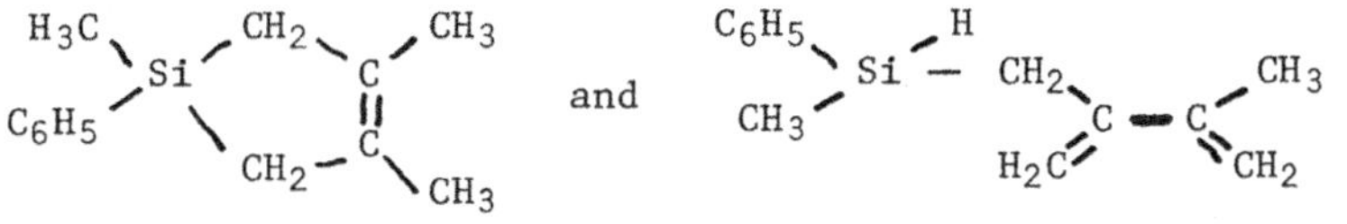

(149)

Apparently, a wide variety of organosilicon compounds can be obtained by the novel reaction pathways made available by UV excitation. The room-temperature mercury $6(^3P_1)$ photosensitization of disilane produces H_2, SiH_4, Si_3H_8, Si_4H_{10} and a solid deposit. Mono- and disilyl radicals are involved. At elevated temperatures the disilyl radical attacks disilane via the displacement reaction:

$$Si_2H_5 + Si_2H_6 \rightarrow Si_3H_8 + SiH_3$$

(150). The $^3P_1$ Hg-sensitized decomposition of SiH_4 yields hydrogen and disilane as well as polymeric silicon hydride (151). Photolysis of acylsilanes in CCl_4 generates silyl radicals which yield chlorosilanes exclusively while photolyses in alcohols yield siloxycarbenes which produce "mixed acetal" insertion products. Photolysis of acylsilanes in non-polar, non-halogenated solvents such as cyclohexane gives complex mixtures of products which result primarily from the attack of silyl radicals on the carbonyl oxygen of an acylsilane followed by disproportionation or recombination reactions. In some instances silyl radicals add to carbon-carbon double bonds (152). Photolysis of $HSi(CH_3)X_2$ ($X = CH_3$ or Cl) with a medium pressure Hg lamp yields $\cdot Si(CH_3)X_2$ radicals which can react with 1,1-di[bis(trifluoromethyl)amino]ethylene to produce silanes of the type $[(CF_3)_2N]_2C{=}CHSi(CH_3)X_2$ (153). The reaction of $GeCl_2$, $SnCl_2$ or Si_2Cl_6 with RLi or $(R'_2N)Li$ [$R = [(CH_3)_3Si]_2CH$, $R' = (CH_3)_3Si$] and subsequent ultraviolet irradiation yield stable metal-centered radicals $R_3Si\cdot$, $R_3M\cdot$

(M = Sn or Ge) and $(R'_2N)_3M\cdot$ (154).

The Hg-sensitized photolysis of mixtures of GeH_4 and NO gases at 2537Å yields Ge_2H_6, $O(GeH_3)_2$, N_2 and H_2 (155). The reaction is a convenient route for the preparation of $O(GeH_3)_2$. The photolysis of potassium trisoxalatogermanate(IV), $K_2Ge(C_2O_4)_3$, in aqueous solution has been shown (156) to yield a Ge(II) oxalate photolysis product. The UV and vacuum-UV photolyses of GeH_4, GeH_3Cl, GeH_2Cl_2, $GeHCl_3$, $GeCl_4$, GeH_3Br and GeH_2Br_2 in argon and carbon monoxide matrices yield a variety of germanium free radical species as a result of primary and secondary photolytic processes (157). The photolysis of solutions containing metal carbonyls and vinyl-substituted germanium derivatives results in the formation of several metal carbonyl derivatives (158). Photolysis of $(CH_3)_2GeM_2$ (M = $Co(CO)_4$, $Fe(CO)_2(C_5H_5)$) proceeds with loss of one carbon monoxide followed by ring closure to yield complexes with the structural unit

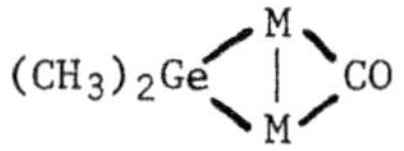

Photolysis of $(CH_3)_2Ge[Mn(CO)_5]_2$ yields the complex $(CH_3)_2GeMn_2(CO)_9$ (159). A number of compounds containing halogen-bridged and germanium-bridged metal-metal bonds can be obtained photochemically (160).

Tin(IV) tetraphenyltetrahydroporphyrin (E_T = 36.5 kcal) has been employed as a triplet photosensitizer for the cis-trans isomerization of thioindigo (161). The flash photolysis of $SnCl_2$ and $SnCl_4$ has been reported (162). $SnCl_2$ is dissociated into SnCl and Cl by absorption of light in the 320 nm region. $SnCl_2$, SnCl and Sn are observed during the flash photolysis of $SnCl_4$. This and other work indicates the synthetic photochemistry of tin compounds should become extensive. The 1,4-concerned addition of $(CH_3)_3SnH$ to singlet excited state penta-1,3-dienes yields cis- and trans-adducts (163). Triphenyltin and tri-n-butyltin radicals, generated photochemically from the corresponding hexaorganoditins, abstract halogen atoms from alkyl halides (164). $Sn[CH(Si\{CH_3\}_3)_2]_2$ upon irradiation with $Cr(CO)_6$ or $Mo(CO)_6$ in hexane affords stannylene complexes (165). The photolyses (2537Å) of azidotriphenyl derivatives of lead, tin and germanium in dichloromethane and tetrahydrofuran have been described. $(C_6H_5)_3PbN_3$ produces $(C_6H_5)_4Pb$ and a polymer in tetrahydrofuran whereas in CH_2Cl_2, $(C_6H_5)_3PbCl$ is obtained (166). Photodecomposition of solid lead chloride by irradiation with a high-pressure Hg lamp results in the formation of elemental lead and chlorine (167). Zinc oxide photosensitized photolysis of lead chloride results in 366 nm light decomposing $PbCl_2$ instead of light with wavelengths below 320 nm.

GROUP VA ELEMENTS

The photolyses of solutions containing the azide ion (168,169), aqueous (170) and solid (171) barium azide and solid lithium (172), zinc (173) and lead (174) azides as well as coordinated azide (74) have been described. Zink (175) has shown how simple MO theory can account for the photochemical behavior of $M(NH_3)_5N_3^{2+}$ (M = Co, Rh, Ir) complexes. The results may be of utility in the design of syntheses. The photolysis of azocyclohexane (>300 nm) yields cyclo-C_6H_{11}-N_2-cyclo-C_6H_{11} while photolysis in the presence of oxygen yields cyclohexyl hydroperoxide and cyclohexanone (176). N,N-Dialkylaromatic amine N-oxides undergo conformationally dependent, homolytic, β cleavages in accord with predictions of the benzene excited state theory (177). The photolysis of methyl nitrite has been thoroughly examined (178). CO and NO are produced and CH_3O radicals are involved in the photoreaction. Photolysis of nitrous acid (330-380 nm) leads to the formation of NO, NO_2 and H_2O (179). The photochemistry of oxides of nitrogen has been extensively examined with regard to air pollution problems (180). The photochemical decomposition of ethylamine (181) and the photo-oxygenation of triethylamine (182) have been characterized.

Irradiation of $NaHPO_3$ solutions with a low-pressure Hg lamp for a few hours leads to more than 10% conversion to $Na_4P_2O_6$. Besides hypophosphate and phosphate, no other products were detected (183). The radiolysis of solutions of simple phosphorus anions (184) has led to the characterization of radicals to be anticipated in photolyses. Since PI_3 and P_2I_4 in cyclohexane have identical spectra and show closely similar photochemical behavior, the inference is that P_2I_4 yields PI_3 on dissolution. Extinction coefficient data is consistent with: $P_2I_4 \rightarrow 1.17\ PI_3 + \frac{1}{6n}(P_5I_3)_n$. Irradiation of solutions of PI_3 and P_2I_4 with 366 nm light yields I_2 and ill-defined polymeric materials (185). Ultraviolet photolysis of P_2F_4 in the presence of C_2H_4, C_3H_6, 2-C_4H_8, C_6H_{10}, C_2F_4 and C_3F_6 results in the formation (186) of $F_2PCH_2CH_2PF_2$, $CH_3CHPF_2CH_2PF_2$, $CH_3CHPF_2CHPF_2CH_3$, $C_6H_{10}(PF_2)_2$, $F_2PCF_2CF_2PF_2$ and $CF_3CFPF_2CF_2PF_2$, respectively. No recoverable amount of analogous products was obtained in similar mixtures of P_2F_4 and 2-C_4F_8 or of P_2F_4 and C_6F_{10}. Photolyses (450 W medium-pressure Hg lamp, Vycor filter) of cyclohexane solutions of phosphoryl azides leads to phosphoramidate and phosphoramide products:

$$(RO)_2PON_3 + \text{cyclohexane} \xrightarrow{h\nu} (RO)_2PONHC_6H_{11} + (RO)_2PONH_2$$

$R = C_2H_5$	$R = C_2H_5$	$R = C_2H_5$
$R = C_6H_5$	$R = C_6H_5$	

The reactive phosphoryl nitrenes, $(RO)_2PON{:}$, are apparently the photochemically produced species (187). UV photolysis of high purity triphenylphosphine crystals gives rise to ESR signals interpreted (188) as being due to diphenylphosphinyl radicals. Photolytic conversion of tetramethyl methylenediphosphonate and dimethyldodecylphosphine to orthophosphate in the presence of hydrogen peroxide has been described (189). Irradiation of triarylphosphate esters with a medium pressure mercury lamp in an argon atmosphere involves the reaction:

$$(ArO)_3PO \xrightarrow{h\nu} Ar\text{-}Ar + ArOPO_3H_2$$

which is of preparative utility (190). The UV photolysis of O,O'-dialkyldithiophosphoric acids (191) gives rise to $(RO)_2P(S)S\cdot$ radicals. Neutral and alkaline solutions of nitrophenylphosphates undergo greatly enhanced hydrolysis when irradiated with ultraviolet and visible light (192). Tetraphenylarsole derivatives of metal carbonyls lose carbon monoxide upon irradiation or heating (193). Triphenylarsine, $Pb(C_6H_5)_4$ and $Sn(C_6H_5)_4$ produce biphenyl when irradiated (2537Å) in benzene (194).

GROUP VIA ELEMENTS

The dioxygenyl salts of arsenic and antimony hexafluoride, O_2AsF_6 and O_2SbF_6, can be prepared (195) by exposure of a mixture of oxygen, fluorine and the appropriate pentafluoride to daylight. The photochemical fluorination of carbonyl fluoride results in the formation of CF_3OF and $(CF_3O)_2$ as the only products (196). Flash photolysis of hydrogen sulfide in the presence of CO_2 affords a variety of intermediates (197) whereas UV photolysis gives hydrogen and sulfur. Photolysis of methanethiol at 254 and 214 nm yields H_2, CH_4 and dimethylsulfide (198). Similarly, ethanethiol photolyses to produce H_2, C_2H_6 and H_2S with ethylene as a minor product (199). Photolysis of H_2S in the presence of $(CH_3)_2SiD_2$ leads to the formation of large yields of D_2. $(CH_3)_2SiDSH$ and $(CH_3)_2SiDS\cdot$ are reactive intermediates (200). Photolysis (200-230 nm) of dimethylsulfide gives CH_4, C_2H_6, CH_3SSCH_3 and CH_3SH as the principal products. The data suggest direct C-S bond scission in the primary process giving rise to "hot" CH_3 and $CH_3S\cdot$ radicals (201). Interestingly, photolysis of crude naphtha (with a high pressure Hg lamp) with aeration can be employed (202) for the desulfurization of naphtha. Irradiation of a solution of OSF_2 in trifluoromethyl hypofluorite, CF_3OF, with a Hg/Xe lamp yields the $\cdot OSF_3$ radical. Irradiation of a solution of SO_2 yields $CF_3OSO_2\cdot$ and $FSO_2\cdot$ radicals (203). The photochemical preparations of peroxydisulfuryl difluoride, fluorine fluorosulfate and peroxysulfuryl difluoride have been described (204). The direct photolysis of aqueous solutions of sulfite yields $\cdot SO_3^-$ and the hydrated electron (205). Poly(biscarbamoyl disulfides) are decomposed by UV radiation to yield carbonyl sulfide and other products (206). Irradiation of dibenzyl diselenide with 350 nm light in the presence of oxygen produces benzaldehyde and elemental selenium whereas irradiation in the absence of oxygen produces dibenzylmonoselenide. Irradiation of dibenzyl triselenocarbonate produces dibenzyldiselenide (207). UV irradiation of xenon bis(pentafluoro-orthochalcogenates) at room temperature results in the quantitative formation of the peroxides $(F_5SeO)_2$ and $(F_5TeO)_2$ (208).

GROUP VIIA ELEMENTS

Photolysis (220-360 nm) of argon matrix samples containing ClF and O_3 produces FClO. Secondary reactions produce $FClO_2$ (209). Flash photolysis of aqueous chlorine dioxide produces Cl_2O_3. The primary act of photodissociation is hypothesized to be: $ClO_2 + h\nu \rightarrow ClO(^2\pi) + O(^3P)$ (210). Gas phase photolysis also produces Cl_2O_3. No thermal reaction occurs between cyanogenchloride and elemental fluorine up to 45°C but, when the gaseous mixture is irradiated with 366 nm light, a reaction takes place, CF_3Cl, CF_2Cl_2 and N_2 being the major products. Fluorine atoms, apparently, are the attacking species (211). Chlorine trifluoride oxide, ClF_3O, can be prepared by photolysis (using a high pressure Hg lamp) of mixtures of the elements, ClF_3-O_2 and ClF-IF_5O. Apparently, photolysis of almost any system capable of producing active oxygen and either ClF,

ClF_3 or ClF_5 will produce ClF_3O. Equipment for performing the syntheses has been described in some detail (212). Studies indicate that both singlet and triplet states of ketones are quenched according to:

$$^{1,3}K + CCl_4 \xrightarrow{313\ nm} K + Cl_3C\cdot + Cl\cdot \qquad (213)$$

This reaction might be useful for <u>in situ</u> production of $Cl\cdot$ and $Cl_3C\cdot$ radicals. The change of ^{35}Cl abundance from 75 to 46% in thiophosgene has been achieved (214) by laser (4705.5Å) excitation of particular isotopic species of thiophosgene followed by chemical reaction of electronically excited phosgene with diethoxyethylene. Flash photolysis of aqueous I_2 solutions produces transient I_2^-. Similar results obtain for methanol and isopropanol solutions of I_2 (215). The photochemistry of I^- in boric acid glass has been studied (216). Irradiation of $HgCl_2$, $HgCl_4^{2-}$, $HgBr_2$, $HgBr_4^{2-}$ and HgI_4^{2-} in aqueous solution gives rise to Cl_2^-, Br_2^- and I_2^- radical anions (217). The photochemical decomposition of the complex formed with n-butylamine and CCl_4 has been examined (218). Photolysis of cyclohexane at 1634Å in the presence of CCl_4 results in the formation of $C_6H_{11}Cl$, $C_6H_{11}CCl_3$ and C_2Cl_6 as well as other products (219). Condensed inert gas-halogen mixtures subjected to argon-ion laser and mercury arc photolysis result in the formation of KrF_2, XeF_2, $XeCl_2$ and XeClF (220).

GROUP VIB ELEMENTS

The photochemistry of this family of elements is indeed extensive and has been the subject of several reviews. Only a very few of the photochemical reactions and syntheses involving Group VIB elements can be considered here.

The primary photoreaction of dibenzenechromium(I) in aqueous solution is decomposition to Cr^+ and two molecules of benzene (221). This result suggests that photolyses in the presence of additional ligands could be employed to prepare species having unusual oxidation states and/or half-sandwich compounds.

The Group VIB hexacarbonyls are superb starting materials for many thermal and photochemical syntheses. A study of the photolysis of the hexacarbonyls in inert-gas matrices at 20°K indicates C_{4v} pentacarbonyl species are formed as intermediates in photochemical reactions (222). Photolysis of $Cr(CO)_6$ in pentane in the presence of excess NO leads to the formation of $Cr(NO)_4$ (223). Crystalline yellow tricarbonyl-Cr(0), -Mo(0) and -W(0) complexes with 2,2-para-cyclophane(22cy), $(CO)_3M(22cy)$, can be prepared (224) by UV irradiation (4 hr) of a boiling solution of 22cy and an excess of $M(CO)_6$ in ligroin (100°C) under nitrogen. Thiazole, isothiazole and 4(5)methyl-isothiazole replace one carbon monoxide in Cr and W hexacarbonyls upon UV irradiation in tetrahydrofuran solution to yield $LM(CO)_5$ complexes. Prolonged irradiation yields disubstituted products (225). The yellow, crystalline diamagnetic $M(CO)_5ts$ (where M = Cr, Mo, W) complexes of thiomorpholin-3-thione(ts) can be prepared by 2 hr u.v. irradiation of a dry ethyl ether solution containing the appropriate metal hexacarbonyl and ts in 1:1 to 1:2 molar ratios (226). Binuclear NCS-bridged metal carbonyl anions can be prepared by the reaction:

$$[M(CO)_5(NCS)]^- + M'(CO)_6 \xrightarrow[\text{THF, 6-8h}]{h\nu} [(OC)_5M\text{-}NCS\text{-}M'(CO)_5]^- + CO$$

(M = Cr, W, M' = Mo, W) (227). Photochemical reaction between $M(CO)_6$ (M = Cr, Mo and W) and SnX_2 (X = Cl, Br, I) or $CsGeCl_3$ in tetrahydrofuran produces $M(CO)_5SnX_2$ and $M(CO)_5GeCl_2$ species (228). Near-UV photolysis (300-380 nm) of $Cr(CO)_6$ in catalytic amounts in neat degassed solutions of a 1,3-diene and a silane having a Si-H bond (∿1:1 mole ratio) provides 1,4-hydrosilation of the diene (229). Photochemical oxidative decarbonylation of tricarbonylbenzenechromium in dry dimethyl disulfide yields Cr(III) methylmercaptide. Chromium(III) ethylmercaptide can be similarly prepared (230). Chromium(III) alkoxides can also be obtained in this manner (231), e.g.,

$$(C_6H_6)Cr(CO)_3 \xrightarrow[h\nu]{CH_3OH} Cr(OCH_3)_3 + C_6H_6 + 3CO$$

The interaction of dimethyl(dimethylarsinothio)arsine sulfide, $(CH_3)_2(S)AsSAs(CH_3)_2$, with u.v.-irradiated tetrahydrofuran solutions of the appropriate metal carbonyls yields the complexes $M(CO)_5L$ (M = Cr, W) and $[MM'(CO)_{10}L]$ (M = M' = Cr; M = Cr, M' = W; M = M' = W; L = dimethyl-(dimethylarsinothio)arsine sulfide. One hour irradiations lead to 40-60% yields of the products (232). Photolysis of THF solutions $Cr(CO)_6$ and diphenyldithiophosphinic acid or dicyclohexyldithiophosphinic acid results, after workup of the photolyzed solution in air, in the formation of the tris-dithio ligand Cr(III) complexes in a much purer form than obtained in the room temperature thermal reactions of the dithio-acids with $CrCl_3 \cdot 6H_2O$. With Mo and W hexacarbonyls a number of (as yet unidentified) products can be obtained (233). $[(CH_3)_2C_6H_5PSCH_3][M(CO)_5I]$ (M = Cr or W) can be obtained thermally and photolytically (234).

The photolysis of tetrabutylammonium hexacyanochromate(III) dissolved in dimethylformamide(DMF) can be used to prepare $Co(CN)_5(DMF)^{2-}$ (235). Irradiation of solid $Cr(NH_3)_6ZnCl_4X$ ($X = NO_3^-$, Cl^-) and $Cr(NH_3)_6X_3$ (X = NO_3, Cl, Br, I) at 366, 404 and 436 nm results in their partial conversion to penta-ammine $Cr(NH_3)_5X^{2+}$ complexes (236). The photoaquation of thiocyanatopentaammine- and chloropentaamine-chromium(III) ions can be sensitized by excited singlet riboflavin (237). Photolysis of $[Cr(RNH_2)_5Cl]^{2+}$ (R = H, CH_3, C_2H_5, n-C_3H_7, n-C_4H_9) in water and 33 w/w% acetone/water yields predominantly cis-$[Cr(RNH_2)_4(H_2O)Cl]^{2+}$ complexes (238). The photolysis of trans-Cr(ethylenediamine)$_2ClF^+$ in aqueous perchloric acid media by irradiation of several of the d-d transitions yields cis-$Cr(en)_2F(H_2O)^{2+}$ and minor amounts of trans-$Cr(en)_2F(H_2O)^+$ as well as a monodentate ethylenediamine complex (239). The photoaquation of $Cr(NH_3)_5X^{2+}$ (X = Cl, Br, $RCOO^-$, NCS^-) complexes yields cis-$Cr(NH_3)_4(H_2O)X^{2+}$ species (240). Biacetyl and acridinium ion sensitize the photoaquation of $Cr(NH_3)_5NCS^{2+}$ in 0.1 N sulfuric acid (241). Trans-$[(H_3N)_5CrOHCr(NH_3)_4Cl](ClO_4)_4$ and trans-$[(H_3N)_5CrOHCr(NH_3)_4(H_2O)](ClO_4)_5$ both yield cis-$[(H_3N)_5CrOHCr(NH_3)_4(H_2O)](ClO_4)_5$ upon irradiation of the d-d bonds. A quantum yield of 0.20 is found independent of the irradiation wavelength (242). Irradiation of a tetrahydrofuran or benzene solution of hexakis-(phenyl isocyanide)chromium(0) and dimethyl fumarate with a high pressure mercury lamp produces pentakis(phenyl isocyanide)(dimethyl fumarate)-

chromium(0). This type of photosubstitution was not observed with hexakis(p-tolyl isocyanide)chromium(0). Similar irradiation of benzene solutions of the isonitrile complexes mentioned above and fumaronitrile yield the pentakis(isonitrile)(fumaronitrile)chromium(0) complexes (243). The oxidation of diethyleneglycol with $K_2Cr_2O_7$ is very photosensitive. Irradiation with a 100 W Xe Lamp of solutions in an ESR cavity has been employed to produce Cr(V) and Cr(III) products. More than one step of the overall reaction is photosensitive. This reaction is of potential in the syntheses of additional Cr(V) complexes as well as Cr(III) compounds (244).

The photochemistry of the octacyanomolybdate(IV) ion has been extensively examined (245-255). A number of products, depending on the photolysis conditions, have been claimed and identified. Photolysis of an ice-cold solution of the acid gives a red anion, $[Mo(CN)_7(H_2O)]^{3-}$, which can be isolated as a silver salt (248). Yellow $K_4[Mo(CN)_7OH]$ can be isolated by addition of the calculated amount of KOH to the red solution of $Mo(CN)_7H_2O^{3-}$ prepared by photolysis of aqueous $H_4[Mo(CN)_8]$ (245). The wavelength dependence of hydrated electron formation from photoexcited $Mo(CN)_8^{4-}$ in 10 N NaOH has been investigated (253). The photohydrolysis of $K_4[Mo(CN)_8]\cdot 2H_2O$ with a Hg lamp can be employed for the preparation of blue $K_3[MoO(OH)(CN)_4]$ and red $K_4MoO_2(CN)_4]\cdot 8H_2O$ (254,255). Photohydrolysis of $K_4W(CN)_8$ can be used for the preparation of yellow-brown $K_4[WO_2(CN)_4]\cdot 6H_2O$ and purple $K_3[WO(OH)(CN)_4]$ (255). The photolysis of octacyanomolybdate(V), $Mo(CN)_8^{3-}$ in aqueous solution (λ = 365 and 436 nm) has been investigated. In neutral or acidic media red $Mo(CN)_7H_2O^{3-}$ is formed. In basic solution yellow $Mo(CN)_7OH^{4-}$ is formed which undergoes fast hydrolysis to $MoO(CN)_4OH^{3-}$. In all cases cyanogen and other nitrogen-containing species are formed (256). It would appear that photolysis of $Mo(CN)_8^{n-}$ (n = 3 or 4) complexes in non-aqueous solvents could lead to the formation of a variety of new materials. Light-induced reactions between $[(\pi\text{-}C_5H_5)Mo(CO)_3]_2$ and Cl^-, Br^-, I^- and NCS^- result in metal-metal bond cleavage (257). The photochromism of the $[\pi\text{-}C_5H_5)Mo(CO)_3]_2$+$Br^-$ system has been investigated (258). Alkyl bromides in benzene react with the complexes $[M(N_2)_2(dpe)_2]$ (M = W or Mo, dpe = 1,2-bis(diphenylphosphino)ethane) on irradiation with visible light to form alkylazo complexes of the type $[MBr(N_2R)dpe)_2]$ (R = CH_3, C_2H_5, t-C_4H_9). In the presence of methyl bromide visible irradiation causes the reaction of tetrahydrofuran with $[M(N_2)_2(dpe)_2]$ complexes to yield tetrahydropyridazido compounds, $[MBr(N_2C_4H_7)(dpe)_2]$ and their hydrobromide salts, $[MBr(N_2C_4H_8)(dpe)_2]Br$ (259). Mono- and di-substituted aminetungsten carbonyls can be prepared (260) by UV photolysis in neat amines in an N_2 atmosphere.

GROUP VIIB ELEMENTS

Solid state photolysis of potassium permanganate with 2537Å light presumably proceeds according to: $KMnO_4 \rightarrow KMnO_2 + O_2$. Above 110°C K_2MnO_2 is formed (261). This would appear to offer another route to Mn(III) compounds. Phosphine and phosphite derivatives of tetrakis(1-pyrazolyl)-boratotricarbonylmanganese(I) and hydrotris(3,5-dimethyl-1-pyrazolyl)-

boratotricarbonylmanganese(I) are air-stable pale yellow to deep orange-yellow crystalline solids. They are obtained by irradiation of benzene or tetrahydrofuran solutions of the parent tricarbonyls and ligands with long-wavelength (3660Å) light (262). Ultraviolet homolysis of the C-Mn bond of $RMn(CO)_5$ ($R = CH_3$ or $CH_2C_6H_5$) occurs in chloroform at ∿-30°C. Irradiation of binuclear metal-metal (M = Mn, Re, Fe, Mo or Co) compounds leads to M-M bond cleavage. Nitrosodurene can be employed as a spin-trap for the radicals so produced thereby affording novel organometallic nitroxide radicals (263). $(CO)_3MnB_3H_8$ can lose CO to form $(CO)_3MnB_3H_8$. This transformation can be brought about by exposing $(CO)_4MnB_3H_8$ to ultraviolet radiation in the gas phase or in CH_2Cl_2 solution at room temperature. The $(CO)_3MnB_3H_8$ reacts with excess CO in CH_2Cl_2 to regenerate $(CO)_4MnB_3H_8$ (264). Photolytic reactions between disulfides and carbonyl derivatives produce unstable $Mn(CO)_5SR$ and $(\pi\text{-}C_5H_5)Ni(CO)SR$ ($R = CF_3$ and C_6F_5) which dimerize via decarbonylation. Photolysis of $(\pi\text{-}C_5F_5)Fe(CO)_2SCF_3$ yields isomers of $[(\pi\text{-}C_5H_5)Fe(CO)SCF_3]_2$ (265). Irradiation of $HRe(CO)_5$ with tributylphosphine(L) in hexane yields $HRe(CO)_4L$ and $HRe(CO)_3L_2$. Similar irradiation of $Re_2(CO)_{10}$ yields $Re_2(CO)_9L$ and and $Re_2(CO)_8L_2$ and other products (266). $Re_2(CO)_{10}$ and $Os_3(CO)_{12}$ when irradiated by 365 nm light initiate the polymerization of methyl methacrylate and acrylonitrile in the absence of organic halide (267). The photochemical reactions of $Mn(NO)(CO)_4$ have been characterized (268). Irradiation of $[Mn(CO)_3LX]_2$ (L = CO, acetonitrile or methyl methacrylate; X = Cl, Br) at 4358Å produces manganese carbonyl, the yield of which is dependent on the nature of L. $Mn(CO)_3L_2X$ also yields manganese carbonyl on photolysis under conditions where the process $Mn(CO)_3L_2X \rightarrow Mn(CO)_3LX + L$ is not inhibited by the presence of L (269). A mixture of $(C_5H_5)Mn(CO)_2CS$ and cis-cyclooctene in n-hexane upon irradiation yields $(C_5H_5)_2Mn(CO)(CS)(C_8H_{14})$ (270). Irradiation of an acetonitrile solution of $[(n\text{-}C_4H_9)_4N]_2[Re_2Cl_8]$ with a 1000-W Hg-Xe lamp equipped with a Pyrex filter results in cleavage of the Re-Re quadruple bond and the formation of orange $[ReCl_3(CH_3CN)_3]$ as well as other products (271). Photolysis of $(NH_4)_2TcCl_6$ dissolved in hydrochloric acid produces $Tc(OH_2)Cl_5^-$ (272). The $^{99m}Tc(CO)_5$ radical, produced in high yield from the β decay of $^{99}Mo(CO)_6$ in the solid state, reacts with photochemically produced $\cdot Mn(CO)_5$ to produce $TcMn(CO)_{10}$ (273).

IRON, RUTHENIUM AND OSMIUM

UV irradiation of sea water solutions of ammonium thiocyanate, glycine and ferrous ammonium sulfate produces insoluble microspheres having peroxidase activity (274). Irradiation of solutions of glycine- and alaline-containing peptides in the presence of $K_3Fe(CN)_6$ or ceric nitrate results in the formation of free radicals (275). Tris(1,10-phenanthroline)iron(III), $Fe(phen)_3^{3+}$, undergoes photoreduction in acidic aqueous media (276). Photolysis of dinuclear $Fe_2O(phen)_4X_4$ (X = Cl, Br) in methanolic solution (20 kK - 40 kK) produces $Fe(phen)_3^{2+}$, Fe^{3+}, formaldehyde and ethylene glycol (277). The low temperature photolysis of $Fe(CN)_6^{3-}$ in aqueous methanol results in the formation of $H_2\dot{C}OH$ and $H\dot{C}O$ radicals as well as $Fe(CN)_5(CH_3OH)^{2-}$ (278). The reaction of photolyzed

(366 nm) aqueous $Na_2Fe(CN)_5NO$ with thiocyanate ion yields blue $Fe(CN)_5SCN^{3-}$ and nitric oxide (279). Photolysis of $Na_2Fe(CN)_5NO$ followed by reaction with thiourea yields blue $Fe(CN)_5(thiourea)^{2-}$ (280). Photolysis of $Fe(H_2O)_6^{3+}$ produces $Fe(H_2O)_6^{2+}$ and OH radicals. Multiple radical scavenging effects of various alcohols on this reaction have been examined (281). The photo-oxidative decarboxylation (at 350 nm) of succinic, gluteric and adipic acids with $FeCl_3$ yields β-chloropropionic acid, γ-butyrolactone and δ-chlorovaleric acid, respectively, as the main products (282). The primary products of the photodecarboxylation of Fe(III) EDTA complexes at 350 nm are CO_2, CH_2O, Fe(II) and ethylenediaminetriacetic acid (283). Photochemical degradation under natural light conditions of the acidic solution monomeric and basic solution monomeric and dimeric complexes formed between Fe^{3+} and nitrilotriacetic acid yields CO_2, H_2CO and iminodiacetic acid and precipitated iron oxide (284). Using a 250-watt General Electric sunlamp Goan et al. (285) were able to show that a variety of iron(III) β-diketonates are readily prepared in good yields by photolysis of $Fe(CO)_5$ in the presence of the β-diketone. Preliminary results of similar reactions with other metal carbonyls were also described. The photoreaction of the iron(III) chelate of 1,1,1,5,5,5-hexafluoro-2,4-pentanedione, $Fe(HFA)_3$ proceeds according to:

$$Fe(HFA)_3 \xrightarrow{h\nu,\ C_2H_5OH} Fe(HFA)_2 \cdot 2C_2H_5OH + HHFA + \text{oxidized solvent}$$

Irradiation at 345 nm results in photoreduction as the primary process (286).

Quantum yields of decomposition have been measured for the photolysis of ferrocene, ruthenocene, nickelocene, cobaltocene, dibenzenechromium and their respective cations. It was found (287) that the "18 electron" compounds were photostable while, the other, all containing unpaired electrons were not. The synthetic implications of this study need be explored. The photooxidation of ferrocene in 50% chloroform-ethanol solution can be sensitized by naphthalene (288). When a solution of azulene, ferrocene and sodium acetate in CCl_4-ethanol (1:1 v/v) is irradiated (365 or 313 nm) ethyl 1-azulenecarboxylate is obtained. In the absence of azulene ethyl ferrocenecarboxylate is the photoproduct (289). Lithium chloride and ferrocene, $FeCp_2$, react in acetone to yield the photosensitive complex $FeCp_2 \cdot LiCl$ (290). Photolysis at 365 nm takes place according to:

$$FeCp_2 \cdot LiCl \xrightarrow[\text{acetone}]{h\nu} Fe^{2+} + 2Cp^- + LiCl$$

and chlorocomplexes of iron(III) are formed as a consequence of secondary thermal reactions. A thermal process involving the ferricenium cation has been suggested and pinacol, indicative of $(CH_3)_2CO^{\bar{\cdot}}$ radicals, was detected as one of the reaction products. Benzoylferrocene photoreacts (466 and 359 nm) with water in dimethylsulfoxide and pyridine to yield cyclopentadienyliron benzoate and free cyclopentadiene (291). Ultraviolet irradiation of formyl-, acetyl- and benzoyl-ferrocene in the presence of hydrogen- and electron donors leads to photoreduction (alcohol production) only with tin(IV) hydrides (292). The photochemistry of benzoylferrocene in hydroxylic solvents has been studied and the benzoylferrocenium cation prepared (293). An improved photochemical method for preparing ferricenium chloride and ferrocenium tetrachloroferrate(III) has been described in some detail (294). UV photolysis of $Fe(CO)_5$ can be used to prepare $Fe(CO)_4$ and $Fe(CO)_x$ ($x < 4$) species (295). Photolysis of $(CH_3OC_6H_4)CH(C_6H_3OCH_3)S\{Fe(CO)_3\}_2$ at 2537Å in degassed ethanol yields the thiolactone $(CH_3OC_6H_4)CH(C_6H_3OCH_3)SC{=}O$ (296). Mixtures of $Fe(CO)_{5-x}(PF_3)_x$ compounds can be prepared from $Fe(CO)_5$ and PF_3 at elevated temperatures and pressures with the result $x \cong 2.5$. To prepare mixtures rich in $Fe(PF_3)_5$ the $x = 2.5$ product and excess PF_3 can be irradiated with a medium pressure Hg lamp to yield a final product with a mean composition about $x = 4.5$ (297). Irradiation of hexane solutions of $Fe(CO)_5$ and various amines with a Hg arc lamp affords $Fe(CO)_4L$ (L = pyridine, α, β or γ-picoline, piperidine) compounds. Similar nitrile complexes, σ-(acrylonitrile)tetracarbonyliron(0) and di-μ-acrylonitrile-bis(tetracarbonyl)-iron(0) can be prepared in the same manner (298). Methyl-substituted butadienetricarbonyliron(0) complexes can be prepared by photochemical reaction of the diene with iron pentacarbonyl in hexane at 0° for several hours. The solvent, excess diene and unreacted $Fe(CO)_5$ can be removed by vacuum distillation at -30°C. Phosphorus trifluoride substitutes photochemically at 0° into the tricarbonyls to yield all possible compounds of the type $(diene)Fe(PF_3)_x(CO)_{3-x}$ (299). Although the dissociation of benzyl isocyanide from bis(benzyl isocyanide)iron(II) phthalocyanine is slow in the dark, it is rapid in fluorescent room lighting. Over a thousandfold increase in rate is observed between dark and full illumination (300). Excitation (254, 280 and 313 nm) of aqueous solutions of cis- and trans-dicyanotetrakis(methyl isocyanide)iron(II) promotes substitution of $CNCH_3$ by water (301).

Halocarbon solvents, e.g., CCl_4, appear to form charge-transfer complexes with ruthenocene. Ruthenocene in halocarbon solvents exhibits a charge-transfer complex band at 285 nm. Irradiation by light, which excites the complexes, but not the components, causes oxidation of ruthenocene to the ruthenocene cation. Yellow crystals of $[Ru(C_5H_5)_2]RuCl_4$ result upon irradiation (280 nm) of ruthenocene in 1,2-dichloroethane solutions under nitrogen. Aqueous HCl can be employed to isolate $Ru(C_5H_5)_2Cl$ from an irradiated solution of $Ru(C_5H_5)_2$ in CCl_4 (302). Tris(2,2'-bipyridine)ruthenium(II), $Ru(bipy)_3^{2+}$, is very useful as a sensitizer and has been employed for sensitization of the photoredox reactions of $Co(C_2O_4)_3^{3-}$ and $Fe(C_2O_4)_3^{3-}$ and photoracemization of $Cr(C_2O_4)_3^{3-}$. In the redox reactions $Ru(bipy)_3^{2+}$ probably behaves as a one-electron reducing agent (303). Ruthenium(II) carbonyloctaethylporphyrin pyridinate in degassed pyridine solution can be photochemically (high pressure Hg lamp, Pyrex filter) converted to ruthenium(II) octaethylporphyrin dipyridinate. Irradiation of the initial carbonyl compound in coordinating solvents leads to the corresponding diadducts (304). The photochemistry of a series of Ru(II)-4-stilbazole complexes (305) and the 185 nm photolysis of Ru(III) ammine complexes (306) have been described.

COBALT, RHODIUM AND IRIDIUM

The photochemistry of compounds of these elements has been reviewed,

Co(III) warranting special attention. The photolysis of alkylcobaloximes and alkylcobinamides is initiated by excitation of the Co-C bond and proceeds by a mechanism related to charge-transfer induced photoreactions of Co(III) complexes (307). Many cobalt corrinoids (derivatives of vitamin B_{12}) are photolabile. Quantum yields at several wavelengths have been determined (308) for the photolysis of methylcobalamin in aqueous solution in the presence of air. The results show that photolysis is initiated by the absorption of a quantum of radiation by the intense bands due to π-π^* transitions localized within the conjugated corrin ring. The initial photolytic step is reversible, i.e.,

$$Co-CH_3 \underset{dark}{\overset{light}{\rightleftharpoons}} Co^{II} + \cdot CH_3$$

The photolysis of Co(III) β-diketonates in various solvents leads to Co(II) and oxidative fragmentation of one β-diketonate group. The reaction is partly reversible in the presence of oxygen (309). The 337 nm pulsed N_2 laser photolysis of $Co(NH_3)_6^{3+}$ demonstrates that the photoaquation ($Co(NH_3)_5OH_2^{3+}$ product) and photoredox modes are uncoupled (310). Biacetyl and quinoline photosensitize the decomposition of $Co(NH_3)_6^{3+}$ to give Co(II) (311). Evidence has been presented (312) which indicates that the photoinsensitivity of the ligand field absorption bands of most Co(III) complexes arises from the relatively rapid rate of ligand field excited singlet state to ground state singlet nonradiative deexcitation compared to ligand field excited singlet state to ligand field excited triplet state intersystem crossing. Flash photolysis studies of $Co(C_2O_4)_3^{3-}$ (313,314) have contributed significantly to the elucidation of the photochemistry of the ion and are suggestive of a number of purely synthetic studies. The low temperature solid state photochemical (∿365 nm) linkage isomerization of $[Co(NH_3)_5NO_2]Cl_2$ apparently involves bidentate(N, O) attachment of the nitrite ligand (315). Finely powdered $[Co(NH_3)_5NO_2]Br_2$ can be converted (316) to the corresponding nitrito isomer by irradiation with a 200W quartz mercury lamp. An interpretation of this behavior has been proposed (317). Photolysis of $[Co(NH_3)_5NO_2]^{2+}$ at 365 nm in 0.1 M perchloric acid containing various amounts of poly-(acrylamide) indicates that viscosity of the medium has no significant effect on the yields of Co(II) and $[Co(NH_3)_5ONO]^{2+}$ (318). The photochemistry of the Co(III) complex with 2,2'-(1,3-diaminopropane)bis(2-methyl-3-butanone oxime) has been described (319). Using the photochemical decomposition of $Co(NH_3)_5O_2CCH_3^{2+}$ as a convenient source of methyl radicals (320) the Co^{II}([14]-4,11-diene N_4) complex in ∿0.1 M $HClO_4$ under a N_2 atmosphere can be converted to $[Co([14]\text{-}4,11\text{-diene } N_4)(OH_2)CH_3](ClO_4)_2$ (321). Photochemical reactions of $(NH_3)_5CoO_2Co(NH_3)_5^{5+}$ (322), $[(H_3N)_3Co(NH_2)O_2(OH)Co(NH_3)_5]^{5+}$ (323) and RCo(salen)·H_2O (R = alkyl) (salen = bis(salicylaldehyde) ethylenediiminato) complexes (324) have been described.

Irradiation of the ligand field absorption bands of $Rh(NH_3)_5Br^{2+}$ yields Br^-, NH_4^+, <u>trans</u>-$Rh(NH_3)_4(OH_2)Br^{2+}$ and $Rh(NH_3)_5OH_2^{3+}$ (325). Irradiation of "d-d" bands of $Rh(NH_3)_5I^{2+}$ produces <u>trans</u>-$Rh(NH_3)_4(OH_2)I^{2+}$ with 90% efficiency (326). Irradiation of solutions of $(-)_{546}$-rhodium(III) D-(-)-1,2-propylenediaminetetraacetate with a low pressure Hg lamp produces a less stable diastereoisomer which returns to the starting material is about three days time (327). The photoequilibrium is established in less than 5 minutes in spite of the inertness of Rh(III) compounds. The reactions $Ir(NO)L_2CO + 2L \xrightarrow{h\nu} IrL_3NCO + L = O$ and

$$Ru(NO)_2L_2 \xrightarrow[\text{or CO, } h\nu]{\text{CO, 100 atm}} Ru(CO)_4L$$

where $L = (C_6H_5)_3P$ have been reported in a preliminary communication (328). Photolysis of aqueous solutions of $[Ir(NH_3)_5N_3](ClO_4)_2$ (>270 nm) in the presence of HCl affords $[Ir(NH_3)_5NH_2Cl](ClO_4)_3$ and N_2. In the presence of sodium iodide, N_2, $Ir(NH_3)_6^{3+}$ and I_2 are formed. Similar irradiation of $[Rh(NH_3)_5N_3](ClO_4)_2$ produces $[Rh(NH_3)_5NH_2Cl](ClO_4)_3$ and $[Rh(NH_3)_6](ClO_4)_3$ (329). Photochemical and photophysical studies of Ir(III)-phenanthroline complexes have been reported (330) recently. Exciplexes of cis-$Ir(phen)_2Cl_2^+$ have been described (331).

NICKEL, PALLADIUM AND PLATINUM

The photochemistry of Pd and Pt compounds is rather extensive compared to that of nickel compounds. Exposure of degassed ethanol solutions of Ni(II) β-diketonates to 254 nm radiation results in deposition of metallic nickel, liberation of β-diketonate and oxidation of the solvent. A black, extremely reactive Ni(I) intermediate has been proposed. The degassed solutions show no detectable photochemical reaction after irradiating for 17.5 hr at 312 nm. The reduction may take place from vibrationally excited ligand π^* states or from ligand states resulting from loss of π-level degeneracy due to chelation. No photochemical reactivity was observed in the presence of oxygen (332). Blue octahedral Ni(2,3,2-tet)$(H_2O)_2^{2+}$ (2,3,2-tet = 1,4,8,11-tetraazaundecane) can be converted to yellow planar Ni(2,3,2-tet)$^{2+}$ by irradiation of an aqueous solution of the chloride salt with 1.06 μm radiation from a neodymium laser (333). Trans-$Ni(gly)_2 \cdot 2H_2O$, trans-$Pd(gly)_2$ and trans-$Pt(gly)_2$ (gly = glycinato) are not very photosensitive to excitation in d-d bands. Trans-$Pd(gly)_2$ and trans-$Pt(gly)_2$ are decomposed upon excitation of electron-transfer bands (334). Cis-$Pt(gly)_2$ changes to the trans-isomer upon excitation in the d-d bands. The latter change have been examined experimentally (334-337) and theoretically (338). Irradiation of aqueous $PtCl_4^{2-}$ (λ > 313 nm) results in photoaquation (339) whereas irradiation at 1849Å results in a redox reaction via photoelectron formation (340). Photolysis of $Pt(NH_3)_4(NH_2)(NO_2)^{2+}$ in aqueous solution results in nitrito-isomerization, photoaquation and oxidation-reduction reactions (341). The photoaquation of hexachloroplatinate(IV) to pentachloroaquoplatinate(IV) has been characterized (342). The photoaquation of $K_2Pt(SCN)_6$ in water-acetonitrile mixtures has been examined (343). Although the photochemistry is non-trivial, some synthetic applications seem indicated.

MISCELLANEOUS

Irradiation of degassed alcohol solutions of Cu(II) β-diketonates with 254 nm radiation results in the production of metallic copper. The photoreduction is reversible in the presence of air. The primary photochemical process was determined to be a reduction of Cu(II) to Cu(I) followed by a thermal reduction of Cu(I) to copper metal. This work shows that unusual reactive Cu(I) species may be easily generated simply by photoreduction of Cu(II) β-diketonates (344). The reaction of porphyrins with Cu(II) β-diketonates in various solvents to form Cu(II) porphyrin compounds is accelerated by light (345). An excited state porphyrin seems to be the reactive species since the photochemical reaction occurs with irradiation by light of some 400 nm wavelength. The photolysis of Cu(II) malonate species in the charge-transfer spectral region leads to acetate ion and CO_2 formations without permanent reduction of the Cu(II) (346). The photoreduction of bis(3,3'-dimethylene-4,4'-dimethyl-2,2'-diquinolyl) copper(II), $Cu(dmch)_2^{2+}$, in alcoholic solution takes place according to

$$[Cu(dmch)_2ROH]^{2+} \xrightarrow{h\nu} [Cu(dmch)_2]^+ + H^+ + RO\cdot$$

upon irradiation of the 27,250 cm^{-1} charge transfer band (347). Similarly, irradiation of mono(2,9-dimethyl-1,10-phenthroline)copper(II) in methanol (348) and the corresponding bis-complex (349) in aqueous or methanol solution yields the Cu(I) complex. The influence of the presence of Cu^{2+} ions on the photochemical reactions of peresters (350) and α,α'-disubstituted stilbenes (351) has been investigated. Copper(II) complexes in amino acid crystals undergo the reaction $Cu(II)+L \xrightarrow{h\nu} Cu(I)+L^+$. The L^+ ligand cation radicals have been detected by ESR spectroscopy (352). The photoredox behavior of $H_2Cu(EDTA)\cdot H_2O$ has been elucidated (353). Pentamminecobalt(III)-cuprous ion dimers, $(NH_3)_5Co^{III}$-acrylato-$Cu^{I\,3+}$ and $(NH_3)_5Co$-fumarato-$Cu^{I\,3+}$, undergo net electron transfer slowly but electron transfer between metal centers can be photoinduced by irradiation of either Co d-d or Cu→ligand charge-transfer bands. Photolytic redox reactions of $(NH_3)_5Co^{III}LCu^I$ (L = alkenoic acids or primary aminoalkenes) have been described (354).

Blue-green crystals of $TiCl_3\cdot 3BuOH$ can be obtained with an N_2 purged pyrex vessel containing a petroleum ether solution of one equivalent of $TiCl_4$ and 3-6 equivalents of n-butanol(BuOH) upon exposure to sunlight for about 4 weeks. $TiCl_3\cdot 3C_2H_5OH$ and $TiCl_3\cdot 3n\text{-}C_3H_7OH$ can be similarly prepared. Exposure to sunlight for similar periods of $Ti(O\text{-}n\text{-}C_3H_7)_4$, $Ti(O\text{-}n\text{-}C_4H_9)_4$ and $Ti(O\text{-}C_2H_5)_4$ solutions in the appropriate alcohol containing about one equivalent of lithium halide (F, Cl or Br) or NaI, yield the green insoluble Ti(III) alkoxides, $Ti(OEt)_3$, $Ti(OPr)_3$ and $Ti(OBu)_3$ (355). $Ti(OC_2H_5)_3$ can be prepared by photolyzing $Ti(OC_2H_5)_4$ at 310 nm in ethanol in the absence of halide ions. The photoreduction of Ti(IV) with 310 and 350 nm light and assorted alcohols has been investigated (356) and suggested as a method of preparation of aldehydes and ketones. The exchange reaction (357):

$$(\pi\text{-}C_5D_5)_2TiCl_2 + (\pi\text{-}C_5H_5)_2TiCl_2 \xrightarrow{313\ nm} 2(\pi\text{-}C_5H_5)(\pi\text{-}C_5D_5)TiCl_2$$

occurs in benzene solution. The alcoholysis of $(\pi\text{-}C_5H_5)_2TiCl_2$ is affected at the same wavelength - the compounds $(\pi\text{-}C_5H_5)(CH_3O)TiCl_2$, $(\pi\text{-}C_5H_5)(C_2H_5O)TiCl_2$ and $(\pi\text{-}C_5H_5)(C_3H_7O)TiCl_2$ have been prepared in this manner (357). Dimethyl derivatives of titanocene, zirconocene and hafnocene readily undergo photolysis with homolytic cleavage of the methyl-metal bond, i.e.,

$$(C_5H_5)_2M(CH_3)_2 \xrightarrow[\text{solvent}]{h\nu} (C_5H_5)_2M + 2CH_3\cdot$$
(M = Ti, Zr, Hf)

These photolytic reactions are solvent dependent. The photolytic reaction:

$$(C_5H_5)_2M(CH_3)_2 + 2PhC{\equiv}CPh \xrightarrow[\text{pentane}]{h\nu} (C_5H_5)_2M$$ 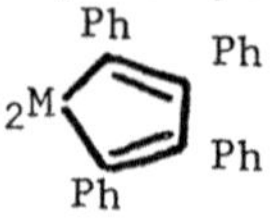

(M = Ti, Zr, Hf) $(Ph{=}C_6H_5)$

has also been examined (358). Chloro-oxobis(2,4-pentanedionato)-vanadium(V), $[VO(acac)_2Cl]$, sensitizes the polymerization of methyl methacrylate when irradiated by wavelengths in the near ultra-violet. Initiation of the polymerization occurs predominantly through scission of chlorine atoms from $[VO(acac)_2Cl]$, i.e.,

$$VO(acac)_2Cl \xrightarrow{365\ nm} VO(acac)_2 + \dot{Cl}$$

This reaction appears to also be useful for the facile introduction of chlorine into various systems (359). Irradiation of vanadium(V) in perchloric acid solution with a argon laser (488 nm) or a high pressure Hg lamp with solution and glass filters yields a solution of oxovanadium(IV) perchlorate. The reaction is presumed to proceed according to: $V(V) + H_2O \xrightarrow{h\nu} V(IV) + \cdot OH + H^+$. An equivalent amount of O_2 is produced (360).

Metal β-diketonates exhibit variable reactivity as quenchers in the photoreduction of benzophenone by benzhydrol. No correlation was found between quenching activity and the magnetic properties of the ground states of the chelates (361). Sensitization of azastilbene photoisomerizations by Zn, Mg and Co porphyrin complexes has been investigated (362).

UV illumination of zinc oxide aqueous suspensions containing assorted organic and inorganic species leads to a variety of photoreductions (363). The photolysis (3261Å) of cadmium vapor and methyl chloride at 400°C leads to the formation of cadmium chloride and a number of organic products (364). UV irradiation of a solution of mercuric acetate in acetic acid with suspended solid sulfur yields SO_4^{2-}, CO_2 and methylmercury acetate (365).

Irradiation of the charge-transfer band of Eu^{3+} in aqueous solution in the presence of scavengers for hydrogen atoms gives rise to formation of Eu^{2+} and evolution of molecular hydrogen (366). Irradiation (λ>365 nm) of a deoxygenated aqueous acid solution containing uranyl ion and excess acetaldehyde yields biacetyl as the main organic product as well as uranium(IV) species (367). The photoreduction of uranyl ion in sulfuric acid with argon and neon laser light and ethanol results in quantum yields of ∿0.6 and 0.5, respectively, for the production of U(IV) (368). $UO_2(pyridine)_2(NO_3)_2$ in absolute ethanol, when irradiated at 254 nm, forms $UO_2(pyridine)(C_2H_5OH)(NO_3)_2$. Pyridine photolysis occurs with the formation of 5-amino-2,4-pentadienal which reacts with $UO_2(NO_3)_2$ to form $UO_2(NO_3)_2(H_2N\text{-}CH{=}CH\text{-}CH{=}CH\text{-}CHO)$ (369). $[(C_2H_5)_4N]_2UCl_6$ in acetonitrile, when irradiated with 254 nm light, yields $UO_2Cl_4^{2-}$ in non-deaerated solution, $UO_2Cl_4^{2-}$ which is photoreduced to U(IV) in deaerated solution and UCl_6^- is formed in the presence of dry HCl. $[(C_2H_5)_4N]_2UO_2Cl_4$ in deaerated acetonitrile can be photoreduced to UO^{2+} and to UCl_6^- in the presence of dry HCl (370). The photochemical preparation of pyridinium oxopentaclorouranate(V), $(pyH)_2UOCl_5$ and $U(OC_2H_5)_5$ from $UO_2(py)_2Cl_2$ has been described (371).

CONCLUDING REMARKS

The purpose of this review was originally to survey inorganic photochemical syntheses. Detailed macroscopic scale photochemical preparations of inorganic compounds, however, are comparatively few. It is hoped that the survey of equipment and techniques and much abbreviated review of photochemical reactions of inorganic compounds will stimulate additional synthetic studies.

The author wishes to acknowledge stimulating conversations with Dr. E. Roland Menzel (Xerox Research Centre of Canada), Mr. D. R. Lorenz and Professor K. Niedenzu (University of Kentucky) which led to a concern with inorganic photochemical syntheses and related problems. The author is also grateful for a "seed" grant from the University of Kentucky Research Foundation for the study of synthetic inorganic photochemistry.

REFERENCES

1. W. A. Noyes, Jr. and V. Boekelheide, "Technique of Organic Chemistry", A. Weissberger, ed., Interscience, N.Y., 1948, Vol. II, p. 79.
2. N. J. Turro, "Molecular Photochemistry", W. A. Benjamin, Inc., N.Y., 1967; R. P. Wayne, "Photochemistry", Elsevier, N.Y., 1970.
3. J. Calvert and J. N. Pitts, "Photochemistry", J. Wiley, N.Y., 1966.
4. A. Cox and T. J. Kemp, "Introductory Photochemistry", McGraw-Hill, N.Y., 1971.
5. J. P. Simons, "Photochemistry and Spectroscopy", J. Wiley, N.Y., 1971; D. R. Arnold, N. C. Baird, J. R. Bolton, J. C. D. Brand, P. W. M. Jacobs, P. de Mayo and W. R. Ware, "Photochemistry", Academic Press, N.Y., 1974.
6. L. R. Koller, "Ultraviolet Radiation", J. Wiley, N.Y., 1965.
7. G. H. Brown, ed., "Photochromism", J. Wiley, N.Y., 1971.
8. A. A. Lamola and N. J. Turro, "Energy Transfer and Organic Photochemistry", J. Wiley, N.Y., 1969.
9. H. Suzuki, "Electronic Absorption Spectra and Geometry of Organic Molecules, Academic Press, N.Y., 1967.
10. R. F. Reinisch, ed., "Photochemistry of Macromolecules", Plenum, N.Y., 1970.
11. G. M. Burnett and A. M. North, eds., "Transfer and Storage of Energy by Molecules", Interscience, N.Y., Vol. 1, 1969.
12. J. B. Birks, "Photophysics of Aromatic Molecules", J. Wiley, N.Y., 1970.
13. V. Balzani and V. Carassiti, "Photochemistry of Coordination Compounds", Academic Press, N.Y., 1970.

14. A. W. Adamson and P. W. Fleischauer, ed., "Concepts of Inorganic Photochemistry", J. Wiley, N.Y., in press.
15. H. D. Breuer, ANGEW. CHEM. INTERNAT. EDN., 13, 370 (1974).
16. R. C. Dougherty, TOPICS CURRENT CHEM., 45, 93 (1974).
17. B. Brocklehurst, RADIATION RES. REVS., 2, 149 (1970).
18. R. M. Hochstrasser in "Excited States of Proteins and Nucleic Acids", R. F. Steiner and I. Weinryb, eds., Plenum Press, N.Y., 1971, Ch. 1.
19. J. F. Ogilvie, J. MOL. STRUCT., 5, 157 (1970).
20. J. C. D. Brand and D. G. Williamson, ADV. PHYS. ORG. CHEM., 1, 365 (1963).
21. M. Ottolenghi, ACCTS. CHEM. RES., 6, 153 (1973).
22. Th. Forster, ANGEW. CHEM. INTERNAT. EDN., 8, 333 (1969).
23. E. H. White, J. D. Miano, C. J. Watkins and E. J. Breaux, ANGEW. CHEM. INTERNAT. EDN., 13, 229 (1974).
24. G. W. Robinson, ANN. REV. PHYS. CHEM., 15, 311 (1964).
25. E. Lippert, ACCTS. CHEM. RES., 3, 74 (1970).
26. P. Seybold and M. Goutermann, CHEM. REVS., 65, 413 (1965).
27. S. K. Lower and M. A. El-Sayed, CHEM. REVS., 66, 199 (1966).
28. O. Buchardt, ANGEW. CHEM. INTERNAT. EDN., 13, 179 (1974).
29. R. Bersohn, ISRAEL J. CHEM., 11, 675 (1973).
30. F. Dorr in "Creation and Detection of the Excited State", A. A. Lamola, ed., M. Dekker, N.Y., 1971, Vol. 1, Part A, Ch. 2.
31. J. Michl, MOL. PHOTOCHEM., 4, 243, 257, 287 (1972). TOPICS CURR. CHEM., 46, 1 (1974); "Chemical Reactivity and Reaction Paths", G. Klopman, ed., J. Wiley, N.Y., 1974, Ch. 8.
32. W. G. Dauben, L. Salem and N. J. Turro, ACCTS. CHEM. RES., 8, 41 (1975).
33. E. W. Schlag, S. Schneider and S. F. Fischer, ANN. REVS. PHYS. CHEM., 22, 465 (1971).
34. W. A. Noyes, Jr. and K. E. Al-Ani, CHEM. REVS., 74, 29 (1974).
35. R. K. Sharma and N. Kharasch, ANGEW. CHEM. INTERNAT. EDN., 7, 36 (1968).
36. P. S. Engel and C. Steel, ACCTS. CHEM. RES., 6, 275 (1973).
37. S. G. Cohen, A. Parola and G. H. Parsons, Jr., CHEM. REVS., 73, 141 (1973).
38. G. G. Spence, E. C. Taylor and O. Buchardt, CHEM. REVS., 70, 231 (1970).
39. J. N. Pitts, Jr., F. Wilkinson and G. S. Hammond, ADV. PHOTOCHEM., 1, 1 (1963).
40. T. R. Evans in "Energy Transfer and Organic Photochemistry", P. A. Leermakers and A. Weissberger, eds., Interscience, N.Y., 1969, Ch. 4.
41. W. M. Riggs in "Analytical Photochemistry and Photochemical Analysis", J. M. Fitzgerald, ed., M. Dekker, Inc., N.Y., 1971, Ch. 4.
42. J. M. Fitzgerald, IBID., Ch. 5.
43. W. W. Wendlandt, IBID., Ch. 8.
44. E. L. Wehry, IBID., Ch. 6.
45. J. N. Demas and G. A. Crosby, J. PHYS. CHEM., 75, 991 (1971).
46. INDUSTRIAL & ENGR. CHEM., 54, 20 (1962).

47. C. L. Bumgardner and E. L. Lawton, ACCTS. CHEM. RES., 7, 14 (1974).
48. J. H. Day, CHEM. REVS., 63, 65 (1963).
49. M. F. Fox, QUART. REVS. CHEM. SOC. (London), 24, 565 (1970).
50. R. B. Bucat and D. W. Watts in "Reaction Mechanisms in Inorganic Chemistry", H. J. Emeleus and M. Tobe, eds., MTP Internat. Rev. Sci., Series One, Vol. 9 (1972). Butterworths, London.
51. M. Halmann, TOPICS PHOSPHORUS CHEM., 4, 49 (1967).
52. H. D. Burrows and T. J. Kemp, CHEM. SOC. REVS., 3, 139 (1974).
53. E. Rabinowitch and R. L. Belford, "Spectroscopy and Photochemistry of Uranyl Compounds", Pergamon Press, N.Y., 1964.
54. R. E. Bozak, ADV. PHOTOCHEM., 8, 227 (1971).
55. A. Schönberg, "Preparative Organic Photochemistry", Springer-Verlag, N.Y., 1968, Ch. 45.
56. W. Strohmeier, ANGEW. CHEM., 76, 873 (1964).
57. M. Wrighton, CHEM. REVS., 74, 401 (1974).
58. E. Koerner von Gustorf and F.-W. Grevels, TOPICS CURR. CHEM., 13, 366 (1969).
59. A. W. Adamson, REC. CHEM. PROGR., 29, 191 (1968).
60. A. W. Adamson, COORD. CHEM. REVS., 3, 169 (1968).
61. V. Balzani, L. Moggi and V. Carassiti, BER. BUNSENGES. PHYS. CHEM., 72, 288 (1968).
62. I. Fischler, J. M. Kelly, P. Kirsch and E. Koerner von Gustorf, MOL. PHOTOCHEM., 5, 497 (1973).
63. W. L. Waltz and R. G. Sutherland, CHEM. SOC. REVS., 1, 241 (1972).
64. E. L. Wehry, QUART. REVS. CHEM. SOC. (London), 21, 213 (1967).
65. A. W. Adamson, W. L. Waltz, E. Zinato, D. W. Watts, P. D. Fleischauer and R. D. Lindholm, CHEM. REVS., 68, 541 (1968).
66. M, K. DeArmond, ACCTS. CHEM. RES., 7, 309 (1974).
67. L. Moggi and V. Balzani, MOL. PHOTOCHEM., 4, 559 (1972).
68. J. F. Endicott and T. Kelly, ANN. SURVEY PHOTOCHEM., 3, 323 (1969).
69. F. E. Lytle, APPL. SPECTRY, 24, 319 (1970).
70. P. D. Fleischauer and P. Fleischauer, CHEM. REVS., 70, 199 (1970).
71. A. D. Kirk, MOL. PHOTOCHEM., 5, 127 (1973).
72. D. Valentine, Jr., ADV. PHOTOCHEM., 6, 123 (1968).
73. V. Balzani, L. Moggi, F. Scandola and V. Carassiti, INORG. CHIM. ACTA REVS., 1, 7 (1967).
74. Z. Dori and R. F. Ziolo, CHEM. REVS., 73, 247 (1973).
75. J. F. Endicott, ISRAEL J. CHEM., 8, 209 (1970).
76. E. L. Simmons and W. W. Wendlandt, COORD. CHEM. REVS., 7, 11 (1971).
77. P. C. Ford, J. D. Petersen and R. E. Hintze, COORD. CHEM. REVS., 14, 67 (1974).
78. P. D. Fleischauer, A. W. Adamson and G. Sartori, PROGR. INORG. CHEM., 17, 1 (1972).
79. H. E. Zimmerman, MOL. PHOTOCHEM., 3, 281 (1971).
80. F. G. Moses, R. S. H. Liu and B. M. Monroe, MOL. PHOTOCHEM., 1, 245 (1969).
81. G. Vesley, MOL. PHOTOCHEM., 3, 193 (1971).
82. C. G. Hatchard and C. A. Parker, PROC. ROY. SOC. (London), 235, 518 (1956).
83. G. J. Fisher, J. C. LeBlanc and H. E. Johns, PHOTOCHEM. PHOTOBIOLOGY, 6, 757 (1967).

84. C. A. Discher, P. F. Smith, I. Lippman and R. Turse, J. PHYS. CHEM., 67, 2501 (1963).
85. E. E. Wegner and A. W. Adamson, J. AMER. CHEM. SOC., 88, 394 (1966).
86. J. N. Pitts, Jr., J. K. S. Wan and E. A. Schuck, J. AMER. CHEM. SOC., 86, 3606 (1964).
87. H. A. Taylor in "Analytical Photochemistry and Photochemical Analysis", J. M. Fitzgerald, ed., M. Dekker, N.Y., 1971, Ch. 3.
88. E. L. Simmons and W. W. Wendlandt, ANAL. CHIM. ACTA, 53, 81 (1971).
89. M. S. Wrighton, D. S. Gindley and D. L. Morse, J. PHYS. CHEM., 78, 2229 (1974).
90. I. E. Den Besten and J. W. Tracy, J. CHEM. EDUC., 50, 303 (1973).
91. B. A. Thompson, R. R. Reeves, Jr. and P. Harteck, J. PHYS. CHEM., 69, 3964 (1965).
92. B. F. Plummer, MOL. PHOTOCHEM., 6, 241 (1974).
93. H. E. Johns in "Creation and Detection of the Excited State", A. A. Lamola, ed., M. Dekker, N.Y., Vol. 1, Part A, 1971, Ch. 3.
94. H. Schultze and H.-R. Vogel, MOL. PHOTOCHEM., 5, 223 (1973).
95. N. R. Greiner, CHEM. PHYS. LETT., 16, 314 (1972).
96. J. Rennert and P. H. Ginsburg, MOL. PHOTOCHEM., 4, 235 (1972).
97. J. W. Pavlik, J. CHEM. EDUC., 46, 568 (1969).
98. A. J. Taggi, J. A. Kapecki, P. H. Mazzocchi and D. A. Seeley, J. CHEM. EDUC., 48, 254 (1971).
99. W. G. Herkstroeter, MOL. PHOTOCHEM., 4, 551 (1972).
100. I. S. Krull, MOL. PHOTOCHEM., 5, 83 (1973).
101. E. J. Smith in "Analytical Phtochemistry and Photochemical Analysis", J. M. Fitzgerald, ed., M. Dekker, N.Y., 1971, Ch. 1.
102. R. Srinivasan in "Organic Photochemical Syntheses", R. Srinivasan and T. D. Roberts, eds., J. Wiley, N.Y., 1971, Vol. 1, Ch. 1.
103. W. M. Hardham in "Creation and Detection of the Excited State", A. A. Lamola, ed., M. Dekker, N.Y., Vol. 1, Part B, 1971, Ch. 13.
104. J. R. McNesby, W. Braun and J. Ball, IBID., Ch. 11.
105. H. E. Gunning and O. P. Strausz, IBID., Ch. 12.
106. E. L. Simmons, J. PHYS. CHEM., 75, 588 (1971); IBID., 78, 1265 (1974).
107. J. M. Thomas, ISRAEL J. CHEM., 10, 573 (1972).
108. E. L. Simmons, COORD. CHEM. REVS., 14, 181 (1974).
109. S. L. Murov, "Handbook of Photochemistry", M. Dekker, N.Y., 1973.
110. A. J. Gordon and R. A. Ford, "The Chemist's Companion", J. Wiley, N.Y., 1972, pp. 348-368.
111. V. S. Letokhov, CHEM. PHYS. LETT., 15, 221 (1972).
112. J. T. Knudtson and E. M. Eyring, ANN. REVS. PHYS. CHEM., 25, 255 (1974).
113. M. Wrighton, H. B. Gray and G. S. Hammond, MOL. PHOTOCHEM., 5, 165 (1973).
114. J. I. Zink, MOL. PHOTOCHEM., 5, 151 (1973).
115. M. J. Incovia and J. I. Zink, INORG. CHEM., 13, 2489 (1974).
116. J. I. Zink, J. AMER. CHEM. SOC., 94, 8039 (1972); IBID., 96, 4464 (1974).
117. J. I. Zink, INORG. CHEM., 12, 1018 (1973); IBID., 12, 1957 (1973).
118. C. Furlani, THEORET. CHIM. ACTA, 34, 233 (1974).
119. W. H. Glaze and T. L. Brewer, J. AMER. CHEM. SOC., 91, 4490 (1969).

120. W. H. Glaze and A. C. Ranade, J. ORG. CHEM., 36, 3331 (1971).
121. S. S. Hixson, J. C. S. CHEM. COMM., 574 (1974).
122. J. B. Smart, R. Hogan, P. A. Scherr, L. Ferrier and J. P. Oliver, J. AMER. CHEM. SOC., 94, 8371 (1972).
123. R. A. Bowie and O. C. Musgrave, PROC. CHEM. SOC., 15 (1964).
124. H. R. Bachmann, H. Nöth, R. Rinck and K. L. Kompa, CHEM. PHYS. LETT. 29, 627 (1974).
125. L. H. Long, PROGR. INORG. CHEM., 15, 1 (1972).
126. D. R. Armstrong, P. G. Perkins and J. J. P. Stewart, J. C. S. DALTON TRANS., 2277 (1973).
127. R. L. Middaugh, INORG. CHEM., 13, 744 (1974).
128. L. G. Sneddon, D. C. Beer and R. N. Grimes, J. AMER. CHEM. SOC., 95, 6623 (1973).
129. F. A. Grimm and R. F. Porter, INORG. CHEM., 7, 706 (1968).
130. M. A. Neiss and R. F. Porter, J. AMER. CHEM. SOC., 94, 1438 (1972).
131. M. Oertel and R. F. Porter, INORG. CHEM., 9, 904 (1970).
132. M. P. Nadler and R. F. Porter, INORG. CHEM., 8, 599 (1969).
133. L. J. Turbini, G. M. Golenwsky and R. F. Porter, INORG. CHEM., 14, 691 (1975).
134. G. H. Lee, II and R. F. Porter, INORG. CHEM., 6, 648 (1967).
135. M. A. Neiss and R. F. Porter, J. PHYS. CHEM., 76, 2630 (1972).
136. A. K. Uriarte and K. G. Hancock, INORG. CHEM., 12, 1428 (1973).
137. K. G. Hancock and D. A. Dickinson, J. AMER. CHEM. SOC., 94, 4396 (1972).
138. K. G. Hancock and D. A. Dickinson, J. AMER. CHEM. SOC., 95, 280 (1973).
139. K. G. Hancock, A. K. Uriarte and D. A. Dickinson, J. AMER. CHEM. SOC., 95, 6980 (1973).
140. J. L. R. Williams, P. J. Grisdale and J. C. Doty, J. AMER. CHEM. SOC., 89, 4538 (1967); J. L. R. Williams, J. C. Doty, P. J. Grisdale, R. Searle, T. H. Regan, G. P. Happ and D. P. Maier, IBID., 89, 5153 (1967).
141. J. J. Eisch, K. Tamao and R. J. Wilcsek, J. AMER. CHEM. SOC., 97, 895 (1974).
142. F. C. Fischer and E. Havinga, REC. TRAV. CHIM., 93, 21 (1974).
143. W. H. Glaze, T. L. Brewer and A. C. Ranade, J. ORGANOMETALLIC CHEM., 25, C6 (1970).
144. J. Furukawa, K. Omura, O. Yamamoto and K. Ishikawa, J. C. S. CHEM. COMM., 77 (1974).
145. G. F. Kirkbright, P. J. Mayne and T. S. West, J. C. S. DALTON TRANS., 1918 (1972).
146. L. D. Weis, T. R. Evans and P. A. Leermakers, J. AMER. CHEM. SOC., 90, 6109 (1968); R. L. Daubendiek and G. R. McMillan, IBID., 95, 1374 (1973).
147. J. K. Hoyano and W. A. G. Graham, INORG. CHEM., 11, 1265 (1972).
148. W. I. Bevan, R. N. Haszeldine, J. Middleton and A. E. Tipping, J. C. S. DALTON TRANS., 2305 (1974).
149. M. Ishikawa, F. Ohi and M. Kumada, J. ORGANOMETALLIC CHEM., 86, C23 (1975).
150. T. L. Pollock, H. S. Sandhu, A. Jodhan and O. P. Strausz, J. AMER. CHEM. SOC., 95, 1017 (1973).

151. H. Niki and G. J. Mains, J. PHYS. CHEM., 68, 304 (1964).
152. A. G. Brook and J. M. Duff, CAN. J. CHEM., 51, 352 (1973).
153. D. H. Coy, F. Fitton, R. N. Haszeldine, M. J. Newlands and A. E. Tipping, J. C. S. DALTON TRANS., 1852 (1974).
154. J. D. Cotton, C. S. Cundy, D. H. Harris, A. Hudson, M. F. Lappert and P. W. Lednor, J. C. S. DALTON TRANS., 651 (1974).
155. R. Varma, K. R. Ramaprasad, A. J. Signorelli and G. K. Sahay, J. INORG. NUCL. CHEM., 37, 563 (1975).
156. E. L. J. Breet and R. van Eldik, INORG. CHIM. ACTA, 9, 177, 183 (1974).
157. W. A. Guillory, R. J. Isabel and G. R. Smith, J. MOL. STRUCT., 19, 473 (1973).
158. R. C. Job and M. D. Curtis, INORG. CHEM., 12, 2510 (1973).
159. R. C. Job and M. D. Curtis, INORG. CHEM., 12, 2514 (1973).
160. M. D. Curtis and R. C. Job, J. AMER. CHEM. SOC., 94, 2153 (1972).
161. G. M. Wyman, B. M. Zarnegar and D. G. Whitten, J. PHYS. CHEM., 77, 2584 (1973).
162. G. A. Oldershaw and K. Robinson, J. CHEM. SOC. (A), 2963 (1971).
163. M. Bigwood and S. Boue, J. C. S. CHEM. COMM., 529 (1974).
164. H. G. Kuivila and C. C. H. Pian, J. C. S. CHEM. COMM., 369 (1974).
165. P. J. Davidson and M. F. Lappert, J. C. S. CHEM. COMM., 317 (1973).
166. J. Jappy and P. N. Preston, INORG. NUCL. CHEM. LETT., 4, 503 (1968).
167. A. Kaldor and G. A. Somorjai, J. PHYS. CHEM., 70, 3538 (1966).
168. K. Shinohara, T. Shida and N. Saito, J. CHEM. PHYS., 37, 2791 (1962).
169. I. Burak and A. Treinin, J. AMER. CHEM. SOC., 87, 4031 (1965).
170. V. R. Pai Verneker and M. Blais, J. PHYS. CHEM., 72, 774 (1968).
171. E. G. Prout and W. G. Sears, J. INORG. NUCL. CHEM., 35, 2163 (1973).
172. E. G. Prout and W. G. Sears, INORG. NUCL. CHEM. LETT., 9, 31 (1973).
173. S. R. Yoganarasimhan and R. K. Sood, INORG. NUCL. CHEM. LETT., 9, 1049 (1973).
174. R. B. Hall and F. Williams, J. CHEM. PHYS., 58, 1036 (1973).
175. J. I. Zink, INORG. CHEM., 14, 446 (1975).
176. J. L. Currie, H. W. Sidebottom and J. M. Tedder, J. C. S. FARADAY TRANS. I, 1851 (1974).
177. V. I. Stenberg and J. E. Schiller, J. AMER. CHEM. SOC., 97, 424 (1975).
178. H. A. Wiebe and J. Heicklen, J. AMER. CHEM. SOC., 95, 1 (1973); H. A. Wiebe, A. Villa, T. M. Hellman and J. Heicklen, IBID., 95, 7 (1973).
179. R. A. Cos, J. PHOTOCHEM., 3, 175 (1974).
180. P. A. Leighton, "Photochemistry of Air Pollution", Academic Press, N.Y., 1961.
181. Y. Kotake, M. Ono and K. Kuwata, BULL. CHEM. SOC. JAPAN, 45, 110 (1972).
182. H. Tsubomura, T. Yagishita and H. Toi, BULL. CHEM. SOC. JAPAN, 46, 110 (1972).
183. A. W. Schwartz and M. van der Veen, INORG. NUCL. CHEM. LETT., 9, 39 (1973).
184. D. Behar and R. W. Fessenden, J. PHYS. CHEM., 76, 1706, 1710 (1972).
185. T. Kennedy and R. S. Sinclair, J. INORG. NUCL. CHEM., 32, 1125 (1970).

186. J. G. Morse and K. W. Morse, INORG. CHEM., 14, 565 (1975).
187. R. Breslow, A. Feiring and F. Herman, J. AMER. CHEM. SOC., 96, 5937 (1974).
188. W. T. Cook, J. S. Vincent, I. Bernal and F. Ramirez, J. CHEM. PHYS., 61, 3479 (1974).
189. R. A. Libby, INORG. CHEM., 10, 386 (1971).
190. R. A. Finnegan and J. A. Matson, J. AMER. CHEM. SOC., 94, 4780 (1972).
191. M. Sato, M. Yanagita, Y. Fujita and T. Kwan, BULL. CHEM. SOC. JAPAN, 44, 1423 (1971).
192. E. Havinga, R. O. de Jongh and W. Dorst, REC. TRAV. CHIM., 75, 378 (1956).
193. E. W. Abel, I. W. Nowell, A. G. J. Modinos and C. Towers, J. C. S. CHEM. COMM., 258 (1973).
194. D. B. Peterson, D. A. G. Walmsley, R. J. Povinelli and M. Burton, J. PHYS. CHEM., 71, 4506 (1967).
195. J. Shamir and J. Binenboym, INORG. CHIM. ACTA, 2, 37 (1968).
196. M. I. Lopez, E. Castellano and H. J. Schumacher, J. PHOTOCHEM., 3, 97 (1974).
197. R. B. Langford and G. A. Oldershaw, J. C. S. FARADAY TRANS. I, 1550 (1972).
198. L. Bridges and J. M. White, J. PHYS. CHEM., 77, 295 (1973).
199. L. Bridges, G. L. Hemphill and J. M. White, J. PHYS. CHEM., 76, 2668 (1972).
200. A. G. Alexander, R. W. Fair and O. P. Strauz, J. PHYS. CHEM., 78, 203 (1974).
201. P. M. Rao and A. R. Knight, CAN. J. CHEM., 50, 844 (1972).
202. Y. Baba, BULL. CHEM. SOC. JAPAN, 47, 204 (1974).
203. J. R. Morton and K. F. Preston, J. CHEM. PHYS., 58, 2657 (1973).
204. M. Gambaruto, J. E. Sicre and H. J. Schumacher, J. FLUORINE CHEM., 5, 175 (1975).
205. O. P. Chawla, N. L. Arthur and R. W. Fessenden, J. PHYS. CHEM., 77, 772 (1973).
206. N. Kobayashi, A. Osawa and T. Fujisawa, CHEM. LETT., 1443 (1974).
207. W. Stanely, M. R. Van DeMark and P. L. Kumber, J. C. S. DALTON TRANS., 700 (1974).
208. K. Seppelt and D. Nothe, INORG. CHEM., 12, 2727 (1973).
209. L. Andrews, F. K. Shi and A. Arkell, J. AMER. CHEM. SOC., 96, 1997 (1974).
210. J. C. Mialocq, F. Barat, L. Gilles, B. Hickel and B. Lesigne, J. PHYS. CHEM., 77, 742 (1973).
211. E. V. Oexler and E. H. Staricco, INORG. NUCL. CHEM. LETT., 8, 469 (1972).
212. D. Pilipovich. H. H. Rogers and R. D. Wilson, INORG. CHEM., 11, 2192 (1972).
213. E. A. Lissi and M. V. Encina, J. PHOTOCHEM., 3, 237 (1974).
214. M. Lamotte, H. J. Dewey, R. A. Keller and J. J. Ritter, CHEM. PHYS. LETT., 30, 165 (1975).
215. P. Fornier de Violet, R. Bonneau and J. Joussot-Dubien, MOL. PHOTOCHEM., 5, 61 (1973); CHEM. PHYS. LETT., 19, 251 (1973).

216. A. Zaliouk-Gitter and A. Treinin, J. CHEM. PHYS., 42, 2019 (1965).
217. M. E. Langmuir and E. Hayon, J. PHYS. CHEM., 71, 3808 (1967).
218. C. J. Biaselle and J. G. Miller, J. AMER. CHEM. SOC., 96, 3813 (1974).
219. J. Nafisi-Movaghar and Y. Hatano, J. PHYS. CHEM., 78, 1899 (1974).
220. W. F. Howard, Jr. and L. Andrews, J. AMER. CHEM. SOC., 96, 7864 (1974).
221. O. Traverso, F. Scandola, V. Balzani and S. Valcher, MOL. PHOTOCHEM., 1, 289 (1969).
222. M. A. Graham, M. Poliakoff and J. J. Turner, J. CHEM. SOC. (A), 2939 (1971); R. N. Perutz and J. J. Turner, INORG. CHEM., 14, 262 (1975).
223. B. I. Swanson and S. K. Satija, J. C. S. CHEM. COMM., 40 (1973).
224. F. Cristiani, D. De Filippo, P. Deplano, F. Devillanova, A. Diaz, E. F. Trogu and G. Verani, INORG. CHIM. ACTA, 12, 119 (1975).
225. K. H. Pannell, C. C.-Y. Lee, C. Parkanyi and R. Redfearn, INORG. CHIM. ACTA, 12, 127 (1975).
226. D. Defilippo. A. Lai, E. F. Trogu, G. Verani and C. Preti, J. INORG. NUCL. CHEM., 36, 73 (1974).
227. H. Behrens, D. Uhlig and E. Lindner, Z. ANORG. ALLGEM CHEM., 394, 8 (1972).
228. D. Uhlig, H. Behrens and E. Lindner, Z. ANORG. ALLGEM. CHEM., 401, 233 (1973).
229. M. S. Wrighton and M. A. Schroeder, J. AMER. CHEM. SOC., 96, 6235 (1974).
230. D. A. Brown, D. Cunningham and W. K. Glass, CHEM. COMM., 736 (1967).
231. D. A. Brown, D. Cunningham and W. K. Glass, CHEM. COMM., 306 (1966).
232. E. W. Ainscough, A. M. Brodie and G. Leng-Ward, J. C. S. DALTON TRANS., 2437 (1974).
233. D. R. Lorenz and J. R. Wasson, Unpublished results.
234. E. W. Ainscough, A. M. Brodie and A. R. Furness, J. C. S. DALTON TRANS., 2360 (1973).
235. H. F. Wasgestian, J. PHYS. CHEM., 76, 1947 (1972).
236. C. D. Flint, P. Greenough and A. P. Matthews, J. C. S. DALTON TRANS., 368 (1973).
237. E. L. Wehry, J. AMER. CHEM. SOC., 95, 2137 (1973).
238. C. F. C. Wong and A. D. Kirk, CAN. J. CHEM., 53, 419 (1975).
239. G. Wirth and R. G. Linck, J. AMER. CHEM. SOC., 95, 5913 (1973).
240. E. Zinato and P. Riccieri, INORG. CHEM., 12, 1451 (1973).
241. J. E. Martin and A. W. Adamson, THEORET. CHIM. ACTA, 20, 119 (1971).
242. P. Riccieri and E. Zinato, INORG. CHIM. ACTA, 7, 117 (1973).
243. K. Iuchi, S. Asada and A. Sugimori, CHEM. LETT., 801 (1974).
244. M. Mitewa, A. Malinovski, P. R. Bontchev and K. Kabassanov, INORG. CHIM. ACTA, 8, 17 (1974).
245. R. P. Mitra and H. Mohan, J. INORG. NUCL. CHEM., 36, 3739 (1974).
246. A. W. Adamson and A. H. Sporer, J. AMER. CHEM. SOC., 80, 3865 (1958).
247. W. L. Waltz, A. W. Adamson and P. D. Fleischauer, J. AMER. CHEM. SOC., 89, 3923 (1967).
248. R. P. Mitra, B. K. Sharma and H. Mohan, CAN.J. CHEM., 47, 2317 (1969).

249. V. Balzani, M. F. Manfrin and L. Moggi, INORG. CHEM., 8, 47 (1969).
250. R. D. Wilson, V. S. Sastri and C. H. Langford, CAN. J. CHEM., 49, 679 (1971).
251. J. R. Perumareddi, Ph.D. thesis, University of Southern California, 1962; University Microfilms, Ann Arbor, Order No. 64-2600.
252. E. Nagorsnik, P. Thomas and E. Hoyer, INORG. NUCL. CHEM. LETT., 10, 353 (1974).
253. M. Shirom and Y. Siderer, J. CHEM. PHYS., 58, 1250 (1973).
254. R. L. Marks and S. E. Popielski, INORG. NUCL. CHEM. LETT., 10, 885 (1974).
255. S. J. Lippard and B. J. Russ, INORG. CHEM., 6, 1943 (1967).
256. G. W. Gray and J. T. Spence, INORG. CHEM., 10, 2751 (1971).
257. A. R. Burkett, T. J. Meyer and D. G. Whitten, J. ORGANOMETAL. CHEM., 67, 67 (1974).
258. J. L. Hughey, IV and T. J. Meyer, INORG. CHEM., 14, 947 (1975).
259. A. A. Diamantis, J. Chatt, G. A. Heath and G. J. Leigh, J. C. S. CHEM. COMM., 27 (1975).
260. S. C. Tripathi, S. C. Srivastava, G. Prasad and R. P. Mani, J. ORGANOMETALLIC CHEM., 86, 229 (1975).
261. E. G. Prout and C. M. Lownds, INORG. NUCL. CHEM. LETT., 9, 377 (1973).
262. A. R. Schoenberg and W. P. Anderson, INORG. CHEM., 11, 85 (1972).
263. A. Hudson, M. F. Lappert, P. W. Lednor and B. K. Nicholson, J. C. S. CHEM. COMM., 966 (1974).
264. D. F. Gaines and S. J. Hildebrandt, J. AMER. CHEM. SOC., 96, 5574 (1974).
265. J. L. Davidson and D. W. A. Sharp, J. C. S. DALTON TRANS., 1957 (1973).
266. B. H. Byers and T. L. Brown, J. AMER. CHEM. SOC., 97, 947 (1975).
267. C. H. Bamford and M. U. Mahmud, J. C. S. CHEM. COMM., 762 (1972).
268. D. P. Keeton and F. Basolo, INORG. CHIM. ACTA, 6, 33 (1972).
269. C. H. Bamford, J. W. Burley and M. Coldbeck, J. C. S. DALTON TRANS., 1846 (1972).
270. A. E. Fenster and I. S. Butler, INORG. CHEM., 13, 915 (1974).
271. G. L. Geoffroy, H. B. Gray and G. S. Hammond, J. AMER. CHEM. SOC., 96, 5565 (1974).
272. M. Koyama, Y. Kanchiku and T. Fujinaga, COORD. CHEM. REVS., 3, 285 (1968).
273. I. G. de Jong and D. R. Wiles, INORG. CHEM., 12, 2519 (1973).
274. A. E. Smith, K. Raab and J. A. Ekpaha-Mensah, EXPERIENTIA, 27, 648 (1971).
275. I. Rosenthal, R. Poupko and D. Elad, J. PHYS. CHEM., 77, 1944 (1973).
276. E. L. Wehry and R. A. Ward, INORG. CHEM., 10, 2660 (1971).
277. P. G. David and E. L. Wehry, MOL.PHOTOCHEM., 5, 21 (1973).
278. M. C. R. Symons, D. X. West and J. G. Wilkinson, J. PHYS. CHEM., 78, 1335 (1974).
279. P. A. Stoeri and D. X. West, J. INORG. NUCL. CHEM., 36, 2347 (1974).
280. P. A. Stoeri and D. X. West, J. INORG. NUCL. CHEM., 36, 3883 (1974).
281. J. H. Carey and C. H. Langford, INORG. NUCL. CHEM. LETT., 10, 591 (1974).

282. J. A. Kuhnle, R. E. Lundin and A. C. Weiss, Jr., J. C. S. CHEM. COMM., 287 (1972).
283. J. H. Carey and C. H. Langford, CAN. J. CHEM., 51, 3665 (1973).
284. T. Trott, R. W. Henwood and C. H. Langford, ENVIRONMENTAL SCI. TECHN., 6, 367 (1972).
285. J. Goan, G. Huether and H. Podall, INORG. CHEM., 2, 1078 (1963).
286. H. D. Gafney, R. L. Lintvedt and I. S. Jaworiwsky, INORG. CHEM., 9, 1728 (1970); H. D. Gafney, Ph.D. thesis, Wayne State University, Detroit, Michigan, 1970; University Microfilms, Ann Arbor, Mich., Order No. 71-17,265.
287. P. Borrell and E. Henderson, INORG. CHIM. ACTA, 12, 215 (1975).
288. O. Traverso, F. Scandola and V. Carassiti, INORG. CHIM. ACTA, 6, 471 (1972).
289. T. Ishigami, T. Akiyama, H. Watanabe, T. Kato and A. Sugimori, J. C. S. DALTON TRANS., 871 (1973).
290. E. Horvath, S. Sostero, O. Traverso and V. Carassiti, GAZZ. CHIM. ITAL., 104, 1003 (1974).
291. L. H. Ali, A. Cox and T. J. Kemp, J. C. S. CHEM. COMM., 265 (1972).
292. J. D. Coyle and Marr, J. ORGANOMETAL. CHEM., 60, 153 (1973).
293. O. Traverso, R. Rossi, S. Sostero and V. Carassiti, MOL. PHOTOCHEM., 5, 457 (1973).
294. O. Traverso, R. Rossi and V. Carassiti, SYN. REACT. INORG. METAL-ORG. CHEM., 4, 309 (1974).
295. M. Poliakoff and J. J. Turner, J. C. S. DALTON TRANS., 1351 (1973); IBID., 2276 (1974); M. Poliakoff, IBID., 210 (1974).
296. H. Alper and A. S. K. Chan, J. AMER. CHEM. SOC., 95, 4905 (1973).
297. H. Mahnke, R. J. Clark, R. Rosanske and R. K. Sheline, J. CHEM. PHYS., 60, 2997 (1974).
298. E. H. Schubert and R. K. Sheline, INORG. CHEM., 5, 1071 (1966).
299. M. A. Busch and R. J. Clark, INORG. CHEM., 14, 219 (1975).
300. D. V. Stynes, J. AMER. CHEM. SOC., 96, 5942 (1974).
301. G. Condorelli, L. Giallongo, A. Giuffrida and G. Romeo, INORG. CHIM. ACTA, 7, 7 (1973).
302. O. Traverso, S. Sostero and G. A. Mazzocchin, INORG. CHIM. ACTA, 11, 237 (1974).
303. J. N. Demas and A. W. Adamson, J. AMER. CHEM. SOC., 95, 5159 (1973).
304. F. R. Hopf, T. P. O'Brien, W. R. Schedit and D. G. Whitten, J. AMER. CHEM. SOC., 97, 277 (1975).
305. P. P. Zarnegar, C. R. Bock and D. G. Whitten, J. AMER. CHEM. SOC., 95, 4367 (1973).
306. J. Siegel and J. N. Armor, J. AMER. CHEM. SOC., 96, 4102 (1974).
307. G. N. Schrauzer, L. P. Lee and J. W. Sibert, J. AMER. CHEM. SOC., 92, 2997 (1970).
308. J. M. Pratt and B. R. D. Whitear, J. CHEM. SOC. (A), 252 (1971).
309. N. Filipescu and H. Way, INORG. CHEM., 8, 1863 (1969).
310. K. M. Cunningham and J. F. Endicott, J. C. S. CHEM. COMM., 1024 (1974).
311. H. D. Gafney and A. W. Adamson, J. PHYS. CHEM., 76, 1105 (1972).
312. P. Natarajan and J. F. Endicott, J. AMER. CHEM. SOC., 94, 3635 (1972).

313. L. Cordemans, J. D'Olieslager, J. Hendrix and S. De Jaegere, J. PHYS. CHEM., 78, 1361 (1974).
314. N. S. Rowan, M. Z. Hoffman and R. M. Milburn, J. AMER. CHEM. SOC., 96, 6060 (1974).
315. D. A. Johnson and K. A. Pashman, INORG. NUCL. CHEM. LETT., 11, 23 (1975).
316. W. W. Wendlandt and J. H. Woodlock, J. INORG. NUCL. CHEM., 27, 259 (1965).
317. J. F. Endicott, INORG. CHEM., 14, 448 (1975).
318. P. Natarajan, J. C. S. CHEM. COMM., 26 (1975).
319. F. Diomedi-Camassei, E. Nocchi, G. Sartori and A. W. Adamson, INORG. CHEM., 14, 25 (1975).
320. T. S. Roche and J. F. Endicott, J. AMER. CHEM. SOC., 94, 8622 (1972).
321. E. R. Kantrowitz, M. Z. Hoffman and J. F. Endicott, J. PHYS. CHEM., 75, 1914 (1971); T. S. Roche and J. F. Endicott, INORG. CHEM., 13, 1575 (1974).
322. J. E. Barnes, J. Barrett, R. W. Brett and J. Brown, J. INORG. NUCL. CHEM., 30, 2207 (1968).
323. J. S. Valentine and D. Valentine, Jr., INORG. CHEM., 12, 1697 (1973).
324. G. Costa, G. Mestroni and G. Pellizer, J. ORGANOMETAL. CHEM., 15, 187 (1968).
325. T. L. Kelly and J. F. Endicott, J. AMER. CHEM. SOC., 94, 278 (1972).
326. T. L. Kelly and J. F. Endicott, J. C. S. CHEM. COMM., 1061 (1971).
327. G. L. Blackmer, J. L. Sudmeier, R. N. Thibedeau and R. M. Wing, INORG. CHEM., 11, 189 (1972).
328. S. Bhaduri, B. F. G. Johnson, C. J. Savory, J. A. Segal and R. H. Walter, J. C. S. CHEM. COMM., 809 (1974).
329. J. L. Reed, F. Wang and F. Basolo, J. AMER. CHEM. SOC., 94, 7173 (1972).
330. R. Ballardini, G. Varani, L. Moggi, V. Balzani, K. R. Olson, F. Scandola and M. Z. Hoffman, J. AMER. CHEM. SOC., 97, 728 (1975).
331. R. Ballardini, G. Varani, L. Moggi and V. Balzani, J. AMER. CHEM. SOC., 96, 7124 (1974).
332. H. D. Gafney and R. L. Lintvedt, J. AMER. CHEM. SOC., 92, 6996 (1970).
333. K. J. Ivin, R. Jamison and J. J. McGarvey, J. AMER. CHEM. SOC., 94, 1763 (1972).
334. V. Balzani, V. Carassiti, L. Moggi and F. Scandola, INORG. CHEM., 4, 1243 (1965).
335. F. Scandola. O. Traverso, V. Balzani, G. L. Lucchini and V. Carassiti, INORG. CHIM. ACTA, 1, 76 (1967).
336. F. Bolletta, M. Gleria and V. Balzani, J. PHYS. CHEM., 76, 3934 (1972); V. Balzani and V. Carassiti, IBID., 72, 383 (1968).
337. F. Bolletta, M. Gleria and V. Balzani, MOL. PHOTOCHEM., 4, 205 (1972).
338. F. S. Richardson, D. D. Shillady and A. Waldrop, INORG. CHIM. ACTA, 5, 279 (1971).
339. F. Scandola, O. Traverso and V. Carassiti, MOL. PHOTOCHEM., 1, 11 (1969).

340. J. L. H. Batiste and R. Rumfeldt, CAN. J. CHEM., 52, 174 (1974).
341. N. Sabbatini, L. Moggi and G. Varani, INORG. CHIM. ACTA, 5, 469 (1971).
342. L. E. Cox, D. G. Peters and E. L. Wehry, J. INORG. NUCL. CHEM., 34, 297 (1972).
343. V. S. Sastri and C. H. Langford, J. INORG. NUCL. CHEM., 36, 2616 (1974).
344. H. D. Gafney and R. L. Lintvedt, J. AMER. CHEM. SOC., 43, 1623 (1971).
345. A. Bluestein and J. M. Sugihara, J. INORG. NUCL. CHEM., 35, 1048 (1973).
346. J. Y. Morimoto and B. A. De Graff, J. PHYS. CHEM., 79, 326 (1975).
347. Ph. Thomas, H. Spindler and D. Rehorek, Z. ANORG. ALLGEM. CHEM., 401, 165 (1973).
348. S. Sundararajan and E. L. Wehry, J. PHYS. CHEM., 76, 1528 (1972).
349. S. Sundararajan and E. L. Wehry, J. INORG. NUCL. CHEM., 34, 3699 (1972).
350. G. Sosnovsky, J. ORG. CHEM., 28, 2934 (1963); TETRAHEDRON, 21, 871 (1965).
351. D. J. Collins, AUST. J. CHEM., 20, 1905 (1967).
352. M. Fujimoto, S. Saito and Y. Tomkiewicz, BIOINORG. CHEM., 2, 341 (1973); M. Fujimoto and Y. Tomkiewicz, J. CHEM. PHYS., 56, 749 (1972).
353. P. Natarajan and J. F. Endicott, J. PHYS. CHEM., 77, 2049 (1973).
354. J. K. Hurst and R. H. Lane, J. AMER. CHEM. SOC., 95, 1703 (1973); J. K. Farr, L. G. Hulett, R. H. Lane and J. K. Hurst, IBID., in press.
355. M. R. Hunt and G. Winter, INORG. NUCL. CHEM. LETT., 6, 529 (1970).
356. F. E. McFarlane and G. W. Tindall, INORG. NUCL. CHEM. LETT., 9, 907 (1973).
357. E. Vitz and C. H. Brubaker, Jr., J. ORGANOMETAL. CHEM., 84, C16 (1974).
358. H. Alt and M. D. Rausch, J. AMER. CHEM. SOC., 96, 5936 (1974).
359. S. M. Aliwi and C. H. Bamford, J. CHEM. SOC. FARADAY TRANS. II, 70, 2092 (1974).
360. B. G. Jeliazkowa, S. Nakamura and H. Fukutomi, BULL. CHEM. SOC. JAPAN, 48, 347 (1975).
361. G. S. Hammond and R. P. Foss, J. PHYS. CHEM., 68, 3739 (1964).
362. D. G. Whitten, P. D. Wildes and C. A. De Rosier, J. AMER. CHEM. SOC., 94, 7811 (1972).
363. J. Cunningham and H. Zainal, J. PHYS. CHEM., 76, 2362 (1972).
364. M. Freiberg, D. Meyerstein and S. Weiss, J. INORG. NUCL. CHEM., 36, 1902 (1974).
365. H. Akagi, E. Takabatake and Y. Fujita, CHEM. LETT., 761 (1974).
366. Y. Haas, G. Stein and R. Tenne, ISRAEL J. CHEM., 10, 529 (1972).
367. S. Sakuraba and R. Matsushima, CHEM. LETT., 911 (1972).
368. J. T. Bell and S. R. Buxton, J. INORG. NUCL. CHEM., 36, 1575 (1974).
369. O. Traverso, V. Carassiti, R. Portanova and P. A. Vigato, INORG. CHIM. ACTA, 9, 227 (1974).
370. G. Condorelli, L. L. Costanzo, S. Pistara and E. Tondello, INORG. CHIM. ACTA, 10, 115 (1974).

371. O. Traverso, R. Portanova and V. Carassiti, INORG. NUCL. CHEM. LETT., 10, 771 (1974).

INDEX

A 5
B 6
C 7
D 8
E 9
F 0
G 1
H 2
I 3
J 4

Printed in Dunstable, United Kingdom

91897630R00219